Sustainable Agriculture Reviews

Volume 21

Series editor
Eric Lichtfouse

Other Publications by Dr. Eric Lichtfouse

Books
Scientific Writing for Impact Factor Journals
Nova Publishers 2013

Sustainable Agriculture
Springer 2009

Sustainable Agriculture Volume 2
Springer 2011

Environmental Chemistry. Green Chemistry and Pollutants in Ecosystems
Springer 2005

Rédiger pour être publié ! Conseils pratiques pour les scientifiques
Springer 2012, 2e édition.

Journals and Series
Agronomy for Sustainable Development
www.springer.com/journal/13593

Sustainable Agriculture Reviews
www.springer.com/series/8380

Environmental Chemistry Letters
www.springer.com/journal/10311

Environmental Chemistry for a Sustainable World
www.springer.com/journal/11480

Blog
Agronomy blog
http://www1.montpellier.inra.fr/agronomy-blog

Magazine
Publier La Science
https://listes.inra.fr/sympa/d_read/veillecaps/

Sustainable agriculture is a rapidly growing field aiming at producing food and energy in a sustainable way for humans and their children. Sustainable agriculture is a discipline that addresses current issues such as climate change, increasing food and fuel prices, poor-nation starvation, rich-nation obesity, water pollution, soil erosion, fertility loss, pest control, and biodiversity depletion.

Novel, environmentally-friendly solutions are proposed based on integrated knowledge from sciences as diverse as agronomy, soil science, molecular biology, chemistry, toxicology, ecology, economy, and social sciences. Indeed, sustainable agriculture decipher mechanisms of processes that occur from the molecular level to the farming system to the global level at time scales ranging from seconds to centuries. For that, scientists use the system approach that involves studying components and interactions of a whole system to address scientific, economic and social issues. In that respect, sustainable agriculture is not a classical, narrow science. Instead of solving problems using the classical painkiller approach that treats only negative impacts, sustainable agriculture treats problem sources.

Because most actual society issues are now intertwined, global, and fast-developing, sustainable agriculture will bring solutions to build a safer world. This book series gathers review articles that analyze current agricultural issues and knowledge, then propose alternative solutions. It will therefore help all scientists, decision-makers, professors, farmers and politicians who wish to build a safe agriculture, energy and food system for future generations.

More information about this series at http://www.springer.com/series/8380

Shivendu Ranjan • Nandita Dasgupta
Eric Lichtfouse
Editors

Nanoscience in Food and Agriculture 2

Springer

Editors
Shivendu Ranjan
Nano-food Research Group, Instrumental and Food Analysis Laboratory, Industrial Biotechnology Division, School of BioSciences and Technology
Vellore Institute of Technology
Vellore, Tamil Nadu, India

Nandita Dasgupta
Nano-food Research Group, Instrumental and Food Analysis Laboratory, Industrial Biotechnology Division, School of BioSciences and Technology
Vellore Institute of Technology
Vellore, Tamil Nadu, India

Eric Lichtfouse
INRA
Dijon, France

ISSN 2210-4410 ISSN 2210-4429 (electronic)
Sustainable Agriculture Reviews
ISBN 978-3-319-39305-6 ISBN 978-3-319-39306-3 (eBook)
DOI 10.1007/978-3-319-39306-3

Library of Congress Control Number: 2016947716

Printed on acid-free paper

This Springer imprint is published by Springer Nature
The registered company is Springer International Publishing AG Switzerland

We dedicate this book to our parents

–The Editors

Nanomaterials: smaller materials – bigger scopes

–Shivendu Ranjan and Nandita Dasgupta

Preface

This book is the second on Nanoscience in Food and Agriculture published in the series Sustainable Agriculture Reviews. With unique properties, nanomaterials are rapidly finding novel applications in many fields such as food, medicine, agriculture and pollution. Such applications include nanoparticles to treat cancer, nanosensors to detect food contamination, nanomaterials for food packaging, nanoencapsulation to preserve nutraceuticals and nanofertilisers for advanced agriculture. After an introductory chapter on property rights of nanomaterials, readers will discover in the following chapters the applications of nanotechnology in food, health, environment, ecotoxicology and agriculture (Fig. 1).

The chapters address four major topics of applied nanotechnology: food, health, pollution and agriculture. Intellectual property rights of nanotechnologies in food and agriculture are reviewed in the first chapter by Chowdhury et al. Then Brandelli et al. explain the use of nano-encapsulated materials for food packaging in Chap. 2. The safety of future food is addressed by Purkayastha and Manhar who review nanosensors for packaging in Chap. 3 (Fig. 2). To improve the quality of food both for nutrition and health, Rao and Naidu discuss the nanoencapsulation of bioactive compounds in Chap. 4. Such use of nutraceuticals is exemplified by Singh et al. who describe food fortification of vitamins and minerals encapsulated in silica in Chap. 5. Nanotechnology for pollutant detection and removal is reviewed in Chap. 6 by Xing-yan Xue et al. The treatment of cancer and other human illnesses using silver nanoparticles is discussed by Kumar and Smita in Chap. 7. Plant illness is then addressed by Kashyap et al. who present the detection of plant pathogens by nanotechnologies in Chap. 8. The ecotoxicology of nanomaterials is addressed by Khan et al. who discuss the toxicity of nanoparticles in soil, microbes, plants and animals

Fig. 1 Nanotechnology will allow to improve the agricultural practices. *Left*: Farmer maintaining the agricultural lands using conventional methods at Village – Salakhua, Chapra, Bihar, India. Copyright: S. Ranjan/VKSMT 2010. *Right*: Farming land at village – Salakhua, Chapra, Bihar, India. Copyright: S. Ranjan/VKSMT 2010

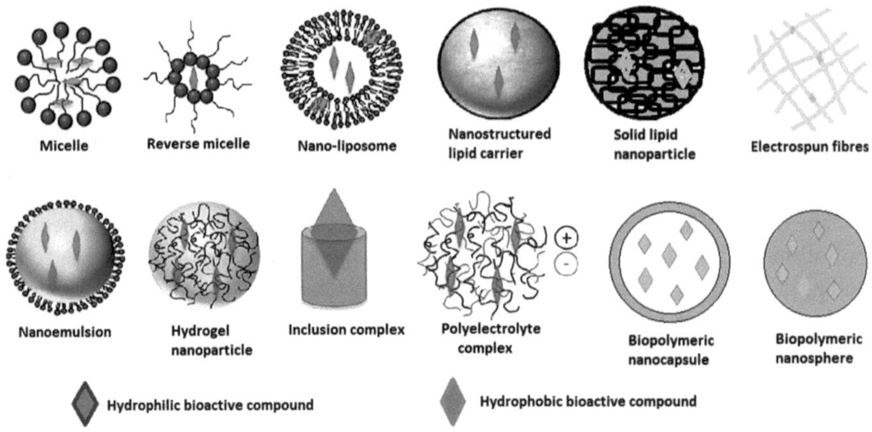

Fig. 2 Nanomaterials to delivery bioactive compounds in food, heath and other fields. Purkayastha and Manhar, Chap. 2

in Chap. 9. Nanoparticles for sustainable agroecosystems are then presented by Belal and El-Ramady in the last chapter.

Thanks for reading

Vellore, India

Dijon, France

Shivendu Ranjan
Nandita Dasgupta
Eric Lichtfouse

Contents

Contributors

Ashwini Angal TATA Chemicals Innovation Centre, Pune, Maharasthra, India

Tanoy Bandyopadhyay Department of Biotechnology, Tocklai Tea Research Institute, Tea Research Association, Jorhat, Assam, India

El-Sayed Belal Agricultural Botany Department, Faculty of Agriculture, Kafr El Sheikh Uni, Kafr El-Sheikh, Egypt

Sangeeta Borchetia Department of Biotechnology, Tocklai Tea Research Institute, Tea Research Association, Jorhat, Assam, India

Adriano Brandelli Laboratório de Bioquímica e Microbiologia Aplicada, Instituto de Ciência e Tecnologia de Alimentos, Universidade Federal do Rio Grande do Sul, Porto Alegre, Rio Grande do Sul, Brazil

Luis Fernando Wentz Brum Laboratório de Bioquímica e Microbiologia Aplicada, Instituto de Ciência e Tecnologia de Alimentos, Universidade Federal do Rio Grande do Sul, Porto Alegre, Rio Grande do Sul, Brazil

Laboratório de Catálise Ziegler-Natta, Instituto de Química, Universidade Federal do Rio Grande do Sul, Porto Alegre, Rio Grande do Sul, Brazil

Hillol Chakdar ICAR-National Bureau of Agriculturally Important Microorganisms, Mau, Uttar Pradesh, India

ICAR-Indian Institute of Wheat and Barley Research (IIWBR), Regional Station, Flowerdale, Shimla, Himachal Pradesh, India

Rong Cheng School of Environment and Natural Resources, Renmin University of China, Beijing, China

Pritom Chowdhury Department of Biotechnology, Tocklai Tea Research Institute, Tea Research Association, Jorhat, Assam, India

Sagarika Das Department of Biotechnology, Tocklai Tea Research Institute, Tea Research Association, Jorhat, Assam, India

Nandita Dasgupta Nano-food Research Group, Instrumental and Food Analysis Laboratory, Industrial Biotechnology Division, School of BioSciences and Technology, Vellore Institute of Technology, Vellore, Tamil Nadu, India

Ashwini K. Dixit Department. of Botany, Guru Ghasidas Vishwavidyalaya, Bilaspur, Chhattisgarh, India

João Henrique Zimnoch dos Santos Laboratório de Catálise Ziegler-Natta, Instituto de Química, Universidade Federal do Rio Grande do Sul, Porto Alegre, Rio Grande do Sul, Brazil

Hassan El-Ramady Soil and Water Department, Faculty of Agriculture, Kafrelsheikh Uni, Kafr El-Sheikh, Egypt

Madhurjya Gogoi Department of Biotechnology, Tocklai Tea Research Institute, Tea Research Association, Jorhat, Assam, India

Pranita Hazarika Department of Biotechnology, Tocklai Tea Research Institute, Tea Research Association, Jorhat, Assam, India

Prem Lal Kashyap ICAR-National Bureau of Agriculturally Important Microorganisms, Mau, Uttar Pradesh, India

ICAR-Indian Institute of Wheat and Barley Research (IIWBR), Regional Station, Flowerdale, Shimla, Himachal Pradesh, India

Naureen S. Khan Department of Botany, Guru Ghasidas Vishwavidyalaya, Bilaspur, Chhattisgarh, India

Brajesh Kumar Centro de Nanociencia y Nanotecnologia, Universidad de las Fuerzas Armadas ESPE, Sangolqui, Ecuador

Department of Chemistry, TATA College, Kolhan University, Chaibasa, Jharkhand, India

Sudheer Kumar ICAR-Indian Institute of Wheat and Barley Research, Karnal, Haryana, India

Eric Lichtfouse INRA, UMR1347 Agroécologie, Dijon, France

Zhong Ma School of Environment and Natural Resources, Renmin University of China, Beijing, China

Ajay Kumar Manhar Department of Molecular Biology and Biotechnology, Tezpur University, Tezpur, Assam, India

Rajendra Mehta Department of Rural Technology, Guru Ghasidas Vishwavidyalaya, Bilaspur, Chhattisgarh, India

Madhav M. Naidu Department of Spice & Flavor Science, Council of Scientific Research Institute- Central Food Technological Research Institute (CSIR-CFTRI), Mysuru, Karnataka, India

K. Pandiyan ICAR-National Bureau of Agriculturally Important Microorganisms, Mau, Uttar Pradesh, India

ICAR-Indian Institute of Wheat and Barley Research (IIWBR), Regional Station, Flowerdale, Shimla, Himachal Pradesh, India

Prabhat Parida TATA Chemicals Innovation Centre, Pune, Maharasthra, India

Manashi Das Purkayastha Department of Horticulture, Assam Agricultural University, Jorhat, Assam, India

Pallavi Rai ICAR-National Bureau of Agriculturally Important Microorganisms, Mau, Uttar Pradesh, India

ICAR-Indian Institute of Wheat and Barley Research (IIWBR), Regional Station, Flowerdale, Shimla, Himachal Pradesh, India

Shivendu Ranjan Nano-food Research Group, Instrumental and Food Analysis Laboratory, Industrial Biotechnology Division, School of BioSciences and Technology, Vellore Institute of Technology, Vellore, Tamil Nadu, India

Pooja J. Rao Department of Spice & Flavor Science, Council of Scientific Research Institute- Central Food Technological Research Institute (CSIR-CFTRI), Mysuru, Karnataka, India

Nikita Rathi TATA Chemicals Innovation Centre, Pune, Maharasthra, India

Debabrata Rautaray TATA Chemicals Innovation Centre, Pune, Maharasthra, India

Shikha Sharma Department of Plant Pathology, Punjab Agricultural University (PAU), Ludhiana, Punjab, India

Lei Shi School of Environment and Natural Resources, Renmin University of China, Beijing, China

Sneha Singh TATA Chemicals Innovation Centre, Pune, Maharasthra, India

Kumari Smita Centro de Nanociencia y Nanotecnologia, Universidad de las Fuerzas Armadas ESPE, Sangolqui, Ecuador

Alok K. Srivastava ICAR-National Bureau of Agriculturally Important Microorganisms, Mau, Uttar Pradesh, India

ICAR-Indian Institute of Wheat and Barley Research (IIWBR), Regional Station, Flowerdale, Shimla, Himachal Pradesh, India

Xing-yan Xue School of Environment and Natural Resources, Renmin University of China, Beijing, China

Afruza Zaman Department of Biotechnology, Tocklai Tea Research Institute, Tea Research Association, Jorhat, Assam, India

Xiang Zheng School of Environment and Natural Resources, Renmin University of China, Beijing, China

About the Editors

Shivendu Ranjan is currently working as DBT-Research Fellow, Department of Biotechnology, Ministry of Science and Technology, Govt of India at VIT University, Vellore, Tamil Nadu, India. He is also serving for a non-government organization as an Honorary-Director, Research Wing, Veer Kunwar Singh Memorial Trust, Chapra, Bihar, India. He is the Founder-Director at Xpert Arena Technological Services Pvt. Ltd., India (www.xpertarena.com); this company is dedicated to serve in online and offline sectors with a vision to simplify the education. He has also founded and drafted the concept for first edition of *VIT Bio Summit* in 2012 and the same has been continued till date by the university.

His area of research is multidisciplinary which are as but not limited to: Nano-food technology, Nano-agri technology, Nanobiotechnology, Nano-toxicology, Natural products technology, Natural products chemistry, Bio-business, Food chemistry and Food Engineering. He has published many scientific articles in international peer reviewed journals and also serving as editorial board member and referee for reputed international peer reviewed journals. He has bagged several awards from different organizations, e.g. Best poster award, achiever award, research award, young researcher award, etc.

Nandita Dasgupta is currently serving as Research Associate at VIT University, Vellore, Tamil Nadu, India. She has exposure of research institutes and industries including CSIR-Central Food Technological Research Institute, Mysore, India, and Uttar Pradesh Drugs and Pharmaceutical Co. Ltd., Lucknow, India. Her areas of interest include toxicological analysis, natural products technology, nanobiotechnology and agri-food technology.

She has published many scientific articles in international peer reviewed journals and also serving as editorial board member and referee for reputed international peer reviewed journals. She has received Elsevier Certificate for "Outstanding Contribution" in reviewing from Elsevier, The Netherlands. She has also been nominated for Elsevier advisory panel for Elsevier, The Netherlands. She was the Guest Editor in Journal of Chemistry for the special issue entitled "Food Nanotechnology Opportunities and Challenges". She has received several awards from different organizations e.g. Best poster award, young researcher award, special achiever award, research award, etc.

Eric Lichtfouse 56, is a soil scientist at the French National Institute for Agricultural Research (INRA). He has invented the C-dating method[1] allowing to measure the dynamics of soil organic molecules[2], thus opening the field of molecular-level investigations of soil carbon sequestration. Chief Editor of the awarded journal Agronomy for Sustainable Development,[3] he has raised the journal rank from 29/53 in 2003, with an impact factor of 0.56, to 2/81 in 2014, with an impact factor of 3.99, in the Agronomy category. He is also Chief Editor and founder of the journal Environmental Chemistry Letters[4] and the book series Sustainable Agriculture Reviews.[5] He is lecturing scientific writing and communication in universities worldwide.[6] His publication assistance service at the INRA has founded the french-english newsletter Publier La Science.[7] He has

[1] http://dx.doi.org/10.1007/s10311-011-0334-2

[2] http://archive.sciencewatch.com/inter/jou/2010/10novAgrSusDev

[3] http://www.springer.com/journal/13593

[4] http://www.springer.com/journal/10311

[5] http://www.springer.com/series/8380

[6] http://fr.slideshare.net/lichtfouse/scientific-writing-and-communication
https://www.youtube.com/playlist?list=PLKEz5Pbi4p3By53Q0gclKPeSBTK2HJGK_

[7] http://www6.inra.fr/caps-publierlascience

published the book Scientific Writing for Impact Factor Journal.[8] This textbook describes in particular the micro-article,[9] a new tool to identify the novelty of experimental results. Further details are available on Slideshare,[10] LinkedIn,[11] ResearchGate,[12] ResearcherID[13] and Orcid.[14]

---

[8] https://www.novapublishers.com/catalog/product_info.php?products_id=42211

[9] http://fr.slideshare.net/lichtfouse/micro-arten

[10] http://fr.slideshare.net/lichtfouse

[11] https://fr.linkedin.com/in/ericlichtfouse

[12] https://www.researchgate.net/profile/Eric_Lichtfouse

[13] http://www.researcherid.com/rid/F-4759-2011

[14] http://orcid.org/0000-0002-8535-8073

Chapter 1
Intellectual Property Rights for Nanotechnology in Agriculture

Pritom Chowdhury, Madhurjya Gogoi, Sagarika Das, Afruza Zaman, Pranita Hazarika, Sangeeta Borchetia, and Tanoy Bandyopadhyay

Abstract Nanoscience studies biological properties of materials at the nanoscale. Nanotechnology research develops improved materials, devices, systems and therapeutics. Nanomaterials can be protected for their intellectual property rights by innovators. However, due to the interdisciplinary nature of nanotechnology, there is a risk of overlapping patent claims and lack of distinction between nano-based and traditional patents. Scientists also must solve ethical and social issues, from health to environmental risk and consumer perception.

This chapter reviews the status of intellectual property rights protection of nanomaterials. The main points are: concerning patent number, there has been recently a tenfold increase of nanomaterial patents; for instance there has been an increase of granted patents from 386 in 2004 to 1106 in 2014 by the United States Patent and Trademark office. Concerning regulatory and policy bodies, the World Intellectual Property Organization and World Health Organization are working to make a comprehensive intellectual property right regulation for nanotechnology products. The United States is the leader of nanotechnology products development, and has made guidelines to make patent search easier for nano-based product. The European Patent Office has also created a new classification for the nano-based inventions. The chapter also gives emphasis on nanotechnology applications in agriculture, such as nanopesticides and nanobiosensors. Environmental, toxicity and health issues associated with nanotechnology products are mentioned.

Keywords Environmental risk • Intellectual property rights • Nanotechnology • Patents • Regulations • World Intellectual Property Organization (WIPO) • Nanopesticide • Nanosensor

---

P. Chowdhury (✉) • M. Gogoi • S. Das • A. Zaman • P. Hazarika • S. Borchetia
T. Bandyopadhyay
Department of Biotechnology, Tocklai Tea Research Institute, Tea Research Association, Jorhat 785008, Assam, India
e-mail: pritomc@gmail.com; madhurjyag@gmail.com; sagarika2608@gmail.com; afruzaz@gmail.com; pranitahazarika@yahoo.in; sborchetia7@gmail.com; tanoy.bandyopadhyay@gmail.com

S. Ranjan et al. (eds.), *Nanoscience in Food and Agriculture 2*, Sustainable Agriculture Reviews 21, DOI 10.1007/978-3-319-39306-3_1

Abbreviations

BIS	Bureau of Indian Standards
CPCB	Central Pollution Control Board
CSIR	Council of Scientific and Industrial Research
CSTP	Committee of Scientific and Technological Policy
DRDO	Defence Research and Development Organization
DST	Department of Science and Technology
EC	European Commission
ECHA	European Chemical Agency
EFSA	European Food Safety Authority
EPA	Environmental Protection Agency
EPO	European Patent Office
EU	European Union
FAO	Food and Agriculture Organization
FD&C Act	Federal Food Drug and Cosmetic Act
FDA	Food and Drug Administration
FIFRA	Federal Insecticide, Fungicide and Rodenticide Act
HEAL	Health and Environment Alliance
IISc	Indian Institute of Science
IIT	Indian Institute of Technology
IPR	Intellectual Property Rights
ISO	International Organization of Standardization
JPO	Japan Patent Office
MSDS	Material Safety Data Sheet
NIFA	National Institute of Food and Agriculture
NSTM	Nano Science and Technology Mission
OECD	Organization for Economic Co-operation and Development
PCT	Patent Cooperation Treaty
PPE	Personal Protective Equipment
REACH	Registration, Evaluation, Authorization and restriction of Chemicals
TRIPS	Trade Related aspects of Intellectual Property Rights
TSCA	Toxic Substances Control Act
UNO	United Nations Organisation
USDA	United States Department of Agriculture
USPTO	United States Patent and Trademark Office
WHO	World Health Organization
WIPO	World Intellectual Property Organization
WPN	Working Party on manufactured Nanomaterials
WTO	World Trade Organization

1.1 Introduction

Nanotechnology is an emerging branch which introduced new terrain on scientific development in a significant way. It deals with interdisciplinary subject with understanding and control of matter at dimension of roughly 100 nm and below. At this scale, the physical, chemical and biological properties of materials differ from the properties of individual atoms and molecules or bulk matter, which enable novel applications. Nanotechnology research and development are directed towards understanding and creating improved materials, devices and systems that exploit these properties as these are discovered and characterized (Raliya et al. 2013). These developments and expectations from nano based applications led governments and industries around the world to invest in nanotechnology research and development. The expectations from nanotechnology have risen to enormous heights, with the American government heralding it as the bringer of the 'Next Industrial Revolution' (Selin 2007). Nanotechnology research and products are reaching newer heights which impends its link up with policy regulation. Intellectual property rights is one such important policy, referring to protection and granting of monopoly rights for limited period to designated innovators by law (Sherman and Bently 1999). Intellectual property rights law comprises of patent, copyright, trademark, design rights and trade secrets. Enforcement of these laws on agriculture sector was traditionally limited because most agricultural research and development was conducted by public sector institutions in both developed and developing countries (Fink and Primo Braga 1999). The development and dissemination of the technologies led to the green revolution which did not pose substantial conflicts around intellectual property rights. However, post 1980s era, role of the private sector in agricultural research and development has increased significantly. In developed countries, almost one-half of agricultural research has been funded by the private sector (Helpman 1993). The emergence of research driven interdisciplinary subjects like nanotechnology has led to increase in reliance on intellectual property rights protection. This advancement for translation of basic scientific research into product for commercialization has accelerated a complexity for intellectual property right protection. Moreover, a research activity in this field has to deal with ethical and social issues, starting from health to environmental risk to consumer perception (Sweeney 2006).

Implementation of new technology in agriculture sector is of extreme importance, particularly to deal with population growth, climate change, pest management and limited availability of nutrient. There are many more important factors, with single aim to solve: how to increase crop production? Nanotechnology may play an effective role for this purpose. Already, there are many important developments on this aspect like enhancement of nutrients absorption by plants, protection of plants, nano-formulated food ingredients and water treatment processes. Nanotechnology derived devices are also explored in the field of plant breeding and genetic transformation. The potential of nanotechnology in agriculture is large, but issues such as increasing the scale of production processes and lowering costs, as

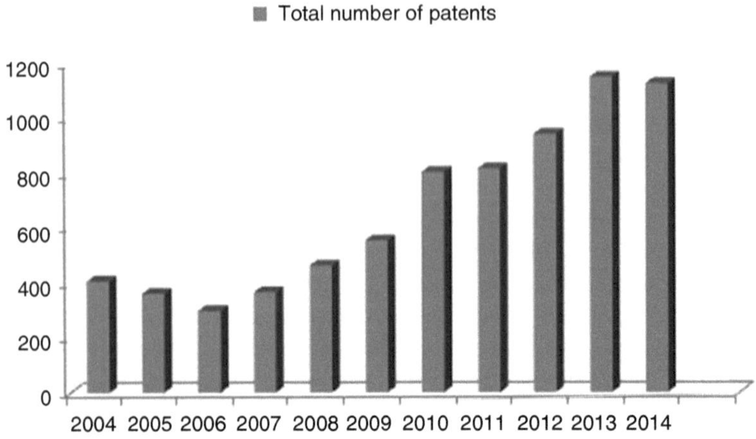

Fig. 1.1 Trend showing increase in the number of nanotechnology patents in the last 10 years as granted by United States Patent and Trademark Office and European Patent Office

well as risk assessment issues are needed to be addressed. This spur in nanotechnology development can be observed by increase in the number of nano related patents (Fig. 1.1) in recent years (Parisi et al. 2014). Patenting on nanotechnology, in general, presents some important concerns as it is an interdisciplinary field with multiple applications and the risk of overlapping patents claim exist. Further, distinction between nano-based patents and traditional patents are not well distinguished. There is a need to understand different regulations and various fields of nanotechnology applications. We have described here some important regulations related to nanotechnology with context to few countries where nanotechnology research is promising. Different application of nanotechnology in agriculture sector is highlighted and risk factor associated with regulatory and environmental risk factor is discussed.

1.2 Intellectual Property Rights Related to Nanotechnology

In the international scenario, nanotechnology inventions are generally protected by patents where other intellectual property rights such as trade secrets might also be applicable (Foster 2007). Some of the international bodies looking into nanotechnology aspects at international forum are discussed here.

1.2.1 United Nations

The United Nations is one of the global governing bodies, and various organizations established under the roof of the United Nations are helping to regulate nanotechnology innovations around the world include:

World Intellectual Property Organization (WIPO): WIPO, established in 1967, is a self-funding agency of the United Nations which acts as a global forum for intellectual property services, policy, information and cooperation. At present, the World Intellectual Property Organization, with headquarters at Geneva, Switzerland, has 188 states as its members. In the milieu of protection of intellectual property rights for new and emerging technologies, the standing committee on the Law of Patents of the World Intellectual Property Organization (SCP 2009) found that the patent system has to constantly conform to the question as to whether a new technological innovation falls within the existing definition of "invention" in the existing patent law. However, it emphasizes the importance of the public welfare perspective. It also states that the regime of intellectual property rights needs to be carefully evaluated to bring an innovation either under laws governing intellectual property rights by creating a new legal mechanism to protect it or alter the existing patent laws to accommodate it. The World Intellectual Property Organization does not grant patents but facilitates international patent protection under the Patent Cooperation Treaty system and complements intellectual property services at the national and/or regional level. An applicant who files a patent application under this treaty can seek protection for an invention in 148 countries throughout the world. The role of the Patent Cooperation Treaty also stretches to help Patent Offices in granting patents and making the wealth of technical information accessible to the public (Havas 2014).

In order to make prior art in various fields accessible, the World Intellectual Property Organization divided the different categories of inventions into different classes. As for nanotechnology, it was tagged as Class B82. This class is further sub divided into (Arnold and Keserü 2013)

1. B82B – Nanostructures formed by manipulation of individual atoms, molecules, or limited collections of atoms or molecules as discrete units; manufacture or treatment thereof.
2. B82Y – Specific uses or applications of nano-structures; measurements or analysis of nano-structures, manufacture or treatment of nano-structures.

Food and Agriculture Organization (FAO): The Food and Agriculture Organization of the United Nations in its expert meeting during 2009 stated that a tiered approach for prioritization of the classes or types of material to be used for an invention in nanotechnology which may help in risk assessment strategies. For this, additional data needed to be generated which would reduce the uncertainties in risk assessment. The need to develop validated testing methods was stressed to address specific data gaps. In 2013, the Food and Agriculture Organization and the World Health Organization published a report, analyzed and summarized the scientific information on nanotechnology and postulated the possible courses of action to be followed by the Food and Agriculture Organization (FAO), the World Health Organization (WHO) and other organizations in this matter (FAO/WHO 2013).

Working Party on manufactured Nanomaterials (WPN): The Organization for Economic Co-operation and Development (OECD) established a working party on manufactured nanotechnology in 2006 under the Committee of Scientific and Technological Policy (CSTP), which works on international co-operation in health-related and environmental safety-related aspects of manufactured nanomaterials (Raimond 2008). According to a Directorate report of the Organization for Economic Co-operation and Development (2007) the three key nanotechnology patent powerhouses were the United States Patent and Trademark Office (USPTO), the European Patent Office (EPO) and the Japan Patent Office (JPO). The report also stated that on the basis of patent applications analyzed between 1978 and 2005, about 33 % of the patent applications were from the United States, followed by Japan and the European Union (EU). The Organization for Economic Co-operation and Development, at present, offers testing guidelines which are globally accepted for hazard recognition and characterization of food chemicals, pesticides, veterinary drugs and other substances to which humans are exposed (FAO 1995).

1.2.2 United States

In the United States nanopesticides are regulated by the Environmental Protection Agency, identifiable in its website since July 2011, which seeks public comment regarding those products regulation and incorporation into their existing Federal Insecticide, Fungicide, and Rodenticide Act (FIFRA) for registration of new pesticide products (Kah and Hofman 2014).

An expert consultation on agri-nanotechnology under the patronage of the World Health Organization and Food and Agriculture Organization suggested a "tiered approach" to risk assessments concerned with engineered nanotechnology materials. In such a tiered approach, priorities could be set based on scientific literature to the product which has risks to public, worker and/or the environment health. Risk assessments could be performed after obtaining product data, leaving low-risk applications further down the priority list. Under the current regulations, companies have the discretion to determine as to whether a nano-scale product, whose macro form is already considered to be safe by the company, may be regarded to be safe in nano form and whether needs to be reported to the Food and Drug Administration or not. However, due to exponentially larger surface-to-mass ratio of the engineered nanotechnology materials which may have different properties as compared to their macro-versions and the determination of acceptable daily intakes would become impossible if companies do not submit data to the regulators for their independent assessment (Suppan 2011). It is critical that there is neither an official nor an unofficial entry of the nano-materials used in consumer or industrial products or industrial processes. Therefore, it is highly necessary that a registry to be formulated which would be an initial step in establishing eventual nanotechnology regulations.

In 2013, the United States Food and Drug Administration (FDA) released a nanotechnology regulatory science research plan which would regulate nanotechnology innovations in an appropriate and balanced manner and at the same time to ensure transparency and enable responsible development of products with new and beneficial properties. The Food and Drug Administration's regulatory approach would have the following attributes (National Science and Technology Council 2014):

1. FDA shall maintain its product-focused, science-based regulatory policy.
2. FDA's approach would respect variations in legal standards for different product-classes.
3. Where pre-market review authority existed, attention to nanomaterials would be incorporated into standing procedures.
4. Where statutory authority did not provide for premarket review, consultation would be encouraged to reduce the risk of unintended harm to human or animal health.
5. FDA would continue constant post-market monitoring.
6. Industry would remain responsible for ensuring that its products met all applicable legal requirements, including safety standards.
7. FDA would collaborate, as appropriate, with domestic and international counterparts on regulatory policy issues.
8. Both for products that are not subject to premarket review and those that were; FDA would offer technical advice and guidance, as needed, to help industry meet its regulatory and statutory obligations.

Since long, the United States has been the leader of nanotechnology (Smith 2013). However, searching for previous patents related to nanotechnology is a tedious job as it is an interdisciplinary field. An additional difficulty in such searches is that all nanotechnology-related inventions may not necessarily contain the word "nano" in its article. In order to overcome such problems, the United States Patent and Trademark Office in October, 2004 established an index to the United States Patent Classification System for Nanotechnology called "Class 977". This would serve as a cross-reference to help examiners and the public to search prior art including issued United States patents and published patent applications (Paradise 2012). The United States Patent and Trademark Office defined certain guidelines (mentioned below) under which nanotechnology innovations must fall to be considered under Class 977 (Shand and Wetter 2007):

1. They must associate with research and technological innovation in a length scale of 1–100 nm in at least one dimension.
2. They must provide a basic idea of the properties of the materials in the nanoscale to create and use structures, devices, and systems that have size-dependent unique characteristics and functions.

Class 977 includes many subclasses (Bhattacharya et al 2012; Kisluik 2010), under which patents were obtained (Fig. 1.2). Apart from Class 977 other important codes

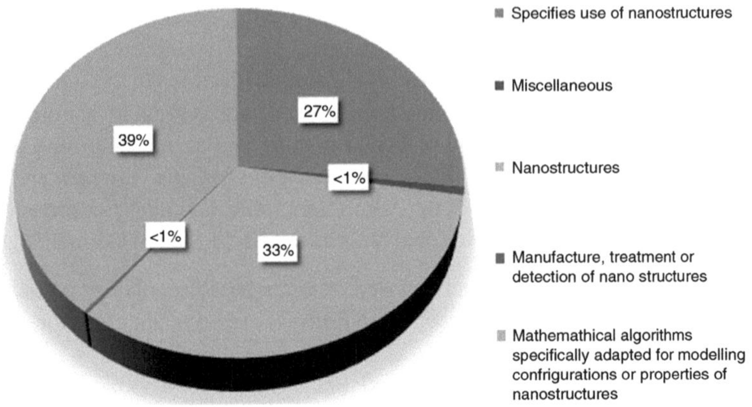

Fig. 1.2 Diagrammatic representation of the percentage of nanotechnology patents obtained by the various sub-classes of Class 977 as of July 2010

are title 35 of the United States Code section 102 which bars the patenting of inventions if those are not novel and can be anticipated by prior art and the title 35 United States Code section 103 which bars patenting obvious inventions (Schauwecker 2011).

1.2.3 European Union

In the European Union nanoparticles are regulated under Registration, Evaluation, Authorization and restriction of Chemicals which in short form is known as REACH. Under REACH, a registrant needs to upgrade his registration for a substance which is already registered in its macro-form but introduced into the market in the nano-form. However, it must fulfilled the following conditions - changes in composition, quantities, uses and safety report to enable appropriate risk management are labeled properly. Evaluation can then be carried out by Member States of REACH and the European Chemicals agency which ask for additional data for the same (European Commission 2011). European Union has also specific regulations dedicated to plant protection products and fertilizers (Bucheli 2014).

The European Commission implemented the communication on the Second Regulatory Review on Nanomaterials in October 2012 (Zornoza 2012) which evaluated the adequacy and execution of European Union legislation for nanomaterials and listed the Commission's proceedings in the field of nanomaterials. Further, the related Staff Working Document (Bergesion and Cole 2015) gave an overview of different nanomaterials in the market and their hazardous properties, along with background information on the definition.

On 13th December 2014, a regulation was applied under the head Article 18 of Regulation (European Union) No. 1169/2011 stating that "All ingredients present in the form of engineered nanomaterials shall be clearly indicated in the list of ingredients. The names of such ingredients shall be followed by the word 'nano' in brackets" (Regulation (EU) 1169/2011). According to article 2 (2) (t) Regulation (European Union) 1169/2011 (Meulen et al 2014):

> "Engineered nanomaterial" means any intentionally produced material that has one or more dimensions of the order of 100 nm or less or that is composed of discrete functional parts, either internally or at the surface, many of which have one or more dimensions of the order of 100 nm or less, including structures, agglomerates or aggregates, which may have a size above the order of 100 nm but retain properties that are characteristic of the nanoscale.

The European Commission provides instructions to the European Food Safety Authority (ESPA), the European Union's risk assessment body for food and feed safety, to provide scientific outputs as technical risk assessment guidelines, scientific opinions on specific products and technical and scientific assistance to the European Commission (EFSA Scientific Committee 2009; EFSA Scientific Committee 2011). Till date, the European Food Safety Authority has already published two scientific opinions on potential risk and risk assessment principles for nanotechnology entering the food and feed chain.

The European Food Safety Authority's major conclusions incorporated that significant level of vagueness and uncertainty persists regarding the characteristics of nanomaterials and involved potential risks which arise in difficulties to recommend any specific and suitable risk assessment strategy for the same. According to the food information to consumers, all ingredients present in the form of engineered nanomaterials must be properly labeled. However, unlike biocides, no labeling is required for nanomaterials in pesticides (Rauscher 2014).

In 2011, the European Food Safety Authority stated that the risk of an engineered nanomaterial in the food and feed category will be determined by its chemical composition, physicochemical properties, interactions with tissues and potential exposure levels but there are currently uncertainties related to the identification, characterization and detection of engineered nanomaterial due to lack of suitable and validated test methods.

The European Union, very similar to the United States Patent and Trademark Office, faced the problems of retrieving nanotechnology-related prior art. In order to solve this problem, the European Patent Office created a new classification for the nano-based inventions, called Y01N. This class was further sub-divided into subclasses as Y01N2 (nanobiotechnology or nano-medicine), Y01N4 (nanotechnology for information processing, storage and transmission), Y01N6 (nanotechnology for materials and surface science), Y01N8 (nanotechnology for interacting, sensing and actuating), Y01N10 (nanotechnology for optics) and Y01N12 fornanomagnetics (McLennan and Rimmer 2012). At the European Patent Office, there are no sectional guidelines for granting a patent. However, novelty and inventiveness are some principles upon which patents are being granted (Escoffier 2006).

1.2.4 India

India is among the fastest growing economies in the world but still lagging behind in research on nanotechnology as evident from the number of patents filed. India is fast adapting to the new trend of nanotechnology and started its nano-mission in 2001 with the project Nano-Science and Technology Initiative by the Department of Science and Technology (DST) which was extended till 2006. Under this project, 19 Centres of excellence were established across the country for research, development and applications of nanotechnology which was followed by the Nano-Science and Technology Mission started by the Department of Science Technology in the year 2007 with a budget of Rs 1000 crores. This mission aimed to provide facilities to scientists, institutions and the industry for promoting basic research. It also helped in developing adequate manpower, international collaborations, developing infrastructure and producing useful products. This mission resulted in about 5000 research articles and useful products like pesticide removal technology from drinking water, nano hydrogel based eye drops, water filters for arsenic and fluoride removal and nano silver based anti-microbial textile coating (Jaiswal 2014). The Union Cabinet gave its clearance on 20th Feb 2014 for the continuation of the Nano-Science and Technology Mission to its second phase in the 12th plan period after success in the first phase where India's place improved by one position in the global listing. For this purpose the Government allocated a grant of Rs 650 crores to the Department of Science and Technology (Kumar 2014; Ali and Sinha 2014).

Discussions on nanotechnology regulations concern on intellectual property rights and risk regulation. Analysis of the regime of intellectual property rights revealed that the requirements of Trade-Related Aspects of Intellectual Property rights (TRIPS) agreement is that all member states of the World Trade Organization (WTO) should adopt minimum standards of intellectual property and allow patents in all fields of technology [Article 27 (1) of TRIPS which provides that patents shall be available for any inventions, whether products or processes, in all fields of technology, provided that they are new, involve an inventive step and are capable of industrial application]. This means that all World Trade Organization members are supposed to provide intellectual property rights in the field of nanotechnology (TERI 2009).

In India, according to Section 2(1)(j) of the Patents (Amendment) Act 2005, an invention is 'a new product or process involving an inventive step and capable of industrial application'. Also, as stated by Section 2 j (a), an 'inventive step' is one which helps to achieve technical advancement and/or having economic importance or both. Additionally, it should also be 'non-obvious' to even a person skilled in the art. However, Section 3(d) adds to the stringency of the patentability of an invention (Chowdhury et al. 2014). According to this section, granting of patents is barred to a derivative form of an already known substance; unless the derivative has exceptionally improved efficacy. It also states that mere discovery of a new property or new use of an already known substance is prohibited. On the other hand, the post-grant opposition clause in Section 25(2) of the Indian patent legislation (2005)

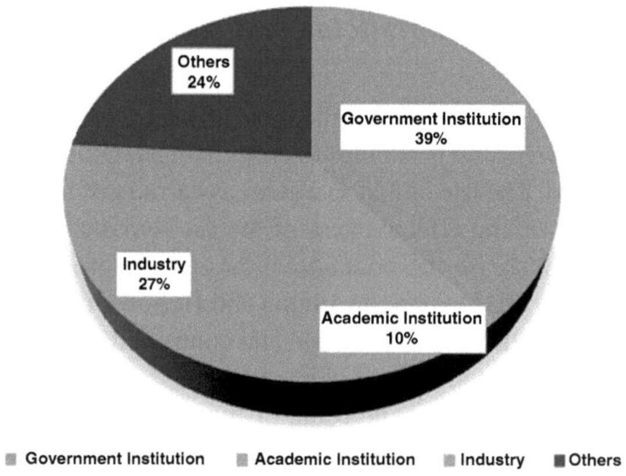

Fig. 1.3 Diagrammatic representation of the percentage of nanotechnology patents attained by various organizations in India till 2007

allows third parties to challenge the validity of a patent after it has been issued. Lack of novelty or non-obviousness, if proven, may lead to amendment or even withdrawal of the patent by the Patent Office. Section 10 of the Patents Act requires detailed information on the invention on the best method of performing the invention known to the applicant to be furnished at the time of application, for claim protection. Insufficient disclosure, if not, can be a valid ground for pre-grant opposition as per Section 25(1)(g) or per Section 25(2)(g) for post-grant opposition (Barpujari 2010).

Regulation is an important part of a technology to be successful but presently there is no strict regulatory framework for nanotechnology in India. It is imperative to devise a regulatory framework which can look into the pros and cons of a particular invention, study public perspective and identify the associated ethical issues. Nanotechnology research community in India is small and limited to a handful of premier institutes (Fig. 1.3) like laboratories of the Council of Scientific and Industrial Research (CSIR), Defense Research and Development Organization (DRDO), Indian Institute of Science (IISc) and Indian Institutes of Technology (IIT). Therefore, it is essential to escalate institutional capacity in order for nanotechnology to blossom in India.

1.3 Nanotechnology Applications

Advances in science and technology can provide potential solution for improvement in the current production system by increasing the resource efficiently (Ditta 2012). This is where various nanotechnology applications (Table 1.1) can play a myriad of roles for improving agricultural output. The nanotechnology application in agricultural sector offers advantage as compared to conventional practices (Aouada and de Moura 2015). Some of the applications of nanotechnology in the agriculture sector have been discussed below.

1.3.1 Nanopesticide and Nanofertilizer

The most common application of nanoscience in agriculture is the application of fertilizers and pesticides to crops using nanodevices (Perez-de-Luque and Hermosín 2013). The use of nanoparticles as carriers of fertilizers and pesticides in agriculture increase the effectiveness of the active materials and reduce their volatilization. It also decrease the contamination of underground water resources (Aouada and de Moura 2015; Perez-de-Luque and Hermosín 2013). One of the major advantages of nanoparticles is the gradual and controlled release of agrochemicals. The controlled release systems allow controlled delivery of active ingredients for a desired period in the vicinity of the roots or the vegetative parts (Aouada and de Moura 2015). Nanotechnology products like nanocapsules, nanoemulsions play an important role in plant protection by acting as smart delivery system of the active ingredients for the control of disease and pest in plants e.g- Neem oil emulsion as larvicidal agent (Parisi et al. 2015).

1.3.2 Soil Improvement

A major application of nanotechnology that co-ordinates with agriculture is the improvement of soil characteristics. The physical and chemical nature of soil largely impacts the efficient growth of the plants. The soil characteristics can be improved by enhancing the water infiltration, aeration, availability of nutrients specially the liquid agrochemical and reducing the shear strength (Bandyopadhyay et al. 2009). Nanotechnology sector has contributed towards the soil improvement by way of producing nanomaterial products like zeolites and nano-clays which help in retention of liquid agrochemicals or water in the soil and allow slow release to the plants (Parisi et al. 2015).

1.3.3 Water Resource Management

Providing access to clean potable water is one of the major global challenges particularly in the developing countries. Water crisis, even in the developed countries, has a tremendous impact not only on health but also on agriculture, manufacture and power industries (OECD 2011). Some of the major water pollutants include microbes, heavy metals, organic chemicals, etc. Moreover, indiscriminate uses of agrochemicals continuously pollute the ground water and shallow water systems (Bhattacharya et al. 2013). The conventional methods of water purification systems are inadequate for the efficient treatment of waste water (Bhati et al. 2015).

Table 1.1 Various nanotechnological application in agricultural sector

Field of application	Products description	References
Nanopesticide and Nanofertilizer	1. Neem oil (Azadirachta indica) nanoemulsion as Larvicidal agent	Anjali et al. (2012), Milani et al. (2012), and Sastry et al. (2009)
	2. Macronutrient Fertilizers Coated with Zinc Oxide	
	3. Pesticides encapsulated in nanoparticles for controlled release	
	4. Ammonia from Buckyballs used as fertilizer	
Soil improvement	Soil-enhancer product, based on a nano clay component, for water retention and release	http://www.geohumas.com/us/products.html
Water resource management	1. Filters coated with TiO2 nanoparticles for the photocatalytic degradation of agrochemicals in contaminated waters	Mc Murray et al. (2006) and Bhati et al. 2015
	2. Carbon nanotube filters for toxin removal	
	3. Gold nanoparticles for pathogen detection	
	4. Nanoparticle for toxic metal detection and removal	
Crop improvement	1. Mesoporus silica nanoparticles transporting DNA to transform plant cells	Torney et al. (2007) and Shrivastava and Dash (2009)
	2. Development of a purple coloured rice variety ‘Khao Kam’ through the drilling of a nano sized hole in the cell wall and membrane of rice cell to insert a nitrogen atom	
	3. Transfection reagents	
	4. DNA loaded supported gold nanoparticles, process for the preparation and use thereof	
Food industry	1. Airtight plastic packaging with silicate nanoparticles	Sastry et al. (2009) and Momin et al. (2013)
	2. Nanoemulsion based ice cream	
	3. Frying oil extender (Oil Fresh) uses nanoproducts to keep frying oil fresh for a longer time	
	4. Polymer composites with nanoclay as improved materials for food packaging	
Nanobiosensor	1. Pesticide detection with a liposome-based nano-biosensor	Vamvakaki and Chaniotakis (2007) and Sastry et al. (2009)
	2. Contamination of packaged food	
	3. Pathogen detection	
	4. Nanosensors linked to a GPS tracking unit for real-time monitoring of soil conditions and crop growth	

Nanoscopic materials such as carbon nanotubes, nanofibres or sponges are used as nanofilters for water purification. For the removal of heavy metals like arsenic, magnetic nanofibres are used for nano-filtration (Bhati et al. 2015; Sekhon 2014).

1.3.4 Crop Improvement

Improvement in the quality of crop plants through conventional breeding approach has been a continuous phenomenon for a long time. The application of genetic engineering to generate transgenic lines is considered to be invaluable in plant genetic research. Nanotechnology also promotes gene transfer by having certain advantages such as small size of nanoparticles, stable integration of gene in the host genome, rapid expression of the transgene, not subjected to microbial attack, easily synthesizable and less toxic. The commonly used nanoparticles as vectors for gene transfer include calcium phosphate, carbon, silica, gold, etc. The nuclear physics laboratory of the Chiang Mai University, Thailand, has developed a new variety of white-grain rice through the application of nanotechnology to modify the traditional purple coloured rice variety (Prassana 2006). Apart from use of nanomaterial as delivery system for genetic modification, the genetically modified plants or microbes can be used for the production of nanomaterials. The agricultural waste products could also be processed to obtain nanomaterials e.g- nanofibres from wheat and soy hulls are used for bio-nanocomposite production (Parisi et al. 2015).

1.3.5 Food Industry

Food industry is a global multi-trillion dollar and multi-technological industry. The application of nanotechnological techniques in the production, processing or packaging of foods are termed as nanofoods (Momin and Joshi 2015). Use of nanoparticles such as micelles, liposomes, nanoemulsions, etc. as well as the development of nanosensors for ensuring food safety has gained tremendous recognition in recent years. The encapsulations of nutraceuticals, flavor enhancers within nanocapsules and nanoparticles that can selectively bind and remove chemicals from food are some of the recently used nanotechnological innovations (Özer et al. 2014). Also fabrication of food grade vitamin E nanoemulsion showed considerable antioxidant and antimicrobial activity (Dasgupta et al. 2015a). Application of nanotechnology has provided great advantages to the food industry and consumers in terms of reduction of preservatives, salts etc. Development of improved tastes and textures through nanoscale processing of foodstuffs and maintaining hygiene during food processing through antibacterial nanocoatings on food preparations are also popular (Özer et al. 2014). Moreover, enzyme assisted synthesis of gold nanoparticles and thermal co-reduction approach for synthesis of silver nanoparticles have been used to bring size variation in nanoparticles (Maddinedi et al. 2015; Dasgupta et al. 2015b). Another major application of nanotechnology is in the food packaging industry where addition of nanoparticles can improve antimicrobial properties etc. Nanoparticles also act as sensors which inform about the quality of the food (Echegoyen 2015; Ranjan et al. 2014; Dasgupta et al. 2015c). Nanobased materials have the ability to considerably enhance the functional properties of packaging materials and improve the shelf life of packaged foods.

1.3.6 Nanobiosensors

A biosensor is a sensing device to screen a particular material by measuring the biological interactions and assessing these interactions into a readable form with the help of electrochemical interpretation. As the name suggests, biosensors are used for detecting a biological specific material such as antibodies, proteins, enzymes, immunological molecules and so on (Malik et al. 2013). Presently, biosensors are applied for the rapid detection of not only body fluids but also food and environmental samples (Thakur et al 2013; Wu et al. 2013). The biosensor technology has shown constant progress in implementing nanotechnology to improve their multidetection capability and sensitivity (Sagadevan and Periasamy 2014). Nanoparticle based biosensors can be easily synthesized without any advanced fabrication processes. Metallic nanoparticles such as gold, silver, platinum, copper etc have been extensively used for their ability to increase an electronic signal. Similarly, magnetic nanoparticles, nanowires, carbon nanomaterials as well as biological nanomaterials have been employed in the synthesis of biosensors (Sagadevan and Periasamy 2014). Environmental monitoring of pollutants present in the atmosphere with help of biosensors is a pre-requisite for the removal of the harmful contaminants from the soil and waste water. Biosensors incorporated with nanomaterials have numerous applications in the real time detection of pesticides, pathogens, toxic materials, odour producing microbes in soil, air, water, etc (Baruah and Dutta 2009). For monitoring plant health and growth condition nano biosensors are being used (Parisi et al. 2015). Nanotechnology has also applications in the dairy industry where it provides thermal insulation and corrosion protection of dairy processing instruments (Baruah and Dutta 2009).

1.4 Agrinanotechnology Regulations in Major Countries

The applications of nanotechnology in the agricultural field are not in the market on a commercial scale as compared to other industrial applications. The inventions of nanotechnology remain confined to academic sector or patent ownership of large companies but fail to capture the market. This is due to the fact that presently huge capital investment is required for nanotechnological inventions which are not cost effective because of regulatory issues and low public acceptance (Parisi et al. 2015). As nanotechnology finds application in various sectors, different regulations are involved in assessing the safety of the nanoproducts. Over the past decade, there has been a rise on the commercialization of nanoscale products. This has led to a change in the regulatory regimes for the efficient application of nanotechnology in agriculture. Adapting to safe manufacturing practices, product testing and assessing potential environmental risks hazard plays a crucial role in enabling the acceptance of nanotechnology in the market (Watson et al. 2011). Different countries impose different regulations on the production and marketing of nano-based products. Here, we have discussed the regulations of few major countries engaged in nanotechnology research.

1.4.1 The United States

US agencies such as the United States Department of Agriculture (USDA), Environment Protection Agency (EPA), Food and Drug Administration, Consumer Product Safety Commission and the cabinet level National Science and Technology Council confers regulations on various aspects of agronanotechnology. The US food and drug administration's nanotechnology task force 33 reported in 2007 that no further legislations are required for the nanomaterials. The Toxic Substances Control Act formed by the California Department of Toxic Substances Control has made it mandatory for the manufacturers to provide information on nanomaterials present in specific chemicals (Watson et al. 2011; Soliman 2012). The National Institute of Food and Agriculture created under the United States Department of Agriculture, manages several nanotechnology grant programs. Environmental Protection Agency under the authority of Fedral Insecticide, Fungicide and Rodenticide Act, registers pesticides and approves nanoscale versions of even conventional substances as newly registered. The Food and Drug Administration under the authorization of the Federal Food, Drug and Cosmetic Act approves food additives, food packaging, processed foods and dietary supplements. Nanoscale materials used as food additives in the processing or packaging of foods require approval from the Food and Drug Administration as safe to be marketed. But items which are "Generally recognized as safe" or are below the threshold of regulation are exempted from the regulatory approval of Food and Drug Administration. On the other hand, certain items such as food products do not have pre-market approval and which impends; Food and Drug Administration and other public to depend on the manufacturers for their safety consent (Watson et al. 2011).

1.4.2 The European Union

European Union regulations on agronanotechnology comprise of several horizontal and vertical legislations. The Food Law Regulation (European Commission) 178/2002 establishes principles and obligations comprising all the stages of food/feed production and distribution. This helps in assessing and managing the risks associated with the use of nanomaterials in food. The Registration, Evaluation, Authorisation and restriction of Chemicals is a horizontal legislation which encompasses both bulk and nanosized materials which needs to be registered at the European Chemical Agency provided they are manufactured or imported in more than one ton per year and are not subjected to other exclusions. Regulation (European Commision) 258/97 is a vertical legislation on the Novel foods and Novel Food ingredients. This regulation makes the pre-market approval of nanoenabled foods mandatory and considers re-evaluation of any nanoscale additives for safety purpose. Directive 91/414/EEC is another vertical legislation regulating the evaluation, marketing and use of pesticides. This legislation is complemented with the Regulation (European Commission) 1107/2009 which follows strict criteria in the approval of substances to be used as pesticides (Watson et al. 2011).

1.4.3 India

The Bureau of Indian Standards lays down the standards for nanotechnology usage with safety standards. The Bureau of Indian Standards follows the guidelines and recommendations of The International Organisation of Standardisation which has developed 42 standards for nanotechnology. The Health and Safety standards mainly deal with the characterization of nanomaterials like gold, silver, zinc oxide, etc. which are used for making water filters in India. The Material Safety Data Sheet for nanomaterials provides information such as physical nature, toxicity, reactivity, health effects, etc. which are considered by The Bureau of Indian Standards for developing Indian standards for nanotechnology in water sector. In the event of an industry which manufactures nanoparticles and discharges its waste into water bodies, the Central Pollution Control Board (an apex body of the Ministry of Environment & Forests) has the authority to inspect the industry, obtain their records and report on the nature of the nanoparticles discharged (Bhati et al. 2015).

1.5 Environmental Implications and Management

Nanotechnology is an evolving technique that has transformed the scenario of global industry but certain risks are associated with the environment and human health. Its regulation and mitigation are also not very clear. The guidelines for the application of nanotechnology are introduced at the regional and national level but there is a need to focus on it at the international platform (Falkner and Jaspers 2012). The main problem for the decision and policy makers is the vagueness associated with the risks of nanotechnology which limits the development of international synchronization.

Nanomaterials can be formed either by natural or synthetic processes. Nanoparticles are said to have many commercial benefits but some are toxic in nature. The physico-chemical nature of nanomaterials can have huge impact on the fate of environment. Insufficient knowledge of nanomaterial toxicity and risks may hinder the pace of industrial application of nano-enabled technologies. Therefore, the potential activities of these materials must be studied comprehensively. In the absence of definitive data, nanomaterial research and regulations could be supported by an efficient characterization of factors leading to toxicity and risks (Linkov et al. 2009a). Studies clustered the nanomaterials into ordered risk categories using Multi-Criteria Decision Analysis by taking into account the variables associated with the toxicity and risk of nanoparticles. However, their research finding does not quantify the environmental risks but attempts were made for recommendations on precise measurements of nanoparticles (Linkov et al. 2009b).

Although nanotechnology offers significant benefits to mankind but have detrimental health effects too. Due to the small size, the nanoparticles can disperse into anatomical barriers to reach the liver, lungs, kidneys, etc. and damage the cells which can cause lesions, granulomas, cancers, Parkinson's disease, Alzheimer's disease, etc. (Chaudhry et al. 2008; Miller and Senjen 2008; Chau et al. 2007; Oberdorster et al. 2005). As a consequence, there is a potential threat to the millions of people working in vast expanses of agricultural lands due to the use of nano fertilizers and pesticides which can enter into the food chain (Rico et al. 2011). Nanomaterials in foods are detrimental for human consumption because these are chemically more reactive and have greater access to our bodies than larger particles which can even compromise with our immune response, impairment of DNA replication and transcription resulting in long term pathological effects (Hoet et al. 2004; Miller and Senjen 2008; Chaudhry et al. 2008; Chaudhry and Castle 2011; Momin et al. 2013).

Detailed understanding of the biological mechanism of nanoparticles is required before application in the field. This will enable the formulation of regulatory rules and help in risk management. Prior to use of products containing nanoparticles in the agricultural sector, a scrupulous analysis is needed for the distribution and amount of particles absorbed by the plants in the food chain. The evolution of a participatory, dynamic and responsive nanotechnology policy is required to develop coordinated risk management strategy in Indian agriculture and food system for the positive economic impacts of this technology to reach the agrarian society (Sastry et al. 2010; Sastry and Rao 2013).

There should be vigorous training of the personals involved in nano agricultural sector. The use of Personal Protective Equipment should be stringent for the workers associated with nanomaterials. Till date, researchers have been trained to cater to the needs of the industry but should be skilled enough to carry out new research experiments, analyse and interpret the results and simultaneously be able to incorporate and implement the theory, tools, and techniques of nanotechnology. There should also be strict rules for the nanotechnology industry to adhere to the engineering controls for effective risk management.

Many countries have recommended various precautionary measures to manage the risk associated with nanoparticles. The Health and Environment Alliance, Belgium recommends a list of measures to manage the risks associated with nanomaterials and stop the commercial sale until appropriate measures are taken. There is a need of a comprehensive policy by uniform international body to alleviate the risk involved with nanotechnology applications.

1.6 Limitation of nanotechnology for Intellectual Property Rights protection

Nanotechnology related inventions are upsurging across all Research & Development and industry sectors. Millions of patents on nanotechnology can be retrieved in the world wide database. For searching prior art on nanotechnology, the patent offices worldwide has classified nanotechnology under the International Patent Classification (WPN 2015). The United States Patent and Trademark Office received around 618330 patent applications in the fiscal year 2014, with a sum total of 24,090 patents related to nanotechnology granted in 2014, which is a growth of more than 12.68 % in the number of nanotechnology patents in comparison to 2013 (USPTO 2014). To grip on the number of patents and scrutinize a nano-invention poses serious concerns among the patent grantees to review that the new patent has additional utility as well as novelty over earlier inventions. Simply submitting a smaller version of a known structure is not patentable. While nanotechnology opens opportunities for new inventions and discoveries, new challenges are confronted at the patent office as the grantees may lack expertise or adequate domain knowledge to assess nanotechnology patent applications leading to disregard the previous inventions (Paradise 2012). The recent increase of patent applications by universities and private sectors are also another potential challenge for IPR protection for rewarding innovations with patent grant (Heller and Eisenberg 1998). The patent owners can exercise their right to exclude others to undergo research work on its invention, which can seriously restrict future research on nano-inventions. Premature disclosure of trade secrets by pressure from the academic circles to publish or report in government funding projects can reduce the ability to secure patents. Shortcuts in drafting a provisional application also increases the possibility of falling short of minimum disclosure requirements causing limitations in scope of claims that can be supported by the final specification (Tullis 2012). The recent euphoria for patenting and the inability of patent offices to handle a large number of patent applications has resulted in the rejection of valid claims (Miller et al. 2004). Commercialization of nanotechnology in the agri-food sector requires a thorough assessment of toxicity and safety concerns before hitting the market (Sastry et al. 2011). A large number of patents have been filed in the area of nano food (eg. nanoemulsions of vitamins, flavours and colorants) and in agricultural system for the delivery of agricultural chemicals and targeted drug delivery (Srinivas et al. 2010; Ehr et al. 2011). Measures should be taken to assess those success patent claims in the field trial before granting patent rights. Regulatory rules to be formulated and thorough investigation is required before these innovations could be transformed into commercialization.

Patent protection is effective only within the issuing country. Thus increased administrative effort is required for securing international patents. Strong patent protection may spur research and invention but also lead to a patent thicket. Though a patent is granted to a nanotechnological invention but becomes invalid due to the operation of a previously granted patent for a similar type of invention leading to expensive litigation. Innovators lacking the resources to litigate patent validity may be forced to license these patents rather than contest them (Barpujari 2010). Delay in commercialization of nanotechnology products will give rise to problems such as unintentional infringement of patents, creating business uncertainty and concerns over patents with some nanotechnology inventions, ending up in courts rather than by the patent office (Clarkson and DeKorte 2006). The nanotechnology IPR landscape is still somewhat chaotic and limitations can be observed when most of the patents do not result in active commercialization.

1.7 Intellectual Property Rights Regulations for Nanoagritechnology

There is no doubt that nanotechnology is the next revolution "in standing" to flourish the agriculture sector. Although patents were granted to nanotechnological applications and processes but more cohesive and clearer policy between different countries is necessary. Nanaotechnology is a multidisciplinary field with application in diverse arena. Therefore, patent granters and examiners need to be trained both in international, national and regional level about its multidimensional aspect. WIPO can play a pivotal role by networking collaboration between different agencies like WHO, World Trade Organization (WTO) where each having distinct, but complementary mandates to work on issues relating to nanotechnology patent issues, public health aspect and trade policy respectively. Most of the developed countries have included the "nanotechnology" term in their existing patent classification system and assigned a separate class. But developing countries like India are yet to assign a separate class and lacking peer policies on nanotechnology. This can be revamped in line of Drug and Cosmetic Amendment Bill 2015, where the word "Drug, Medical technology, Clinical trials" etc. are included and regulations are clearly mentioned.

Another way of promoting nanotechnology research and smoothening patent grants is "stronger public-private partnership". One way to achieve this is cohesive licensing model where incentive to research and development driven companies may be given to promote nanotechnology products. It will help to control price in future which is one of the major hurdle in other fields like drug industry. Continuous discussions are being made in different platforms and policies are being amended but concentrating solely on intellectual property rights will not help for success stories. Different international and national regulatory agencies on social, health and economics should come under same platform to help on policies formulation for promoting innovation on nanotechnology. Countries should increase their gross domestic product investment in nanotechnology research to promote innovation especially in agriculture sector, where nanotechnology holds a potential for "Agriculture Revolution".

References

Ali A, Sinha K (2014) Prospects of nanotechnology development in the health sector in India. Int J Health Sci 2(2):109–125

Anjali CH, Sharma Y, Mukherjee A, Chandrasekaran N (2012) Neem oil (Azadirachta indica) nanoemlsion – a potent larvicidal agent against Culex quinquefasciatus. Pest Manag Sci 68(2):158–163. doi:10.1002/ps.2233

Aouada FA, de Moura MR (2015) Nanotechnology applied in agriculture: controlled release of agrochemicals. In: Rai M, Ribeiro C, Mattoso L, Duran N (eds) Nanotechnologies in food and agriculture. Springer International Publishing, Cham, pp 103–118. doi:10.1007/978-3-319-14024-7_5

Arnold B, Keserü (2013) Nanotechnology and law in the 21st century. In: Cofola 2013: the conference proceedings, Brno, pp 178–188

Bandyopadhyay KK, Hati KM, Singh R (2009) Management options for improving soil physical environment for sustainable agricultural production: a brief review. J Agric Phys 9:1–8

Barpujari I (2010) The patent regime and nanotechnology: issues and challenges. J Intellect Property Rights 15:206–213

Baruah S, Dutta J (2009) Nanotechnology applications in pollution sensing and degradation in agriculture: a review. Environ Chem Lett 7(3):191–204. doi:10.1007/s10311-009-0228-8

Bergesion LL, Cole MF (2015) Regulatory implications of nanotechnology. In: Sutariya VB, Pathak Y (eds) Biointeractions of nanomaterials. CRC Press, Boca Raton, pp 315–346

Bhati M, Kumar SR, Bansal P (2015) Nanotechnology application in water sector from the beginning to end of the value chain: an Indian perspective. ISTIP Policy Bulletin No. 6. Available online at: http://ifwn.org/wp-content/uploads/2015/04/Nanotechnology-Application-in-Water-Sector-NISTADS-report.pdf

Bhattacharya S, Shilpa, Pushkaran JA (2012) Nanotechnology research and innovation in India: drawing insights from bibliographic and innovation indicators. CSIR-NISTADS strategy paper II

Bhattacharya S, Saha I, Mukhopadhyay A, Chattopadhyay D, Ghosh UC, Chatterjee D (2013) Role of nanotechnology in water treatment and purification: potential applications and implications. Int J Chem Sci Technol 3(3):59–64

Bucheli T (2014) Agricultural applications of nanotechnology. In: Parisi C, Vigani M, Rodriguez Cerozo E (eds).Proceedings of a workshop on nanotechnology for the agricultural sector: from research to the field, Seville, European Commission Joint Research Centre, Institute for Prospective Technological Studies, Luxembourg, pp 10–12

Chau CF, Wu SH, Yen GC (2007) The development of regulations for food nanotechnology. Trends Food Sci Technol 18(5):269–280. doi:10.1016/j.tifs.2007.01.007

Chaudhry Q, Castle L (2011) Food applications of nanotechnologies: an overview of opportunities and challenges for developing countries. Trends Food Sci Technol 22(11):595–603. doi:10.1016/j.tifs.2011.01.001

Chaudhry Q, Scotter M, Blackburn J, Ross B, Boxall A, Castle L, Aitken R, Watkins R (2008) Applications and implications of nanotechnologies for the food sector. Food Add Contam Part A 25(3):241–258. doi:10.1080/02652030701744538

Chowdhury P, Khan SA, Dutta P, Mahanta J (2014) Pharmaceutical and biotech product patents in India: doldrums or blissfulness? Nat Biotechnol 32:133–134. doi:10.1038/nbt.2809

Clarkson G, DeKorte D (2006) The problem of patent thickets in convergent technologies. Ann NY Acad Sci 1093:180–200. doi:10.1196/annals.1382.014
Dasgupta N, Ranjan S, Mundra S, Chidambaram R, Kumar A (2015a) Fabrication of food grade vitamin E nanoemulsion by low energy approach, characterization and its application. Int J Food Prop. doi:10.1080/10942912.2015.1042587
Dasgupta N, Ranjan S, Rajendran B, Manickam V, Ramalingam C, Avadhani GS, Kumar A (2015b) Thermal co-reduction approach to vary size of silver nanoparticle: its microbial and cellular toxicology. Environ Sci Pollut Res Int. doi:10.1007/s11356-015-4570-z
Dasgupta N, Ranjan S, Mundekkad D, Ramalingam C, Shanker R, Kumar A (2015c) Nanotechnology in agro-food: from field to plate. Food Res Int 69:381–400
Ditta A (2012) How helpful is nanotechnology in agriculture? Adv Nat Sci Nanosci Nanotechnol 3(3):1–10. doi:10.1088/2043-6262/3/3/033002
Drug and Cosmetic act bill, India, amended 2015
Echegoyen Y (2015) Nano-developments for food packaging and labeling applications. In: Rai M, Ribeiro C, Mattoso L, Duran N (eds) Nanotechnologies in food and agriculture. Springer International Publishing, Cham, pp 141–166. doi:10.1007/978-3-319-14024-7_7
EFSA Scientific Committee (2011) Scientific opinion on guidance on the risk assessment of the application of nanoscience and nanotechnologies in the food and feed chain. EFSA J 9(5):2140 [36 pp.] doi:10.2903/j.efsa.2011.2140
Ehr RJ, Thomas HK, Dale CS, Mike PT, Kerrm YY, Qiang Z et al (2011) Mesosized capsules useful for the delivery of agricultural chemicals. US Patent Number: US 20110045975A1
Escoffier L (2006) Nanotechnology: the revolutionary technology seen from the U.S. and European perspective. DIME working papers on Intellectual Property Rights, Paper No. 14, Birkbeck Centre for Innovation Management Research, Luca Escoffier, Queen Mary IP Reserach Institute London. Available online at: http://www.bbk.ac.uk/innovation/publications/dime/docs/WP14-IPR.pdf
European Commission (2011) Commission recommendation of 18 October 2011 on the definition of nanomaterial (2011/696/EU). Off J Eur Union L275:38–40
Falkner R, Jaspers N (2012) Regulating nanotechnologies: risk, uncertainty and the global governance gap. Global Environ Politics 12(1):30–55. doi:10.1162/GLEP_a_00096
FAO (1995) Application of risk analysis to food standards issues: report of the joint FAO/WHO expert consultation, Geneva, Switzerland, Ch-5
FAO/WHO (2013) State of the art on the initiatives and activities relevant to risk assessment and risk management of nanotechnologies in the food and agriculture sectors. FAO/WHO technical paper. Available online at: http://www.fao.org/docrep/018/i3281e/i3281e.pdf
Fink C, Primo Braga CA (1999) How stronger protection of intellectual property rights affects international trade flows. World Bank Policy Research Working Paper No. 2051. The World Bank, Washington, DC
Foster L E (2007) Nanotechnology, 3rd edn. Pearson Education, India, pp 106–108
Havas A (2014) Trapped by the high-tech myth: the need and chances for a new policy rationale. In: Hirsch-Kreinsen H, Schwinge I (eds) Knowledge-intensive entrepreneurship in low-tech industries. Edward Elgar Publishing Limited, Glos, UK, pp 193–217
Heller MA, Eisenberg RS (1998) Can patents deter innovation? The anticommons in biomedical research. Science 280(5364):698–701. doi:10.1126/science.280.5364.698
Helpman E (1993) Innovation, imitation, and intellectual property rights. Econometrica 61(6):1247–1280
Hoet PH, Bruske-Hohlfeld I, Salata OV (2004) Nanoparticles – known and unknown health risks. J Nanobiotechnol 2(1):12. doi:10.1186/1477-3155-2-12
Jaiswal RR (2014) Phase II of nano-mission approved. The Science: Science and Technology for IAS. Available online at http://thescience.co.in/phase-ii-of-nano-mission-approved/
Kah M, Hofman T (2014) Nanopesticide research: current trends and future priorities. Environ Int 63:224–235
Kisluik B (2010) Nanotechnology-related issues at the United States Patent and Trademark Office. Available online at http://www.nist.gov/tpo/upload/BruceKisliuk.pdf

Kumar A (2014) Nanotechnology development in India: an overview. Research and Information System for Developing Countries (RIS) research paper # 193

Linkov I, Steevens J, Chappel M, Tervonen T, Figueira JR, Merad M(2009a) Classifying nanomaterial risks using multi-criteria decision analysis. In: Linkov I, Steevens J(eds) Nanomaterials: risks and benefits. NATO Science for Peace and Security Series C: Environmental Security, pp 179–191. doi: 10.1007/978-1-4020-9491-0_13

Linkov I, Steevens J, Adlakha-Hutcheon G, Bennett E, Chappell M, Colvin V, Davis JM, Davis T, Elder A, Foss Hansen S, Hakkinen PB, Hussain SM, Karkan D, Korenstein R, Lynch I, Metcalfe C, Ramadan AB, Satterstrom FK (2009) Emerging methods and tools for environmental risk assessment, decision-making, and policy for nanomaterials: summary of NATO Advanced Research Workshop. J Nanopart Res 11(3):513–527. doi:10.1007/s11051-008-9514-9

Maddinedi SB, Mandal BK, Ranjan S, Dasgupta N (2015) Diastase assisted green synthesis of size-controlled gold nanoparticles. RSC Adv 5:26727–26733. doi:10.1039/C5RA03117F

Malik P, Katyal V, Malik V, Asatkar A,Inwati G, Mukherjee TK (2013) Nanobiosensors: concepts and variations. ISRN Nanomaterials, volume 2013. http://dx.doi.org/10.1155/2013/327435

Mc Murray TA, Dunlop PSM, Byrne JA (2006) The photocatalytic degradation of atrazine on nanoparticulate TiO2 films. J Photochem Photobiol A Chem 182(1):43–51

McLennan A, Rimmer M (2012) Cosmo, cosmolino: patent law and nanotechnology. In: Rimmer M, McLennan A (eds) Intellectual properties and emerging technologies: the new biology. Edward Elgar Pub, Glos, UK, pp 255–290

Meulen BVD, Bremmers H, Purnhagen K, Gupta N, Bouwmeester H, Geyer LL (2014) Governing nano foods: Principles-based responsive regulation. EFFoST critical reviews # 3. Academic, London, p 70

Milani N, McLaughlin MJ, Stacey SP, Kirkby JK, Hettiarachchi GM, Beak DG, Cornelis G (2012) Dissolution kinetics of macronutrient fertilizers coated with manufactured zinc oxide nanoparticles. J Agric Food Chem 60:3991–3998. doi:10.1021/jf205191y

Miller G, Senjen R (2008) Out of the laboratory and on to our plates: nanotechnology in food and agriculture. Friends of the Earth Australia, Europe and USA. Available online at: http://www.denix.osd.mil/cmrmd/upload/Nanotech_Food_Agric.pdf

Miller JC, Serrato R, Represas-Cardenas JM, Kundahl G (2004) The handbook of nanotechnology: business, policy and intellectual property law. Wiley, Hoboken

Momin JK, Joshi BH (2015) Nanotechnology in foods. In: Rai M, Ribeiro C, Mattoso L, Duran N (eds) Nanotechnologies in food and agriculture. Springer International Publishing, Cham, pp 3–24. doi:10.1007/978-3-319-14024-7_1

Momin JK, Jayakumar C, Prajapati JB (2013) Potential of nanotechnology in functional foods. Emir J Food Agric 25(1):10–19. doi:10.9755/ejfa.v25i1.9368

Oberdorster G, Oberdorster E, Oberdorster J (2005) Nanotoxicology: an emerging discipline evolving from studies of ultrafine particles. Environ Health Perspect 113(7):823–839. doi:10.1289/ehp.7339

OECD (2011) Fostering nanotechnology to address global challenges: water. OECD, Paris, Available online at: http://www.oecd.org/sti/nano/47601818.pdf

Özer EA, Özcan M, Didin M (2014) Nanotechnology in food and agriculture industry. In: Malik A, Erginkaya Z, Ahmad S, Erten H (eds) Food processing: strategies for quality assessment. Food engineering series. Springer, New York, pp 477–497. doi: 10.1007/978-1-4939-1378-7_19

Paradise J (2012) Claiming nanotechnology: improving USPTO efforts at classification of emerging nano-enabled pharmaceutical technologies. Northwest J Technol Intellect Prop 10(3):169–208

Parisi C, Vigani M, Rodríguez-Cerezo E (2014) Proceedings of a workshop on "Nanotechnology for the agricultural sector: from research to the field". JRC scientific and policy reports

Parisi C, Vigani M, Rodríguez-Cerezo E (2015) Agricultural nanotechnologies: what are the current possibilities? Nanotoday 10(2):124–127. doi:10.1016/j.nantod.2014.09.009

Perez-de-Luque A, Hermosín MC (2013) Nanotechnology and its use in agriculture. In: Bagchi D, Bagchi M, Moriyama H, Shahidi F (eds) Bio-nanotechnology: a revolution in food, biomedical and health sciences. Blackwell Publishing Ltd., Oxford. doi:10.1002/9781118451915.ch20

Prassana BM (2006) Nanotechnology in agriculture, ICAR National Fellow, Division of Genetics, IARI, New Delhi- 110012. Available online at: http://www.iasri.res.in/ebook/EBADAT/6-Other%20Useful%20Techniques/10-nanotech_in_Agriculture__BM_Prasanna__1.2.2007.pdf

Raimond RR (2008) Ethical considerations regarding the international development and application of nanotechnology and nanoscale materials. In: 37th annual conference on environmental law, Keystone, Colorado

Raliya R, Tarafdar JC, Gulecha K, Choudhury K, Ram R, Mal P, Saran RP (2013) Review article; Scope of nanosciences and nanotechnology in Agriculture. J Appl Biol Biotechnol 1(3):41–44. doi:10.7324/JABB.2013.1307

Ranjan S, Dasgupta N, Chakraborty AR, Samuel SM, Ramalingam C, Shanker R, Kumar A (2014) Nanoscience and nanotechnologies in food industries: opportunities and research trends. J Nanopart Res 16:2464. doi:10.1007/s11051-014-2464-5

Rauscher H (2014) Role of JRC-IHCP in Nanotechnology: overview of policy support activities on nanomaterials safety. In: Proceedings of a workshop on "Nanotechnology for the agricultural sector: from research to the field". pp 23. Available online at: https://ec.europa.eu/jrc/sites/default/files/ipts_jrc_89736_(online)__final.pdf

Rico CM, Majumdar S, Duarte-Gardea M, Peralta-Videa JR, Gardea-Torresdey JL (2011) Interaction of nanoparticles with edible plants and their possible implications in the food chain. J Agric Food Chem 59(8):3485–3498. doi:10.1021/jf104517j

Sagadevan S, Periasamy M (2014) Recent trends in nanobiosensors and their applications – a review. Rev Adv Mater Sci 36:62–69

Sastry RK, Rao NH (2013) Emerging technologies for enhancing Indian agriculture-case of nano-biotechnology. Asian Biotechnol Dev Rev 15(1):1–9

Sastry KR, Rashmi HB, Rao NH, Ilyas SM (2009) Nanotechnology and agriculture in India: the second green revolution? In: Invited paper presented in: presented in session IV, OECD conference on potential environmental benefits of nanotechnology. Fostering Safe Inovation- Led Growth, OECD Conference Centre, Paris, France

Sastry RK, Rashmi HB, Rao NH, Ilyas SM (2010) Integrating nanotechnology into agri-food systems research in India: a conceptual framework. Technol Forecast Soc Change 77(4):639–648. doi:10.1016/j.techfore.2009.11.008

Sastry RK, Rashmi HB, Rao NH (2011) Nanotechnology for enhancing food security in India. Food Policy 36(3):391–400

Schauwecker M (2011) Nanotechnology inventions in U.S. patent law. TTLF working paper

Scientific Committee EFSA (2009) Scientific opinion of the scientific committee on a request from the European Commission on the potential risks arising from nanoscience and nanotechnologies on food and feed safety. EFSA J 958:1–39

SCP (2009) Exclusions from patentable subject matter and exceptions and limitations to the rights. WIPO Doc. WIPO SCP/13/3

Sekhon BS (2014) Nanotechnology in agri-food production: an overview. Nanotechnol Sci Appl 7:31–53. doi:10.2147/NSA.S39406

Selin C (2007) Expectations and the emergence of nanotechnology. Sci Technol Hum Values 32:196–220

Shand H, Wetter KJ (2007) Trends in intellectual property and nanotechnology: implications for the global south. J Intellectual Property Rights 12:111–117

Sherman B, Bently L (1999) The making of modern intellectual property law: the British experience 1760–1911. Cambridge University Press, Cambridge

Shrivastava S, Dash D (2009) Agrifood nanotechnology: a tiny revolution in food and agriculture. J Nano Res 6:1–14

Smith EG (2013) U.S.-based inventors lead world in nanotechnology patents: study. Reuters. Available online at http://www.reuters.com/article/2013/02/14/us-patents-nanotechnology-idUSBRE91D0YL20130214

Soliman A (2012) The need for stronger nanotechnology regulation., Food Safety News. Available online at, http://www.foodsafetynews.com/2012/10/why-we-should-have-more-regulations-on-nanotechnology/#.VbhnQ3qqo7s

Srinivas PR, Philbert M, Vu TQ, Huang Q, Kokini JL, Saos E et al (2010) Nanotechnology research: applications in nutritional sciences. J Nutr 140:119–124. doi:10.3945/jn.109.115048

Suppan S (2011) Racing ahead U.S. Agri-Nanotechnology in the absence of regulation. Institute for Agriculture and Trade Policy, Minneapolis

Sweeney AE (2006) Social and ethical dimensions of nanoscale science and engineering research. Sci Eng Ethics 12:435–464

National Science and Technology Council (2014) National nanotechnology initiative: strategic plan

TERI (2009) Regulatory challenges posed by Nanotechnology Developments in India TERI project: capability, governance, and nanotechnology developments – a focus on India. The Energy and Resources Institute, New Delhi. [Project Report No. 2006ST21: D6]

Thakur MS, Ranjan R, Vinayaka AC, Abhijith KS, Sharma R (2013) Nanoparticles and biophotonics as efficient tools in resonance energy transfer-based biosensing for monitoring food toxins and pesticides. In: Park B, Apell M (eds) Advances in applied nanotechnology for agriculture, vol 1143, ACS symposium series., pp 5–84. doi:10.1021/bk-2013-1143.ch004

Torney F, Trewyn BG, Lin VSY, Wang K (2007) Mesoporous silica nanoparticles deliver DNA and chemicals into plants. Nat Nanotechnol 2(5):295–300. doi:10.1038/nnano.2007.108

Tullis TK (2012) Current intellectual property issues in nanotechnology. Nanotechnol Rev 1(2):189–205. doi:10.1515/ntrev-2012-0501

USPTO (2014) Performance & accountability report., Available online at, http://www.uspto.gov/about/stratplan/ar/USPTOFY2014PAR.pdf

Vamvakaki V, Chaniotakis NA (2007) Pesticide detection with a liposome- based nano- biosensor. Biosens Bioelect 22(12):2848–2853

Watson SB, Gergely A, Janus ER (2011) Where is "Agronanotechnology" heading in the United States and European Union? Nat Resour Environ 26(1):8

WPN (2015) Policy environments and governance for innovation and sustainable growth through nanotechnology. DSTI/STP/NANO (2013)13/FINAL. Available online at:http://www.oecd.org/officialdocuments/publicdisplaydocumentpdf/?cote=DSTI/STP/NANO(2013)13/FINAL&doclanguage=en

Wu X, Chen J, Park B, Huang YW, Zhao Y (2013) The use of silver nanorod array based surface enhanced raman scattering sensor for food safety applications. In: Park B, Appell M (eds) Advances in applied nanotechnology for agriculture, vol 1143, ACS symposium series., pp 85–108. doi:10.1021/bk-2013-1143.ch005

Zornoza L (2012) European commission opens second regulatory review of nanomaterials., Available online at, http://www.raps.org/focus-online/news/news-article-view/article/2350/

Chapter 2
Nanobiotechnology Methods to Incorporate Bioactive Compounds in Food Packaging

Adriano Brandelli, Luis Fernando Wentz Brum, and João Henrique Zimnoch dos Santos

Abstract Plastics have been successfully used as packaging materials, but increased environmental concerns lead to a growing interest on alternative biocompatible materials. The development of bionanocomposites for application in the food packaging sector is a major emergent fields in nano-research. Encapsulation can be done using procedures such as spray drying, spray cooling and chilling, fluidized bed coating, coacervation, liposome entrapment, co-crystallization, nanoemulsion, air suspension, freeze-drying, interfacial polymerization, molecular inclusion and sol–gel. Lipid-based nanostructures such as liposomes and solid-lipid nanoparticles are used for the delivery of bioactive molecules showing antimicrobial and antioxidant activities. Fat-soluble vitamins can be incorporated in nanoemulsions, which, added to polymeric nanoparticles, are suitable vehicles to phytochemicals like quercitin and curcumin. Nanofibers can also be used to incorporate bioactive substances in food packaging. This chapter presents methods for encapsulation of bioactive compounds into nanostructures, which can then be incorporated into packaging materials to improve food quality and safety.

Keywords Active packaging • Antimicrobial • Antioxidant • Bioactive molecules • Encapsulation • Liposomes • Nanocomposites • Nanoemulsion • Nanofibers • Nanostructures

---

A. Brandelli (✉)
Laboratório de Bioquímica e Microbiologia Aplicada, Instituto de Ciência e Tecnologia de Alimentos, Universidade Federal do Rio Grande do Sul, Porto Alegre, Brazil
e-mail: abrand@ufrgs.br

L.F.W. Brum
Laboratório de Bioquímica e Microbiologia Aplicada, Instituto de Ciência e Tecnologia de Alimentos, Universidade Federal do Rio Grande do Sul, Porto Alegre, Brazil

Laboratório de Catálise Ziegler-Natta, Instituto de Química, Universidade Federal do Rio Grande do Sul, Porto Alegre, Brazil

J.H.Z. dos Santos
Laboratório de Catálise Ziegler-Natta, Instituto de Química, Universidade Federal do Rio Grande do Sul, Porto Alegre, Brazil

S. Ranjan et al. (eds.), *Nanoscience in Food and Agriculture 2*, Sustainable Agriculture Reviews 21, DOI 10.1007/978-3-319-39306-3_2

2.1 Introduction

A package can be defined as the material unit that functions as barrier between the contents and the exterior atmosphere. Packaging materials should provide containment, protection, convenience and communication, the four basic functions of traditional packaging, but also create appropriate conditions that are essential for obtaining a suitable shelf life and maintaining quality and safety during transportation and storage of products. The basic assets of packaging materials are suitable mechanical, optical, and thermal properties, but an adequate food package should also avoid gain or loss of moisture, prevent microbial contamination, avoid loss of nutrients and act as a barrier against permeation of water vapor, oxygen, carbon dioxide, and other volatile compounds such as flavors and taints (Rhim et al. 2013). The most innovative advances in packaging are associated with the development of biodegradable/biocompatible materials and "active" and "intelligent" packaging. Active packaging aims the shelf life extension and also the preservation or improvement of product quality, whereas the purpose of intelligent packaging is to provide indication and monitor the freshness of the product (Dainelli et al. 2008; Vanderroost et al. 2014).

Plastics are remarkable materials for packaging due to their low price, ease of processing and modeling, lightness and flexibility of thermal and mechanical properties. Food packaging represents the largest demanding request for plastics in the form of films, sheets, bottles, cups, tubs, trays, among others (Lagaron and Lopez-Rubio 2011; Rhim et al. 2013). The most frequently used polymers are petroleum-based materials like polypropylene (PP), polyvinyl chloride (PVC), polystyrene (PS), polyethylene terephthalate (PET), and various grades of polyethylene (PE) (Marsh and Bugusu 2007). Although packaging can help to reduce organic waste by preserving foods, the substantial increase in the use of synthetic plastics raised several environmental concerns (Siracusa et al. 2008). Consequently, development and use of biodegradable and/or eco-friendly materials has received increasing attention. Unfortunately, biopolymers present relatively poor mechanical and barrier properties for packaging applications, which presently limit their industrial use. Since biopolymers are essentially hydrophilic, the improvement of moisture barrier characteristics becomes a special challenge. The functional properties of natural polymers have been improved by reinforcement of the biopolymer matrix with nanoparticles, resulting in a new class of materials called bionanocomposites (Lagaron and Lopez-Rubio 2011; Vanderroost et al. 2014).

These advances on bionanocomposites may be helpful for the development of eco-friendly active packaging. Active packaging is designed to intentionally incorporate compounds that would be released into or absorb substances from the packaged food or the food surrounding environment (Silvestre et al. 2011). An active compound can be incorporated inside the packaging material or onto its surface, in multilayer structures or in particular elements associated with the packaging such as sachets, labels or bottle caps (Dainelli et al. 2008; Vanderroost et al. 2014). Various applications of nanoparticles in the food industries are focused in (i) sensory improvements (flavor/color enhancement, texture modification), (ii) increased absorption and targeted delivery of nutrients and bioactive compounds, (iii) stabilization of active ingredients such as nutraceuticals in food structures, (iv) packaging and product innovation to increase shelf life, (v) sensors to assess the safety of food, (vi) as an antimicrobial against food borne pathogenic bacteria (Ranjan et al. 2014).

All of the agricultural activities and the corresponding manufacturing ones generate considerable quantities of different types of by-products and waste. There is a worldwide concern regarding the use of industrial waste. Such waste is the part of raw material that is rejected during processing in the food industry. These residues are rich in nutrients as bioactive compounds; thus, they can be improved and incorporated into human diet or other industries processes (da Silva and Jorge 2014; Naziri et al. 2014).

Through encapsulation, bioactive compounds such as antimicrobials, antioxidants and enzymes can be protected against hash processing conditions of package manufacture and may have improvement in controlled release behavior during storage (Sant'Anna et al. 2011; McClements 2014). Thus, the objective of this paper is to discuss different methods for encapsulation of bioactive compounds into nanostructures, which can be incorporated into packaging materials to improve food quality and safety.

2.2 Encapsulation of Bioactive Molecules

Figure 2.1 illustrates a survey of the last 10 years considering the keywords *encapsulation*, *bioactive* and *food* and shows an increasing trend with regard to the number of publications, of which 23 articles have already been published in the six first months of 2015.

Encapsulation is a useful tool to improve delivery of bioactive molecules (e.g. antioxidants, antimicrobials, minerals, vitamins, phytosterols, lutein, fatty acids, lycopene) and living cells, e.g. probiotics, into foods (Nedovic et al. 2011). The addition of bioactive ingredients or living cells should not cause undesirable effects on the sensory properties, color or flavor of food products. A number of encapsulation techniques can be used to produce nano or microparticulate systems, being emulsification, evaporation, spray-drying and coacervation the most extensively used (Champagne and Fustier 2007; Gómez-Mascaraque et al. 2015).

In general, encapsulation of active agents aims at: (i) stabilizing the material through the formation of enveloping wall; (ii) tuning the structure of the encapsulating material to promote the suitable leaching/leakage and to avoid the insertion/migration of unwanted components within the encapsulating device. The development of edible nano- or microencapsulation matrices has been envisaged as a plausible option to protect these biologically active compounds against adverse conditions. The classical systems developed in nano- or microencapsulation are based on reservoir or matricial particles (Fang and Bhandari 2010; Gómez-Mascaraque et al. 2015). These systems may constitute a physico-chemical barrier

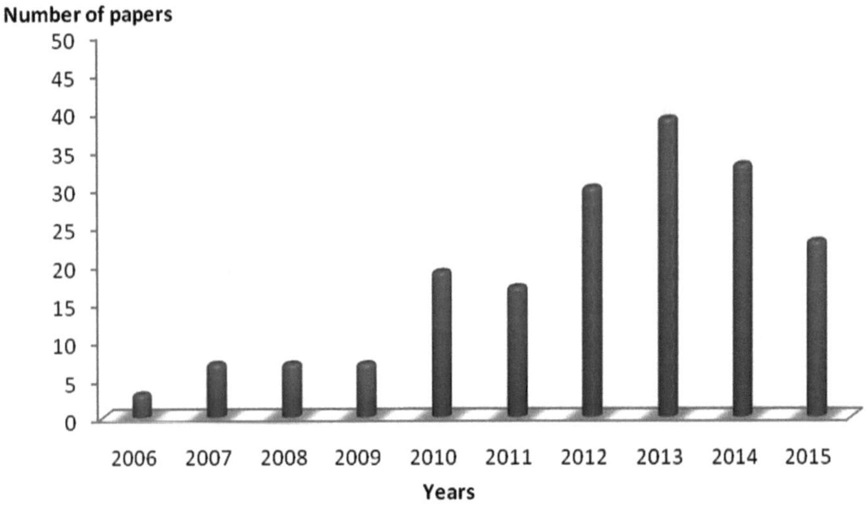

Fig. 2.1 Survey of the last 10 years considering the keywords *encapsulation*, *bioactive* and *food*. An increasing trend in publications can be observed from 2010. Values for 2015 correspond to the six initial months (Database: Scopus. Available at: www.scopus.com)

against prooxidant elements such as free radicals, oxygen or UV. Encapsulation may also represents a mean to improve biological efficiencies such as shelf life, control of delivery of active components, and could prevent the onset of side effects (Gonnet et al. 2010).

Many encapsulation procedures have been proposed but none of them can be considered as a universally applicable technique for a bioactive or food component. This is because individual food components have their own typical molecular structure. Encapsulation processes should lead to particles with a high encapsulation degree, good polydispersity index (PDI) and elevated shelf life (Gonnet et al. 2010; Fang and Bhandari 2010). Sophisticated shell materials and technologies have been developed and an extremely wide variety of functionalities can now be achieved through microencapsulation. In most cases, encapsulation refers to a technology in which the components are completely enveloped, covered and protected by a physical barrier, without any protrusion of the components. Also, encapsulation has been defined as a technology of packaging solids, liquids, or gaseous materials into small capsules that release their contents at controlled rates over prolonged periods and under specific conditions (Gouin 2004; Nedovic et al. 2011).

Most microcapsules are small spheres with diameters ranging from few micrometers to few millimeters. However many of these microcapsules bear little resemblance to those of simple spheres. In fact, both size and shape of formed microparticles depend on the materials and methods used to prepare them (Champagne and Fustier 2007; Gharsallaoi et al. 2007). As particle size affects texture, the addition of large particles is undesirable in most cases, explaining why encapsulation becomes crucial in the food sector. It must be stressed, however, that the food industry is severely limited with respect to the compounds that can be used for encapsulation. Compatibility with the food components is not the only requirement that an encapsulation procedure has to meet. It also should have specific characteristics to withstand influences from the environment (de Vos et al. 2010).

The current encapsulation systems include different types of micro- or nanoparticles (capsules and spheres) produced from a wide range of wall materials (monomers and/or polymers) by a large number of different processes. These include spray drying, spray cooling and chilling, extrusion, fluidized bed coating, coacervation, liposome entrapment, centrifugal suspension separation, co-crystallization, nanoemulsion, air suspension, freeze-drying, interfacial polymerization, molecular inclusion, sol–gel, among other. However, many techniques used in the non-food sectors cannot be simply extended to food ingredients. Thus, encapsulation techniques can be performed considering the limited options of approved food-grade compounds (Gharsallaoi et al. 2007; Champagne and Fustier 2007; Fang and Bhandari 2010).

When encapsulation is aimed to prevent excessive degradation of a sensitive ingredient or to reduce flashing off volatile flavors, the cost-in-use must be lower than the non-encapsulated ingredient. However, if encapsulation provides the ingredient a special property, the cost-in-use can be slightly higher than the non-encapsulated ingredient. To protect food from microbial spoilage and prevent food borne outbreaks one possible approach is the application of encapsulated antimicrobial coatings on food contact materials or incorporation of antimicrobial compounds in food packaging to kill or inhibit the contaminating microorganisms and ensure the safety of the food during storage as well as distribution (Gouin 2004; Gharsallaoi et al. 2007).

Figure 2.2 illustrates in more detail the bioactive delivery characteristics of some of the more commonly used nanosystems, namely liposomes, nanocapsules, nanospheres and micelles or nanoemulsions. Bioactive compounds are encapsulated in different ways according to the synthesized frame: through hydrophobic and hydrophilic regions of the liposomes; the formation of a spherical wall for the nanocapsules; aggregation of the components within or on the surface of nanospheres; or stabilization liquid and solid agents or hydrophilic and hydrophobic within the micelles. Particle size is crucial in determining the particle's kinetics and biodistribution. Particle diameters of 10–100 nm accumulate more readily within cells than larger particles (Brandelli 2012; Brandelli and Taylor 2015).

In the following sections, some strategies for incorporating bioactive compounds into nanostructures are presented. Diverse nanostructures, such as nanoliposomes, polymeric nanoparticles, nanoemulsions, nanofibers and nanocomposite polymers are available for encapsulation of antimicrobial, antioxidant and other bioactive molecules.

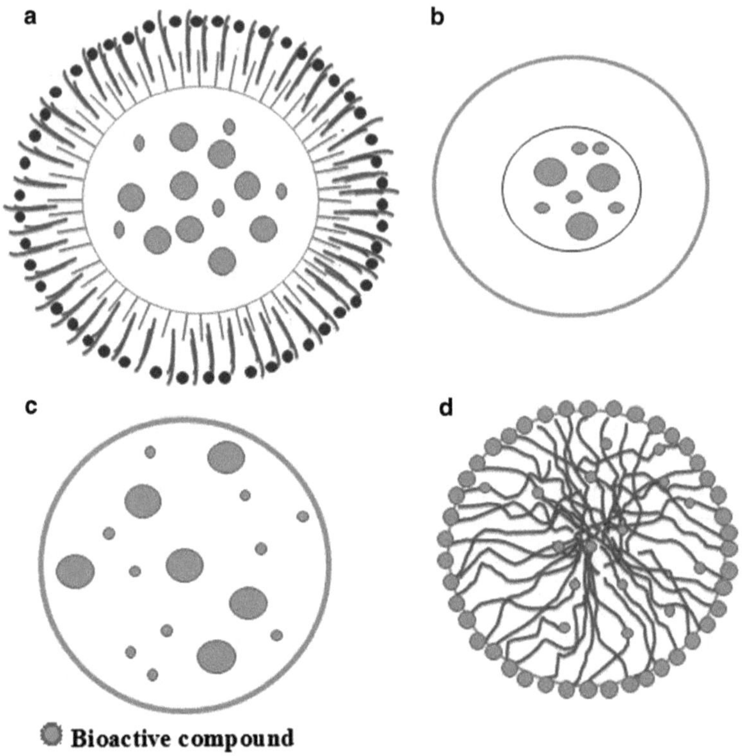

Fig. 2.2 Schematic representation of structural characteristics of some of the more commonly used nanosystems, namely: (**a**) liposomes; (**b**) nanocapsules; (**c**) nanospheres and (**d**) micelles or nanoemulsions

2.3 Lipid Nanoparticles

2.3.1 Liposomes

Liposomes are spherical structures in the colloidal system where phospholipids are dispersed in aqueous solution followed by forming the bilayer membrane of particles through self-assembling. These structures may be used for a wide variety of applications, including the entrapment and controlled release of drugs or nutraceuticals compounds and as model cells or membranes (Thompson et al. 2006; Tan et al. 2013).

The main methods for liposomes preparation have been reviewed (Patil and Jadhav 2014). The authors divided methods in eight major groups listed in Table 2.1. Several liposomal formulations are already in the market, while quite a few are still in the pipeline. Conventional techniques for liposome preparation and size reduction remain popular as these are simple to implement and do not require

Table 2.1 Methods for preparation of liposomes[a]

Major methods	Specific techniques
(1) Conventional methods for preparing giant unilamellar vesicles	(A) Gentle hydration of a phospholipid film
	(B) Electro-formation
	(C) Coalescence of small vesicles
(2) Conventional methods for preparing multilamellar vesicles	(A) Hydration of a phospholipid film under hydrodynamic flow
	(B) Solvent spherule method
	(C) Hydration of proliposomes
(3) Conventional methods for preparing small and large unilamellar vesicles	(A) Reverse phase evaporation
	(B) Injection of organic solvent with dissolved phospholipids into an aqueous phase
	(C) Detergent dialysis
	(D) Reduction of size and lamellarity of multilamellar vesicles
(4) Microfluidic methods for liposome formation	(A) Micro hydrodynamic focusing
	(B) Microfluidic droplets for formation of giant vesicles
	(C) Pulsed jet flow microfluidics for giant vesicles
	(D) Thin film hydration in microtubes
(5) Supercritical fluids for liposome formation	
(6) Modified electroformation for preparation of giant vesicles	
(7) Size reduction of multilamellar vesicles and giant unilamellar vesicles	
(8) Other novel methods for preparing liposomes	(A) Freeze drying of double emulsions
	(B) Membrane contactor for preparation of liposomes
	(C) Hydration of phospholipids deposited on nanostructured materials
	(D) Liposome preparation by curvature-tuning

[a]Compiled from Patil and Jadhav (2014)

sophisticated equipment. However, not all laboratory scale techniques are easy to scale-up for industrial liposome production (Patil and Jadhav 2014).

Liposomes have been used as carrier systems to deliver drugs and other bioactive substances in animals and humans. The main advantages of using liposomes as delivery systems are the potential reducing of undesirable secondary-effects and ensuring good cellular uptake, since liposomes resemble the structure of biomembranes (Chang et al. 2008). The low physical stability and high semi-permeability of membranes are challenging tasks for the incorporation of liposomes into foods. For an efficient application of liposomal carriers as food delivery systems, it is necessary to obtain deeper insights into the solubilization site, stability and impact of encapsulated compounds on bilayer membrane properties, such as fluidity,

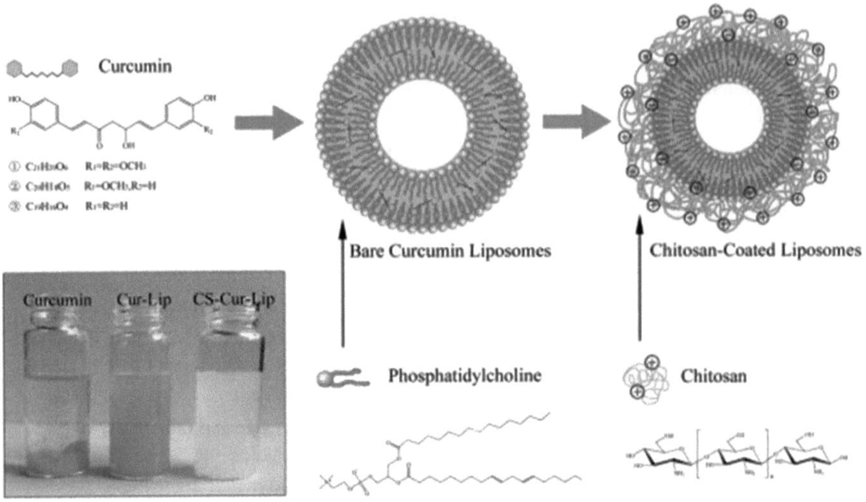

Fig. 2.3 Schematic representation of the production of liposomes incorporating curcumin and chitosan-coated liposomes. The inset photograph shows images of free curcumin, liposomes incorporating curcumin and liposomes incorporating curcumin coated with chitosan (Reprinted with permission from reference Liu et al. (2015)). *Cur-Lip* liposomes incorporating curcumin, *CS-Cur-Lip* liposomes incorporating curcumin coated with chitosan

micropolarity and molecule dynamics. The reactions occurring in phospholipid vesicles (liposomes) have received considerable attention because both stability and reactivity of the liposome are potentially regulated by keeping an intact molecular structure. However, the use of liposomal systems exhibits additional advantages amongst simple micelles and emulsions, when those systems are applied as delivery systems in food (Yoshimoto et al. 2007; Frenzel and Steffen-Heins 2015).

Liu et al. (2015) presented the structure stability of chitosan-coated curcumin liposome and the *in vitro* release of curcumin as shown in Fig. 2.3. The ethanol injection method was used to prepare curcumin liposome. The chitosan solution (1 %) was prepared by dissolving chitosan in distilled water containing 0.1 % acetic acid and then added drop-wisely into bare curcumin liposome suspension. The in vitro release study showed that the cumulative release rate was faster with temperature increase and chitosan could decrease the release rate.

Bioactive and nutraceutical substances have been incorporated into liposomes for delivery in food systems. Table 2.2 presents some examples of studies in recent years that have used liposomes as the primary method of encapsulation of bioactive compounds.

According to Marsanasco et al. (2015), liposomes provide a significant protective effect to thermolabile folic acid, remaining half of this vitamin active after pasteurization. The phosphatidylcholine based liposomes revealed a low peroxidative trend, and liposomes incorporating vitamins remained stable after pasteurization. Furthermore, adding vitamins to phosphatidylcholine:stearic acid and

Table 2.2 Examples of bioactive compound encapsulated by liposomes

Bioactive	Liposome	Reference
ω-3 and ω-6 fatty acids, folic acid, vitamin E and vitamin C	Phosphatidylcholine	Marsanasco et al. (2015)
Quercetin	Soy phospholipid and whey protein isolate	Frenzel and Steffen-Heins (2015)
Bioactive peptides	Phosphatidylcholine	Mosquera et al. (2014)
Pediocin	Soybean phosphatidylcholine	de Mello et al. (2013)
Tea polyphenol extract	Milk phospholipids	Gülseren et al. (2012)
Nisin	Phospholipid	Taylor et al. (2005) and Malheiros et al. (2010a)

phosphatidylcholine:calcium stearate systems favored liposomal aggregation and membrane rigidity, respectively.

Physical stability of liposomes benefits from whey protein isolate coating, as indicated by prolonged shelf life, elimination of osmotic effects in the presence of salts or sugars, and a lower sensitivity towards low pH values during simulated *in vitro* gastric digestion. The use of whey protein isolate as liposome coating material is a new approach in food science, while chitosan has already been investigated as a coating biopolymer for liposomes and is permitted for food applications (Frenzel and Steffen-Heins 2015).

In addition to physical characterization, the influence of encapsulated tea polyphenol on the viability of adenocarcinoma cell line HT-29 was investigated and compared to that of soy phospholipid liposomes (Gülseren et al. 2012). The average size of the milk phospholipid liposomes was lower than 200 nm and all liposomes were visibly stable over the experimental time. Both liposome types were negatively charged, which has not only implications for their colloidal stability, but could potentially influence their absorption. The work clearly demonstrated that tea polyphenols can be encapsulated in liposome preparations, and milk phospholipids represent an appropriate delivery vector. Although both vesicles showed high bioefficacy and delivery of polyphenols, there were differences in their behavior. It was concluded that milk phospholipids could represent an alternative source to soy phospholipids for liposome preparation and that the differences in composition and charge may allow further fine tune the cellular uptake of the bioactive.

A peptide fraction from sea bream scales hydrolysate showing both antioxidant and antihypertensive activity was encapsulated into phosphatidylcholine liposomes. The liposomes showed a mean diameter of 90 nm and the biological activities were maintained after encapsulation (Mosquera et al. 2014). A low-molecular-weight peptide fraction (lower than 1 kDa) showing antihypertensive activity, obtained from hydrolyzed squid tunics, was encapsulated into phosphatidylcholine liposomal nanovesicles. The peptide concentration affected the encapsulation efficiency and the stability of the resulting liposomes, and the peptide amount that provided maximum stability was established as 1.75 g/L. Liposomes obtained with this peptide concentration showed an average diameter of 70.3 nm and proved to be stable in the pH range 3–7 at 4 °C. These liposomes were incorporated into fish gelatin without detriment of rheological properties and thermal stability of the gels, indicating an interesting characteristic for application in edible films and coatings (Mosquera et al. 2015).

Nisin shows a broad antimicrobial activity against Gram-positive bacteria and is approved for food use in many countries. However, its activity may be impaired in food systems due to undesirable interaction with some food components or enzyme inactivation (Cleveland et al. 2001). Liposomal systems containing nisin have been shown to be effective against food pathogens (Malheiros et al. 2010a). Some examples of application in food systems were described and phosphatidylcholine liposomes containing nisin were able to control *Listeria monocytogenes* in milk (Malheiros et al. 2010b) and cheese (Malheiros et al. 2012). Liposomes containing nisin were embedded into hydroxypropyl cellulose films, using a film forming solution containing both free and encapsulated nisin. This film inhibited the growth of *L. monocytogenes*, representing an innovative concept of biodegradable active films that may be useful for improving food safety (Imran et al. 2012).

Pediocin is a broad-spectrum bacteriocin from lactic acid bacteria that belongs to the Class II bacteriocins, a group known as "antilisterial" or "*Listeria*-active" peptides (Papagianni and Anastasiadou 2009). This bacteriocin was incorporated into liposomes and the effect of free and nanovesicle-encapsulated pediocin was evaluated against *Listeria* species. Encapsulation of pediocin into PC liposomes was carried out by the thin-film hydration method with bath-type sonicator. The efficiency of encapsulation was 80 % and low polydispersity index values were observed, indicating an adequate size distribution of liposomes. The stability of liposome vesicles showed potential for application in foods. The antimicrobial activity of the pediocin was preserved by encapsulation. Pediocin-loaded nanovesicles may provide an important tool for controlling spoilage and pathogenic organisms in food and may improve pediocin stability and efficacy in food matrices (de Mello et al. 2013).

2.3.2 Solid Lipid Nanoparticles

Chevalier and Bolzinger (2013) presented the classical description of well-established Pickering emulsions. Pickering emulsions are emulsions of any type, either oil-inwater (o/w), water-in-oil (w/o), or even multiple, stabilized by solid particles in the place of surfactants. Solid stabilizing particles are necessarily smaller than emulsion droplets. Solid particles of nanometric size (or sub-micron, about 100 nm) allow the stabilization of droplets as small as few micrometers diameter; stabilization of larger droplets is possible as well. The solid particles adsorbed at the oil–water interface stabilize the droplets in place of the surfactant molecules. Figure 2.4 shows the structure of Pickering emulsions in comparison to the classical structure of emulsions.

Solid lipid nanoparticles are interesting systems for delivery of hydrophobic bioactive compounds that can be prepared using food grade lipids. Qian et al. (2013)

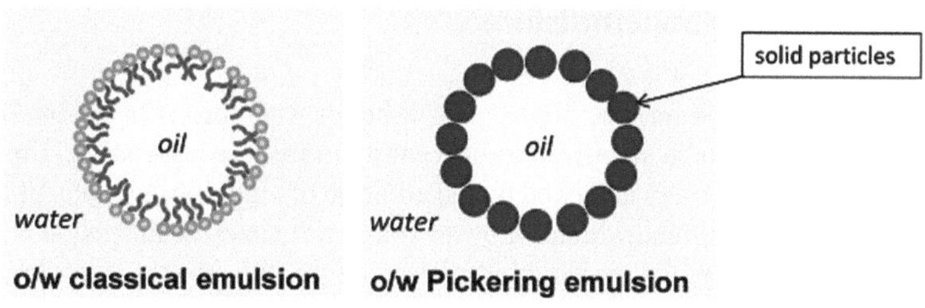

Fig. 2.4 Drawing of a classical (surfactant-based) emulsion and a Pickering emulsion (Reprinted with permission from reference Chevalier and Bolzinger (2013))

aimed to produce solid lipid nanoparticles that had a more disordered crystal structure by mixing two lipid phases with different melting characteristics together: cocoa butter and hydrogenated palm oil. They hypothesized that by blending these two types of lipids it would be able to produce physically stable solid lipid nanoparticles with improved ability to inhibit the chemical degradation of an encapsulated lipophilic component (β-carotene). Solid lipid nanoparticles and liquid lipid nanoparticles were prepared using a hot high-pressure homogenization method. Liquid lipid nanoparticles were produced by directly cooling the hot nanoemulsions since this temperature is above the crystallization temperature of the mixed lipid systems. Solid lipid nanoparticles were produced by cooling the hot nanoemulsions since this is below the crystallization temperature of the mixed lipid phase, and then heating the sample back since this is below the melting temperature of the mixed lipid phase. Liquid lipid nanoparticles had better stability to droplet aggregation and β-carotene degradation than solid lipid nanoparticles after storage for 8 days: the droplet diameter increased around 35 % for solid lipid nanoparticles and around 1 % for liquid lipid nanoparticles. The high stability of this system can be attributed to the ability of the non-ionic surfactant used to generate a strong steric repulsion between the liquid lipid nanoparticles and also to the ability of the fat crystals within the lipid nanoparticles to promote partial coalescence and expulsion of carotenoids to the particle exterior.

Helgason et al. (2009) examined the effect of surfactant concentration on stability and crystal structure of suspensions of solid lipid nanoparticles composed of a model triacylglycerol (tripalmitin) stabilized by a non-ionic surfactant (Tween 20). Emulsion droplets after homogenization had a mean particle diameter of 134 nm and a polydispersity index of 0.08. Solid lipid nanoparticle dispersions rapidly gelled due to aggregation of particles driven by hydrophobic attraction between insufficiently covered lipid crystal surfaces. The Tween 20 concentration in the aqueous phase decreased after tripalmitin crystallization suggesting additional surfactant adsorption onto solid surfaces. At higher Tween 20 concentrations, solid lipid nanoparticles have increasingly complex crystal structures. The results suggested that surfactant coverage at the interface might influence crystal structure and stability of solid lipid nanoparticles via surface-mediated crystal growth.

2.4 Micro and Nanoemulsions

Emulsions can be in the form of oil drops in water or water drops in oil, but in both cases the drops need to be stabilized to prevent them from re-coalescing. Emulsion droplet stabilization is often achieved by the addition of amphiphilic molecules such as surfactants or emulsifiers, which act by decreasing the interfacial tension between the phases, increasing the steric hindrances and/or the electrostatic repulsion between the droplets (Rayner et al. 2014). One of the main advantages of nanoemulsions is that considerably less surfactant is required to their formation when compared with conventional emulsions. Food-grade nanoemulsions can be prepared by high-energy methods, such as high-pressure homogenization or sonication, or low-energy methods, such as phase inversion temperature, spontaneous emulsification, or emulsion phase inversion Such nanoemulsions have good stability against droplet aggregation because the range of attractive forces acting between the droplets decreases with decreasing particle size, whilst the steric repulsion is less dependent on particle size (Gulotta et al. 2014; Ranjan et al. 2014). Table 2.3 presents examples of recent studies that have used micro or nanoemulsion as the primary method for encapsulation of bioactive compounds.

Zeeb et al. (2015) studied the stability of protein-based nanoemulsions fabricated using a high-pressure emulsification approach and the influence of small chain alcohols on their formation. The addition of alcohol, regardless of type, had a major impact on the final droplet size. Presence of 5–10 % alcohol led to the formation of smaller oil droplets compared to the equivalent alcohol-free emulsions - a phenomena that might be attributed to changes in the interfacial tension. The mean particle diameter decreased with increasing alcohol concentrations from 0 to 10 %, but extensive droplet aggregation occurred at higher levels. This phenomenon was attributed to enhanced protein–protein interactions between the adsorbed emulsifier

Table 2.3 Examples of bioactive compound encapsulated by micro or nanoemulsion

Bioactive	Micro/nanoemulsion	Reference
None	(1) Caseinate; (2) Fish gelatin; and (3) Whey protein isolate. With or without alcohol	Zeeb et al. (2015)
Vitamin D3 and ω-3 fatty acid	Flaxseed oil stabilized with calcium caseinate with or without lecithin	Mehmood (2015)
Essential oils	Sodium alginate	Salvia-Trujillo et al. (2015)
Quercitin	Blend of lipids: glyceryl monostearate, glycerol monolaurate and caprylic capric triglyceride	Ni et al. (2015)
Quercitin	Ethyl acetate-water	Gonçalves et al. (2015)
Curcumin	Whey protein concentrate	Sari et al. (2015)
Vitamin E acetate	Edible mustard oil and Tween 80	Dasgupta et al. (2016)
β-carotene	Anhydrous milk fat	Zhang et al. (2013)

molecules in the presence of alcohol leading to droplet flocculation. This study demonstrated that the addition of alcohol might be a useful tool for producing protein-stabilized nanoemulsions suitable as delivery systems for lipophilic bioactives, and the addition of short-chain alcohols to such emulsions prior to homogenization can reduce the droplet size, thereby facilitating nanoemulsion formation.

Nanoemulsions were developed for delivery of vitamin D3 and ω-3 fatty acid in cheddar cheese. Recovery level of vitamin D3 in cheese was 91 and 84 % in the presence or absence of lecithin, respectively. The fortification of foods with encapsulated vitamins and other bioactive molecules may have a positive impact on the composition, yield and chemical stability of the resultant product (Mehmood 2015). Vitamin E acetate nanoemulsion was produced using mustard oil and Tween 80 and floculation was not observed in 15 days. This nanoemulsion showed improved antioxidant and antimicrobial bioactivities and could be potentially applied to increase the shelf life of fruit juices (Dasgupta et al. 2016).

The formation of transparent dispersions of nanostructured lipid carriers using the phase-inversion temperature method was studied by Zhang et al. (2013). To form a coarse emulsion, anhydrous milk fat was heated for complete melting, followed by mixing with warm aqueous phase under agitation. β-carotene was dissolved in warm anhydrous milk fat before emulsification. The degradation of β-carotene encapsulated in nanostructured lipid carriers was significantly reduced when compared to its encapsulation in the soybean-oil-based nanoemulsion, which is associated to the lower mobility in the amorphous lipid core. β-Carotene degraded much faster in melted anhydrous milk fat (79.0 %) than in soybean oil (28.9 %) at 65 °C after 16 days, suggesting soybean oil having better antioxidant properties. At 21 °C, β-carotene did not show obvious degradation in both lipids during storage. The degradation tests in bulk lipids support the conclusion that enhanced stability of β-carotene in nanostructured lipid carriers is due to the amorphous lipid structures of anhydrous milk fat core.

Microfluidized nanoemulsions incorporating essential oils were characterized in terms of droplet size, size distribution, viscosity and color, besides in vitro antimicrobial activity against *Escherichia coli* in comparison with conventional emulsions. Further processing of blends by microfluidization after a previous high shear homogenization rendered nanoemulsions with a reduced droplet size. Moreover, strong differences between emulsion droplet size, size distribution and viscosity of coarse emulsions were observed, depending on the essential oil used in the formulation. The ζ-potential of nanoemulsions indicated a strong electrostatic repulsion of the dispersed oil droplets in the aqueous phase. Emulsions and nanoemulsions containing lemongrass, clove, thyme or palmarosa essential oils showed the strongest antimicrobial activity (Salvia-Trujillo et al. 2015). Nanoemulsions of clove bud and oregano essential oils were incorporated into water-soluble methylcellulose films and their antimicrobial activity was tested in sliced bread. Antimicrobial films containing emulsified essential oils caused a reduction in the counts of yeasts and molds in sliced bread during 15 days (Otoni et al. 2014).

Quercetin is a polyphenolic compound that has been associated with multiple biological effects, including antioxidant, radical scavenging and antimicrobial

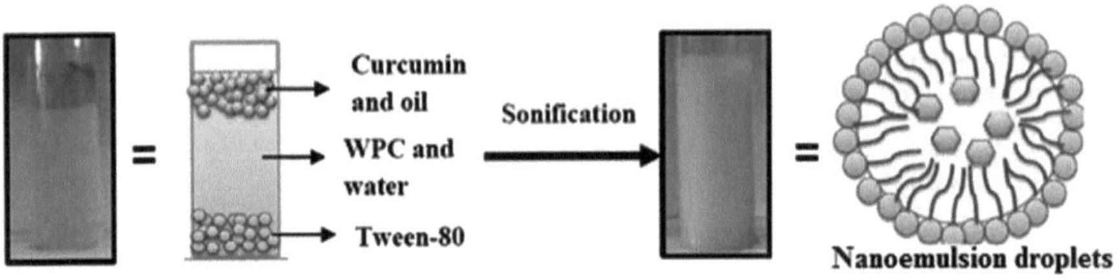

Fig. 2.5 Methodology for preparation of curcumin encapsulated nanoemulsion. A suspension of curcumin dispersed in carrier oil was mixed with an aqueous solution of whey protein concentrate, and sonicated in presence of the surfactant Tween 80, resulting in homogeneous nanoemulsion (Reprinted with permission from reference Sari et al. (2015)). *WPC* whey protein concentrate

activities. An in vitro digestion model consisting of mouth, gastric and small intestinal phases was used to determine the potential biological fate of the quercitin-nanostructured lipid carrier. The in vitro digestion of free or encapsulated quercitin was performed under stirring and maintained at 37 °C throughout the digestion process. The model indicated that bioaccessibility of quercitin under simulated small intestine conditions was improved by encapsulation. The formulated quercitin-nanostructured lipid carrier showed high encapsulation efficiency (93.5 %) and the stability analysis indicated that the nanocarriers remained stable during 60 days of storage. The possible reason for higher encapsulation efficiency of quercitin in nanostructured lipid carriers could be the hydrophobic character of this polyphenol, leading to higher drug partitioning into the lipid matrix of nanoparticle and lesser into the external aqueous phase (Ni et al. 2015).

Water-soluble formulations of quercetin through pressurized ethyl acetate-water emulsion technique have been also developed. The dissolution is achieved by mixing a quercitin suspension in pressurized ethyl acetate at ambient temperature, with a stream of preheated and pressurized ethyl acetate, using a T-mixer. Formulations of quercetin carried out with lecithin yielded the best results with encapsulation efficiencies around 76 % and a micellar particle size in the range of nanometers. The aqueous suspensions of emulsified quercetin presented concentrations of this flavonoid up to 630 ppm, being 315-fold higher soluble when compared with pure quercetin in water (Gonçalves et al. 2015).

The preparation, physicochemical characterization and in vitro digestion kinetics of curcumin nanoemulsion was studied by Sari et al. (2015). The coarse emulsion was prepared using magnetic stirring at room temperature for different time intervals, and fine emulsion was prepared by ultrasound processing the coarse emulsion as shown in Fig. 2.5.

Curcumin, being highly unstable and hydrophobic, is difficult to incorporate in aqueous food systems. For this reason, curcumin was encapsulated inside carrier oil in an emulsion form. An encapsulation efficiency of 90 % was achieved, and the more stable formulation maintained the initial properties for 27 days storage at 25 °C. In vitro release kinetics of curcumin from nanoemulsion by simulated gastrointestinal studies showed that the curcumin nanoemulsion was relatively resistant to pepsin digestion, but pancreatin causes release of curcumin from nanoemulsion. The digestion of encapsulated bioactives in the gastrointestinal tract is a complex process and its impact on the release of bioactive component plays a major role in the uptake, distribution as well as bioavailability of the component (Sari et al. 2015).

Similarly, it can be assumed that addition of nanoemulsified bioactives like quercetin and curcumin to films and coatings should improve stability and controlled release, representing a beneficial to the food packaged in these systems.

2.5 Polymeric Nanoparticles

Molecules composed of regions, typically one hydrophobic and another hydrophilic monomer, with opposite affinities for an aqueous solvent can form polymeric nanoparticles. Protein-based nano-encapsulated materials are particularly interesting because they are relatively easy to prepare and can form complexes with polysaccharides, lipids, or other biopolymers. They are called nanocapsules when used as vehicles for delivery of essential nutrients or pharmaceuticals. There are six classical methods for the preparation of nanocapsules: nanoprecipitation, emulsion-diffusion, double emulsification, emulsion-coacervation, polymer-coating and layer-by-layer. Polymeric nanoparticles are named nanocapsules when they contain a polymeric wall composed of non-ionic surfactants, macromolecules, phospholipids and an oil core. These are prepared mostly by two techniques: the interfacial polymerization and interfacial nano-deposition. A wide variety of nutrients can also be incorporated (Kothamasu et al. 2012; Ranjan et al. 2014; Dasgupta et al. 2015).

Although the use of polymeric nanoparticles for delivery of bioactive molecules has been described (Zhang et al. 2010), the focus of this section is mainly on the nanoparticles developed from biopolymers. Diverse examples of nanoparticles based on protein and/or polysaccharides with potential food applications are discussed below.

Table 2.4 illustrates some studies in recent years that have used polymeric nanoparticles as the primary method of encapsulation of bioactive compounds.

Nanoparticles of whey protein isolate and beet pectin were prepared for anthocyanin encapsulation. Biopolymers were assembled using a combination of two physicochemical phenomena: thermal treatment of protein-polysaccharide mixtures to induce thermal denaturation and aggregation of the globular protein molecules as nanoparticles and pH adjustment to promote coating of protein nanoparticles by oppositely charged polysaccharide molecules. The biopolymer nanoparticles were loaded with an anthocyanin-rich extract. This process led to the formation of relatively small anionic biopolymer nanoparticles (200 nm) at pH 4. A higher loading efficiency was observed when anthocyanin was added before heating the whey protein isolate-beet protein solution, which was attributed to increased protein-polyphenol interactions. This study has shown that biopolymer particles can be produced by thermal processing and electrostatic complexation of whey protein–pectin mixtures. However, the biopolymer particles were only physically stable over

Table 2.4 Examples of bioactive compound encapsulated by polymeric material.

Bioactive	Polymeric material	Reference
Anthocyanin	Whey protein isolate and beet pectin	Arroyo-Maya and McClements (2015)
–	Zein and alginate	Hu and McClements (2015)
Chitosan/fucoidan	Polystyrene	Pinheiro et al. (2015)
Trimetilsilila (TMCS)	Silica aerogel	Oh et al. (2015)
Curcumin	k-carrageenan	Xu et al. (2014)
Quercetin	Chitosan/lecithin	Souza et al. (2014)
Linoleic acid	Chitosan/β-lactoglobulin	Ha et al. (2013)
Curcumin	Casein	Pan et al. (2013)
–	Gelatin and shellac	Patel et al. (2013)
Lipase	Silica shell	Macario et al. (2013)
Quercetin	Polyvinyl alcohol and Eudragit®	Wu et al. (2008)

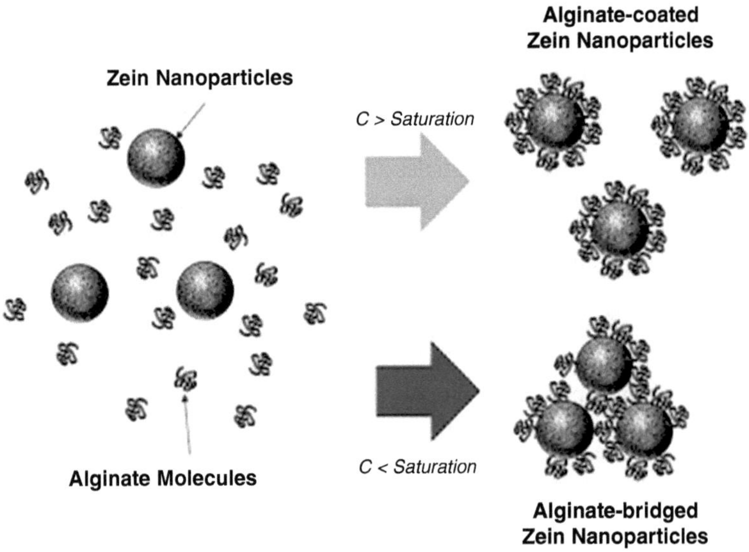

Fig. 2.6 Proposed mechanism of sodium alginate stabilization of zein nanoparticles. The nanoparticles had a core diameter of about 80 nm and a shell thickness of about 40 nm. The amount of alginate required to saturate the surfaces of zein nanoparticles was 2.0 mg/m^2. Alginate-bridged zein nanoparticles are formed with lower alginate concentration (Reprinted with permission from reference Hu and McClements (2015))

a limited range of pH conditions (pH ≈ 3.8 to 4.2), which was attributed to changes in electrostatic, hydrophobic, bridging, and depletion interactions at lower and higher pH values (Arroyo-Maya and McClements 2015).

Protein-polysaccharide nanoparticles from zein and alginate were prepared using a combined antisolvent precipitation/electrostatic deposition method (Fig. 2.6). In addition, the influence of pH, ionic strength, and heating on the stability of these nanoparticles was investigated to assess their potential for commercial applications. Zein nanoparticles were highly susceptible to aggregation when they were prepared in the absence of non-ionic surfactant (Tween 80), which can be attributed to hydrophobic attraction between non-polar patches on zein surfaces. Their stability was greatly improved in the presence of Tween 80 since the non-polar tails of the surfactant molecules adsorb to these non-polar patches, while the polar head groups generate a steric repulsion. The yield of the zein-alginate particles was 95 %, which mean that the majority of the zein added to the water phase formed nanoparticles, also exhibiting superior storage stabilities during room temperature and refrigerator storage. These core/shell biopolymer nanoparticles have potential to be used as nano-delivery systems for bioactive molecules in food and pharmaceutical formulations (Hu and McClements 2015).

Fucoidan is mainly composed of fucose and uronic acids, whereas chitosan is mainly composed by D-glucosamine and is known by its antimicrobial activity. Layer-by-layer deposition of chitosan/fucoidan on a sacrificial template allows the production of biodegradable hollow nanocapsules with potential antioxidant and antimicrobial activity and with great potential to act as a controlled delivery system for bioactive compounds. The resulting nanocapsules also showed a high antioxidant activity at a concentration of 1 mg/mL fucoidan. Thus, chitosan/fucoidan nanocapsules are a promising delivery system for water soluble bioactive compounds, showing a great potential for application in food and pharmaceutical industries (Pinheiro et al. 2015).

Encapsulation efficiency of quercetin into chitosan/lecithin nanoparticles presented values higher than 95 % for all tested concentrations. Moreover, encapsulated quercetin showed improved antioxidant properties. Nanoparticles presented a spherical morphology and stability studies showed that nanoparticles are stable to temperatures ranging between 5 and 70 °C and a pH variation from 3.3 to 5.0. The results suggest that quercetin-loaded chitosan/lecithin nanoparticles can be used in the manufacture of functional foods (Souza et al. 2014). Other authors also describe the development of nanoparticulate systems for encapsulation of quercetin, which has been used as a model molecule for encapsulation studies. Ha et al. (2013) produced linoleic acid-modified chitosan/β-lactoglobulin nanoparticles by the modified ionic gelation method. Spherical shape nanoparticles in the range of 170 and 350 nm were successfully formed, and encapsulation efficiency of quercetin was enhanced with increased amount of linoleic acid and decreased temperature. A novel quercetin nanoparticle system was prepared by Wu et al. (2008) with a simple nanoprecipitation technology with Eudragit® and polyvinyl alcohol as carriers. The utilization of a weight ratio quercitin: Eudragit:polyvinyl alcoohol at 1:10:10 resulted an encapsulation efficiency were over 99 %. In addition, the antioxidant activity of the encapsulated quercitin was more effective than pure quercitin on 2,2-diphenyl-1-picrylhydrazyl scavenging, superoxide anion scavenging, and lipid peroxidation assays. This study also established that the release mechanisms of quercitin were attributed to the reduction of drug particle size, formation of high-energy amorphous state, and intermolecular hydrogen bonding.

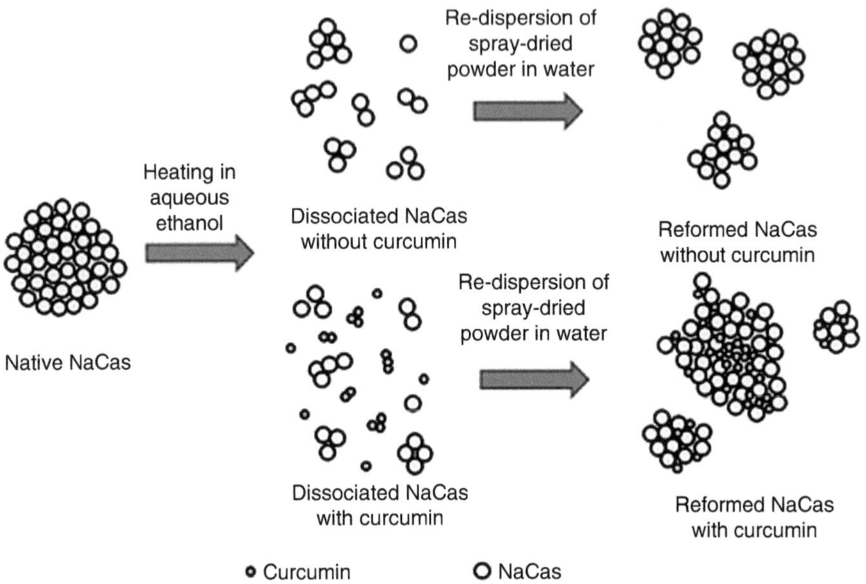

Fig. 2.7 Schematic mechanism of the structural changes of sodium caseinate subjected to heating in aqueous ethanol and spray-drying for encapsulation of curcumin. The increased nanoparticle dimension, together with fluorescence and FTIR spectroscopy results, suggested that curcumin was entrapped in the nanoparticle core through hydrophobic interactions (Reprinted with permission from reference Pan et al. (2013). NaCas, sodium caseinate)

The characteristics of curcumin encapsulation and protection in a k-carrageenan and lysozyme system have been studied. The results indicated that the spherical shaped k-carrageenan/lysozyme complex with 1 μm diameter was spontaneously formed by one spot incubation. The encapsulation efficiency of curcumin by the k-carrageenan/lysozyme complex (2:1 ratio) were 71 %. The results demonstrated that the protein/polysaccharide complex has great potential to deliver sensitive amphiphilic bioactive compounds, protecting them against unfriendly environment (Xu et al. 2014).

Encapsulation of curcumin was also studied by spray-drying dispersions with casein according to the diagram in Fig. 2.7. The spray-dried powder contained 16.7 % (dry basis) curcumin and the encapsulation efficiency was about 83 %. The encapsulation caused the loss of crystallinity of curcumin because it was entrapped in the nanoparticle core through hydrophobic interactions. Free curcumin has a potentially high antioxidant activity because its phenolic -OH groups play a major role in the oxidation reactions. However, pristine curcumin has a low antioxidant activity when dissolved in 95 % ethanol followed by dilution in water, because the low solubility limits the amount of dissolved curcumin molecules to react with free radicals in the aqueous phase. The curcumin encapsulated in casein nanoparticles had higher biological activity, as assessed by antioxidant and cell proliferation assays, than pristine curcumin, likely due to the improved dispersibility (Pan et al.

2013). This simple approach may be applied to encapsulate various lipophilic bioactive compounds.

Patel et al. (2013) successfully demonstrated the generation of novel microcapsules from two natural products gelatin and shellac, both of which are edible and approved by the Food and Drug Administration. Shellac, a natural biodegradable resin from insect origin (*Kerria lacca*), is a complex mixture of polar and non-polar components consisting of polyhydroxy polycarboxylic esters, lactones and anhydrides with the main acid components being aleuritic and terpenic acids. The all-natural microcapsules were generated using a simple extrusion method wherein the gelatin-shellac was dropped in acidic medium resulting in an instantaneous generation of spherical microcapsules that retained their shape on drying. The formation of the microcapsules was basically due to the strong interactions between two oppositely charged polymers and the instant precipitation of acid-resistant shellac.

Macario et al. (2013) reported the synthesis of organic–inorganic nanoparticles with spherical morphology where lipase was encapsulated as active compound. These nanospheres are composed of a purely organic internal liposomal phase in which the bioactive molecule (lipase) is encapsulated in a microaqueous environment. The liposomal phase is covered with porous inorganic silica shell. The synthesis of hybrid nanospheres occurs in two stages: preparation of liposome nanospheres encapsulating the enzyme and formation of an inorganic porous silica shell around the organic nanospheres. Maintaining the ratio between liposome and silica equal to 1, the highest enzyme immobilization efficiency (93–95 %) was achieved. Silica can stabilize the internal liposomal phase and, consequently, isolate and protect the bioactive molecules. The tests showed excellent catalytic performance, near to those of free lipase. This suggests that the enzyme has been immobilized in its active form and the silica shell protects the organic phase, because its activity is preserved after the second reaction cycle. This fact may be associated to the higher density of the external microporous silica shell, which could avoid the decomposition of the enzyme after successive catalytic runs.

Gram-negative *Salmonella typhimurium* LT2 and *Salmonella typhimurium* 14082 s and Gram-positive *Listeria innocua* NADC 2841 were utilized by Oh et al. (2015) in order to study the interactions of the developed food-contact surfaces with bacteria through dip inoculation. Silica (SiO_2) aerogel was synthesized by the sol–gel polymerization of tetraethylorthosilicate via hydrolysis and condensation reaction using ammonium hydroxide as catalyst. The reaction was allowed to take place for 24 h and the silica aerogel formed was dried using supercritical carbon dioxide. This resulted in hydrophilic silica aerogel, which was submerged in 6 % trimethylsilyl chloride solution to functionalize silica surfaces with trimethylsilyl chloride. Figure 2.8 illustrates the change of hydrophilic to hydrophobic behavior of silica aerogel. Compared with the negative control, the positive control and silica aerogel led to decreased counts of *salmonellae* by 1.2 ± 0.1 and 3.1 ± 0.1 log units, corresponding to 93.2 and 99.9 % reduction, respectively. The log reductions in the number of *L. innocua* were 1.3 ± 0.0 and 3.0 ± 0.0 for the positive control and silica aerogel, which represents a reduction of 94.8 % and 99.9 %, respectively. Overall, bacterial anti-adhesion property as well as other distinctive properties such as

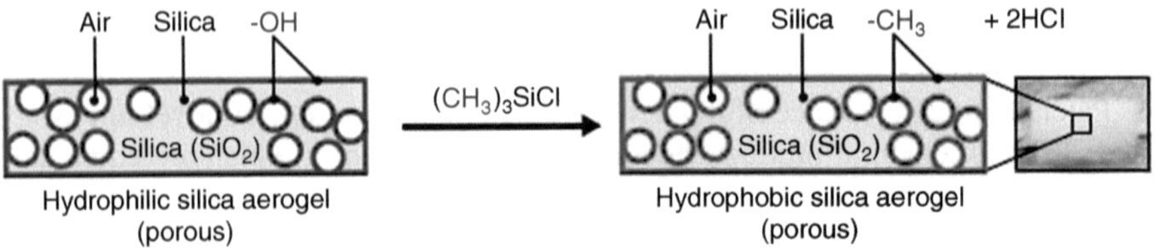

Fig. 2.8 Schematic of representation of the modification of hydrophilic to hydrophobic character of silica (SiO_2) aerogel via methylation reaction using trimethylsilyl chloride (Adapted with permission from reference Oh et al. (2015))

superior thermal insulation and ultra-lightweight make hydrophobically-modified silica aerogel an attractive candidate as a novel food-contact surface.

2.6 Nanofibers

Recently, electrospinning has received great attention in functional food and active food packaging systems. This simple technique allows the production of fibrous polymeric unwoven membranes formed by polymer fibers with nanometric diameters. Electrospinning process takes place at ambient conditions, and electrospun fibers are more suitable to encapsulate thermally labile substances than fibers prepared by conventional melt spinning or films produced by extrusion process or other encapsulation methods (Wei 2012). Figure 2.9 illustrates the electospinning process and drug loading on nanofiber surface. Bioactive compounds can be loaded on the surface of nanofibers by simple physical adsorption or adsorption to surface functionalized nanofibers (Brandelli 2012; Mascheroni et al. 2013). Table 2.5 presents some recent studies that have used nanofibers as the primary method of encapsulation of bioactive compounds.

Aceituno-Medina et al. (2013) evaluated the feasibility of producing electrospun fibers from different amaranth protein isolate:pullulan blends and the influence of Tween 80 (non-ionic surfactant) on the morphology and molecular organization of the electro-deposited material. Addition of Tween 80 significantly improved the fiber morphology for the blends with greater protein content, resulting in defect-free smooth electrospun fibers for the 70:30 and 80:20 amaranth protein isolate:pullulan compositions. Presence of pullulan in the blends resulted in increased viscosity and lower conductivity of the solutions, related to a better chain entanglement and decrease in the polyelectrolyte protein character, respectively, both factors needed for fiber formation. The solutions were stable, which could be attributed to the formation of protein-polysaccharide soluble complexes through the formation of hydrogen bonds (C=O···HO, OH···OH, HO···HN), hydrophobic interactions, and/or ionic bonds.

Aceituno-Medina et al. (2015a) investigated the potential of amaranth-based structures for encapsulation and protection of two potent antioxidant compounds,

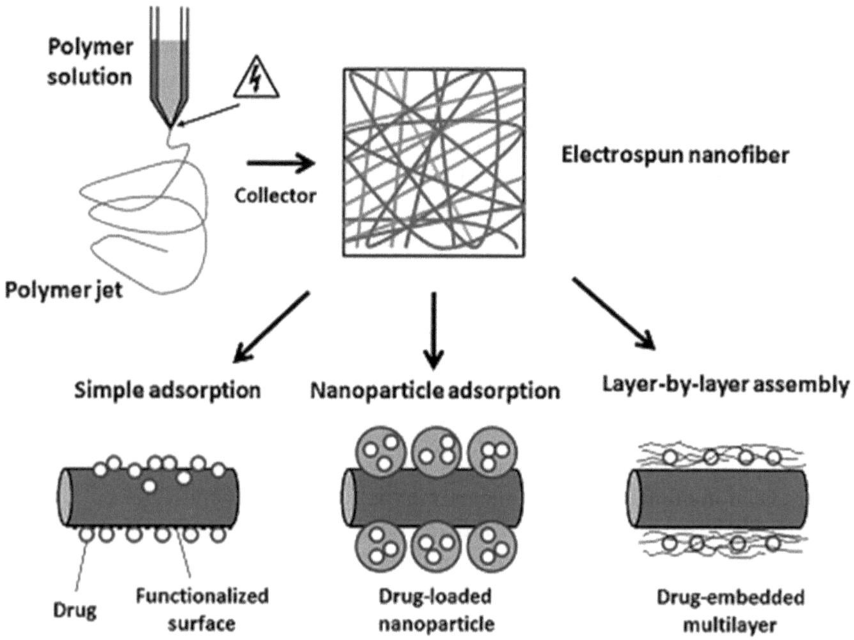

Fig. 2.9 Schematic representation of electrospinning and methods of drug loading on the surface electrospun nanofibers. The bioactive molecules can be loaded to the nanofibers by simple physical adsorption, by assembly of nanoparticles on the nanofiber surface, multilayer assembly, or through chemical immobilization (Reprinted with permission from reference Brandelli (2012))

Table 2.5 Examples of bioactive compound encapsulated with nanofibers

Bioactive	Nanofiber	Reference
Folic acid	Amaranth protein isolate and pullulan	Aceituno-Medina et al. (2015a)
Quercetin and ferulic acid	Amaranth protein isolate and pullulan	Aceituno-Medina et al. (2015b)
Plantaricin 423	Polyethylene oxide and poly(lactic acid)	Heunis et al. (2011)
Nisin	Polyethylene oxide and poly(lactic acid)	Heunis et al. (2013)

quercetin and ferulic acid, to be incorporated into functional foods. In order to develop the electrospun fibers for encapsulation of the bioactives, 50:50 and 80:20 amaranth protein isolate:pullulan blends with the surfactant Tween 80 were prepared using 95 % formic acid as solvent. Smooth ultrathin electrospun fibers were obtained in which the antioxidants were homogeneously distributed. Thermal analysis showed that the degradation temperature within the nanofibers is still far from the usual temperatures used for food processing. Smooth, defect-free structures were obtained, which contained the antioxidants evenly distributed along the fibers.

Efficiency obtained for quercetin was higher, but it should be considered that the concentration of bioactive in this case was half the concentration of ferulic acid used. Therefore, the developed electrospun structures show promissory results in terms of bioactive protection for application in functional foods.

Same authors also investigated the potential of amaranth-based structures for encapsulation and photoprotection of folic acid for food fortification (Aceituno-Medina et al. 2015b). In order to develop the electrospun fibers for encapsulation of folic acid, a blend of 80:20 amaranth protein isolate:pullulan with the surfactant Tween 80 was prepared. Addition of 100 mg of folic acid per g of biopolymer to the polymeric solution used for electrospinning resulted in increased apparent viscosity and, thus, in thicker electrospun fibers. Folic acid was encapsulated through electrospinning in amaranth protein isolate:pullulan ultrathin fibers with encapsulation efficiency higher than 95 %. Even though no specific chemical interactions were shown to be established between the vitamin and the matrix materials. Therefore, the API:pullulan structures have demonstrated a great potential for encapsulation and protection of photosensitive bioactives for food-related applications.

Nanofibers have been used as a delivery platform for bacteriocins as well. A gelatin nanofiber incorporating nisin produced by electrostatic spinning inhibited *Lactobacillus plantarum*, *S. aureus* and *L. monocytogenes*. Although the nanofiber mats were easily broken and dissolved in water, cross-linking with glutaraldehyde strengthen the mat. This antimicrobial material was stored for up to 5 months at 25 °C, maintaining the antibacterial activity after this time (Dheraprasart et al. 2009). Another antimicrobial nanofiber was generated by electrospinning nisin (Nisaplin) into polyethylene oxide and poly(lactic acid) (50:50) blend nanofibers. Active nisin diffused from the nanofiber wound dressings for at least 4 days *in vitro*, as shown by consecutive transfers onto plates seeded with strains of methicillin-resistant *S. aureus* Heunis et al. (2013). Same authors successfully used this same polymer blend to incorporate other antimicrobial peptides, namely plantaricin 423 and bacteriocin ST4SA into nanofiber mats Heunis et al. (2011).

Nanofibers can be very interesting nanostructures for developing functional high-performance packaging materials. The ability to produce non-woven mats composed of nanofibers with diameters below 100 nm from biopolymer solutions can be used to produce food packaging materials with superior mechanical properties and very large specific surface area (Kriegel et al. 2008). These nanofibers based on biopolymers can be loaded with large amounts of bioactive compounds, such as antimicrobials, antioxidants, enzymes, and reinforcing materials resulting in eco-friendly active packaging. For instance, starch-chitosan hybrid films are entirely green since they are produced from renewable and fully degradable biopolymers. Addition of chitosan in packaging can enable to increase the food shelf life due to its antimicrobial activity, and incorporation of cellulose nanofiber into the film matrix caused an increase in tensile strength of starch-chitosan films. In addition, antimicrobial analysis showed that the addition of cellulose nanofiber could increase the effect towards Gram positive bacteria (Salehudin et al. 2014).

2.7 Gels and Films

An edible coating or film has been defined as a thin, continuous layer of edible material formed or placed on or between foods or food components. Furthermore, acting as protective barriers, edible films can be used as carriers of bioactive compounds, thus enhancing the functional properties of the food product by promoting health benefits. Edible films and coatings generally exhibit lower moisture barriers due to their hydrophilicity. To improve moisture barrier characteristics, hydrophobic compounds such as surfactants could be added to hydrophilic materials. Edible films and coatings made from carbohydrates or chitosan have been shown to improve physicochemical and microbiological quality of fresh-cut fruits and vegetables (Cé et al. 2012; Ramos et al. 2012). Table 2.6 illustrates some recent studies dealing with films and coatings as the primary method of encapsulation of bioactive compounds.

Arcan and Yemenicioglu (2014) aimed at developing a scientific basis for edible packaging materials having controlled release properties for multiple active compounds capable of showing antimicrobial and antioxidant effects not only in packed food (active packaging) but also in the human physiological system after consumption of the enriched packed food (bioactive packaging). The films were produced by slowly mixing zein and glycerol with a magnetic stirrer. The mixture was then heated under continuous mixing until it started to boil. The blend zein films has lower initial release rates for the model active compounds, lysozyme and (+)-catechin, than the zein control films, respectively. The change of fatty acid chain length affected both catechin and lysozyme release rates while the change of fatty acid double bond number affected only the catechin release rate. The film morphologies suggested that the blend films owe their controlled release properties mainly to the microspheres formed within their matrix and encapsulation of active compounds.

Wichchukit et al. (2013) evaluated the physical/mechanical properties of a release system made from whey protein/alginate gel beads in a viscous medium that mimics a food beverage. Four polymer solutions were made by mixing Na-alginate with denatured whey protein isolate. Riboflavin was then added at a concentration of 0.25 g/50 g to those gel solutions under minimum light exposure to prevent degradation. Viscoelastic characteristics were not observed in pure alginate solution; it behaved as a Newtonian solution with no measurable elastic behavior. In contrast, the combination of whey protein and alginate resulted in shear thinning behavior of solutions. Solutions with high apparent viscosity resulted in larger beads. Mechanical

Table 2.6 Examples of bioactive compound encapsulated with gels and films.

Bioactive	Components	Reference
Cinnamaldehyde	Chitosan	Rieger et al. (2015)
Lysozyme, catechins, and a group of flavonoids	Zein and glycerol	Arcan and Yemenicioglu (2014)
Riboflavin	Whey protein isolate and alginate gel	Wichchukit et al. (2013)

relaxation tests of the four types of gel beads quantified the viscoelastic property. Pure whey protein beads tended to be more solid and relaxed the least, while pure alginate beads lack sufficient structural integrity.

Rieger et al. (2015) confirmed that cinnamaldehyde has a strong antimicrobial effect against *E. coli*, *S. aureus*, and *P. aeruginosa*. Increased encapsulation of cinnamaldehyde was enabled using a model surfactant, sorbitan monooleate (SpanVR80). Chitosan/ cinnamaldehyde films were prepared by adding the various amounts of cinnamaldehyde to the chitosan solution. Qualitatively, films exhibited well-defined structural color, which quantitatively ranged from 145 to 345 nm thick. These results suggest that nanostructured chitosan- cinnamaldehyde coatings hold potential to delay bacterial colonization on a range of surfaces, from indwelling medical device to food processing surfaces. Films fabricated using SpanVR80 demonstrated an increased release that paralleled the increased cinnamaldehyde incorporation. NMR indicated that a majority of the cinnamaldehyde was physically incorporated into the hydrophilic chitosan films. Spin-coating can be used to incorporate and deliver high-loadings of a model essential oil, cinnamaldehyde, from ultrathin chitosan films. These natural bioactive films hold potential for use as bioactive coatings in food packaging.

2.8 Nanocomposite Packaging

2.8.1 Active Nanocomposite Packaging

Active packaging is intentionally designed to incorporate components that release, absorb or modify substances into or from the packaged food or the environment surrounding the food. The major developments on active nanocomposite packaging are described on antimicrobial films, oxygen scavenging systems and enzyme immobilization systems (Silvestre et al. 2011; Ranjan et al. 2014).

Several antimicrobial substances, including natural products like bacteriocins, essential oils, isothiocyanates and sorbic acid, antimicrobial enzymes, metallic nanoparticles, and some modified clay minerals have been used to develop antimicrobial nanocomposite packaging materials (Azeredo 2013; Rhim et al. 2013). Diverse food packaging systems incorporated with silver nanoparticles have been described. The incorporation of silver nanoparticles in biopolymer films such as cellulose, starch, chitosan and alginate, exhibit consistent antimicrobial activity against both Gram-positive and Gram-negative bacteria (Unalan et al. 2014). The bacteriocin nisin has been also tested in several antimicrobial packaging systems. This bacteriocin was adsorbed by coating method on poly(lactic acid) containing cellulose nanocrystals and applied on sliced cooked ham to control *Listeria monocytogenes* (Salmieri et al. 2014). Recent studies showed efficient inhibition of *L. monocytogenes* by incorporation of nisin into poly(butylene adipate-*co*-terephthalate) films (Zehetmeyer et al. 2016), and starch nanocomposites containing halloysite nanoclay as reinforcement (Meira et al. 2016).

Protection against oxidative damage can be achieved by developing food packaging materials containing oxygen scavengers, which results into lowering the oxygen level (Rhim et al. 2013). Active packaging based on nanocomposite films were developed for selective control of oxygen diffusion. Different polymers containing nanostructured antioxidants or oxygen scavengers can be used as active packaging for sliced meat, poultry, beverages, cooked pasta, ready-to-eat snacks, and fish (Neethirajan and Jayas 2011).

The immobilization and incorporation of enzymes in packaging materials is an alternative to its direct use in food matrix since the enzymatic activity can be inactivated by some processing conditions and/or interaction with other compounds. Nanoscale immobilization can increase the available surface area and enhance performance by improving stability to pH and temperature and resistance to proteases or other denaturing compounds, besides allowing repeated use or controlled release into food (Rhim et al. 2013; Ranjan et al. 2014).

2.8.2 *Smart Nanocomposite Packaging*

Intelligent packaging systems are intended to include innovative functions such as detection, recording, tracing and communication. The condition of packaged food is monitored during storage and transportation through indicators and/or sensors in the form of a package label or a print on packaging films. In addition, such devices can inform the supplier or consumer that foodstuffs are still fresh, or has spoiled, kept at the appropriate temperature along the supply chain, or whether the packaging has been opened (Silvestre et al. 2011; Ranjan et al. 2014).

Nanoparticles have been used as reactive units in composite packaging materials to inform about the condition of the package. These nanosensors are designed to detect selected chemical compounds, such as pesticides, allergens and toxins, and are able to respond to microbial contamination, spoilage products, and environmental changes on temperature, humidity, and oxygen level. In this way, nanosensors inform the consumer, providing real-time status of food freshness and eliminating the need for inaccurate expiration dates. Nanosensors can be a valuable tool for the manufacturers, protecting against food mishandling after dispatch, ensuring that the product reaches the consumer in a suitable condition (Silvestre et al. 2011).

2.8.3 *Nanoreinforcements*

Significant improvements in mechanical and barrier properties, dimensional stability, and solvent resistance can be achieved by including some nanofillers into biopolymer matrixes intended for packaging purposes. Nanoreinforcement materials can include clay and silicate nanoplatelets, silica nanoparticles, carbon nanotubes, cellulose nanofibers or nanowhiskers, and chitin or chitosan nanoparticles

(Sorrentino et al. 2007). Currently, only the inorganic clays have attracted some attention by the packaging industry, due to their availability, low cost, significant improvements, and easy to process.

Mineral clays are considered environmentally friendly, naturally abundant and inexpensive, and regarded as safe food additives according to Food and Drug Administration and European Food Safety Authority (Ibarguren et al. 2014). Montmorillonite, a member of the smectite group, is one of the most widely used natural clays, particularly exploited for the fabrication of polymer-clay nanocomposites (Azeredo 2013). The improved barrier properties of polymer-clay nanocomposites appear to be due to an increased tortuous path for a diffusive molecule, forcing them to travel a longer way to diffuse through the film. Montmorillonite layers were efficiently dispersed and distributed within soy-protein-based films, generating a significant strengthening of the nanolayer, leading to resistance to breakage, and decrease in extension, moisture content, solubility, and permeability to water vapor (Echeverría et al. 2014). Lee et al. (2014) showed that two types of nanoclays incorporated to sesame seed meal protein nanocomposite film enhanced their physical properties (tensile strength, elongation and water vapor permeability), indicating that the films could be applied in food packaging.

Nanocellulose structures are described as a multi-performance material with potential for application in cellulose-based food packaging, showing promising properties like firmness, biocompatibility and suitable water vapor permeability. Cellulose nanocrystal obtained from the paper-mulberry bast pulp was tested as a reinforcing agent for the preparation of agar based bio-nanocomposites. Properties of agar films like mechanical and water vapor barrier properties were significantly improved by blending with cellulose nanocrystals (Reddy and Rhim 2014). The tensile strength and tensile modulus of agar films increased by 25 % and 40 %, respectively, in the composite film with 5 % nanocellulose, whereas the water vapor permeability decreased by 25 % in nanocomposite with 3 % nanocellulose.

2.9 Conclusion

In the near future, many nanostructured bioactive compounds could be developed with efficacy. Advances in the strategies for incorporation of bioactive molecules into nanostructures and development of nanocomposite materials for food packaging constitute a field of great promise in the food sector. Development of integrated nanomaterial-bioactive systems for food processing, packaging, proper delivery of nutraceuticals and quality control is a topic of utmost interest to the worldwide community. Therefore, production of food grade nanostructures conjugated with bioactive compounds or plant extracts and their incorporation in food packaging for enhanced properties using nanotechnology are some of the future directions in the field. The use of agroindustrial byproducts as a source of natural polymers for development of such nanostructures and packaging materials constitute an interesting economical and eco-friendly alternative.

Acknowledgements A. Brandelli and J.H.Z. Santos are research awardees of Conselho Nacional de Desenvolvimento Científico e Tecnológico (Brasília, Brazil). L.F.W. Brum is a PhD studentship from Coordenação de Aperfeiçoamento de Pessoal de Nível Superior (Brasília, Brazil).

References

Aceituno-Medina M, Mendoza S, Lagaron JM, López-Rubio A (2013) Development and characterization of food-grade electrospun fibers from amaranth protein and pullulan blends. Food Res Int 54:667–674. doi:10.1016/j.foodres.2013.07.055

Aceituno-Medina M, Mendoza S, Rodríguez BA, Lagaron JM, López-Rubio A (2015a) Improved antioxidant capacity of quercetin and ferulic acid during *in-vitro* digestion through encapsulation within food-grade electrospun fibers. J Funct Foods 12:332–341. doi:10.1016/j.jff.2014.11.028

Aceituno-Medina M, Mendoza S, Langaron JM, López-Rubio A (2015b) Photoprotection of folic acid upon encapsulation in food-grade amaranth (*Amaranthus hypochondriacus* L.) protein isolate and pullulan electrospun fibers. LWT Food Sci Technol 62:970–975. doi:10.1016/j.lwt.2015.02.025

Arcan I, Yemenicioglu A (2014) Controlled release properties of zein-fatty acid blend films for multiple bioactive compounds. J Agric Food Chem 62:8238–8246. doi:10.1021/jf500666w

Arroyo-Maya IJ, McClements DJ (2015) Biopolymer nanoparticles as potential delivery systems for anthocyanins: Fabrication and properties. Food Res Int 69:1–8. doi:10.1016/j.foodres.2014.12.005

Azeredo HMC (2013) Antimicrobial nanostructures in food packaging. Trends Food Sci Technol 30:56–69. doi:10.1016/j.tifs.2012.11.006

Brandelli A (2012) Nanostructures as promising tools for delivery of antimicrobial peptides. Mini-Rev Med Chem 12:731–741. doi:10.2174/138955712801264774

Brandelli A, Taylor TM (2015) Nanostructured and nanoencapsulated natural antimicrobials for use in food products. In: Taylor TM (ed) Handbook of natural antimicrobials for food safety and quality. Elsevier, Oxford, pp 229–257. doi:10.1016/B978-1-78242-034-7.00011-6

Cé N, Noreña CPZ, Brandelli A (2012) Antimicrobial activity of chitosan films containing nisin, peptide P34, and natamycin. J Food 10:21–26. doi:10.1080/19476337.2010.537371

Champagne CP, Fustier P (2007) Microencapsulation for the improved delivery of bioactive compounds into foods. Curr Opin Biotechnol 18:184–190. doi:10.1016/j.copbio.2007.03.001

Chang CC, Liu DZ, Lin SY, Liang HJ, Hou WC, Huang WJ, Chang CH, Ho FM, Liang YC (2008) Liposome encapsulation reduces cantharidin toxicity. Food Chem Toxicol 46:3116–3121. doi:10.1016/j.fct.2008.06.084

Chevalier Y, Bolzinger MA (2013) Emulsions stabilized with solid nanoparticles: pickering emulsions. Coll Surf A Physicochem Eng Aspects 439:23–34. doi:10.1016/j.colsurfa.2013.02.054

Cleveland J, Montville TJ, Nes IF, Chikindas ML (2001) Bacteriocins: safe, natural antimicrobials for food preservation. Int J Food Microbiol 71:1–20. doi:10.1016/S0168-1605(01)00560-8

da Silva AC, Jorge N (2014) Bioactive compounds of the lipid fractions of agro-industrial waste. Food Res Int 66:493–500. doi:10.1016/j.foodres.2014.10.025

Dainelli D, Gontard N, Spyropoulosc D, Zondervan-van den Beuken E, Tobback P (2008) Active and intelligent food packaging: legal aspects and safety concerns. Trends Food Sci Technol 19:S103–S112. doi:10.1016/j.tifs.2008.09.011

Dasgupta N, Ranjan S, Mundekkad D, Ramalingam C, Shanker R, Kumar A (2015) Nanotechnology in agro-food: from field to plate. Food Res Int 69:381–400. doi:10.1016/j.foodres.2015.01.005

Dasgupta N, Ranjan S, Mundra S, Chidambaram R, Kumar A (2016) Fabrication of food grade vitamin E nanoemulsion by low energy approach, characterization and its application. Int J Food Prop 19:700–708. doi:10.1080/10942912.2015.1042587 de Mello MB, Malheiros PS, Brandelli A, da Silveira NP, Jantzen MM, da Motta AS (2013) Characterization and antilisterial effect of phosphatidylcholine nanovesicles containing the antimicrobial peptide pediocin. Probiotics Antimicrob Proteins 5:43–50. doi:10.1007/s12602-013-9125-3

de Vos P, Faas MM, Spasojevic M, Sikkema J (2010) Encapsulation for preservation of functionality and targeted delivery of bioactive food components. Int Dairy J 20:292–302. doi:10.1016/j.idairyj.2009.11.008

Dheraprasart C, Renggpipat S, Supaphol P, Tattiyakul J (2009) Morphology, release characteristics, and antimicrobial effect of nisin-loaded electrospun gelatin fiber mat. J Food Prot 72:2293–2300. doi:10.1016/j.polymdegradstab.2012.01.003

Echeverría I, Eisenberg P, Mauri AN (2014) Nanocomposites films based on soy proteins and montmorillonite processed by casting. J Memb Sci 449:15–26

Fang Z, Bhandari B (2010) Encapsulation of polyphenols e a review. Trends Food Sci Technol 21:510–523. doi:10.1016/j.tifs.2010.08.003

Frenzel M, Steffen-Heins A (2015) Impact of quercetin and fish oil encapsulation on bilayer membrane and oxidation stability of liposomes. Food Chem 185:48–57. doi:10.1016/j.foodchem.2015.03.121

Gharsallaoi A, Roudaut G, Chambin O, Voilley A, Saurel R (2007) Applications of spray-drying in microencapsulation of food ingredients: an overview. Food Res Int 40:1107–1121. doi:10.1016/j.foodres.2007.07.004

Gómez-Mascaraque LG, Lagarón JM, López-Rubio A (2015) Electrosprayed gelatin submicroparticles as edible carriers for the encapsulation of polyphenols of interest in functional foods. Food Hydrocoll 49:42–52. doi:10.1016/j.foodhyd.2015.03.006

Gonçalves VSS, Rodríguez-Rojo S, De Paz E, Mato C, Cocero MJ (2015) Production of water soluble quercetin formulations by pressurized ethyl acetate-in-water emulsion technique using natural origin surfactants. Food Hydrocoll 51:295–304. doi:10.1016/j.foodhyd.2015.05.006

Gonnet M, Lethuaut L, Boury F (2010) New trends in encapsulation of liposoluble vitamins. J Control Release 146:276–290. doi:10.1016/j.jconrel.2010.01.037

Gouin S (2004) Microencapsulation: industrial appraisal of existing technologies and trends. Trends Food Sci Technol 15:330–347. doi:10.1016/j.tifs.2003.10.005

Gulotta A, Saberi AH, Nicoli MC, McClements DJ (2014) Nanoemulsion-based delivery systems for polyunsaturated (ω-3) oils: formation using a spontaneous emulsification method. J Agric Food Chem 62:1720–1725, doi: 10.1021/jf4054808

Gülseren I, Guri A, Corredig M (2012) Encapsulation of tea polyphenols in nanoliposomes prepared with milk phospholipids and their effect on the viability of HT-29 human carcinoma cells. Food Dig 3:36–45. doi:10.1007/s13228-012-0019-8

Ha H-K, Kim JW, Lee M-R, Lee W-J (2013) Formation and characterization of quercetin-loaded chitosan oligosaccharide/β-lactoglobulin nanoparticle. Food Res Int 52:82–90. doi:10.1016/j.foodres.2013.02.021

Helgason T, Awad TS, Kristbergsson K, McClements DJ, Weiss J (2009) Effect of surfactant surface coverage on formation of solid lipid nanoparticles (SLN). J Colloid Interface Sci 334:75–81. doi:10.1016/j.jcis.2009.03.012

Heunis T, Bshena O, Klumperman B, Dicks LMT (2011) Release of bacteriocins from nanofibers prepared with combinations of poly(D, L-lactide) (PDLLA) and poly(ethylene oxide) (PEO). Int J Mol Sci 12:2158–2173. doi:10.3390/ijms12042158

Heunis TDJ, Smith C, Dicks LMT (2013) Evaluation of a nisin-eluting nanofiber scaffold to treat *Staphylococcus aureus*-induced skin infections in mice. Antimicrob Agents Chemother 57:3928–3935. doi:10.1128/AAC.00622-13

Hu K, McClements DJ (2015) Fabrication of biopolymer nanoparticles by antisolvent precipitation and electrostatic deposition: Zein-alginate core/shell nanoparticles. Food Hydrocoll 44:101–108. doi:10.1016/j.foodhyd.2014.09.015

Ibarguren C, Naranjo PM, Stötzel C, Audisio MC, Sham EL, Torres EMF, Müller FA (2014) Adsorption of nisin on raw montmorillonite. Appl Clay Sci 90:88–95. doi:10.1016/j.clay.2013.12.031

Imran M, Revol-Junelles AM, René N (2012) Microstructure and physico-chemical evaluation of nano-emulsion-based antimicrobial peptides embedded in bioactive packaging films. Food Hydrocoll 29:407–419. doi:10.1016/j.foodhyd.2012.04.010

Kothamasu P, Kanumur H, Ravur N, Maddu C, Parasuramrajam R, Thangavel S (2012) Nanocapsules: the weapons for novel drug delivery systems. BioImpacts 2:71–81. doi:10.5681/bi.2012.011

Kriegel C, Arecchi A, Kit K, McClements DJ, Weiss J (2008) Fabrication, functionalization, and application of electrospun biopolymer nanofibers. Crit Rev Food Sci Nutr 48:775–797. doi:10.1080/10408390802241325

Lagaron JM, Lopez-Rubio A (2011) Nanotechnology for bioplastics: opportunities, challenges and strategies. Trends Food Sci Technol 22:611–617. doi:10.1016/j.tifs.2011.01.007

Lee JH, Song NB, Jo WS, Song KB (2014) Effects of nano-clay type and content on the physical properties of sesame seed meal protein composite films. Int J Food Sci Technol 49:1869–1875. doi:10.1111/ijfs.12496

Liu Y, Liu D, Zhu L, Gan Q, Le X (2015) Temperature-dependent structure stability and in vitro release of chitosan-coated curcumin liposome. Food Res Int 74:97–105. doi:10.1016/j.foodres.2015.04.024

Macario A, Verri F, Diaz U, Corma A, Giordano G (2013) Pure silica nanoparticles for liposome/lipase system encapsulation: application in biodiesel production. Catal Today 204:148–155. doi:10.1016/j.cattod.2012.07.014

Malheiros PS, Daroit DJ, Brandelli A (2010a) Food applications of liposome-encapsulated antimicrobial peptides. Trends Food Sci Technol 21:284–292. doi:10.1016/j.tifs.2010.03.003

Malheiros PS, Daroit DJ, Silveira NP, Brandelli A (2010b) Effect of nanovesicle-encapsulated nisin on growth of *Listeria monocytogenes* in milk. Food Microbiol 27:175–178. doi:10.1016/j.fm.2009.09.013

Malheiros PS, Sant'Anna V, Barbosa MS, Brandelli A, Franco BDGM (2012) Effect of liposome-encapsulated nisin and bacteriocin-like substance P34 on *Listeria monocytogenes* growth in Minas Frescal cheese. Int J Food Microbiol 156:272–277. doi:10.1016/j.ijfoodmicro.2012.04.004

Marsanasco M, Marquez AL, Wagner JR, Chiaramoni NS, Alonso SV (2015) Bioactive compounds as functional food ingredients: characterization in model system and sensory evaluation in chocolate milk. J Food Eng 166:55–63. doi:10.1016/j.jfoodeng.2015.05.007

Marsh K, Bugusu B (2007) Food packaging – roles, materials, and environmental issues. J Food Sci 72:38–55. doi:10.1111/j.1750-3841.2007.00301.x

Mascheroni E, Fuenmayor CA, Cosio MS, Di Silvestro G, Piergiovanni L, Mannino S, Schiraldi A (2013) Encapsulation of volatiles in nanofibrous polysaccharide membranes for humidity-triggered release. Carbohydr Polym 98:17–25. doi:10.1016/j.carbpol.2013.04.068

McClements DJ (2014) Nanoparticle- and microparticle-based delivery systems: encapsulation, protection and release of active compounds. CRC Press, Boca Raton, 572p

Mehmood T (2015) Optimization of the canola oil based vitamin E nanoemulsions stabilized by food grade mixed surfactants using response surface methodology. Food Chem 183:1–7. doi:10.1016/j.foodchem.2015.03.021

Meira SMM, Zehetmeyer G, Scheibel JM, Werner JO, Brandelli A (2016) Starch-halloysite nanocomposites containing nisin: characterization and inhibition of *Listeria monocytogenes* in cheese. LWT Food Sci Technol. doi:10.1016/j.lwt.2015.12.006

Mosquera M, Giménez B, Silva IM, Boelter JF, Montero P, Gómez-Guillén C, Brandelli A (2014) Nanoencapsulation of an active peptidic fraction from sea bream scales collagen. Food Chem 156:144–150. doi:10.1016/j.foodchem.2014.02.011

Mosquera M, Giménez B, Montero P, Gómez-Guillén MC (2015) Incorporation of liposomes containing squid tunic ACE-inhibitory peptides into fish gelatin. J Sci Food Agric. doi:10.1002/jsfa.7145

Naziri E, Nenadis N, Mantzouridou FT, Tsimidou MZ (2014) Valorization of the major agrifood industrial by-products and waste from Central Macedonia (Greece) for the recovery of compounds for food applications. Food Res Int 65:350–358. doi:10.1016/j.foodres.2014.09.013

Nedovic V, Kalusevic A, Manojilovic V, Levic S, Bugarski B (2011) An overview of encapsulation technologies for food applications. Procedia Food Sci 1:1806–1815. doi:10.1016/j.profoo.2011.09.266

Neethirajan S, Jayas DS (2011) Nanotechnology for the food and bioprocessing industries. Food Bioprocess Technol 4:39–47. doi:10.1007/s11947-010-0328-2

Ni S, Sun R, Zhao G, Xia Q (2015) Quercetin loaded nanostructured lipid carrier for food fortification: preparation, characterization and in vitro study. J Food Process Eng 38:93–106. doi:10.1111/jfpe.12130

Oh JK, Perez K, Kohli N, Kara V, Li J, Min Y, Castillo A, Taylor M, Jayaraman A, Cisneros-Zevallos L, Akbulut M (2015) Hydrophobically-modified silica aerogels: novel food-contact surfaces with bacterial anti-adhesion properties. Food Control 52:132–141. doi:10.1016/j.foodcont.2014.12.029

Otoni CG, Pontes SF, Medeiros EA, Soares NF (2014) Edible films from methylcellulose and nanoemulsions of clove bud (*Syzygium aromaticum*) and oregano (*Origanum vulgare*) essential oils as shelf life extenders for sliced bread. J Agric Food Chem 62:5214–5219. doi:10.1021/jf501055f

Pan K, Zhong Q, Baek SJ (2013) Enhanced dispersibility and bioactivity of curcumin by encapsulation in casein nanocapsules. J Agric Food Chem 61:6036–6043. doi:10.1021/jf400752a

Papagianni M, Anastasiadou S (2009) Pediocins: the bacteriocins of Pediococci. Sources, production, properties and applications. Microbial Cell Factories 8, doi: 10.1186/1475-2859-8-3

Patel AR, Remijn C, Cabero A-iM, Heussen PCM, Hoorn JWMS, Velikov KP (2013) Novel all-natural microcapsules from gelatin and Shellac for biorelated applications. Adv Funct Mater 23:4710–4718. doi:10.1002/adfm.201300320

Patil YP, Jadhav S (2014) Novel methods for liposome preparation. Chem Phys Lipids 177:8–18. doi:10.1016/j.chemphyslip.2013.10.011

Pinheiro AC, Bourbon AI, Cerqueira MA, Maricato E, Nunes C, Coimbra MA, Vicente AA (2015) Chitosan/fucoidan multilayer nanocapsules as a vehicle for controlled release of bioactive compounds. Carbohydr Polym 115:1–9. doi:10.1016/j.carbpol.2014.07.016

Qian C, Decker EA, Xiao H, McClements DJ (2013) Impact of lipid nanoparticle physical state on particle aggregation and β-carotene degradation: potential limitations of solid lipid nanoparticles. Food Res Int 52:342–349. doi:10.1016/j.foodres.2013.03.035

Ramos OL, Fernandes JC, Silva SI, Pintado ME, Malcata FX (2012) Edible films and coatings from whey proteins: a review on formulation, and on mechanical and bioactive properties. Crit Rev Food Sci Nutr 52:533–552. doi:10.1080/10408398.2010.500528

Ranjan S, Dasgupta N, Chakraborty AR, Samuel SM, Ramalingam C, Shanker R, Kumar A (2014) Nanoscience and nanotechnologies in food industries: opportunities and research trends. J Nanopart Res 16:2464. doi:10.1007/s11051-014-2464-5

Rayner M, Marku D, Eriksson M, Sjöö M, Dejmek P, Wahlgren M (2014) Biomass-based particles for the formulation of Pickering type emulsions in food and topical applications. Coll Surf A Physicochem Eng Aspects 458:48–62. doi:10.1016/j.colsurfa.2014.03.053

Reddy JP, Rhim JW (2014) Characterization of bionanocomposite films prepared with agar and paper-mulberry pulp nanocellulose. Carbohydr Polym 110:480–488. doi:10.1016/j.carbpol.2014.04.056

Rhim JW, Park HM, Ha CS (2013) Bio-nanocomposites for food packaging applications. Prog Polym Sci 38:1629–1652. doi:10.1016/j.progpolymsci.2013.05.008

Rieger KA, Eagan NM, Schiffman JD (2015) Encapsulation of cinnamaldehyde into nanostructured chitosan films. J Appl Polymer Sci 132, doi 10.1002/APP.41739

Salehudin MH, Salleh E, Mamat SNH, Muhamad II (2014) Starch based active packaging film reinforced with empty fruit bunch (EFB) cellulose nanofiber. Procedia Chem 9:23–33. doi:10.1016/j.proche.2014.05.004
Salmieri S, Islam F, Khan RA, Hossain FM, Ibrahim HMM, Miao C, Hama WY, Lacroix M (2014) Antimicrobial nanocomposite films made of poly(lactic acid)-cellulose nanocrystals (PLA-CNC) in food applications: part A – effect of nisin release on the inactivation of *Listeria monocytogenes* in ham. Cellulose 21:1837–1850. doi:10.1007/s10570-014-0406-0
Salvia-Trujillo L, Rojas-Graü A, Soliva-Fortuny R, Martín-Belloso O (2015) Physicochemical characterization and antimicrobial activity of food grade emulsions and nanoemulsions incorporating essential oils. Food Hydrocoll 43:547–556. doi:10.1016/j.foodhyd.2014.07.012
Sant'Anna V, Malheiros PS, Brandelli A (2011) Liposome-encapsulation protects bacteriocin-like substance P34 against inhibition by Maillard reaction products. Food Res Int 44:326–330. doi:10.1016/j.foodres.2010.10.012
Sari TP, Mann B, Kumar R, Singh RRB, Sharma R, Bhardwaj M, Athira S (2015) Preparation and characterization of nanoemulsion encapsulating curcumin. Food Hydrocoll 43:540–546. doi:10.1016/j.foodhyd.2014.07.011
Silvestre S, Duraccio D, Cimmino S (2011) Food packaging based on polymer nanomaterials. Prog Polym Sci 36:1766–1782. doi:10.1016/j.progpolymsci.2011.02.003
Siracusa V, Rocculi P, Romain S, Dalla Rosa M (2008) Biodegradable polymers for food packaging: a review. Trends Food Sci Technol 19:634–643. doi:10.1016/j.tifs.2008.07.003
Sorrentino A, Gorrasi G, Vittoria V (2007) Potential perspectives of bio-nanocomposites for food packaging applications. Trends Food Sci Technol 18:84–95. doi:10.1016/j.tifs.2006.09.004
Souza MP, Vaz AFM, Correia MTS, Cerqueira MA, Vicente AA, Carneiro-da-Cunha MG (2014) Quercetin-loaded lecithin/chitosan nanoparticles for functional food applications. Food Bioprocess Technol 7:1149–1159. doi:10.1007/s11947-013-1160-2
Tan C, Xia S, Xue J, Xie J, Feng B, Zhang X (2013) Liposomes as vehicles for lutein: preparation, stability, liposomal membrane dynamics, and structure. J Agric Food Chem 61:8175–8184. doi:10.1021/jf402085f
Taylor TM, Davidson PM, Bruce BD, Weiss J (2005) Ultrasonic spectroscopy and differential scanning calorimetry of liposomal-encapsulated nisin. J Agric Food Chem 53:8722–8728. doi:10.1021/jf050726k
Thompson AK, Hindmarsh JP, Haisman D, Rades T, Singh H (2006) Comparison of the structure and properties of liposomes prepared from milk fat globule membrane and soy phospholipids. J Agric Food Chem 54:3704–3711. doi:10.1021/jf052859b
Unalan UI, Cerri G, Marcuzzo E, Cozzolino CA, Farris S (2014) Nanocomposite films and coatings using inorganic nanobuilding blocks (NBB): current applications and future opportunities in the food packaging sector. RSC Adv 4:29393–29428. doi:10.1039/C4RA01778A
Vanderroost M, Ragaert P, Devlieghere P, De Meulenaer B (2014) Intelligent food packaging: the next generation. Trends Food Sci Technol 39:47–62. doi:10.1016/j.tifs.2014.06.009
Wei Q (2012) Functional nanofibers and their applications. Woodhead Publishing, Oxford
Wichchukit S, Oztop MH, McCarthy MJ, McCarthy KL (2013) Whey protein/alginate beads as carriers of a bioactive component. Food Hydrocoll 33:66–73. doi:10.1016/j.foodhyd.2013.02.013
Wu T-H, Yen F-L, Lin L-T, Tsai T-R, Lin C-C, Cham T-M (2008) Preparation, physicochemical characterization, and antioxidant effects of quercetin nanoparticles. Int J Pharm 346:160–168. doi:10.1016/j.ijpharm.2007.06.036
Xu W, Jin W, Zhang C, Li Z, Lin L, Huang Q, Ye S, Li B (2014) Curcumin loaded and protective system based on complex of k-carrageenan and lysozyme. Food Res Int 59:61–66. doi:10.1016/j.foodres.2014.01.059
Yoshimoto M, Sakamoto H, Yoshimoto N, Kuboi R, Nakao K (2007) Stabilization of quaternary structure and activity of bovine liver catalase through encapsulation in liposomes. Enzym Microb Technol 41:849–858. doi:10.1016/j.enzmictec.2007.07.008

Zeeb B, Herz E, McClements DJ, Weiss J (2015) Reprint of: Impact of alcohols on the formation and stability of protein-stabilized nanoemulsions. J Colloid Interface Sci 449:13–20. doi:10.1016/j.jcis.2014.07.034

Zehetmeyer G, Meira SMM, Scheibel JM, Oliveira RVB, Brandelli A, Soares RMD (2016) Influence of melt processing on biodegradable nisin-PBAT films intended for active food packaging applications. J Appl Polym Sci. doi:10.1002/app.43212

Zhang L, Pornpattananangkul D, Hu CMJ, Huang CM (2010) Development of nanoparticles for antimicrobial drug delivery. Curr Med Chem 17:585–594. doi:10.2174/092986710790416290

Zhang L, Hayes DG, Chen G, Zhong Q (2013) Transparent dispersions of milk-fat-based nanostructured lipid carriers for delivery of β-carotene. J Agric Food Chem 61:9435–9443. doi:10.1021/jf403512c

Chapter 3
Nanotechnological Applications in Food Packaging, Sensors and Bioactive Delivery Systems

Manashi Das Purkayastha and Ajay Kumar Manhar

Abstract The development of nanomaterial-based products has boosted research in food science. Packaging has been the core subject of research in food nanotechnology. The mechanical and barrier performance of packaging are improved by nanocomposites. Intelligent packaging with communication functions is currently being developed using indicators of oxygen, humidity and freshness. However the concern of nanoparticle migration into the food must be addressed prior to commercial use. For biosensors, functionalized nanomaterials act as catalytic or immobilization platform or as electro-optical labels for enhancing the sensitivity and selectivity of the detection. Other functions of nanomaterials include encapsulation, protection, and enhancement of the solubility and bioavailability of active food ingredients. This review describes nanocomposites as high barrier packaging material, nano-sensors to achieve safer foods with lower incidence of chemical contaminants or adulterants or pathogenic microorganisms, and nanoencapsulation strategies as innovative delivery systems of bioactive compounds. The technical aspects of these topics are discussed with respect to synthesis, mode of action and functional performance.

Keywords Nanomaterial • Food packaging • Biosensor • Bioactive compounds • Nanoencapsulation

M. Das Purkayastha (✉)
Department of Horticulture (Food Science & Technology Programme), Assam Agricultural University, 785013 Jorhat, Assam, India
e-mail: manashidas@aau.ac.in

A.K. Manhar
Department of Molecular Biology and Biotechnology, Tezpur University, 784028 Tezpur, Assam, India
e-mail: akmanhar@tezu.ernet.in

S. Ranjan et al. (eds.), *Nanoscience in Food and Agriculture 2*, Sustainable Agriculture Reviews 21, DOI 10.1007/978-3-319-39306-3_3

3.1 Introduction

Nanotechnology is a new and fast emerging field in the research realm. The interest of scientists is unequivocally burgeoning in the area of nanotechnology-derived food ingredients, additives and barrier materials. Nanofood is a term used to describe edible stuff or ingredient that use nanotechnology techniques, tools or engineered nanoparticles that have been added during cultivation, production, processing or packaging. This innovative technology also holds a key role in fabrication of sensors and nanomaterial-based assays for the detection of food-relevant analytes (gases, organic compounds and pathogens). Basically nanotechnology deals with investigation and employment of materials that have at least one dimension (or contain components having at least one dimension) that is approximately lower than 100 nm (de Azeredo 2009). Particles of such small size exhibit physico-chemical properties that are significantly different from their macroscale counterparts. Size usually relates to functionality of food materials; smaller size meaning a bigger surface area, is desirable for purposes such as improved water absorption, flavour release, bioavailability and faster rate of catalysis. For example, reduction of particle size of Djulis (*Chenopodium formosanum*) to 20–75 nm after nanogrinding with yttria-stabilized zirconia increased its antioxidant (radical scavenging) activity (Tsai et al. 2011).

Research involving nanomaterial in the arena of food technology has skyrocketed over the last decade; however the domain of commercially viable nanotechnology-based products is still in its infancy. This is because several studies have indicated that consumers prefer nanoparticles in an "out-of-food" application (Bieberstein et al. 2013). The main concern stems from the unpredictable interaction(s) of nano-sized materials at the molecular or cellular level and their possible ill impact on consumer's health and environment (Frewer et al. 2014). Nevertheless, the potential use of nanoparticles is being assessed in almost every segment of the food industry. It is expected that nanotechnology would bring a range of benefits to the food sectors, including new flavours, taste, enhanced nutrient absorption, improved packaging materials and enhanced agricultural produce (Dasgupta et al. 2015).

The review discusses some of the most promising applications of nanotechnology, with special emphasis on nanoparticle-based food packaging, sensors and assays for detecting chemical contaminants and pathogens at surprisingly low levels or monitoring changes in food during packaged condition, and encapsulation of flavours, bioactive compounds or textural enhancers. These areas of food nanotechnology are still a lesser-known subfield of the greater nanotechnology spectrum, and the present paper anticipates addressing this knowledge deficit. Such food-related aspects are most likely to receive consumer attention in the near future.

3.2 Types of Nanomaterials

Nano-sized entity may include any of the following forms: nanoparticles, nanotubes, fullerenes, nanofibres, nanowhiskers, nanosheets or nanolaminate. Nanotube has a cylindrical lattice structure; fullerene has a spherical shape; and nanofibres have a length to diameter ratio of at least 3:1; in nano scale. Nanowhiskers are long fibres having a cross-sectional diameter of 5–20 nm. Nanosheet or nanolaminate is the arrangement of the above-mentioned entities into fine film whose dimension is in the nano range. There are also inorganic nanoparticles which include quantum dots, magnetic and metallic nanoparticles. Quantum dots or nanodots are photostable fluorescent particles with size lower than 10 nm. These are used for bioimaging and anti-counterfeiting agents (Das Purkayastha et al. 2014a). Metallic nanoparticles include gold, silver, iron, zinc, nickel, aluminium, titanium dioxide, etc. and are used a labels for biosensors. They can exist in different geometries such as nanospheres, nanoshells, nanorods, or nanocages. Magnetic nanoparticles are spherical nanocrystals comprised of an iron core and are used to label biomolecules in bioassays and drug targeting. Many of these different nano forms are either in use or under investigation for use within the food industry.

3.3 Manufacture of Nanomaterials

Synthesis of nanomaterial follows two building strategies: 'top-down' and 'bottom-up' approaches. 'Top-down' approach involves breaking down larger particles to nanoscale dimension by physical or chemical means, for example, size reduction by use of force such as compression, impact and shear. Currently, commercial production of nanomaterials primarily involves the 'top-down' approach, which is executed by employing processes such as grinding, milling, etching, nanolithography and precision engineering (Chiang et al. 2012; Sanguansri and Augustin 2006). By contrast, in the 'bottom-up' approach, nanometric structures are built from individual atoms or molecules capable of self-assembling, which in-turn relies on balancing attraction and repulsion forces between molecules as building blocks to form functional nanoscopically structured entity. This can be achieved by changing the temperature, concentration, pH and ionic strength of the system. Methods of bottom up manufacture include crystallisation, layer-by-layer deposition, solvent extraction/evaporation, self-assembly, etc. Biological nanostructures rely on molecular recognition for self-assembly and the templating of atomic and molecular structures. Examples of such self-assembled nanostructures in food include micelle complexation, protein-polysaccharide coacervates and liposomes. Application of the 'bottom-up' approach in industry is expected to increase in upcoming years, because top-down approaches are often accompanied by heat generation and do not provide consistently impressive results especially in terms of size distribution of the synthesized particle, e.g. grinding and milling.

3.4 Application in Food Packaging

Due to the public preference for "natural" food products and the fear of consumers about "nanofoods", currently the most active area of food research is packaging. Food after production must be contained in a suitable package material that serves numerous functions. The prime duties of a packaging material in a food processing industry are to maintain the quality and safety of the product during storage and transportation, to extend the produce's shelf-life, should hinder gain or loss of moisture, carbon dioxide, oxygen and other volatile compounds, to prevent microbial contamination and must provide good mechanical, optical and thermal properties. In addition to these basic properties, it is desirable for the packaging material to biodegrade in a suitable time period after their use, without causing environmental problems (Duncan 2011). Among the conventional food packaging materials such as paper, cardboard, plastic, glass, ceramic and metal, petroleum-based plastic has been widely used. Lately several research groups started the preparation and examination of various kinds of biopolymer-based packaging materials, as they have been considered as an environmentally-friendly substitute for the use of non-biodegradable and non-renewable plastics. Though bio-derived polymer packaging have garnered considerable attention due to biodegradability, their poor mechanical, barrier and high hydrophilic properties with low thermal processability severely limits their usefulness in industries and overshadow their other benevolent features. To this end, complex multi-layer films and polymer blends by coextrusion technique are being tested, as no single polymer can exhibit all the desired properties for every conceivable food packaging application. Furthermore, multi-layer films and biopolymer blending possess high production and material costs, require the use of additional additives and adhesives, and reduce optical property, thus causing a major limitation in their industrial use. It has been suggested that inherent shortcomings of such biopackaging materials may be overcome by nanocomposite technology, wherein biopolymer-based packaging materials act as excellent vehicle for incorporating nano form additives (nanoparticle-reinforced packaging). Bio-nanocomposites consist of one or more biopolymer matrices reinforced with a nanoscale filler material (usually $\leq 5\%$ w/w), such as clay nanoplatelets (Avella et al. 2005), silica (SiO_2) nanoparticles (Tang et al. 2008), carbon nanotubes or graphene (Coleman et al. 2006), starch nanocrystals (Kristo and Biliaderis 2007), cellulose nanofibers or whiskers (Fortunati et al. 2013a), chitin or chitosan nanoparticles (Sriupayo et al. 2005), and metal oxide nanoparticles (Ag, Cu, CuO, TiO_2, ZnO, Pd, Fe) to name a few examples. Addition of nanofibers and nano-rods (e.g. carbon nanotubes, cellulose nanofibers/whiskers) can confer useful physical properties like strength and stiffness that far exceed conventional materials (Chen et al. 2005). Fibers with aspect ratio higher than 300 are particularly useful (Bradley et al. 2011). Nanofillers dispersed within a biocompatible polymeric matrix even show faster rate of biodegradation than control polymer containing no nanoscale fillers (Das Purkayastha et al. 2014a). This in-turn offers environmental benefits over conventional plastics or composites. Many authors reviewed the state-of-the-art of biopolymer-based nanocomposites that are related to the extension of shelf life of packaged foods and enhancement of packaging performance (de Azeredo 2009; Lagaron and Lopez-Rubio 2011; Mihindukulasuriya and Lim 2014; Peelman et al. 2013; Ranjan et al. 2014; Rhim et al. 2013; Sanchez-Garcia et al. 2010a; Sorrentino et al. 2007).

The manner in which the polymer and nanoplatelets is fabricated can play a significant role in modulating the distribution of nanomaterial throughout the matrix,

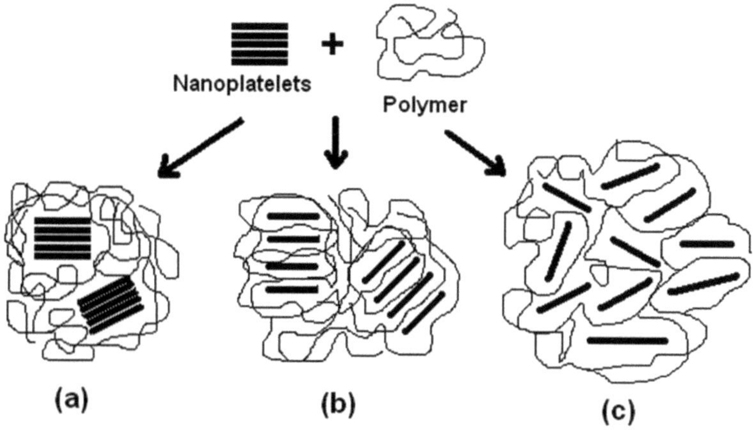

Fig. 3.1 Different morphologies of polymer-nanoplatelets interaction: (**a**) tactoid, (**b**) intercalated, (**c**) exfoliated (Adapted from Ranjan et al. 2014). Tactoid conformation results in a microcomposite, while intercalated and exfoliated arrangements give nanocomposites

and therefore, the barrier properties (passage of water vapour, oxygen, aroma and tainting compounds) of the resulting composite. In conventional composites (microcomposites), nanoplatelets (nanoclay) tend to stick together, particularly when dispersed in non-polar polymer environment. Such agglomeration of platelets lead to a 'tactoid' structure with reduced aspect ratio and barrier efficiencies (Fig. 3.1). When polymer-nanoplatelets interactions are more favourable or the platelets are disintegrated by high shear force (ultrasonication), intercalated or exfoliated structures are formed. Intercalated morphologies are characterized by the penetration of polymer chains into the interlayer region of the platelets, resulting in an ordered stack with alternating polymer and platelet layers. In exfoliated configuration, individual platelets are well separated and randomly dispersed in the polymer matrix with extensive polymer penetration (Jin and Zhong 2012). Generally intercalation and exfoliation (or delamination) is done by using one of the following two approaches: in the first approach (also called *in situ* polymerization), liquid monomer is inserted into the gallery volume of the nano-filler and subsequently polymer formation occurs by heat or radiation or a suitable initiator. In the second approach, the nanofiller is mixed directly with the polymer matrix which is in either the molten state or solubilized in a solvent. Depending on the nanofiller-polymer compatibility, the polymer chains crawl into the interlayer space. Theoretically, the best gas barrier properties would be obtained in polymer nanocomposites with fully exfoliated morphology with large aspect ratio. This is because the filler material (nanoplatelets) are essentially impermeable inorganic crystals, so gas molecules have to diffuse around them rather than a short straight line path which lies perpendicular to the film surface (Fig. 3.2). This results in a longer mean path for gas diffusion to occur through the film, in presence of fillers. Essentially, the presence of layered nanostructures delays the diffusion of molecules by making the pathway more 'tortuous'. This is

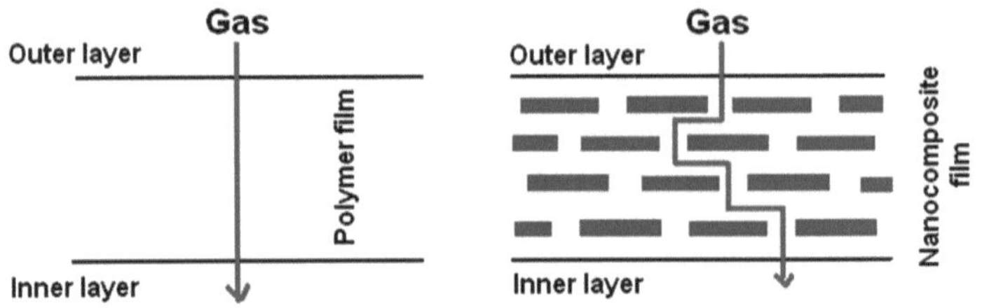

Fig. 3.2 Graphical illustration of the 'tortuous' pathway created by exfoliation (Adapted from Duncan 2011). Presence of layered nanoparticles increases the tortuosity

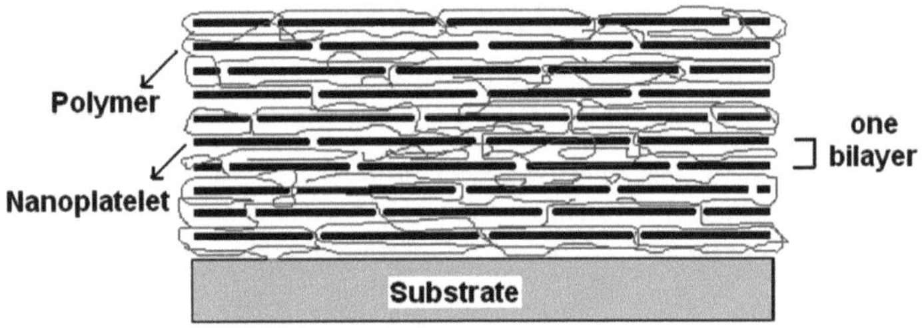

Fig. 3.3 Layer-by-layer deposition of polymer and nanoplatelets to form 'brick wall' nanocomposite (Adapted from Duncan 2011). Nanoreinforcement imparts high mechanical and barrier properties to the nanocomposites

the most widely accepted theory for explaining the improved barrier properties of polymer-organoclay nanocomposites.

Attainment of complete exfoliation of clay nanoplatelets is often difficult. So, a bottom-up strategy called layer-by-layer assembly was developed to fabricate multilayer films of nanoscale thickness. As the name suggests, layer-by-layer assembly is constructed by submerging the substrate in a solution of positively charged polymer, followed by drying and then submerging in a solution of negatively charged nanoplatelets. Electrostatic attraction between the oppositely charged layers forms a bilayer, also known as nanoplatelet-polymer bilayer. The cycle of alternate submerging is repeated until the desired numbers of bilayers are deposited on the substrate (Duncan 2011). The most fascinating attributes of such packaging material is its extremely high gas barrier property and strength. These impressive features are due to the "brick wall"-like fabrication of the nanocomposite, wherein the nanoplatelets and the polymer are arranged in distinct monolayers in an alternating fashion (Fig. 3.3). In other words, layer-by-layer assembly promotes extensive exfoliation and increases the tortuosity of the pathway (Duncan 2011).

As the risk of anaerobic microbial growth becomes the major worry when high barrier packaging materials are applied, multilayer nanocomposite technique can be explored (Fig. 3.4). For example, a barrier layer (polymer-clay nanocomposite) may be glued with other structural layers having oxygen scavenging or antimicrobial property (Rhim et al. 2013). Hence, hybrid nanocomposite coating or layering is presumed to be the wave of the future.

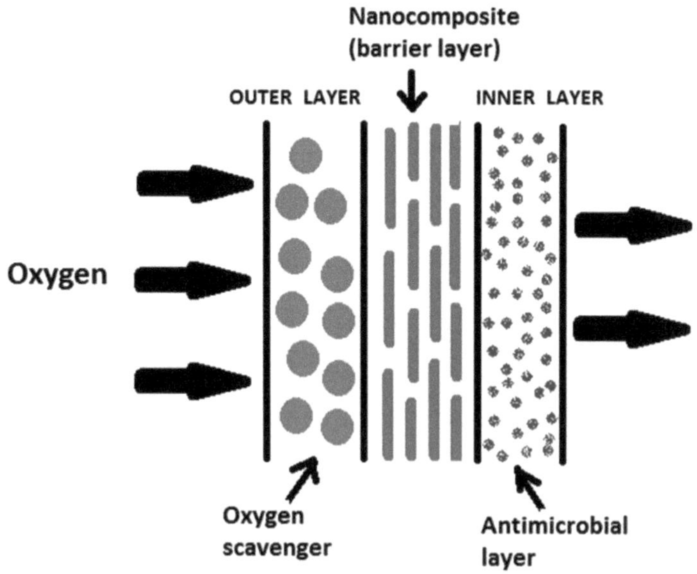

Fig. 3.4 Diagrammatic depiction of multilayer nanocomposite packaging. Such nanocomposite formulations are expected to considerably enhance the shelf-life of many types of food

Though many research papers focused on improvement of the properties of nanocomposite-based packaging, in reality, their optical properties are still not satisfactory (especially when carbon nanofiller are added) and find difficulty in 'clarity'-related purposes (Sanchez-Garcia et al. 2010b). Moreover, nano-level fillers complicate their regulation by federal agencies, i.e., the issues of non-intended migration and potential toxicity, and entail added difficulty when it comes to infant and geriatric food contact materials. Food packaging materials which are designed to release particulate nanomaterials into food are likely to receive more pessimistic consumer perception, than the ones from which no nanoparticles migrate. The latter may have a marginal 'active' effect limited only to the food contact material surface, and the former will be regarded as a food additive under regulation. Such particulate nanomaterials need to ensure their position on the food additives' positive list before use.

3.4.1 *Types of Packaging*

On the basis of the manufacturing processes, polymers used for packaging can be segregated into different groups: (i) polymers derived from chemical synthesis, such as polylactide, polyvinyl alcohol, polycaprolactone, polyvinyl chloride, polyethylene, etc. Such polymers can be further sub-divided into biodegradable and non-biodegradable classes; (ii) biopolymers derived from plant and animal sources, such as starch, chitosan, alginate, carrageenan, gluten, protein isolates, collagen, casein, waxes, fatty acid derivatives, etc.; (iii) biopolymers produced by microbial fermentation, such as polyhydroxyalkanoates, pullulan, curdlan, etc (Rhim et al. 2013).

The possible role of nanostructures in polymer-based packaging materials can be broadly divided into following categories:

- *Nanocomposites or bio-nanocomposites*: Nanomaterial loaded polymer or biopolymer matrix, which are either cast into film or applied as coating (inside or outside surface or sandwiched as a layer in a laminate). Nanocomposites are found to display higher barrier, mechanical, rheological and thermal properties compared to the neat polymers and conventional composites, due to the high aspect ratio and surface area of the nanoparticles like nano-titanium nitride, montmorillonite, etc. (Fabra et al. 2013).
- *Active packaging*: Packaging material incorporated with nanoparticles, also known as nano-additives, having antimicrobial properties or biocides; or antioxidative potential or oxygen scavenging property; or UV protection, e.g. nano-TiO_2; or ethylene absorbing ability, e.g. palladium doped zeolite nanoparticles. The sole motive for addition of these ingredients is to prolong the shelf-life of the packaged food. Among these, food contact materials imbued with antimicrobial nanoparticles, such as oxides of silver, zinc, copper, magnesium and carbon, or the film enriched with aluminium, iron, silicate nanoparticles and nanoclay for selective control of oxygen, aroma and moisture transmission, are of current interest. The migration of the cations from the polymer matrices is the key point to determine their antimicrobial effectiveness (Conte et al. 2013). Besides this general agreement, a mechanistic explanation of its bactericidal activity is still elusive. Among the current hypothesized mechanisms, physical cell membrane damage causing rupture, generation of reactive oxygen species and oxidative stress caused by nanoparticles are the most developed paradigms (Das Purkayastha et al. 2014b). Detailed description of metallic-based antimicrobial nanocomposites may be found in numerous review papers and books (Llorens et al. 2012; Sung et al. 2013a; Rhim et al. 2013; Vermeiren et al. 1999). Although there are many benefits of antimicrobial nanoparticles, there is a concern over their potential ill-effect on the gut natural microflora (Chaudhry et al. 2008).

Another new emergent at R&D stage include the development of hydrophobic or dirt-repellent nano-coating known as 'self-sanitizing surface'. Hydrophobic wax nanocrystals cause water to form droplets which roll off the surface and pick up loose dirt along their path, mimicking the "Lotus effect" (Chaudhry et al. 2008), and the photocatalytic action of nano-TiO_2 kills the surface microorganisms, producing a self-cleaning action (Kuan et al. 2012). This breakthrough can help to maintain hygiene in abattoirs, kitchen utensils, fish and meat processing machineries (Ravichandran 2010). Representative polymer based nanocomposites and their enhanced material properties are tabulated in Table 3.1.

- *Intelligent* or *"smart" packaging*: Incorporating nanosensors or indicators to monitor and report the condition of the packaged food. Indicators usually pro-

Table 3.1 Some representatives of polymer based nanocomposites and their enhanced material properties

Enhanced properties	Carrier/Polymer used	Nanomaterial	Reference
Thermal and mechanical strength	Poly(vinyl alcohol)	Cellulose nanocrystals	Fortunati et al. (2013b)
	Low-density polyethylene	Nanoclay	Hemati and Garmabi (2011)
	Polyethylene	silicate	Hyun et al. (2003)
	Polypropylene	Nano silica	Wu et al. (2005a)
	Low-density polyethylene	Organic montmorillonite	Xie et al. (2012)
	Polyvinyl chloride	Organoclay	Mingliang and Demin (2008)
	Polyvinyl chloride/ethylene vinyl acetate	Organic montmorillonite	Chuayjuljit et al. (2008)
	Poly(ethylene-co-vinyl acetate)	Nanosilica	Dasan et al. (2010)
	Unsaturated polyester and vinyl ester oligomer	Organic montmorillonite	Sharmila et al. (2010)
	Poly(methyl methacrylate)	Layered silicate	Mohanty and Nayak (2010)
	Polyimide	Nanoclay	Park and Chang (2009)
	Polystyrene	Organomontmorillonite-Cloisite 20A	Nayak and Mohanty (2009)
	Polyethylene terephthalate	Nanoclay	Ghanbari et al. (2013)
	Poly(vinyl alcohol)/ Poly(vinyl pyrrolidone)	Sodium montmorillonite	Mondal et al. (2013)
	Polypropylene	montmorillonite	Zhu et al. (2014)
	Polyethylene	Layered silicate	Chrissopoulou et al. (2005)
	Polyimide	Nanoclay	Gao et al. (2010)
	Polybutylene succinate	Nanoclay	Lee et al. (2002)
	Corn starch	Montmorillonite	Huang et al. (2005)
	Soy protein	Montmorillonite	Chen and Zhang (2006)
	Starch	Cellulose whiskers	Dufresne et al. (2000)

(continued)

Table 3.1 (continued)

Enhanced properties	Carrier/Polymer used	Nanomaterial	Reference
	Poly(lactic acid)/ Polycaprolactone blend	Nanoclay	Cabedo et al. (2006)
	Starch	Nano silica	Xiong et al. (2008)
	Polyvinyl alcohol	Nano silica	Jia et al. (2007)
	Pullulan	Starch nanocrystals	Kristo and Biliaderis (2007)
	Chitosan	Chitin whiskers	Sriupayo et al. (2005)
	Hydroxypropyl methylcellulose	Chitosan/tripolyphosphate	De Moura et al. (2009)
	Polyvinyl alcohol	Silver	Mbhele et al. (2003)
	Polylactic acid	Bentonite	Petersson and Oksman (2006)
	Pea starch	Cellulose whiskers	Chen et al. (2009)
	Polyurethane	Cellulose nanofibrils	Wu et al. (2007)
	Poly(ethylene-2,6-naphtalene)	Carbon nanotubes	Kim et al. (2008)
	Polyvinyl alcohol	Carbon nanotubes	Bin et al. (2006)
	Polypropylene	Carbon nanotubes	López Manchado et al. (2005)
	Polyamide	Carbon nanotubes	Zeng et al. (2006)
	Polypropylene	Layered silica	Wu et al. (2002)
Biodegradability	Polylactic acid	Layered silicate	Sinha Ray and Okamoto (2003)
	Polylactic acid	Montmorillonite	Sinha Ray et al. (2003)
	Polylactic acid	Nanoclay	Nieddu et al. (2009); Paul et al. (2005)
	Polyhydroxybutyrate	Layered silicate	Maiti et al. (2007)
	Soy protein	Furfural modified organoclay	Sasmal et al. (2009)
	Starch/polyvinyl alcohol	Layered silicate	Tang et al. (2008)
	Corn starch	Montmorillonite	Huang and Yu (2006)
Gas and water vapour barrier properties	Poly(e-caprolactone)	Cellulose nanocrystals	Follain et al. (2013)

	Polypropylene	Montmorillonite	Frounchi et al. (2006)
	Ethylene-vinyl alcohol/Poly(lactic acid)	Nanoclay	Lagaron et al. (2005)
	Low-density polyethylene	Nanoclay	Arunvisut et al. (2007)
	Low-density polyethylene	Potassium permanganate	Khosravi et al. (2013)
	Polypropylene	Nanoclay	Manikantan and Varadharaju (2011)
	Polyimide	Nanoclay	Khayankarn et al. (2003)
	Polyimide	NanoClay	Kim and Chang (2013)
	High-density polyethylene	Montmorillonite	Horst et al. (2014)
	High-density polyethylene	α-cellulose nanofibres	Fendler et al. (2007)
	Ethylene-vinyl alcohol	Organomontmorillonite	Kim and Cha (2014)
	Polyimide	NanoClay	Yano et al. (1997)
	Agar	NanoClay	Rhim (2011)
	Thermoplastic starch	NanoClay	Park et al. (2003)
	Whey protein	NanoClay	Sothornvit et al. (2009)
	Soy protein	Montmorillonite	Kumar et al. (2010)
	Wheat gluten	Montmorillonite	Tunc et al. (2007)
	Polycaprolactone	Montmorillonite	Gorrasi et al. (2002)
	Poly(butylene succinate)/thermoplastic starch	Organoclay	Park et al. (2007)
	Cellulose acetate	Organoclay	Park et al. (2004)
	Polylactic acid	Layered silicate	Koh et al. (2008)
	Polyethylene	Chitosan/bentonite	Vartiainen et al. (2010)
	Polylactic acid	Nanoclay	Chowdhury (2008)

(continued)

Table 3.1 (continued)

Enhanced properties	Carrier/Polymer used	Nanomaterial	Reference
	Chitosan and methyl cellulose	Mica	Lagaron and Fendler (2009)
	Polyethylene terephthalate	Cellulose whiskers	Sanchez-Garcia et al. (2007)
	Polyethylene terephthalate	Montmorillonite	Jang et al. (2008)
	Maleic anhydride modified polypropylene	Nano silica	Vladimiriov et al. (2006)
	Soy protein	Chitin whiskers	Lu et al. (2004)
Antimicrobial property	Polyethylene	Silver	Del Nobile et al. (2004)
	Chitosan	Silver/clay	Rhim et al. (2006)
	Polyvinylpyrrolidone	Silver	An et al. (2008)
	Cellulose	Silver	Pinto et al. (2009); Fernandez et al. (2010)
	Zein, agar, polycaprolactone	Silver-montmorillonite	Incoronato et al. (2010)
	Cellulose	Copper	Mary et al. (2009)
	Chitosan	Copper	Cardenas et al. (2009)
	Ethylene vinyl alcohol	Titanium dioxide	Cerrada et al. (2008)
	Stainless steel	Titanium dioxide	Verran et al. (2010); Chorianopoulos et al. (2011)
	Polylactic acid	Zinc oxide	Jin and Gurtler (2011)
	Low-density polyethylene	Silver, titanium dioxide, zinc oxide	Emamifar et al. (2010)
	Polycaprolactone	Magnesium aluminium hydrocalcite	Meera et al. (2012); Sorrentino et al. (2005)
	Chitosan	Silver zeolite	Rhim et al. (2006)
	Polystyrene	Silver silicate	Egger et al. (2009)
	Agar	Silver montmorillonite	Valodkar et al. (2010)
	Chitosan/starch	Silver	Yoksan and Chirachanchai (2010)

	Chitosan	Silver, zinc oxide	Li et al. (2010)
	Glass	zinc oxide	Applerot et al. (2009)
	Polypropylene	Titanium dioxide	Chawengkijwanich and Hayata (2008)
	Polyamide 6	Silver	Damm et al. (2007)
	Polyvinyl chloride	Silver/ titanium dioxide	Cheng et al. (2006)

vide immediate visual, quantitative information, unlike a sensor, about the packaged food by means of a colour change or an increase in colour intensity or diffusion of a dye. These can be a gas indicator (informs about altered gas concentrations such as CO_2, O_2, water vapor, ethanol, etc.), freshness indicator (detects reactions between microbial growth metabolites and integrated indicators; estimate the remaining shelf life of perishable products) or time-temperature indicator (warns about temperature abuse for chilled or frozen food products during distribution and storage). The later is achieved by thermochromic ink which changes colour with exposure to different temperatures (Vanderroost et al. 2014). A large body of scientific publications on indicators (labels or nanosensors) is available due to exhaustive research efforts undertaken during the last decade, and is discussed in the next section.

3.5 Application in Biosensor or Nanosensor

A biosensor is defined as a sensing element for selective detection of target and a method to transduce the interaction as a measuring signal (Warriner et al. 2014). The potential benefit of this emerging technology is receiving growing interest because nanotechnology-based sensors can rapidly detect pathogens, gases, aromas, freshness of processed product, flavour and food contaminants or toxins, and thereby reduce health risks and frequency of food-borne illness. The nanomaterial used in biosensor, such as magnetic nanoparticle, carbon nanotube, nanorods, quantum dots, nanowires, nanochannels, etc., have high capacity for charge transfer, large surface-to-volume ratio for immobilization of bioaffinity agents, high quantum yield and resistance to photo-degradation, i.e., excellent optical properties. All these tuneable properties of nanomaterial contribute to the improved performance of biosensors, making them suitable to reach lower detection limits and higher sensitivity values. Beyond the benefits afforded to miniaturization, reduced sample volume and detection time, the main focus of nano-biosensor, also known as nanosensor, a modern and efficient class of detection system that can give a fast 'yes-no' response or ensure a similar simple communication with the end-user, is to enhance traceability and reflect the actual food quality in real time.

Biosensors encountered in the food sector are derived from those devices initially fabricated to meet the health care needs. Appropriately, the use of biosensors in food pathogen detection has been continuously growing in the last decade, with *Salmonella*, *Escherichia coli* and *Listeria monocytogenes* among the most studied microbial contaminants. The majority of sensors developed for pathogen detection have been based on immunoassay coupled with immunocapture (Immunomagnetic separation) or flow through microfluidics with the specific antibodies immobilized on the nanoparticle or sensor surface (Fig. 3.5). In Immunomagnetic separation, nanoscale magnetic particles, attached with selective antibodies (recognition element), allow large analyte capture efficiency (due to specific antibody-antigen interaction). The captured (target) analyte can then be easily purified (selective separation by a magnet field) from the food matrix, and finally this hybrid is monitored through

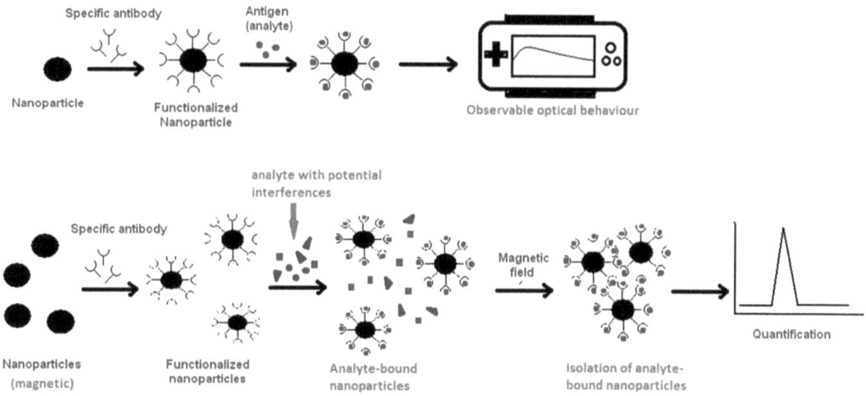

Fig. 3.5 Schematic representation of detection of analyte using immunocapture or immunomagnetic separation (Adapted from Duncan 2011)

a signal transduction mechanism, which explicitly detect the interaction between the pathogen and the recognition element improving the sensitivity of the analysis (Su and Li 2004).

Development of biosensor for toxin or allergen detection has been a major challenge as compared to microbial detection, due to greater degree of diversity in the nature of toxin molecules, types and cumbersomeness faced in extracting low levels of microbial/chemical toxin (in the order of $\leq$ng) from food matrices. Because of the latter, most reports on biosensors for toxin estimation have either been limited to using artificially spiked samples as opposed to considering direct extraction from original food sample, or necessitate pre-extraction step(s) to concentrate the analyte up to a detectable level. Separation from the sample matrix can be achieved by nanoparticle-based Magnetic Solid Phase Extraction, akin to Immunomagnetic separation (Duncan 2011). Other strategies devised for the convenient detection of a variety of food adulterants or allergens or contaminants include colorimetric (absorptive colour change), fluorimetric (change in fluorescence signal) and electrochemical (change in material's conductivity when the target analyte binds to the conductive nanoparticle) methods. Compared to optical (colorimetric and fluorimetric) methods, electrochemical approach may prove to be more useful for food matrices because the problem of light scattering and absorption from various food ingredients can be avoided (Duncan 2011). There is a pressing need for developing sensitive and selective method for analysis of residual pesticides in agricultural products owing to their high level of toxicity. As organophosphates are powerful neurotoxins, biosensors based on the inhibition of acetylcholine esterase (AChE) enzyme have been widely used for their detection (Table 3.2). Apart from AChE, fluorescence assay based on specific recognition of organophosphates by organophosphorus hydrolase (OPH) enzyme has also been developed. Analytes are not limited to harmful substances: several studies showed that nanoparticle-based detection can be used to measure sugar (Li et al. 2009a), polyphenols (Carralero-Sanz et al. 2005), vitamins (Liu et al. 2006), gliadin (Staiano et al. 2009), alcohol

Table 3.2 Some examples of nanomaterial-based biosensors applied in food analysis

Pathogen	Food matrix	Detection limit	Reference	Description of sensing technique
Escherichia coli 0157:H7	Ground beef	1.2×10^3 cells	Varshney et al. (2007)	Streptavidin-coated magnetic nanopartciles was conjugated with biotin-labeled polyclonal goat anti-*E. coli* antibody, which was used to capture target bacteria and was injected through the microfluidic flow cell (having embedded gold inter-digitated array microelectrode) for impedance measurement.
E. coli 0157:H7	Ground beef	1 cell	Zhao et al. (2004)	Tris(2,2'-bipyridyl) dichlororuthenium (II) hexahydrate (RuBpy) encapsulated in silica nanoparticle was conjugated to anti-*E.coli* antibodies for isolation of target pathogen and then detected by fluorescence signals.
E. coli 0157:H7	Ground beef	6 log cfu/g	Ochoa and Harrington (2005)	Magnetic beads coated with antibodies, which were specific to antigens of serotype 0157, were used to isolate the bacteria. Matrix-assisted laser desorption/ ionization (MALDI) Time-of-flight (TOF)-MS furnished bacterial mass spectra for organism identification.
E. coli 0157:H7	Wash water of green leafy vegetables	0.39 log cfu/ml	Magana et al. (2013)	Wash water was filtered to concentrate the target bacteria. The concentrated sample was subjected to electrochemiluminescence assay using antibodies labelled with silica encapsulated ruthenium (II) tri-bipyridal.

E. coli 0157:H7	Ground beef homogenate and milk	5 log cfu/ml	Aydin et al. (2014)	Magnetic-bead based immunoassay was used to isolate the pathogen, followed by tyramide-signal amplification (TySA) enzyme-linked immunosorbent assay (ELISA). TySA is based on horseradish peroxidase-tyramide conjugate (as fluorescent label) for increasing sensitivity and specificity.
E. coli	Phosphate buffer	10^4 cells/ml	El-Boubbou et al. (2007)	Surface of magnetic nanoparticles were functionalized with D-mannose and fluorescein-labelled concanavalin A. These functionalized nanoparticles were incubated with the pathogen and magnetic field was applied to separate the MNP-*E. coli* aggregate. The captured bacteria were imaged by fluorescent dye.
E. coli	Drinking water	3–5 cells	Kalele et al. (2006)	Antibody-anchored silver nano-shells were used to isolate the pathogen. *E. coli*-conjugated antibody-anchored nano-shells reduced the intensity of Surface Plasmon Resonance, which provided rapid detection. Nanorod attachment to the bacterial cell surface, followed by exposure to NIR resulted in a significant reduction in cell viability.
E. coli	–	–	Singh et al. (2009)	Antibody-conjugated gold nanoparticles are used to isolate the bacteria and the bound complex increases the two-photon Resonance Rayleigh scattering intensity for detection.

(continued)

Table 3.2 (continued)

Pathogen	Food matrix	Detection limit	Reference	Description of sensing technique
E. coli	Ground beef	–	Varshney et al. (2005)	Pathogen isolated by immunomagnetic separation using specific antibody functionalized iron oxide nanoparticles and detection by PCR analysis.
E. coli	Phosphate buffer	10^4 cells/ml	El-Boubbou et al. (2007)	Pathogen was isolated using magnetic iron oxide nanoparticles having sugar molecules adsorbed on their surfaces. Isolated bacteria were subsequently detected using fluorescent staining.
E. coli	-	10^3–10^7 cfu/ml	Su and Li (2004)	Pathogen separated by immunomagnetic separation and detected by fluorescent tags using quantum dots.
E. coli	Milk	50 cfu/strip	Lin et al. (2005)	Immunosensor strip was fabricated with carbon ink modified with gold nanoparticles (13 nm diameter) and ferrocene-dicarboxylic acid to detect horse-peroxidase activity, which in-turn amplifies the amperometric signal.
E. coli	water	3 cfu/10 ml	Zhang et al. (2009)	Copper having anti-*E. coli* antibody labels, is deposited on gold nanoparticles. Specific antigen-antibody binding releases Cu^{2+} ions which are detected by anodic stripping voltammetry.
E. coli, *Salmonella typhimurium*	2 % milk and spinach extract	10^4–10^5 cfu/ml	Ravindranath et al. (2009)	Magnetic iron oxide nanoparticles were functionalized with species- and strain-specific antibodies to isolate the target organism and then optically detected by FTIR/mid IR spectroscopy.

Salmonella sp.	Pork	10^2 cfu/ml	Gong-Jun et al. (2009)	Specific monoclonal antibodies were immobilized on gold nanaoprticles, which interacted with the pathogen to produce a change in impedance or capacitance.
Salmonella typhimurium	Phosphate buffer	98.9 cfu/ml	Dungchaia et al. (2008)	Monoclonal antibodies were immobilized on polystyrene microwells and allowed to bind with the bacteria. Pathogen specific antibody-colloidal gold conjugate was added to interact with the bound bacteria, followed by the addition of ascorbic acid and copper sulphate into the polystyrene microwells. The ascorbic acid reduced Cu^{2+} ions to Cu^0, which deposited onto gold nanoparticle tags and produced voltammetric changes (anodic stripping voltammetry).
Salmonella typhimurium	Chicken carcass wash water	10^3 cells/ml	Yang and Li (2005)	Pathogen was separated using anti-Salmonella antibody coated magnetic nanobeads and was reacted to secondary biotin-labelled anti-Salmonella antibody. Fluorescent quantum dots coated with streptavidin was allowed to react with biotin on the secondary antibody. Intensity of fluorescence provided a quantitative method for microbial detection.

(continued)

Table 3.2 (continued)

Pathogen	Food matrix	Detection limit	Reference	Description of sensing technique
Salmonella typhimurium	Phosphate buffer (pH 7.4)	30 cfu/ml	Wang et al. (2011a)	The indicator probe was designed through bioconjugating the *Salmonella* sequence-specific oligonucleotide with the fluorescent $Ru(bpy)_3{}^{2+}$ -doped silica nanoparticles that were prepared by W/O microemulsion method. Through a sandwich-type DNA hybridization procedure, the target *Salmonella* DNA was captured and the indicator probe was assembled onto solid matrix so that the *Salmonella* DNA can be measured by the fluorescent signals of assembled indicator probes.
E. coli 0157:H7, *Listeria monocytogenes*, *Salmonella* sp.	Rinse solutions from ground beef, poultry and lettuce	20–50 cfu/ml	Wang et al. (2011b)	Streptavidin-conjugated magnetic nanobeads (MNB) of 30 and 150 nm diameter, and quantum dots (QDs) (530, 580 and 620 nm emission wavelength) were separately coated with biotinylated anti-*Salmonella*, anti-*E. coli* and anti-*Listeria* antibodies. After being magnetically separated from the food sample, the MNB-cell conjugates were mixed with the immune-QDs to form the MNB-cell-QDs complexes, and the fluorescence intensity of which was measured at 530, 580 and 620 nm to determine the population of *Salmonella*, *E. coli* and *Listeria*, respectively.

Listeria monocytogenes	Beef, milk, lettuce rinse solution	4 log cfu/g	Kanayeva et al. (2012)	Pathogen captured by Immunomagnetic separation using nanobeads, which are flowed over a series of interdigitated electrodes that results in a change in impedance (impedimetric microfluidic system).
Listeria monocytogenes	Milk	10^2 cfu/0.5 ml	Yang et al. (2007)	Attachment of antibodies selective for the pathogen onto functionalized magnetic iron oxide nanoparticles (IMS) for separation of target bacteria and then detecting with real –time PCR analysis.
Listeria monocytogenes	Buffer	4.7×10^2 cfu/ml	Wang et al. (2008)	Conductive TiO_2 nanowire bundles, coated with pathogen-selective antibodies, were deposited between two gold electrodes. Bacteria bind to the antibodies and cause measurable change in impedance.

(continued)

Table 3.2 (continued)

Pathogen	Food matrix	Detection limit	Reference	Description of sensing technique
Listeria monocytogenes	Poultry products, pork and ready-to-eat foods	50 cfu/ml	Wang et al. (2010a)	FITC-IgG complexes were selected as core materials to prepare stable core–shell (FITC-IgG)-doped fluorescent silica nanoparticles (fsNPs) by a reverse microemulsion method. Then the prepared fsNPs were conjugated with polyclonal rabbit anti-L. monocytogenes antibody (pAb) and used as indicator probe. A sandwich-type immune affinity reaction between polyclonal rabbit anti-*L. monocytogenes* antibody coated onto microplate wells, target bacteria and the fsNPs–antibody conjugates subsequently was conducted to detect target bacteria and assemble the indicator probe onto the wells. The target bacteria were measured by the fluorescent signals of the assembled indicator probes.
Staphylococcus aureus	Milk	1.5×10^5 cfu	Sung et al. (2013b)	A nanocomposite was synthesized by coating the magnetic nanoparticles with bovine serum albumin (BSA) and then adsorbing the gold nanoparticles (AuNPs) and anti-*S. aureus* antibodies on its surface. The cells were captured by a magnet and concentrated by passing through a micro-filter that retains nanoparticle bound cells with unbound ones passing through. Fluorescence of the AuNPswas used to quantify the number of captured cells.

Pseudomonas aeruginosa	0.85 % NaCl solution	–	Norman et al. (2008)	Surface of gold nanorods were functionalized to bind with anti-*P. aeruginosa* primary antibodies. The antibody conjugated nanorods were mixed with bacteria and exposed to NIR light. Following irradiation, bacteria were stained to determine cell viability.
Mycobacterium avium paratuberculosis	Milk	10^8 cells/ml	Yakes et al. (2008)	Monoclonal antibodies are immobilized on the surface of gold nanoparticles (extrinsic Raman labels), that targets a specific surface protein of the pathogenic microorganism. Capturing of antigen produces large Surface-enhanced Raman scattering (SERS) signals.
Microbial toxin				
Aflatoxin (AF)	Drinking water, Red wine	0.0012 ng/ml (for AF B1, B2 and G1) and 0.0031 ng/ml for AF G2	McCullum et al. (2014)	Surface of ferric oxide nanobeads was modified with polydopamine onto which the aflatoxin in the sample binds (adsorbent). Separation of the hybrid from the sample was achieved by magnetic field (MSPE) followed by quantification with HPLC-MS/MS.

(continued)

Table 3.2 (continued)

Pathogen	Food matrix	Detection limit	Reference	Description of sensing technique
Aflatoxin-B1	Peanut, rice, milk, flour and soybean	1 ng/g	Zhou et al. (2012)	Graphene/conducting polymer/gold nanoparticle/ionic liquid composite film were deposited on gold electrode surface. Ionic liquid provided a benign microenvironment for the antibody. Presence of grapheme and gold nanoparticles aided enhanced electron transfer of the conducting polymer film. Upon binding of the aflatoxin target, there was a concentration-dependent charge transfer resistance, which formed the basis of the impedimetric sensor.
Aflatoxin-B1 (AFB1)	Milk	0.01 ng/ml	Jin et al. (2009)	Piezoelectric quartz crystal immunosensor was prepared by immobilization of BSA-AFB1 on the probe surface. By competitive immunoreactions of AFB1 with monoclonal anti-AFB1 antibody. The probe was incubated with gold nanoparticles-labelled secondary antibody, and the increased weight helps in detection.

Aflatoxin-B1	Pig feed	5 μg/kg	Delmulle et al. (2005)	It is an immunoassay-based lateral flow dip-stick technique. The dipstick membrane was biofunctionalized with aflatoxin B1-BSA conjugate and rabbit anti-mouse antibodies as capture reagent. The detector reagent consisted of colloidal gold particles coated with monoclonal anti-aflatoxin B1 antibodies. Analyte present in the sample competes with aflatoxin-B1 immobilized on the membrane for binding with antibodies (pink-coloured band).
Aflatoxin-B2	Peanut, Hazelnut, pistachios, almonds	0.9 ng/ml	Tang et al. (2009)	Membrane-based lateral-flow immune-dipstick assay, where detector reagent consisted of magnetic nanogold microspheres with nano-Fe_2O_3 as core and gold nanoparticles as shell and biofunctionalized with monoclonal anti-aflatoxin B2 antibodies.

(continued)

Table 3.2 (continued)

Pathogen	Food matrix	Detection limit	Reference	Description of sensing technique
Botulinum neurotoxin A and B	Milk, apple juice, orange juice	5–25 ng/ml	Ching et al. (2012)	Lateral Flow device consists of a nitrocellulose membrane to which an anti-mouse immunoglobulin and specific capture antibodies are immobilized. To the lower end of the membrane, a conjugation pad containing immobilized anti-Botulinum gold-conjugated antibody. Sample flows by capillary action through the conjugation pad. Toxin binds with the gold conjugated antibody and together they bind with the capture antibody, resulting in a colorimetric change.
Botulinum neurotoxin A	Phosphate buffer	5–31 pM	Warner et al. (2009)	It involves sandwich immunoassay where the primary antibody specific for neurotoxin is attached to a solid support (such as Sepharose bead and surface of a 96-well plate). The toxin fragment is bound in an incubation step to the antibody. A secondary antibody attached to the reporter quantum dots, is then bound to the toxin fragment-primary antibody complex. The fluorescence signal from the complete sandwich complex is collected by a fibre optic system.

Citrinin (CIT)	Rice	0.5 ng/ml	Arevalo et al. (2011)	CIT mycotoxin was immobilized on a gold surface electrodeposited on a glassy carbon electrode modified with a cysteamine self-assembled monolayer having monoclonal mouse anti-CIT immunoglobulin (Ig) labelled with horseradish peroxidise (sec Ab-HRP) reacts with mAb-CIT on the electrode surface. The HRP in presence of H_2O_2 catalyzes the oxidation of catechol. This electrochemical reduction is detected by amperometric measurement.
Deoxynivalenol (DON)	Wheat, maize	500 µg/kg	Speranskaya et al. (2014)	Hydrophobic cadmium selenide-based core-shell quantum dots were dispersed in aqueous solution by maleic anhydride-based polymer encapsulation. The obtained hydrophilic encapsulation. The obtained hydrophilic quantum dots covered with polymer retained 90 % of their fluorescence, and were modified with anti-DON antibodies (biolabels) to detect the target mycotoxin, using fluorescence-labeled immunosorbent assay (FLISA)
Cholera toxin, *E. coli* heat labile toxin and *Staphylococcus aureus* toxins (enterotoxin A and B, and shock syndrome toxin)	Drinking water, milk, meat extract	0.1–1 pg/ml	Shlyapnikov et al. (2012)	Assay consists of 3 steps: electrophoretic collection of toxins on antibody microarray, labelling of captured antigens with secondary biotinylated antibodies, and detection of biotin labels by scanning the microarray surface with streptavidin-coated magnetic nanobeads in a shear-flow.

(continued)

Table 3.2 (continued)

Pathogen	Food matrix	Detection limit	Reference	Description of sensing technique
Microcystin (toxin from cyanobacteria)	Drinking water	0.6 ng/ml	Wang et al. (2009)	Antibody to the microcystin-LR (MC-LR) was dispersed with single-walled carbon nanotubes (SWNTs). The change in electrical conductivity of anti-MCLR-coated SWNTs was used for estimation.
Ochratoxin-A	50 mM phosphate buffer (pH 7.0) with 10 % methanol	0.25 ng/dl	Kaushik et al. (2009)	Cerium oxide nanoparticles – chitosan based nanobiocomposite film was deposited onto indium-tin-oxide coated glass substrate, which has been used to coimmobilize rabbit immunoglobin and BSA for mycotoxin detection. The electrochemical signal is enhanced by the presence of Nano-CeO_2
Staphylococcus aureus Enterotoxin B	–	10–35 fm	Mishra et al. (2008)	Nanowire field effect transistors (nano-FET) were lithographically fabricated using 50 nm doped polysilicon nanowires attached to two small gold terminals separated from each other by a approximately 150 nm gap to serve as the basis for electronic detection of bacteria toxins.

Cholera toxin	–	10^{-16} g (equivalent to 100 μl of 10^{-15} g/ml)	Viswanathan et al. (2006)	Potassium ferrocyanide-encapsulated and ganglioside (GM1)-functionalized liposomes act as highly specific recognition labels for the amplified detection of cholera toxin. The sensing interface consists of monoclonal antibody against the B subunit of cholera toxin (CT) that is linked to poly(3,4-ethylenedioxythiophene) coated on Nafion-supported multiwalled carbon nanotube caste film on a glassy carbon electrode. The CT is detected by a "sandwich-type" assay on the electronic transducers, where the toxin is first bound to the anti-CT antibody and then to the GM1-functionalized liposome. The potassium ferrocyanide molecules are released from the bounded liposomes on the electrode by lyses with methanolic solution of Triton X-100. The released electroactive marker is measured by adsorptive square-wave stripping voltammetry.
Pesticides/insecticides/herbicides				
Methyl parathion and chlorpyrifos	0.002 M phosphate buffer (pH 7.0)	1×10^{-12} M	Viswanathan et al. (2009)	Acetylcholine esterase was immobilized on polyaniline, which in-turn was deposited on vertically assembled single-walled carbon nanotubes, wrapped single stranded DNA.

(continued)

Table 3.2 (continued)

Pathogen	Food matrix	Detection limit	Reference	Description of sensing technique
Chlorpyrifos	Drinking water, surface water and agricultural runoff water	3.8 ng/ml	Chen et al. (2010a)	Competitive fluorescence-linked immunosorbent assay (cFLISA) based on quantum dot-streptavidin conjugate (QDs-SA) was developed. The QDs-SA conjugate, which consists of 3-mercaptopropyl acid-stabilized CdTe nanoparticle QDs and streptavidin (SA) made through the active ester method, was employed to improve the sensitivity of QDs-SA-cFLISA. The QDs-SA conjugate was used as a fluorescence signal system and the biotin-streptavidin system (B-S system) was used for signal amplification.
Parathion	Buffer (pH 2.0–7.0)	3 ng/ml	Wang and Li (2008)	ZrO_2/gold nanocomposite film electrode showed strong affinity towards the phosphate group of parathion and the detection was studied with square wave voltammetry.
Paraoxon, Methyl parathion	0.05 M phosphate buffer (pH 7.4)	0.15 μM (paraoxon) and 0.8 μM (methyl parathion)	Deo et al. (2005)	Amperometric biosensor based on a carbon nanotube-organophosphorus hydrolase layer, which detects the enzymatically generated *p*-nitrophenol product.
Paraoxon	CH_3OH/H_2O (1:1 v/v)	10^{-8} M	Ji et al. (2005)	CdSe/ZnS core-shell quantum dots (QDs) were modified with organophosphorus hydrolase (OPH). Photoluminescence intensity of the OPH-QDs bioconjugate was quenched in presence of the insecticide.

Paraoxon	20 mM glycine buffer (pH 9.0)	–	Simoniana et al. (2005)	Organophosphate hydrolase (OPH)/gold nanoparticle bioconjugate was prepared, and then incubated with a fluorescent enzyme inhibitor or decoy. The fluorescence intensity of the decoy was sensitive to the proximity of the gold nanoparticles, and thus could be used to indicate that the decoy was bound to the OPH (competitive inhibitor of the enzyme). Then different concentrations of paraoxon were introduced to the OPH-AuNP-conjugate-decoy mixture, and normalized ratio of fluorescence intensities was measured.
2,4-Dichlorophenoxyacetic acid	50 mM phosphate buffer (pH 7.4)	250 pg/ml	Vinayaka et al. (2009)	It is based on competitive fluoroimmunoassay, where CdTe quantum dots are immobilized with anti-2,4-D antibodies.
Dichlorodiphenyltrichloroethane (DDT)	Grape, mango juice, milk and cauliflower	1000–27 ng/ml	Lisa et al. (2009)	Anti-DDT antibodies conjugated gold nanoparticles were treated with DDT to form an immunocomplex, which was further reacted with DDA (1,1,1-trichloro-2,2-bis-chlorophenyl acetic acid)-BSA conjugate immobilized on nitrocellulose membrane. The binding of nanoparticles conjugated anti-DDT antibodies to the immobilized DDA-BSA developed red colour.

(continued)

Table 3.2 (continued)

Pathogen	Food matrix	Detection limit	Reference	Description of sensing technique
Atrazine (herbicide)	Drinking water	1.0 ppb	Kaur et al. (2007)	It is based on dip-stick competitive immunoassay where hapten (a derivative of atrazine)-protein-gold nanoparticle conjugate competes with the free antigen (atrazine) present in the sample. The colour developed due to conjugate is correlated with concentration of atrazine.
Allergen/food colourant/adulterant				
Peanut allergen-Ara h1	Peanut milk beverage	0.041 fM	Sun et al. (2015)	A stem-loop DNA probe biosensor was prepared using multilayer graphene–gold nanocomposite which was electrodeposited onto a glassy carbon electrode and then a thiolated hairpin DNA-biotin probe was immobilized onto the modified electrode surface. The multilayer graphene-gold composite amplify the electrochemical signal and form the anchoring site. The probe was switched to an "off" state in the presence of target DNA.
Melamine	Raw milk, infant formula	2.5 ppb	Ai et al. (2009)	Gold nanoparticles functionalized with cyanuric acid selectively bind to the melamine. When bound to melamine, aggregated nanoparticles (blue) exhibit different absorptive properties than unbound nanoparticles (red).

Melamine	Raw milk	6 ppb	Kuang et al. (2011)	Colorimetric detection using crown-ether modified gold nanoparticles.
Melamine	Raw milk and milk powder	0.2 nM	Ni et al. (2014)	The AuNPs can conjugate with melamine to form AuNPs–melamine aggregates which could selectively improve the ability of bare AuNPs to catalyze 3,3',5,5'-tetramethlybenzidine (TMB) in the presence of H_2O_2 and lead to the significant enhancement of the absorbance.
Gliadin (for Celiac patients)	Flour extract	60 ng/ml	Staiano et al. (2009)	Gliadin was first captured on glass surface or silver island films (SIF) coated with anti-Gliadin antibodies, and then incubated with fluorescence (rhodamine)-labelled anti-Gliadin antibodies. The signal from the fluorophore is detected by total internal reflection fluorescence mode, which was found to be higher in case of SIFs compared to that from glass substrate not modified with a SIF.
Cyanide	Groundwater, tap water, pond water, and lake water	200 nM	Liu et al. (2010)	Cyanide etching-induced fluorescence quenching of gold nanoclusters, in presence of Elsner reaction between cyanide and the gold atoms of gold nanoclusters.
Sulfite	Red and white wines	6 mg/l	Rawal et al. (2012)	Sulfite oxidase was immobilized onto carboxylated gold coated magnetic nanoparticles, deposited onto the surface of gold electrode. Oxidation of sulphite was detected by amperometric response.

(continued)

Table 3.2 (continued)

Pathogen	Food matrix	Detection limit	Reference	Description of sensing technique
Sulfite	White wine	2.0×10^{-8} M	Zhang et al. (2013a)	In a flow injection chemiluminescence (CL) method, CoFe2O4 nanoparticles were found to possess intrinsic oxidase-like activity and could catalyze luminol oxidation by dissolved oxygen to produce intensified CL. When sulfite binds to the luminol – CoFe2O4 system, CL can either decrease (at low sulfite concentration) or increase (at high sulfite concentration).
ß-adrenergic agonists or growth promoters (Rectopamine, clenbuterol, salbutamol)	Meat (pork)	1.52 pg/ml (rectopamine) 1.44 pg/ml (salbutamol) 1.38 pg/ml (clenbuterol)	Wang et al. (2013a)	Surface of reduced grapheme oxide was immobilized with artificial antigens (anti-RAC, anti-SAL, anti-CLB) and silver-palladium alloy nanoparticles were used to label antibodies. Individual screen-printed carbon electrode and competition strategy was used for detection by capturing the free nanoparticles by the corresponding antigen-antibody reaction.
Amaranth (synthetic azo dye)	Soft drinks	35 nM	Wang et al. (2010b)	At a multiwall carbon nanotube film-modified glassy carbon electrode (denoted as MWNT/GCE), an irreversible and greatly improved oxidation peak is observed by cyclic voltammetry, suggesting that MWNT film possesses a notable enhancement effect on the oxidation of amaranth.

Polycyclic Aromatic Hydrocarbons (PAHs)	Edible oils	0.34 – 2.9 ng/g	Zhao et al. (2011)	Magnetic multiwalled carbon nanotubes were applied to magnetic solid-phase extraction (MSPE) of eight heavy molecular weight PAHs including chrysene, benzo[*a*]anthracene, benzo[*b*] fluoranthene, benzo[*k*]fluoranthene, benzo[*a*]pyrene, indeno[1,2,3-*cd*]pyrene, dibenzo[*a*,*h*]anthracene and benzo[*g*,*h*,*i*] perylene, and then coupled with gas chromatography–mass spectrometry (GC – MS) for their detection.
Lasalocid (LAS) and salicylate (SAL) (veterinary drugs)	Tap water, poultry feed premix, and egg samples	1.0 ng/ml (LAS) 4.0 ng/ml (SAL)	Castillo-García et al. (2012)	Tb_4O_7 nanoparticles as analytical reagents was used for checking their capability to obtain terbium sensitized luminescence in the presence of LAS and SAL.
Sulfamethazine (SM2) (veterinary drug)	Chicken muscle tissue	1.0 ng/ml	Ding et al. (2006)	Competitive fluorescence-linked immunosorbent assay (cFLISA) method for the detection of SM2 was demonstrated using quantum dots (QDs) as the fluorescence label coupled with secondary antibody.
Triclosan (antimicrobial)	Toothpaste samples	16.5 μg/l (about 57 nM)	Yang et al. (2009)	Multiwalled carbon nanotubes (MWCNT) film was constructed on the surface of a glassy carbon electrode (GCE) via the method of dip-coating and evaporating. At the MWCNT film, the oxidation signals of triclosan remarkably increase, suggesting that the MWCNT film exhibits a considerable enhancement effect with triclosan.

(continued)

Table 3.2 (continued)

Pathogen	Food matrix	Detection limit	Reference	Description of sensing technique
Ponceau 4R and Allura Red (synthetic colourants)	Soft drinks	15 μg/l (Ponceau 4R) 25 μg/l (Allura Red)	Zhang et al. (2010)	At a multiwalled carbon nanotubes film-modified glassy carbon electrode (denoted as MWNT/GCE), two sensitive and well-shaped oxidation peaks are observed for Ponceau 4R and Allura red.
Ponceau 4R and Amaranth (synthetic colourants)	Soft drinks	2 nmol/l (Ponceau 4R) 36 nmol/l (Amaranth)	Zhang et al. (2014)	Expanded graphite paste electrode (EGPE) was prepared by mixing expanded graphite with solid paraffin. EGPE showed higher electrocatalytic activity towards the oxidization of Ponceau 4R and Amaranth
Ponceau 4R and Amaranth	Fruit drink samples	5×10^{-10} M (Amaranth) 1×10^{-9} M (Ponceau 4R)	Wang et al. (2015)	polypyrrole was used to decorated onto the surface of carbon nanotubes (CNT–ppy) through in-situ chemical oxidative polymerization in the presence of $FeCl_2$ and H_2O_2. Compared with individual CNT modified glass carbon electrode (CNT/GCE), the CNT–ppy modified GCE (CNT–ppy/GCE) enhanced the electrocatalytic oxidation signals of Ponceau 4R and Amaranth.

Sudan I	Red chili powder	11.4×10^{-7} M/l	Prabakaran and Pandian (2015)	Silver nanoparticles decorated graphene oxide was deposited on glassy carbon electrode to modify the electrode surface in order to enhance the electrocatalytic reduction process of Sudan I. AgNPs@GO/GCE catalyzes the reduction of azo group, –N=N-followed by electrochemical oxidation of -OH group present in Sudan I dye molecule. Quantitative detection of dye was carried out by amperometry method.
Sudan I	Hot chilli powder and ketchup	0.03 μmol/l	Mo et al. (2010)	Electrode was fabricated with hydrophobic ionic liquid (trihexyltetradecylphosphonium bis(trifluoromethylsulfonyl)imide), multi-walled carbon nanotubes and cationic gemini surfactant ($C_{12}H_{25}N(CH_3)_2$–C_4H_8–$N(CH_3)_2C_{12}H_{25}Br_2$). Electrochemical behaviour of Sudan I was detected by cyclic voltammetry.
Orange II	0.1 M acetate buffer (pH 4.5)	3.4×10^{-10} M	Yun et al. (2015)	Electrochemically reduced graphene oxide was grafted with 5-amino-1,3,4-thiadiazole-2-thiol-Pt nanoparticles (denoted as ERGO-ATDT-Pt) onto a glassy carbon electrode (GCE). ERGO-ATDT-Pt/GCE showed improved sensitivity to the electrocatalytic determination.

(continued)

Table 3.2 (continued)

Pathogen	Food matrix	Detection limit	Reference	Description of sensing technique
Food constituent				
Polyphenols	Wine	2.1×10^{-7} M (phenol); 1.5×10^{-7} M (catechol); 6.6×10^{-7} M (caffeic acid); 6.2×10^{-7} M (chlorogenic acid); 70×10^{-7} M (gallic acid); 20×10^{-7} M (3,4-dihydroxybenzaldehyde)	Carralero-Sanz et al. (2005)	Glassy carbon electrode, modified with electrodeposited gold nanoparticles, was immobilized with tyrosinase enzyme (Try-nAu-GCE), by cross-linking with glutaraldehyde. Oxidation of the phenols by enzyme to *o*-quinone is detected by amperometric response, which was proportional to the substrate concentration.
Glucose	20 mM phosphate buffer (pH 7.4)	0.5 mM	Li et al. (2009a)	Photoluminescent CdTe quantum dots and glucose oxidase are integrated into a multilayer film by electrostatic layer-by-layer self-assembly technique. H_2O_2 produced by the oxidation of glucose with the enzyme etches the quantum dots surface, and hence quenches its photoluminescence.
Glucose	0.2 M acetate buffer (pH 4.0)	3×10^{-5} M	Wei and Wang (2008)	Oxidation of glucose by glucose oxidase enzyme liberates H_2O_2. Fe_3O_4 magnetic nanoparticles catalyze the oxidation of 2,2'-azino-bis(3-ethylbenzothiazoline-6-sulfonic acid)diammonium salt (ABTS) in presence of H_2O_2. Concentration of glucose is determined by the colorimetric change cause by coloured oxidized ABTS.

Glucose	Commercial beverages (pomegranate, peach, orange and mixed fruit)	0.05 mM	Ozdemir et al. (2010)	Pyranose oxidase was immobilized on gold nanoparticles-polyaniline/silver chloride/gelatine nanocomposite. This hybrid was coated on glass carbon electrode, which detects the consumed oxygen during enzymatic reaction via amperometric detection.
Glucose	50 mM phosphate buffer (pH 7.4)	1 μM	Lu et al. (2007)	Exfoliated graphite nanoplatelets were decorated with platinum and palladium nanoparticles. These metal decorated nanoplatelets were suspended in Nafion (a sulfonate tetrafluoroethylene copolymer). Glucose oxidase was immobilized on Nafion layer and the glucose concentration was detected by cyclic voltammetry of the redox process.
Glucose	0.1 M phosphate buffer (pH 7.4)	20 μM	Zhao et al. (2007)	Glucose oxidase was immobilized onto the ZnO:Co nanocluster-assembled thin film through Nafion-assisted cross-linking technique. The redox process of the enzymatic reaction was analyzed by amperometric response.
Fructose	Honey	1×10^{-6} M	Antiochia et al. (2004)	The sensor is made up of carbon nanotubes paste electrode modified with an electropolymerized film of 3,4-dihydroxybenzaldehyde and its activity is based on D-fructose dehydrogenase enzyme immobilized on the top of the electrode. The current from the redox reaction was detected by amperometric response.

(continued)

Table 3.2 (continued)

Pathogen	Food matrix	Detection limit	Reference	Description of sensing technique
Fructose	0.1 M phosphate buffer (pH 5.0)	5 mmol/dm^3	Tominaga et al. (2009)	Multi-walled carbon nanotubes were synthesized on platinum plate electrode by chemical vapour deposition. D-fructose dehydrogenase enzyme, immobilized onto the electrode surface, produced catalytic oxidation current which was used for calibrating fructose concentration.
Maltose	Buffer	–	Sandros et al. (2006)	Protein attached Ru (II) complex is assembled with ZnS coated CdSe nanoparticles. This maltose-binding protein changes the fluorescence intensity of the associated nanoparticle.
Vitamin B1	0.1 M citrate buffer (pH 4.0)	0.9 ng/ml	Liu et al. (2006)	Interaction of gold nanoparticles with vitamin B1 enhanced the Resonance Rayleigh Scattering intensity, which was directly proportional to the concentration of the vitamin.
Soy protein	Yoghurt	0.05 mg/l	Godoy-Navajas et al. (2011)	A conjugate composed of anti-soy protein antibodies bound to Nile Blue-doped silica nanoparticles was synthesized and these nanobeads were linked with anti-soy protein antibodies. Soy protein was immobilized onto ELISA microplates and the fluorescence was measured during analysis.

Casein	Cheese	5×10^{-8} g/ml	Cao et al. (2011)	The electrode was made of film having polyarginine modified carbon nanotubes and anti-casein antibodies immobilized gold nanoparticles. Formation of antigen-antibody complex reduced the peak current of ferricyanide redox pair which was detected by differential pulsed voltammetry.

dehydrogenase activity (Wang et al. 2013b), synthetic antioxidant like propyl gallate (Xu et al. 2015), etc.

Another area of nanotechnology-based sensing in relation to food safety or quality includes accurate shelf-life prediction of packaged food. Produce which are either spoiled or unpalatable, exhibiting unacceptable odors, colours or other sensory defects can now be discerned with the help of user-friendly biosensors/devices, and the consumers need not have to rely on expiry dates to assess the quality of the product, especially when the packaging material prevents extensive sensory exposure. In industrial processing line, packaged food are tested randomly during a production run and the assays require destruction of the package, which is time-consuming and costly, and yet does not ensure that the unsampled packages meet the requisite quality and safety standards (Duncan 2011). To this end, real-time monitoring of specific degradation product(s) or headspace gas(es) of a package using nanosensor supported noninvasive technique has been devised to assess the quality of the contained food long after it has left the prediction facility. Such easy-to-read shelf-life prediction sensors are close to commercialization. A major thrust in this direction is the development of intelligent (smart) packaging; the embedded nanosensor in the packaging film triggers a colour change to alert the consumer that the shelf-life is either ending or has ended. A promising in-package oxygen sensor which utilizes UV-active TiO_2 or SnO_2 nanoparticles and a redox-active dye (methylene blue), is devised by Mills and Hazafy (2008). It changes colour in response to even minute quantities of oxygen and thus can be used to detect the presence of leaks or possible compromised seal integrity in every packets. Similarly package moisture can be quickly determined without invasive sampling by colorimetric change of a swelled polymer strip containing carbon-coated copper nanoparticle (Luechinger et al. 2007). A non-invasive method of measuring in-package CO_2 level is based upon the half-life analysis of fluorophore (luminescent dye) encapsulated in polymer nanobeads (von Bultzingslöwen et al. 2002). A real-time NH_3 sensor, based on nanoporous silica-alumina coated silicon-on-insulator micro-ring resonator, showed a detection limit of 5 ppm (Yebo et al. 2012). Inkjet-printed film of carbon nanotube bundles was blueprinted on cellulosics (paper and cloth) for detecting aggressive oxidizing vapors such as nitrogen dioxide and chlorine at 250 and 500 ppb, respectively, at room temperature in ambient air without the aid of a vapor concentrator (Ammu et al. 2012). Indicators of meat and fish spoilage include detection of gaseous amines using nanofibrils of perylene-based fluorophores (Che et al. 2008; Che and Zang 2009; Hernández-Jover et al. 1996) or xanthine sensor (xanthine is a breakdown product of purine) based on chitosan modified ZnO nanoparticles on a multi-walled carbon nanotube scaffold within a polyaniline matrix onto which the xanthine oxidase is bound (Devi et al. 2012, 2013).

Nanotechnology can help food industries in providing authentication and avoiding counterfeiting; preventing adulteration and allowing the brand owners to monitor their supply chains without having to share company information (Neethirajan and Jayas 2011). Nam et al. (2003) have made nanodisks of gold and nickel, functionalized with chromophoric dye that emit a unique light spectrum when illuminated with a laser beam, to encrypt information (DNA) to be used as biological tags for tracking food products. Bloch et al. (2014) added DNA (food tracer) encapsulated silica nanoparticles into a milk processing chain, thereby the milk-derived products (yoghurt and cheese) could be uniquely labeled and monitored. Li et al. (2005) have created a nanobarcode detection system that fluoresces under ultraviolet light, also called security inks. Similarly, a forgery-proof biodegradable packaging films, embedded with oilseed cake-derived fluorescent carbon nanodots, has been reported by our group (Das Purkayastha et al. 2014a). The inherent photoluminescence property of the carbon nanodots was easily discernible under the UV light and showed radical scavenging property and hemocompatibility. Presence of nanodots imparted key novelty to the nanocomposite by minimizing the ingress of oxygen, retarding the oxidation of oil samples packed within the film pouches, and ameliorating the biodegradability rate of the film material (Das Purkayastha et al. 2014a).

Further progress in the field of nanosensor includes the so-called 'Electronic tongue' technology where sensor arrays are fabricated having artificial taste receptors, e.g. for sweetness or bitterness, which are immobilized on the surface of polypyrrole modified carbon nanotubes. Upon binding of specific compounds from the sample, associated with either sweetness or bitterness, there is a re-orientation of the polypyrrole layer which is transduced as an electronic signal (Apetrei et al. 2010). Likewise, polyaniline boronic acid conducting polymer-nanoparticle hybrid is the basis of 'Electronic nose', which has been demonstrated for monitoring spoilage of stored grain (Neethirajan et al. 2013), progress of wine fermentation, cheese ripening and meat spoilage (Sliwinska et al. 2014). Over the years, devices based on competitive immunoassay in 'dip-stick format', also known as Lateral Flow technique, are expected to gain interest. By controlling the target antibody, this assay can be extended for use with various food mycotoxins and pesticides. Among the near-market improvement is Radio Frequency Identification Display (RFID) technology, incorporating nanoscale organic thin-film concept that will enable the food or its ingredients to be traced back to the source of origin, assist quick distribution and reports for anomalies such as temperature and humidity variation, short lifespan, etc. (Chaudhry et al. 2008). RFID tags for food contact materials is discussed in detail by Vanderroost et al. (2014). Table 3.2 provides a non-exhaustive list of some of the nanomaterial-based sensors applied in food analysis, along with brief illustration of the working principle, as it is out of the scope of the article to go into their respective instrumentation details. An overview of the different sensing technology involved in biosensor design has been reviewed by Warriner et al. (2014), and Pérez-López and Merkoçi (2011). Few biosensors have been commercialized for food testing, but most of the trials on nanosensors are still in the early stages of development and require further intensive research prior to integrating them in food packaging, because there may be possible hazardous effect of the sensor's biological and nano entities like antibodies, immunoglobulins, etc. on the contained food (Vanderroost et al. 2014).

3.6 Application in Bioactive Delivery System or Nanoencapsulation

Another significant contribution of nanotechnology to the food sector is the development of nanometric delivery system, which allows encapsulation of sensitive bioactive compounds like polyphenols, micronutrients, enzymes, functional peptides, antioxidants and nutraceuticals, in nanosized particles mainly to preserve their functionality and biological activity against adverse environment, such as oxidation or degradation during processing or during passage through harsh acidic environment of the stomach, until their release at targeted sites (Ezhilarasi et al. 2013). This is also termed as precision targeting. The concept of 'nanoencapsulation' stems from the already well established 'microencapsulation' technology. Nanoencapsulation is defined as a process by which core materials, i.e. bioactive or functional ingredients, are packed within a wall material of nanoscale range to form nano-capsules or nanospheres. Nanocapsule is a core-shell structure in which the bioactive material is confined to a cavity enclosed by a thick-walled polymer membrane, while nanosphere is a matrix system where bioactive compounds are distributed throughout the continuous polymer network (Fig. 3.6) (Ezhilarasi et al. 2013; Hamidi et al. 2008). The membrane can be made up of protein, lipid, carbohydrate or other biodegradable polymer. Additional functions of nanoencapsulation are to enhance solubility and bioavailability; i.e., the increased surface area per unit volume of the entrapped bioactive ameliorates the solubility and allows its easy passage from the gut to the bloodstream, aiding in bioavailability, and enable controlled release and sometimes masking the unpleasant taste of bioactive ingredients, bitter taste of functional peptides, etc.

Next to the above, localization of the bioactive influences the nanoparticle structure and release characteristics: the active compounds can be distributed throughout the polymer matrix (dispersion type) or confined to the core surrounded by polymer

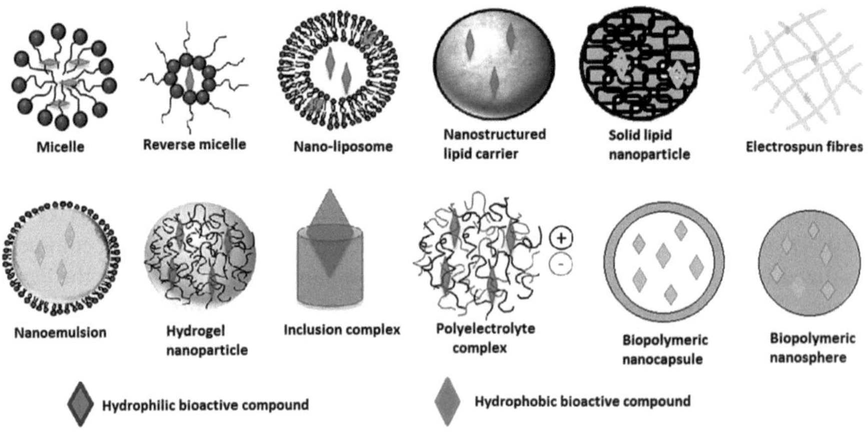

Fig. 3.6 Different types of bioactive delivery systems (Adapted from Joye et al. 2014)

shell, known as enriched core-type, or may be concentrated in the outer film surrounding the core, known as enriched shell-type. The bioactive release is governed by different mechanisms depending on the nature of nanoparticle used (Joye and McClements 2013; Joye et al. 2014): diffusion through intact particle matrix; release due to particle swelling by solvent or particle disintegration or erosion by chemical or enzymatic degradation. The former two mechanisms are more likely to occur in polymer-based nanoparticles, e.g. polylactide-co-glycolide, and results in sustained or delayed release; whereas the latter two phenomena give rise to burst release and mostly occur in hydrophilic carriers, e.g. carbohydrate and protein-based wall materials. Moreover, entrapment of active compounds in the shell or particle surface, instead of the core, is likely to reduce the possibility of a controlled release mechanism and the drug can be damaged by environmental agents. In spite of the increasing number of reports, analysis and patent applications on nanotechnology, studies relating to release kinetic modeling of nanoencapsulated food ingredient are very scarce.

Nanoencapsulates are also known as nanocarriers, which are mostly lipid-based or natural biodegradable polymers. Such polymers include albumin, gelatin, alginate, collagen, chitosan, whey protein, α-lactalbumin, etc. Different excipients, i.e. polymer or surfactant, can be applied to design edible nanoparticles for releasing their payload at appropriate site of action in the gastro-intestinal tract: nanoparticles made of digestible protein shall dissociate in the stomach due to protease activity; lipidic nanoparticles will mainly dissociate in the small intestine due to lipase activity, and dietary fibers can be used to fabricate nanoparticles that will remain intact until they reach the colon (Joye et al. 2014) Despite of applicability of protein and carbohydrate-based nanocarriers in food systems, lipid-based ones are more preferred as they bear advantage of higher encapsulation efficacy and better processability; albeit encapsulation load higher than 50 % tends to increase the risk of bioactive leakage and surface defects (Fathi et al. 2012). Moreover, lipophilic systems are highly permeable through the intestine via active transport and facilitated diffusion (Ezhilarasi et al. 2013). Lipid-based nanocarriers are basically of four types: nanoliposomes or liposomes, solid lipid nanoparticles, nanostructured lipid carrier and nanoemulsion (Fig. 3.6). Apart from these, other types of nano-delivery systems in food include archaeosomes and nano-cochleates (Ranjan et al. 2014). Archaeosomes are liposomes made from Archaeobacteria, and are more thermostable and stress resistant than liposomes, and can protect thermolabile antioxidants during processing. While nano-cochleates are usually 50 nm in size, delivers protein, peptides, DNA, and are resistant to degradation in the gastrointestinal tract.

The choice of nanoencapsulation techniques will depend on the physico-chemical properties of the core and wall materials. Based on the solubility in water, nutraceuticals to be nanoencapsulated can be either hydrophilic, such as ascorbic acid, polyphenols, vitamin-B, etc., or lipophilic, such as lycopene, ß-carotene, lutein, phytosterols, docosahexaenoic acid, etc. Emulsification, coacervation, spray drying, freeze drying and supercritical fluid technique are used for the encapsulation of both hydrophilic and lipophilic compounds; whereas inclusion complex, emulsification-solvent evaporation and nanoprecipitation methods are mostly used for lipophilic compounds (Ezhilarasi et al. 2013). A state-of-the-art of current and projected processes used for nanoencapsulation is briefly discussed below. Readers interested in a more detailed account of specific technique should consult the review articles by Ezhilarasi et al. (2013); Fathi et al. (2012, 2014), and Joye et al. (2014).

3.6.1 Biopolymer-Based Nanoparticle Fabrication:

Biopolymer-based nanoparticles can be made by various methods, the common ones are presented below:

3.6.1.1 Nanoprecipitation or Anti-solvent Precipitation or Solvent Displacement Technique

In this method, water miscible organic solvent containing the dissolved biopolymer like polycaprolactone, polylactic acid, polylactide-co-glycolide, or polyalkylcyanoacrylate, is mixed with the aqueous phase, in such a way that the solvent surrounding the biopolymer rapidly diffuses into the aqueous phase, thereby promoting polymer particle formation by precipitation of polymer through nucleation and growth (Sanna et al. 2015). The resulting nanopcapsules or nanospheres are often around 100 nm and exhibit good stability against degradation, higher cellular uptake and encapsulation efficiency (Anand et al. 2010; Gou et al. 2011; Ribeiro et al. 2008; Patel et al. 2010). Efficiency of entrapment of bioactives within the polymer particle depends on the similarity in polarity of the matrix material (biopolymer) and the molecule of interest, i.e., hydrophobic polymers can encapsulate lipophilic bioactives and hydrophilic ones can entrap hydrophilic compounds. The major limitation of this technique is its dependency on water-miscible solvents, and selection of appropriate food-grade anti-solvent phase compatible with the polymer (Ezhilarasi et al. 2013).

3.6.1.2 Emulsification-Solvent Evaporation Method

It is a 2-step process where the polymer, encapsulant and organic solvent are emulsified into an aqueous phase, in presence of an emulsifier (by high speed homogenization or ultrasonication). In the second step, the solvent is evaporated under reduced pressure to induce polymer precipitation as nanospheres. Similar to nanoprecipitation technique, emulsification-solvent evaporation method also remains to be an efficient method of producing particles less than 100 nm diameter, with high drug loading capacity (75–96 %) and stability (Ezhilarasi et al. 2013). This procedure of nanoparticle making has been successfully implemented to produce curcumin-loaded nanocarriers with improved solubility and resistance to degradation (Dandekar et al. 2009; Mukerjee and Vishwanatha 2009; Sowasod et al. 2008).

3.6.1.3 Supercritical Fluid Technology

Supercritical fluid technique has attracted increasing attention in recent years because it requires minimum organic solvents and is suitable for encapsulation of heat-labile and oxidation-prone compounds because it provides an inert medium with supercritical fluids like carbon dioxide or water or nitrogen, and uses low critical temperature (Fathi et al. 2014); nonetheless, the high capital cost of the equipment and the need to maintain the extractant in the supercritical state are the prime disadvantages of this process. The basic principle of this technology comprises of saturating the supercritical fluid (maintained at high pressure) with organic solvent or emulsion (containing biopolymer and encapsulant), and then expanding it through a nozzle leading to sudden pressure drop and evaporation of supercritical fluid along with organic solvent and consequent precipitation of solutes into nano-sized particles (Mattea et al. 2009; Santos et al. 2012).

3.6.1.4 Electrospinning and Electrospraying

These are electrohydrodynamic processes for obtaining fibers or particles of nanometer scale from a polymer solution, under the application of high potential electric field (1–30 kV). Though both the processes are considered as 'sister' technologies because of their similar fundamental features, there are certain aspects which discern the two processes (Bhushani and Anandharamakrishnan 2014). In electrospinning, the polymer solution is pumped through a capillary or fine conductive needle to form a droplet, which in-turn elongates as a fibre under high voltage and concomitant solvent evaporation occurs. On the other hand, the polymer solution is atomized into fine droplets under high voltage in electrospraying. Evaporation of the solvent occurs during the flight of the droplets and then contract into solid polymeric particles instead of fibers (Fathi et al. 2014). The recent advancement in this technique is the addition of a concentric or dual nozzle spinneret to the existing setup which helps to incorporate bioactive ingredients in the inner capillary (to form core material) and the wall material in the outer capillary, enabling the production of core-shell fibers or particles (Sakuldao et al. 2011). These techniques have been successfully employed for encapsulation of *Bifidobacteria* (López-Rubio et al. 2012), vanillin (Kayaci and Uyar 2012), epigallocatechin-3-gallate (Sun et al. 2011), ß-carotene (Fernandez et al. 2009), curcumin (Brahatheeswaran et al. 2012), α-tocopherol (Wongsasulak et al. 2014), etc. In addition to the above, co-electrospinning and coaxial electrospinning have been used for immobilization of enzymes like glucose oxidase (Ge et al. 2012), cellulase (Wu et al. 2005b), ß-galactosidase (El-Aassar 2013) and lipase (Xie and Hsieh 2003). Protein, polysaccharides and their composite blends with synthetic polymers have been reported as encapsulation matrices for electrospun and electrosprayed products. Lately, electrospun fibers, produced as interwined non-woven mat, have been extended for utilization as filtering aid in beverage clarification (Veleirinho and Lopes-da-Silva 2009). Like any other encapsulation system, electrospinning can mar the viability of encapsulated probiotics or bioactive ingredients due to high voltage applied (Heunis et al. 2010).

3.6.2 *Emulsification (Nanoemulsion)*

Nanoemulsion remains to be one of the most promising techniques of generating versatile nanocarriers, and oil-in-water (o/w) nanoemulsions have been shown to possess great prospective for application within several food products. Nanoemulsion is an extension of microemulsion process with 10–100 nm sized oil droplets being surrounded by a thin interfacial layer of emulsifier and dispersed uniformly in aqueous medium. As nanoemulsions are thermodynamically unfavourable systems, presence of an emulsifier such as cationic, anionic or non-ionic surfactants, or hydrocolloid like polysaccharides or protein, is a must to stabilize such system; thereby reducing the interfacial tension between the two phases. This nano-delivery system has several advantages over conventional emulsions, such as improved solubility of bioactives, mainly for poorly water-soluble substances); better stability to particle aggregation and gravitational separation; and optical transparency, which scatters light waves weakly due to very small size, and hence are suitable for incorporation into clear or slightly turbid beverages (McClements and Rao 2011); this being the most critical benefit. Nanoemulsions can either be used directly in the liquid state or be dried to powder using spray drying or freeze drying methods after emulsification (Ezhilarasi et al. 2013). Conversely, the stability of bioactives (resveratrol, quercetin, ß-carotene, omega-3 fatty acids, vitamin E) in the lipid core or in the interfacial region of the nanoemulsion can be improved either by coating them with one or more biopolymer(s) or by enzymatic cross-linking of the interfacial layer (Joye et al. 2014). Recently, multiple emulsions such as water-in-oil-in-water (w/o/w) emulsion for encapsulating flavonoids (Akhtar et al. 2014) and resveratrol (Matos et al. 2014) have been developed. These have the added benefit of encapsulating both hydrophilic and lipophilic bioactives within a single delivery system.

In general, both high-energy and low-energy approaches are widely used to prepare nanoemulsions. In high-energy methods, strong disruptive forces are generated with mechanical devices (like high shear stirring, high speed or high pressure homogenization, ultrasonication and microfluidization), which are capable of breaking and dispersing oil droplets into the aqueous phase; on the other hand, low-energy approach relies on the spontaneous formation of tiny droplets of dispersed phase within the continuous phase when the composition or environmental conditions are altered (e.g., phase-inversion, solvent demixing and spontaneous-emulsification). In industrial food operations, high-energy techniques are mostly used, as they are capable of preparing nanoemulsions from a variety of different starting materials; however most of the energy provided is dissipated in the form of heat, which in-turn may destroy certain heat-sensitive components. Regarding the formation of nanoemulsions by low-energy approach, it is fairly simple and does not require expensive equipments but uses relatively higher concentration of synthetic surfactants for tailoring the size distribution, composition and other attributes of the nano-droplets (Dasgupta et al. 2016). This may limit their use for many food applications. Formulation, fabrication, properties, performance, biological fate and potential toxicity of food-grade nanoemulsion has been comprehensively reviewed by (McClements and Rao 2011)

3.6.3 Coacervation

This method involves trapping of the bioactive within a particle (hydrocolloid shell) formed by electrostatic complexation of oppositely charged biopolymer(s). Formation of coacervates (biopolymeric shells) can also be induced by hydrophobic interaction and hydrogen bonding. Depending on the number of polymer type used, the process can be simple coacervation (one polymer type) or complex coacervation (two or more types of polymers). Alternatively, such shell can be formed by cross-linking with an enzyme (transglutaminase) or suitable chemicals (glutaraldehyde, tannins, tripolyphosphates), followed by an appropriate drying technique like vacuum drying, freeze drying or spray drying. Gelatin, acacia and chitosan have been found to be promising wall materials for this technique and the produced encapsulates, mainly lying within the range of 100–600 nm, improved the thermal stability of the entrapped functional ingredients (Chen et al. 2010b).

3.6.4 Micelles and Reverse Micelles

Micelles are self-assembled spherical entities ($d < 20$ nm) with a hydrophobic core made of non-polar surfactant tails, and a hydrophilic shell comprised of polar surfactant head groups (Zhang et al 2006). The scenario is completely opposite in case of a reverse micelle, where the surfactant head groups align to form the hydrophilic core and the non-polar tails are directed outward. Lipophilic components are encapsulated within the hydrophobic core of micelles, while hydrophilic compounds can be trapped in the core of reverse micelles. Micelles are composed of only surfactants; however co-solvent or co-surfactant (along with biopolymer) may be added to alter their dimension and properties (Vrignaud et al. 2012). In general, high amount of surfactant is required for creating stable micelles, which may limit their application in certain foods.

3.6.5 Molecular Inclusion or Inclusion Complexation

This refers to the association of an ingredient (generally termed as ligand) into a molecular entity bearing molecular level cavities (such as α- and ß-cyclodextrin, ß-lactoglobulin) (Ezhilarasi et al. 2013). Here the 'host' molecule traps a part of the 'guest' molecule by physico-chemical forces like hydrogen bonding, van der Waals attraction and hydrophobic interaction and in-turn liberates a part of the water from the host cavity (Joye et al. 2014). Thus the geometric compatibility between the two molecules is one of the main factors governing the efficacy of inclusion process. Unfortunately, presence of such food-grade host molecules with central cavity is very rare. For enhancing stability, such complexes can further be incorporated into suitable carriers such as liposomes (Kayaci et al. 2013a). Inclusion complexation technique is used in encapsulation of volatile organic molecules (essential oils and oleoresins like eugenol) (Kayaci et al. 2013b), particularly for masking odors and off-flavour, and preserving aroma (Ezhilarasi et al. 2013).

3.6.6 Nano-liposomes or Liposomes

These liquid nano-carriers are mainly made of phospholipids (derived from soy, egg, dairy or sunflower lecithin) and water (aqueous medium), where the polar head of phospholipids are directed towards aqueous phase of the inner and outer media, and the hydrophobic tails are oriented into the bilayer. Phospholipids are heated above their thermotropic phase transition temperature to align themselves in vesicular structures, which then self-assembles into onion-skin like spherical structures in aqueous environment upon energy input (Joye et al. 2014). Thus liposomes have aqueous interior and hydrophobic inner core of the bilayer membrane (Fig. 3.6). The resulting spherical core-shell structure can be classified as unilamellar vesicles, with one lipid bilayer; multilamellar vesicles, having more than one concentric lipid bilayers; or multivesicular vesicles, having small non-concentric vesicles entrapped within a single lipid bilayer. On the basis of sizes, they can be further categorized into small unilamellar vesicle having diameter within 20–100 nm and formed by delamination of outer layer of multivesicular vesicles; or large unilamellar vesicle with size up to few micrometer. Usually high energy/mechanical methods, such as extrusion through fine pore filters or membrane extrusion, untrasonication, high pressure homogenization, microfluidization and colloid mill, result in unilamellar vesicles; on the contrary, low energy/non-mechanical methods like reversed-phase evaporation, solvent evaporation and depletion of mixed detergent-lipid micelles, give multilamellar or multivesicular vesicles (Joye et al. 2014).

The structural properties of the polymer surrounding the outer surface of the vesicles govern the function of the liposomes; for e.g., thermosensitive/temperature-sensitive liposomes are either polymer-coated or lipid bilayer modified with specific polymer. These are destabilized above a critical temperature, leading to release of the encapsulated material. These kinds of carrier, as nano-additives in ready-to-eat or ready-to-cook meals, are ideal for flavour release during high cooking temperature (Fathi et al. 2012). Likewise, pH-sensitive liposomes can be prepared using amphiphilic lipid molecules like saturated phosphatidylethanolamine and oleic acid or polymethacrylic acid-co-stearyl methacrylate (Cho et al. 2009). Such stimulus-responsive liposomes can find significant potential in liberating biostatic or biocidal compounds upon pH change as a result of increased microbial activity in fermented foods. Again, presence of chitosan coating or long chain saturated acyl chains (disteroylphosphatidylcholine or hydrogenated soy phosphatidylcholine) or cholesterol increases the longevity, and therefore, the release time of liposomes (Li et al. 2009b; Laye et al. 2008; Zaru et al. 2009)

The advantages of nano-liposomes over nanoemulsion include their ability to entrap hydrophilic (in the interior core), and lipophilic and amphiphilic molecules (in the lipid bilayer) simultaneously, and impart higher protection to sensitive bioactives (ascorbic acid, glutathione) (Suntres and Shek 1996).

3.6.7 Solid Lipid Nanoparticles

Solid lipid nanoparticles are nano-size carriers consisting of a matrix in which the lipid phase is fully or partially solidified with an ordered crystalline matrix. Solid lipid nanoparticles can be prepared from a single pure lipid or a mixture of several lipids (Tamjidi et al. 2013) and is created from an oil-in-water nanoemulsion (by high pressure homogenization) during which the lipid phase is maintained above its melting point (5–10 °C above) and then cooled (by lyophilization) to induce solidification. This method of solid lipid nanoparticles preparation is known as hot homogenization technique and its major demerit is the use of high temperature that destroys the heat-labile bioactive compounds and the loss of hydrophilic functional ingredients to the aqueous (water) phase (Fathi et al. 2012). In order to overcome these shortcomings, cold homogenization can be used, where the bioactives are incorporated into the lipid melt and cooled rapidly for crystallization, followed by homogenization at low temperature (0 °C). The solid state of the matrix minimizes partitioning of the bioactive to the water phase. Apart from homogenization, solid lipid nanoparticles can also be synthesized by emulsification-solvent evaporation technique, where lipid in organic solvent is mixed with bioactives and emulsifier, followed by ultrasonication (to form nanoemulsion) at a temperature above the melting point of lipid. The formed nanoemulsion is quickly added to chilled water to allow lipid precipitation (as nanospheres) and subsequent solvent evaporation is performed (Fathi et al. 2012). Uniform distribution of bioactive compounds in the solid lipid nanoparticles matrix is obtained in cold homogenization method and also when highly hydrophobic functional ingredients are added during hot homogenization method.

The prime advantage of solid lipid nanoparticles include the high drug loading efficiency, controlled (prolonged) and targeted drug release especially when bioactives are concentrated in the core of solid lipid nanoparticles, non-toxicity of the carrier, feasibility of large-scale production and greater protection of the encapsulated compounds from degradation (lower thermal/molecular diffusion process into the solid matrix). However, like any other encapsulation techniques, solid lipid nanoparticles also suffer from certain disadvantages such as accumulation of bioactives towards the outer shell of solid lipid nanoparticles leads to burst release behavior, which makes it improper for certain food applications. Another problem of solid lipid nanoparticles is its inability to trap the drug during storage: polymorphic transition of some lipids tends to attain highly crystalline particles (α-crystal → ß' crystal → ß-crystal) with a perfect crystal lattice, leading to expulsion of the drug. So, use of oils forming less-ordered crystalline phase and cryoprotectants (e.g. sorbitol, mannose, trehalose, etc.) can enhance drug loading capacity (Fathi et al. 2012).

3.6.8 *Nanostructured Lipid Carrier*

As majority of drugs have higher solubility in liquid phase, increased solidification (crystallization) of fat in solid lipid nanoparticles results in drug expulsion with very poor encapsulation load. To overcome this limitation, solid lipid nanoparticles are modified into nanostructured lipid carriers in which the lipid is made of both sold and liquid phases at room temperature. Nanostructured lipid carrier contains an oil-loaded core surrounded by an amorphous solid lipid. This leads to improved physical protection of drug and its controlled release, duc to drug immobilization within the outer solid fat matrix (Teeranachaideekul et al. 2007; Zhang et al. 2013b). Nanostructured lipid carrier augments the loading efficiency by two ways: the inner liquid (oil) serves as a suitable medium for dissolving drugs and the selected lipid blend shows polymorphic transition ($\alpha \rightarrow$ ß' $\rightarrow \alpha$ fat crystal). It therefore seems that nanostructured lipid carrier may prove to be a boon for the encapsulation of lipophilic bioactive compounds and can dispel the drawbacks of other lipid nanocarriers. Based on structural organization, Tamjidi et al. (2013) have categorized nanostructured lipid carrier into three groups which are as follows:

3.6.8.1 Imperfectly Structured Solid Matrix

This type of nanostructured lipid carrier is prepared by mixing glycerides having very different fatty acid composition (mixture of saturated and unsaturated fatty acids). This incompatibility between the lipids results in imperfections/holes in the lipid crystal (presence of amorphous clusters) where drug accommodation occurs (drug payload increases).

3.6.8.2 Structureless Solid Amorphous Matrix

This kind of nanostructured lipid carrier is produced by mixing solid lipids with hydroxyoctacosanylhydroxystearate, isopropylmyristate, etc., as a result of which the core congeals into an amorphous lipid state with negligible crystal holes.

3.6.8.3 Multiple Oil-in-Fat-in-Water Carriers

In multiple type nanostructured lipid carrier, the solid matrix contains tiny liquid nano-compartments of oil, where solubility of drug is higher. Such nano-compartments are found when higher amount of oil is mixed with solid matrix. The solid matrix surrounding the nano-compartments prevents drug leakage and facilitates controlled release.

3.6.9 Hydrogel Nanoparticles (Nanogel)

Hydrogels are physically or chemically cross-linked three dimensional (bio-)polymeric network with the capability of imbibing appreciable amount of water or fluid. Nanogel presents the benevolent feature of high water content (hydrophilicity) in a nano-sized form. Hydrogel nanoparticles can be created by desolvation, emulsion-templating, reverse micellar method and polyelectrolyte complexation or self-assembly (electrostatic attraction between biopolymers with oppositely charged groups), wherein the water or aqueous phase is gelled by acidification (Nag et al. 2011), addition of multivalent ions (Singh and Kim 2005) or cross-linking agent (Ethirajan et al. 2008) or temperature adjustment (Burey et al. 2008). The resulting particle contains bioactive or particulate substances (e.g. lipid) trapped within the polymer matrix and are known as filled hydrogel nanoparticles (Aouada et al. 2011). Gel particles can be further coated with other polymers to increase stability and reduce the leaching of bioactive compounds. At present, hydrogel nanoparticles are not much explored in the field of food science and their utility has exclusively been reported for drug delivery in pharmaceutical applications (Hamidi et al. 2008); nevertheless their role in food industries can be envisioned in near future.

3.7 Natural Food Constituents Processed/Formulated into Nanostructures (Nano-Nutraceuticals)

A major focus of the present nanotechnology application is to reduce the particle size of natural food elements like vitamins, carotenoids, phytosterols, antioxidants, colour and flavour, with the sole aim of improving the uptake and bioavailability of nutrients and dietary supplements (Chiang et al. 2014; Sonkaew et al. 2012; Yu et al. 2006). A number of processed nanostructured or nanotextured food products are presently at R&D or near-market stage. Production of nutraceuticals or natural food components at the nanoscale will be of significant interest to the consumers and their introduction to the market is expected to be relatively easy for a nano ingredient because the origin of these nanostructures are considered to be of food grade and so they are likely to be solubilized or digested in the gastro-intestinal tract.

Nanostructured ingredients claim to offer improved rheology, texture, mouthfeel and stability to the processed food-stuff, as compared to their conventional bulk equivalent. Commercial success in this area has been achieved by nanotextured mayonnaise, spreads and ice cream, where nanodroplets of water negates the need of high quantity of fat and surfactant during emulsification (fat reduction from

8–16 % to 1 %) (Cushen et al. 2012). This offers a healthier and tasteful alternative to the consumer. Aligned with increased interest to eliminate the use of synthetic surfactant in emulsion system, Pickering emulsion, stabilized by lactoferrin nanoparticles showed extended stability under oral condition (artificial saliva), which was linked to emulsion sensorial perception and product acceptability (Shimoni et al. 2013). A different endeavor was made to improve the texture of shrimps using nano-sized chitin and chitosan (polysaccharide existing naturally in exoskeleton of crustaceans) (Chantarasataporn et al. 2013). Chitin and chitosan whiskers swell the gaps inside the myofibrillar fibers of the shrimp flesh, resulting in a layer structure similar to that of the fresh shrimp. Other examples emanating from this novel technique would be nanostructured ß-carotene (Astete et al. 2009), lycopene (Limpens et al. 2006) and coenzyme Q10 (Ankola et al. 2007).

Several nanostructured food ingredients can also be used as carrier of bioactive components to execute multifarious functions. For example, casein nanostructure imparted colloidal stability and prevented oxidation of ω-3 fatty acids (Zimet et al. 2011). Self-assembled nanotubes of zein protein and α-lactalbumin improved the aqueous dispersibility of water-insoluble essential oil (Wu et al. 2012). In the same way, self-assembled nanostructures of chitosan and polyglutamic acid was designed to trap tea catechins for enhancing its assimilation in the human body (Chen et al. 2010b).

At the moment, there are very few commercial products proclaiming to have nanostructured natural ingredients (Chaudhry and Castle 2011). Maintaining nanoscale nutraceuticals in a stable state throughout the production chain is invariably challenging. So encapsulation of such ingredients is often necessary. Nano-sizing may lead to changes in bioassimilation profile: an increased bioavailability of vitamins and minerals may not always be beneficial for consumer health and may exceed the ADI (acceptable daily intake) value for that additive (Chaudhry and Castle 2011). For example, excessive accumulation of vitamin A may induce liver problem. Also nano-additive may undergo various transformations in complex food and gastro-intestinal tract due to agglomeration, aggregation, binding with other food components or form reaction byproducts with stomach acid, enzymes, etc. (Wang et al 2014). This notion is vouched by the findings of Peters et al. (2012), where the authors found large agglomerates of nano-silica under gastric digestion stage. Such alterations may cause nanomaterials to lose their 'nano' configuration and characteristics and perhaps carry insoluble materials to the circulatory system.

3.8 Conclusion

Despite the strong upsurge in the investigation of nanoparticulate system, the road to successful implementation of nanotechnology in food industries is still long. The biological fate of nanoparticles during digestion, absorption and excretion is not yet fully understood. It is apparent from a number of studies that inorganic nanoparticles have negative implication on biomolecules and cellular components (Smolkova et al. 2015). Many industries are not willing to disclose their work in nanotechnology (Duncan 2011) and because of this knowledge gap, estimates of the current and future market share of nanotechnology-enabled products is still illusive. In a recent report by Chen et al. (2013) and Weir et al. (2012), it has been shown that the number of food products containing nano-TiO_2 is much larger than known, and consumers have already often been exposed to engineered nanoparticles in daily life. This aggravates the public fear about nanofoods (Siegrist et al. 2008). Available literature suggests that there are uncertainties over the adequacy of current global regulations, as they are not sufficient to cope with the challenges posed by nanotechnology in food related products. Difficulty in assessing the characteristics and precise detection of nanomaterials in complex food matrices, and absence of detailed toxicological data, are impeding the implementation of effective legislation (Chaudhry et al. 2008; Cushen et al. 2012). So, efforts are underway to establish clear and consistent guidelines for risk assessment of nanoparticles in food, together with their appropriate permissible limit in food items.

Though promising results are expected from the plethora of published research papers, there is no comprehensive information on determining the capacity of these nanoparticles or nano-delivery systems to withstand realistic food processing conditions. Advances in the laboratory do not always translate into commercial success: many approaches developed within research laboratories may not be suitable for scale-up to industrial production facilities. It is imagined that the near future application of food nanotechnology are likely to be for high-value products, in particular the packaging, and nano-carriers for delivery of drugs and dietary supplements. Ideally, consumers would like to buy nanomaterials made from acceptable food grade ingredients and it would be prudent to adopt proper labeling of nano-food products to inform the consumer. This might help in presenting the image of nanotechnology in a favourable or positive way. Another area where there seems to be a lot of scope in the coming years is nanotechnology-based food sensors, some of which are already in use in biomedical applications.

Virtually, nanotechnology has penetrated into every aspect of the food sector in some way. The dispute on the benefits and risks of applying nanotechnology in the food sector will last long; nonetheless there is optimism that novel nano-scale materials are expected to evolve in food industries.

References

Ai K, Liu Y, Lu L (2009) Hydrogen-bonding recognition-induced color change of gold nanoparticles for visual detection of melamine in raw milk and infant formula. J Am Chem Soc 131(27):9496–9497

Akhtar M, Murray BS, Afeisume EI, Khew SH (2014) Encapsulation of flavonoid in multiple emulsion using spinning disc reactor technology. Food Hydrocoll 34:62–67

Ammu S, Dua V, Agnihotra SR, Surwade SP, Phulgirkar A, Patel SK, Manohar SK (2012) Flexible, all-organic chemiresistor for detecting chemically aggressive vapors. J Am Chem Soc 134(10):4553–4556

An J, Zhang M, Wang S, Tang J (2008) Physical, chemical and microbiological changes in stored green asparagus spears as affected by coating of silver nanoparticles-PVP. LWT 41:1000–1007

Anand P, Nair HB, Sung B, Kunnumakkara AB, Yadav VR, Tekmal RR, Aggarwal BB (2010) Design of curcuminloaded PLGA nanoparticles formulation with enhanced cellular uptake, and increased bioactivity *in vitro* and superior bioavailability *in vivo*. Biochem Pharmacol 79(3):330–338

Ankola DD, Viswanad B, Bhardwaj V, Ramarao P, Kumar MN (2007) Development of potent oral nanoparticulate formulation of coenzyme Q10 for treatment of hypertension: can the simple nutritional supplements be used as first line therapeutic agents for prophylaxis/therapy? Eur J Pharm Biopharm 67:361–369

Antiochia R, Lavagnini I, Magno F (2004) Amperometric mediated carbon nanotube paste biosensor for fructose determination. Anal Lett 37(8):1657–1669

Aouada FA, de Moura MR, Orts WJ, Mattoso LHC (2011) Preparation and characterization of novel micro- and nanocomposite hydrogels containing cellulosic fibrils. J Agric Food Chem 59:9433–9442

Apetrei C, Apetrei IM, Villanueva S, de Sajab JA, Gutierrez-Rosalesc F, Rodriguez-Mendez ML (2010) Combination of an e-nose, an e-tongue and an e-eye for the characterisation of olive oils with different degree of bitterness. Anal Chim Acta 663:91–97

Applerot G, Perkas N, Amirian G, Girshevitz O, Dedanken A (2009) Coating of glass with ZnO via ultrasonic irradiation and a study of its antibacterial properties. Appl Surf Sci 256:S3–S8

Arevalo FJ, Granero AM, Fernandez H, Raba J, Zon MA (2011) Citrinin (CIT) determination in rice samples using a microfluidic electrochemical immunosensor. Talanta 83(3):966–973

Arunvisut S, Phummanee S, Somwangthanaroj A (2007) Effect of clay on mechanical and gas barrier properties of blown film LDPE/clay nanocomposites. J Appl Polym Sci 106:2210–2217

Astete CE, Sabliov CM, Watanabe F, Biris A (2009) Ca^{2+} crosslinked alginic acid nanoparticles for solubilization of lipophilic natural colorants. J Agric Food Chem 57:7505–7512

Avella M, De Vlieger JJ, Errico ME, Fischer S, Vacca P, Volpe MG (2005) Biodegradable starch/clay nanocomposite films for food packaging applications. Food Chem 93:467–474

Aydin M, Herzig GPD, Jeong KC, Dunigan S, Shah P, Ahn S (2014) Rapid and sensitive detection of *Escherichia coli* O157:H7 in milk and ground beef using magnetic bead based immunoassay coupled with tyramide signal amplification. J Food Protect 77(1):100–105

Bhushani JA, Anandharamakrishnan C (2014) Electrospinning and electrospraying techniques: potential food based applications. Trends Food Sci Tech 38:21–33

Bieberstein A, Roosen J, Marette S, Blanchemanche S, Vandermoere F (2013) Consumer choices for nano-food and nano-packaging in France and Germany. Eur Rev Agric Econ 40(1):73–94

Bin Y, Mine M, Koganemaru A, Jiang X, Matsuo M (2006) Morphology and mechanical and electrical properties of oriented PVA–VGCF and PVA–MWNT composites. Polymer 47:1308–1317

Bloch MS, Paunescu D, Stoessel PR, Mora CA, Stark WJ, Grass RN (2014) Labeling milk along its production chain with DNA encapsulated in silica. J Agric Food Chem 62:10615–10620

Bradley EL, Castle L, Chaudhry Q (2011) Applications of nanomaterials in food packaging with a consideration of opportunities for developing countries. Trends Food Sci Tech 22:604–610

Brahatheeswaran D, Mathew A, Aswathy RG, Nagaoka Y, Venugopal K, Yoshida Y, Maekawa T, Sakthikumar D (2012) Hybrid fluorescent curcumin loaded zein electrospun nanofibrous scaffold for biomedical applications. Biomed Mater 7(4):045001

Burey P, Bhandari BR, Howes T, Gidley MJ (2008) Hydrocolloid gel particles: formation, characterization, and application. Crit Rev Food Sci Nutr 48:361–377

Cabedo L, Feijoo JL, Villanueva MP, Lagarón JM, Giménez E (2006) Optimization of biodegradable nanocomposites based on PLA/PCL blends for food packaging applications. Macromol Symp 233:191–197

Cao Q, Zhao H, Yang Y, He Y, Ding N, Wang J, Wu Z, Xiang K, Wang G (2011) Electrochemical immunosensor for casein based on gold nanoparticles and poly(L-Arginine)/multi-walled carbon nanotubes composite film functionalized interface. Biosens Bioelectron 26(8):3469–3474
Cardenas G, Diaz J, Melendrez MF, Cruzat C, Garcia Cancino A (2009) Colloidal Cu nanoparticles/chitosan composite film obtained by microwave heating for food package applications. Polym Bull 62:511–524
Carralero-Sanz V, Luz-Mena M, González-Cortes A, Yañez-Sedeño P, Pingarrón J (2005) Development of a tyrosinase biosensor based on gold nanoparticles-modified glassy carbon electrodes. Application to the measurement of a bioelectrochemical polyphenols index in wines. Anal Chim Acta 528:1–8
Castillo-García ML, Aguilar-Caballos MP, Gómez-Hens A (2012) Application of Tb_4O_7 nanoparticles for lasalocid and salicylate determination in food analysis. J Agric Food Chem 60:11741–11747
Cerrada ML, Serrano C, Sanchez-Chaves M, Fernandez-Garcia M, Fernandez-Martin F, De Andres A, Riobóo RJJ, Kubacka A, Ferrer M, Fernández-García M (2008) Self-sterilized EVOH-TiO_2 nanocomposites: interface effects on biocidal properties. Adv Funct Mater 18:1949–1960
Chantarasataporn P, Yoksan R, Visessanguan W, Chirachanchai S (2013) Water-based nano-sized chitin and chitosan as seafood additive through a case study of Pacific white shrimp (*Litopenaeus vannamei*). Food Hydrocoll 32:341–348
Chaudhry Q, Castle L (2011) Food applications of nanotechnologies: an overview of opportunities and challenges for developing countries. Trends Food Sci Tech 22:595–603
Chaudhry Q, Scotter M, Blackburn J, Ross B, Boxall A, Castle L, Aitken R, Watkins R (2008) Applications and implications of nanotechnologies for the food sector. Food Addit Contam Part A 25(3):241–258
Chawengkijwanich C, Hayata Y (2008) Development of TiO_2 powder coated food packaging film and its ability to inactivate *Esherrichia coli in vitro* and in actual test. Int J Food Microbiol 123:288–292
Che Y, Zang L (2009) Enhanced fluorescence sensing of amine vapor based on ultrathin nanofibers. Chem Commun 14(34):5106–5108
Che Y, Yang X, Loser S, Zang L (2008) Expedient vapor probing of organic amines using fluorescent nanofibers fabricated from an n-type organic semiconductor. Nano Lett 8:2219–2223
Chen P, Zhang L (2006) Interaction and properties of highly exfoliated soy protein/montmorillonite nanocomposites. Macromolecules 7:1700–1706
Chen W, Tao X, Xue P, Cheng X (2005) Enhanced mechanical properties and morphological characterizations of poly(vinyl alcohol)-carbon nanotube composite films. Appl Surf Sci 252:1404–1409
Chen Y, Liu C, Chang PR, Cao X, Anderson DP (2009) Bionanocomposites based on pea starch and cellulose nanowhiskers hydrolyzed from pea hull fibre: effect of hydrolysis time. Carbohydr Polym 76(4):607–615
Chen Y, Ren HL, Liu N, Sai N, Liu X, Liu Z, Gao Z, Ning BA (2010a) A fluoroimmunoassay based on quantum dot-streptavidin conjugate for the detection of chlorpyrifos. J Agric Food Chem 58:8895–8903
Chen Y-C, Yu S-H, Tsai G-J, Tang D-W, Mi F-L, Peng Y-P (2010b) Novel technology for the preparation of self-assembled catechin/gelatin nanoparticles and their characterization. J Agric Food Chem 58:6728–6734
Chen X-X, Cheng B, Yang Y-X, Cao A, Liu J-H, Du L-J, Liu Y, Zhao Y, Wang H (2013) Characterization and preliminary toxicity assay of nano-titanium dioxide additive in sugar-coated chewing gum. Small 9(9–10):1765–1774
Cheng Q, Li C, Pavlinek V, Saha P, Wang H (2006) Surface-modified antibacterial TiO_2/Ag^+ nanoparticles: preparation and properties. Appl Surf Sci 252:4154–4160
Chiang L-H, Chen S-H, Yeh A-I (2012) Preparation of nano/submicrometer Yam and its benefits on collagen secretion from skin fibroblast cells. J Agric Food Chem 60:12332–12340

Chiang L-H, Chen S-H, Yeh A-I (2014) Preparation of nano/submicrometer *Ganoderma tsugae* and its mutagenic potencies and cytotoxicity. J Agric Food Chem 62:12244–12255
Ching KH, Lin A, McGarvey JA, Stanker LH, Hnasko R (2012) Rapid and selective detection of botulinum neurotoxin serotype-A and -B with a single immunochromatographic test strip. J Immunol Methods 380(1–2):23–29
Cho EC, Lim HJ, Kim HJ, Son ED, Choi HJ, Park JH, Kim J-W, Kim J (2009) Role of pH-sensitive polymer-liposome complex in enhancing cellular uptake of biologically active drugs. Mater Sci Eng C 29(3):774–778
Chorianopoulos NG, Tsoukleris DS, Panagou EZ, Falaras P, Nychas GJE (2011) Use of titanium dioxide (TiO_2) photocatalysts as alternative means for *Listeria monocytogenes* biofilm disinfection in food processing. Food Microbiol 28:164–170
Chowdhury SR (2008) Some important aspects in designing high molecular weight poly(L-lactic acid)-clay nanocomposites with desired properties. Polym Int 57:1326–1332
Chrissopoulou K, Altintzi I, Anastasiadis SH, Giannelis EP, Pitsikalis M, Hadjichristidis N, Theophilou N (2005) Controlling the miscibility of polyethylene/layered silicate nanocomposites by altering the polymer/surface interactions. Polymer 46:12440–12451
Chuayjuljit S, Thongraar R, Saravari O (2008) Preparation and properties of PVC/EVA/organo-modified montmorillonite nanocomposites. J Reinf Plast Comp 27:431–442
Coleman JN, Khan U, Blau WJ, Guńko YK (2006) Small but strong: a review of the mechanical properties of carbon nanotube-polymer composite. Carbon 44:1624–1652
Conte A, Longano D, Costa C, Ditaranto N, Ancona A, Cioffi N, Scrocco C, Sabbatini L, Contò F, Del Nobile MA (2013) A novel preservation technique applied to fiordilatte cheese. Innov Food Sci Emerg Technol 19:158–165
Cushen M, Kerry J, Morris M, Cruz-Romero M, Cummins E (2012) Nanotechnologies in the food industry–recent developments, risks and regulation. Trends Food Sci Tech 24:30–46
Damm C, Münstedt H, Rösch A (2007) Long-term antimicrobial polyamide 6/ silver-nanocomposites. J Mater Sci 42(15):6067–6073
Dandekar P, Jain R, Kumar C, Subramanian S, Samuel G, Venkatesh M, Patravale V (2009) Curcumin loaded pH sensitive nanoparticles for the treatment of colon cancer. J Biomed Nanotechnol 5(5):445–455
Das Purkayastha M, Manhar AK, Das VK, Borah A, Mandal M, Thakur AJ, Mahanta CL (2014a) Antioxidative, hemocompatible, fluorescent carbon nanodots from an 'End-of-pipe" agricultural waste: exploring its new horizon in the food-packaging domain. J Agric Food Chem 62:4509–4520
Das Purkayastha M, Manhar AK, Mandal M, Mahanta CL (2014b) Industrial waste-derived nanoparticles and microspheres can be potent anti-microbial and functional ingredients. J Appl Chem 171427
Dasan KP, Purushothaman E, Unnikrishnan G (2010) Analysis of physical and solvent transport behavior of poly (ethylene-covinyl acetate)/silica composites. J Reinf Plast Comp 29:238–246
Dasgupta N, Ranjan S, Mundekkad D, Ramalingam C, Shanker R, Kumar A (2015) Nanotechnology in agro-food: from field to plate. Food Res Int 69:381–400
Dasgupta N, Ranjan S, Mundra S, Ramalingam C, Kumar A (2016) Fabrication of food grade vitamin E nanoemulsion by low energy approach, characterization and its application. Int J Food Prop 19:700–708
de Azeredo HMC (2009) Nanocomposites for food packaging applications. Food Res Int 42:1240–1253
De Moura MR, Aouada FA, Avena-Bustillos RJ, McHugh TH, Krochta JM, Mattoso LHC (2009) Improved barrier and mechanical properties of novel hydroxypropyl methylcellulose edible films with chitosan/tripolyphosphate nanoparticles. J Food Eng 92:448–453
Del Nobile MA, Cannarsi M, Altieri C, Sinigaglia M, Favia P, Iacoviello G, D'Agostino R (2004) Effect of Ag-containing nanocomposite active packaging system on survival of *Alicyclobacillus acidoterrestris*. J Food Sci 6:379–383

Delmulle BS, De Saeger SM, Sibanda L, Barna-Vetro I, Van-Peteghem CH (2005) Development of an immunoassay based lateral flow dipstick for the rapid detection of aflatoxin B1 in pig feed. J Agric Food Chem 53:3364–3368

Deo PR, Wang J, Block I, Mulchandani A, Joshi KA, Trojanowicz M, Scholz F, Chen W, Lin Y (2005) Determination of organophosphate pesticides at a carbon nanotube/organophosphorus hydrolase electrochemical biosensor. Anal Chim Acta 530:185–189

Devi R, Yadav S, Pundir CS (2012) Amperometric determination of xanthine in fish meat by zinc oxide nanoparticle/chitosan/multiwalled carbon nanotube/polyaniline composite film bound xanthine oxidase. Analyst 137(3):754–759

Devi R, Batra B, Lata S, Yadav S, Pundir CS (2013) A method for determination of xanthine in meat by amperometric biosensor based on silver nanoparticles/cysteine modified Au electrode. Process Biochem 48(2):242–249

Ding S, Chen J, Jiang H, He J, Shi W, Zhao W, Shen J (2006) Application of quantum dot-antibody conjugates for detection of sulfamethazine residue in chicken muscle tissue. J Agric Food Chem 54:6139–6142

Dufresne A, Dupeyre D, Vignon MR (2000) Cellulose microfibrils from potato tuber cells: processing and characterization of starch–cellulose microfibril composites. J Appl Polym Sci 76:2080–2092

Duncan TV (2011) Applications of nanotechnology in food packaging and food safety: barrier materials, antimicrobials and sensors. J Colloid Interface Sci 363:1–24

Dungchaia W, Siangprohb W, Chaicumpac W, Tongtawed P, Chailapakula O (2008) Salmonella typhi determination using voltammetric amplification of nanoparticles: a highly sensitive strategy for metalloimmunoassay based on a copper-enhanced gold label. Talanta 77:727–732

Egger S, Lehmann RP, Height MJ, Loessner MJ, Schuppler M (2009) Antimicrobial properties of novel silver-silica nanocomposite material. Appl Environ Microbiol 75:2973–2976

El-Aassar MR (2013) Functionalized electrospun nanofibers from poly (AN-co-MMA) for enzyme immobilization. J Mol Catal B Enzym 85:140–148

El-Boubbou K, Gruden C, Huang X (2007) Magnetic glyco-nanoparticles: a unique tool for rapid pathogen detection, decontamination, and strain differentiation. J Am Chem Soc 129:13392–13393

Emamifar A, Kadivar M, Shahedi M, Soleimanian-Zad S (2010) Evaluation of nanocomposites packaging containing Ag and ZnO on shelf-life of fresh orange juice. Innov Food Sci Emerg Technol 11:742–748

Ethirajan A, Schoeller K, Musyanovych A, Ziener U, Landfester K (2008) Synthesis and optimization of gelatin nanoparticles using the miniemulsion process. Biomacromolecules 9:2383–2389

Ezhilarasi PN, Karthik P, Chhanwal N, Anandharamakrishnan C (2013) Nanoencapsulation techniques for food bioactive components: a review. Food Bioprocess Technol 6:628–647

Fabra MJ, Busolo MA, Lopez-Rubio A, Lagaron JM (2013) Nanostructured biolayers in food packaging. Trends Food Sci Tech 31:79–87

Fathi M, Mozafari MR, Mohebbi M (2012) Nanoencapsulation of food ingredients using lipid based delivery systems. Trends Food Sci Tech 23:13–27

Fathi M, Martín Á, McClements DJ (2014) Nanoencapsulation of food ingredients using carbohydrate based delivery systems. Trends Food Sci Tech 39:18–39

Fendler A, Villanueva MP, Gimenez E, Lagaron JM (2007) Characterization of the barrier properties of composites of HDPE and purified cellulose fibers. Cellulose 14:427–438

Fernandez A, Torres-Giner S, Lagaron JM (2009) Novel route to stabilization of bioactive antioxidants by encapsulation in electrospun fibers of zein prolamine. Food Hydrocoll 23(5):1427–1432

Fernandez A, Picouet P, Lloret E (2010) Reduction of the spoilage-related microflora in absorbent pads by silver nanotechnology during MAP packaging of beef meat. J Food Protect 73:2263–2269

Follain N, Belbekhouche S, Bras J, Siqueira G, Marais S, Dufresne A (2013) Water transport properties of bionanocomposites reinforced by Luffa cylindrica cellulose nanocrystals. J Membr Sci 427:218–229

Fortunati E, Peltzer M, Armentano I, Jiménez A, Kenny JM (2013a) Combined effects of cellulose nanocrystals and silver nanoparticles on the barrier and migration properties of PLA nanobiocomposites. J Food Eng 118:117–124

Fortunati E, Puglia D, Luzi F, Santulli C, Kenny JM, Torre L (2013b) Binary PVA bionanocomposites containing cellulose nanocrystals extracted from different natural sources: part I. Carbohydr Polym 97:825–836

Frewer LJ, Gupta N, George S, Fischer ARH, Giles EL, Coles D (2014) Consumer attitudes towards nanotechnologies applied to food production. Trends Food Sci Tech 40:211–225

Frounchi M, Dadbin S, Salehpour Z, Noferesti M (2006) Gas barrier properties of PP/EPDM blend nanocomposites. J Membr Sci 282:142–148

Gao Y, Choudhury NR, Dutta NK (2010) Systematic study of interfacial interactions between clays and an ionomer. J Appl Polym Sci 117:3395–3405

Ge L, Zhao YS, Mo T, Li JR, Li P (2012) Immobilization of glucose oxidase in electrospun nanofibrous membranes for food preservation. Food Control 26(1):188–193

Ghanbari A, Heuzey M-C, Carreau PJ, Ton-That M-T (2013) Morphological and rheological properties of PET/clay nanocomposites. Rheol Acta 52:59–74

Godoy-Navajas J, Caballos MPA, Gomez-Hens A (2011) Heterogeneous immunoassay for soy protein determination using nile blue-doped silica nanoparticles as labels and front-surface long wavelength fluorimetry. Anal Chim Acta 701(2):194–199

Gong-Jun Y, Jin-Lin H, Wen-Jing M, Ming S, Xin-An J (2009) A reusable capacitive immunosensor for detection of Salmonella spp. based on grafted ethylene diamine and self-assembled gold nanoparticle monolayers. Anal Chim Acta 647:159–166

Gorrasi G, Tortora M, Vittoria V, Galli G, Chiellini E (2002) Transport and mechanical properties of blends of poly(epsilon-caprolactone) and a modified montmorillonite-poly(epsilon-caprolactone) nanocomposite. J Polym Sci B Polym Phys 40:1118–1124

Gou M, Men K, Shi H, Xiang M, Zhang J, Song J, Long J, Wan Y, Luo F, Zhao X, Qian Z (2011) Curcumin-loaded biodegradable polymeric micelles for colon cancer therapy *in vitro* and *in vivo*. Nanoscale 3(4):1558–1567

Hamidi M, Azadi A, Rafiei P (2008) Hydrogel nanoparticles in drug delivery. Adv Drug Deliv Rev 60:1638–1649

Hemati F, Garmabi H (2011) Compatibilised LDPE/LLDPE/nanoclay nanocomposites: I. Structural, mechanical, and thermal properties. Can J Chem Eng 89:187–196

Hernández-Jover T, Izquierdo-Pulido M, Veciana-Nogués MT, Vidal-Carou MC (1996) Ion-pair high-performance liquid chromatographic determination of biogenic amines in meat and meat products. J Agric Food Chem 44(9):2710–2715

Heunis TDJ, Botes M, Dicks LMT (2010) Encapsulation of *Lactobacillus plantarum* and its bacteriocin in nanofibers. Probiotics Antimicrob Proteins 2(1):46–51

Horst MF, Quinzani LM, Failla MD (2014) Rheological and barrier properties of nanocomposites of HDPE and exfoliated montmorillonite. J Thermoplast Compos Mater 27:106–125

Huang M, Yu J (2006) Structure and properties of thermoplastic corn starch/montmorillonite biodegradable composites. J Appl Polym Sci 99:170–176

Huang M, Yu J, Ma X (2005) High mechanical performance MMT-urea and formamide-plasticized thermoplastic corn starch biodegradable nanocomposites. Carbohydr Polym 62:1–7

Hyun K, Chong W, Koo M, Chung IJ (2003) Physical properties of polyethylene/silicate nanocomposite blown films. J Appl Polym Sci 89:2131–2136

Incoronato AL, Buonocore GG, Conte A, Lavorgna M, Del Nobile MA (2010) Active systems based on silver-montmorillonite nanoparticles embedded into bio-based polymer matrices for packaging applications. J Food Protect 73:2256–2262

Jang WS, Rawson I, Grunlan JC (2008) Layer-by-layer assembly of thin film oxygen barrier. Thin Solid Films 516(15):4819–4825

Ji X, Zheng J, Xu J, Rastogi VK, Cheng T-C, DeFrank JJ, Leblanc RM (2005) (CdSe)ZnS quantum dots and organophosphorus hydrolase bioconjugate as biosensors for detection of paraoxon. J Phys Chem B 109:3793–3799

Jia X, Li Y, Cheng Q, Zhang S, Zhang B (2007) Preparation and properties of poly(vinyl alcohol)/ silica nanocomposites derived from copolymerization of vinyl silica nanoparticles and vinyl acetate. Eur Polym J 43:1123–1131

Jin T, Gurtler JB (2011) Inactivation of *Salmonella* in liquid egg albumen by antimicrobial bottle coatings infused with allyl isothiocyanate, nisin and zinc oxide nanoparticles. J Appl Microbiol 110:704–712

Jin M, Zhong Q (2012) Structure modification of montmorillonite nanoclay by surface coating with soy protein. J Agric Food Chem 60:11965–11971

Jin X, Jin X, Chen L, Jiang J, Shen G, Yu R (2009) Piezoelectric immunosensor with gold nanoparticles enhanced competitive immunoreaction technique for quantification of aflatoxin B1. Biosens Bioelectron 24(8):2580–2585

Joye IJ, McClements DJ (2013) Production of nanoparticles by anti-solvent precipitation for use in food systems. Trends Food Sci Tech 34:109–123

Joye IJ, Davidov-Pardo G, McClements DJ (2014) Nanotechnology for increased micronutrient bioavailability. Trends Food Sci Tech 40:168–182

Kalele SA, Kundu AA, Gosavi SW, Deobagkar DN, Deobagkar DD, Kulkarni SK (2006) Rapid detection of *Escherichia coli* by using antibody-conjugated silver nanoshells. Small 2(3):335–338

Kanayeva DA, Wang RH, Rhoads D, Erf GF, Slavik MF, Tung S, Li Y (2012) Efficient separation and sensitive detection of Listeria monocytogenes using an impedance immunosensor based on magnetic nanoparticles, a microfluidic chip, and an interdigitated microelectrode. J Food Protect 75(11):1951–1959

Kaur J, Singh KV, Boro TKR, Raje M, Varshney GC, Suri CR (2007) Immunochromatographic dipstick assay format using gold nanoparticles labeled protein-hapten conjugate for the detection of atrazine. Environ Sci Technol 41:5028–5036

Kaushik A, Solanki PR, Pandey MK, Ahmad S, Malhotra BD (2009) Cerium oxide-chitosan based nanobiocomposite for food borne mycotoxin detection. Appl Phys Lett 95:173703

Kayaci F, Uyar T (2012) Encapsulation of vanillin/cyclodextrin inclusion complex in electrospun polyvinyl alcohol (PVA) nanowebs: prolonged shelf-life and high temperature stability of vanillin. Food Chem 133:641–649

Kayaci F, Umu OCO, Tekinay T, Uyar T (2013a) Antibacterial electrospun poly(lactic acid) (PLA) nanofibrous webs incorporating triclosan/cyclodextrin inclusion complexes. J Agric Food Chem 61:3901–3908

Kayaci F, Ertas Y, Uyar T (2013b) Enhanced thermal stability of eugenol by cyclodextrin inclusion complex encapsulated in electrospun polymeric nanofibers. J Agric Food Chem 61:8156–8165

Khayankarn O, Magaraphan R, Schwank JW (2003) Adhesion and permeability of polyimide clay nanocomposite films for protective coatings. J Appl Polym Sci 89:2875–2881

Khosravi R, Hashemi SA, Sabet SA, Rezadoust AM (2013) Thermal, dynamic mechanical, and barrier studies of potassium permanganate-LDPE nanocomposites. Polym-Plast Technol Eng 52:126–132

Kim SW, Cha S-H (2014) Thermal, mechanical, and gas barrier properties of ethylene-vinyl alcohol copolymer-based nanocomposites for food packaging films: effects of nanoclay loading. J Appl Polym Sci 131:40289–40297

Kim Y, Chang J-H (2013) Colorless and transparent polyimide nanocomposites: thermo-optical properties, morphology, and gas permeation. Macromol Res 21:228–233

Kim JY, Han S, Hong S (2008) Effect of modified carbon nanotube on the properties of aromatic polyester nanocomposites. Polymer 49:3335–3345

Koh HC, Parka JS, Jeong MA, Hwang HY, Hong YT, Ha SY, Nam SY (2008) Preparation and gas permeation properties of biodegradable polymer/layered silicate nanocomposite membranes. Desalination 233:201–209

Kristo E, Biliaderis CG (2007) Physical properties of starch nanocrystal reinforced pullulan films. Carbohydr Polym 68:146–158

Kuan C-Y, Yee-Fung W, Yuen K-H, Liong M-T (2012) Nanotech: propensity in foods and bioactives. Crit Rev Food Sci Nutr 52(1):55–71

Kuang H, Chen W, Yan W, Xu L, Zhu Y, Liu L, Chu H, Peng C, Wang L, Kotov NA, Xu C (2011) Crown ether assembly of gold nanoparticles: melamine sensor. Biosens Bioelectron 26(5):2032–2037

Kumar P, Sandeep KP, Alavi S, Truong VD, Gorga RE (2010) Preparation and characterization of bio-nanocomposite films based on soy protein isolate and montmorillonite using melt extrusion. J Food Eng 100:480–489

Lagaron JM, Fendler A (2009) High water barrier nanobiocomposites of methyl cellulose and chitosan for film and coating applications. J Plast Film Sheet 25(1):47–59

Lagaron JM, Lopez-Rubio A (2011) Nanotechnology for bioplastics: opportunities, challenges and strategies. Trends Food Sci Tech 22:611–617

Lagaron JM, Cabedo L, Cava D, Feijoo JL, Gavara R, Gimenez E (2005) Improving packaged food quality and safety. Part 2: nanocomposites. Food Addit Contam 22:994–998

Laye C, McClements DJ, Weiss J (2008) Formation of biopolymer-coated liposomes by electrostatic deposition of chitosan. J Food Sci 73:7–15

Lee SR, Park HM, Lim HT, Kang KY, Li L, Cho WJ, Ha CS (2002) Microstructure, tensile properties, and biodegradability of aliphatic polyester/clay nanocomposites. Polymer 43:2495–2500

Li Y, Yht CU, Luo D (2005) Multiplexed detection of pathogen DNA with DNA-based fluorescence nanobarcodes. Nat Biotechnol 23:885–889

Li X, Zhou Y, Zheng Z, Yue X, Dai Z, Liu S, Tang Z (2009a) Glucose biosensor based on nanocomposite films of CdTe quantum dots and glucose oxidase. Langmuir 25(11):6580–6586

Li N, Zhuang C, Wang M, Sun X, Nie S, Pan W (2009b) Liposome coated with low molecular weight chitosan and its potential use in ocular drug delivery. Int J Pharm 379:131–138

Li LH, Deng JC, Deng HR, Liu ZL, Li XL (2010) Preparation, characterization and antimicrobial activities of chitosan/Ag/ZnO blend films. Chem Eng J 160:378–382

Limpens J, Schröder FH, de Ridder CM, Bolder CA, Wildhagen MF, Obermüller-Jevic UC, Krämer K, van Weerden WM (2006) Combined lycopene and vitamin E treatment suppresses the growth of PC-346C human prostate cancer cells in nude mice. J Nutr 136:1287–1293

Lin G, Lee YR, Liu YC, Wu YG (2005) Ortho effects for inhibition mechanisms of butyrylcholinesterase by o-substituted phenyl N-butylcarbamates and comparison with acetylcholinesterase, cholesterol esterase, and lipase. Chem Res Toxicol 18:1124–1131

Lisa M, Chouhan RS, Vinayaka AC, Manonmani HK, Thakur MS (2009) Gold nanoparticles based dipstick immunoassay for the rapid detection of dichlorodiphenyltrichloroethane: an organochlorine pesticide. Biosens Bioelectron 25:224–227

Liu S, Chen Y, Liu Z, Hu X, Wang F (2006) A highly sensitive resonance rayleigh scattering method for the determination of Vitamin B1 with gold nanoparticles probe. Microchim Acta 154:87–93

Liu Y, Ai K, Cheng X, Huo L, Lu L (2010) Gold-nanocluster-based fluorescent sensors for highly sensitive and selective detection of cyanide in water. Adv Funct Mater 20(6):951–956

Llorens A, Lloret E, Picouet PA, Trbojevich R, Fernandez A (2012) Metallic-based micro and nanocomposites in food contact materials and active food packaging. Trends Food Sci Tech 24:19–29

López Manchado MA, Valentini L, Biagotti J, Kenny JM (2005) Thermal and mechanical properties of single-walled carbon nanotubes–polypropylene composites prepared by melt processing. Carbon 43:1499–1505

López-Rubio A, Sanchez E, Wilkanowicz S, Sanz Y, Lagaron JM (2012) Electrospinning as a useful technique for the encapsulation of living bifidobacteria in food hydrocolloids. Food Hydrocoll 28:159–167

Lu Y, Weng L, Zhang L (2004) Morphology and properties of soy protein isolate thermoplastics reinforced with chitin whiskers. Biomacromolecules 5:1046–1051

Lu J, Do I, Drzal LT, Worden RM, Lee I (2007) Nanometaldecorated exfoliated graphite nanoplatelet based glucose biosensors with high sensitivity and fast response. Biosens Bioelectron 23:1825–1832

Luechinger NA, Loher S, Athanassiou EK, Grass RN, Stark WJ (2007) Highly sensitive optical detection of humidity on polymer/metal nanoparticle hybrid films. Langmuir 23(6):3473–3477

Magana S, Schlemmer SM, Leskinen SD, Kearns EA, Lim DV (2013) Automated dead-end ultrafiltration for concentration and recovery of total coliform bacteria and laboratory-spiked Escherichia coli O157:H7 from 50-liter produce washes to enhance detection by an electrochemiluminescence immunoassay. J Food Protect 76(7):1152–1160

Maiti P, Batt CA, Giannelis EP (2007) New biodegradable polyhydroxybutyrate/layered silicate nanocomposites. Biomacromolecules 8:3393–3400

Manikantan MR, Varadharaju N (2011) Preparation and properties of polypropylene-based nanocomposite films for food packaging. Packag Technol Sci 24:191–209

Mary G, Bajpai SK, Chand N (2009) Copper (II) ions and copper nanoparticles-loaded chemically modified cotton cellulose fibers with fair antibacterial properties. J Appl Polym Sci 113:757–766

Matos M, Gutiérrez G, Coca J, Pazos C (2014) Preparation of water-in-oil-in-water (W1/O/W2) double emulsions containing trans-resveratrol. Colloids Surf A Physicochem Eng Asp 442:69–79

Mattea F, Martín Á, Matías-Gago AN, Cocero MJ (2009) Supercritical antisolvent precipitation from an emulsion: ß-carotene nanoparticle formation. J Supercrit Fluids 51:238–247

Mbhele ZH, Salemane MG, van Sittert CGCE, Nedeljkovic JM, Djokovic V, Luyt AS (2003) Fabrication and characterization of silver–polyvinyl alcohol nanocomposites. Chem Mater 15(26):5019–5024

McClements DJ, Rao J (2011) Food-grade nanoemulsions: formulation, fabrication, properties, performance, biological fate, and potential toxicity. Crit Rev Food Sci Nutr 51(4):285–330

McCullum C, Tchounwou P, Ding LS, Liao X, Liu YM (2014) Extraction of aflatoxins from liquid foodstuff samples with polydopamine-coated superparamagnetic nanoparticles for HPLCMS/MS analysis. J Agric Food Chem 62(19):4261–4267

Meera AP, Thomas PS, Thomas S (2012) Effect of organoclay on the gas barrier properties of natural rubber nanocomposites. Polym Compos 33:524–531

Mihindukulasuriya SDF, Lim L-T (2014) Nanotechnology development in food packaging: a review. Trends Food Sci Tech 40:149–167

Mills A, Hazafy D (2008) A solvent-based intelligence ink for oxygen. Analyst 133(2):213–218

Mingliang GE, Demin J (2008) Influence of organoclay prepared by solid state method on the morphology and properties of polyvinyl chloride/organoclay nanocomposites. J Elastom Plast 40:223–235

Mishra NN, Maki WC, Cameron E, Nelson R, Winterrowd P, Rastogi SK, Filanoski B, Maki GK (2008) Ultra-sensitive detection of bacterial toxin with silicon nanowire transistor. Lab Chip 8(6):868–871

Mo Z, Zhang Y, Zhao F, Xiao F, Guo G, Zeng B (2010) Sensitive voltammetric determination of Sudan I in food samples by using gemini surfactant–ionic liquid–multiwalled carbon nanotube composite film modified glassy carbon electrodes. Food Chem 121(1):233–237

Mohanty S, Nayak SK (2010) Effect of organo-modified layered silicates on the properties of poly (methyl methacrylate) nanocomposites. J Thermoplast Compos Mater 23:623–645

Mondal D, Mollick MMR, Bhowmick B, Maity D, Bain MK, Rana D, Mukhopadhyay A, Dana K, Chattopadhyay D (2013) Effect of poly (vinyl pyrrolidone) on the morphology and physical

properties of poly (vinyl alcohol)/sodium montmorillonite nanocomposite films. Prog Nat Sci Mater Int 23:579–587
Mukerjee A, Vishwanatha JK (2009) Formulation, characterization and evaluation of curcumin loaded PLGA nanosphere for cancer therapy. J Anticancer Res 29(10):3867–3875
Nag A, Han K-S, Singh H (2011) Microencapsulation of probiotic bacteria using pH-induced gelation of sodium caseinate and gellan gum. Int Dairy J 21:247–253
Nam JM, Thaxton CS, Mirkin CA (2003) Nanoparticlebased bio-bar codes for the ultrasensitive detection of proteins. Science 301(5641):1884–1886
Nayak SK, Mohanty S (2009) Dynamic mechanical, rheological, and thermal properties of intercalated polystyrene/ organomontmorillonite nanocomposites: Effect of clay modification on the mechanical and morphological behaviors. J Appl Polym Sci 112:778–787
Neethirajan S, Jayas DS (2011) Nanotechnology for the Food and Bioprocessing Industries. Food Bioprocess Technol 4:39–47
Neethirajan S, Freund M, Jayas DS (2013) Poly(aniline boronic acid) polymers and methods of use. United States Patent 8454819B2
Ni P, Dai H, Wang Y, Sun Y, Shi Y, Hu J, Li Z (2014) Visual detection of melamine based on the peroxidase-like activity enhancement of bare gold nanoparticles. Biosens Bioelectron 60:286–291
Nieddu E, Mazzucco L, Gentile P, Benko T, Balbo V, Mandrile R, Ciardelli G (2009) Preparation and biodegradation of clay composite of PLA. React Funct Polym 69:371–379
Norman RS, Stone JW, Gole A, Murphy CJ, Sabo-Attwood TL (2008) Targeted Photothermal Lysis of the pathogenic bacteria, *Pseudomonas aeruginosa*, with gold nanorods. Nano Lett 8:302–306
Ochoa ML, Harrington PB (2005) Immunomagnetic isolation of enterohemorrhagic *Escherichia coli* O157:H7 from ground beef and identification by matrix-assisted laser desorption/ionization time-of-flight mass spectrometry and database searches. Anal Chem 77(16):5258–5267
Ozdemir C, Yeni F, Odaci D, Timur S (2010) Electrochemical glucose biosensing by pyranose oxidase immobilized in gold nanoparticle-polyaniline/AgCl/gelatinnanocomposite matrix. Food Chem 119:380–385
Park JS, Chang JH (2009) Colorless polyimide nanocomposite films with pristine clay: thermal behavior, mechanical property, morphology, and optical transparency. Polym Eng Sci 49:1357–1365
Park HM, Lee WK, Park CY, Ha CS (2003) Environmentally friendly polymer hybrids. Part I mechanical, thermal, and barrier properties of thermoplastic starch/clay nanocomposites. J Mater Sci 38:909–915
Park HM, Mohanty A, Misra M, Drzal LT (2004) Green nanocomposites from cellulose acetate bioplastic and clay: effect of ecofriendly triethyl citrate plasticizer. Biomacromolecules 5:2281–2288
Park HM, Kim GH, Ha CS (2007) Preparation and characterization of biodegradable aliphatic polyester/thermoplastic starch/organoclay ternary hybrid nanocomposites. Compos Interf 14:427–438
Patel AR, Hu YC, Tiwari JK, Velikov KP (2010) Synthesis and characterisation of zein-curcumin colloidal particles. Soft Matter 6:6192–6199
Paul MA, Delcourt C, Alexandre M, Degée P, Monteverde F, Dubois P (2005) Polylactide/montmorillonite nanocomposites: study of the hydrolytic degradation. Polym Degrad Stabil 87:535–542
Peelman N, Ragaert P, Meulenaer BD, Adons D, Peeters R, Cardon L, Impe FV, Devlieghere F (2013) Application of bioplastics for food packaging. Trends Food Sci Tech 32:128–141
Pérez-López B, Merkoçi A (2011) Nanomaterials based biosensors for food analysis applications. Trends Food Sci Tech 22:625–639
Peters R, Kramer E, Oomen AG, Rivera ZEH, Oegema G, Tromp PC, Fokkink R, Rietveld A, Marvin HJP, Weigel S, Peijnenburg AACM, Bouwmeester H (2012) Presence of nano-sized

silica during *in vitro* digestion of foods containing silica as a food additive. ACS Nano 6(3):2441–2451
Petersson L, Oksman K (2006) Biopolymer based nanocomposites: comparing layered silicates and microcrystalline cellulose as nanoreinforcement. Compos Sci Technol 66:2187–2196
Pinto RJB, Marques PAAP, Neto CP, Trindade T, Daina S, Sadocco P (2009) Antibacterial activity of nanocomposites of silver and bacterial or vegetable cellulosic fibers. Acta Biomater 5:2279–2289
Prabakaran E, Pandian K (2015) Amperometric detection of Sudan I in red chili powder samples using Ag nanoparticles decorated graphene oxide modified glassy carbon electrode. Food Chem 166:198–205
Ranjan S, Dasgupta N, Roy Chakraborty A, Samuel SM, Ramalingam C, Shanker R, Kumar A (2014) Nanoscience and nanotechnologies in food industries: opportunities and research trends. J Nanopart Res 16:2464
Ravichandran R (2010) Nanotechnology applications in food and food processing: innovative green approaches, opportunities and uncertainties for global market. Int J Green Nanotechnol Phys Chem 1(2):P72–P96
Ravindranath SP, Mauer LJ, Deb-Roy C, Irudayaraj J (2009) Biofunctionalized magnetic nanoparticle integrated mid-infrared pathogen sensor for food matrixes. Anal Chem 81(8):2840–2846
Rawal R, Chawla S, Pundir CS (2012) An electrochemical sulfite biosensor based on gold coated magnetic nanoparticles modified gold electrode. Biosens Bioelectron 31(1):144–150
Rhim J-W (2011) Effect of clay contents on mechanical and water vapor barrier properties of agar-based nanocomposite films. Carbohydr Polym 86:691–699
Rhim J-W, Hong SI, Park H-M, Ng PKW (2006) Preparation and characterization of chitosan-based nanocomposite films with antimicrobial activity. J Agric Food Chem 54:5814–5822
Rhim J-W, Park H-M, Ha C-S (2013) Bio-nanocomposites for food packaging applications. Prog Polym Sci 38(10–11):1629–1652
Ribeiro HS, Chua BS, Ichikawab S, Nakajima M (2008) Preparation of nanodispersions containing β-carotene by solvent displacement method. Food Hydrocoll 22(1):12–17
Sakuldao S, Yoovidhya T, Wongsasulak S (2011) Coaxial electrospinning and sustained release properties of gelatin-cellulose acetate core-shell ultrafine fibres. Sci Asia 37(4):335–343
Sanchez-Garcia MD, Gimenez E, Lagaron JM (2007) Comparative barrier performance of novel PET nanocomposites with biopolyester nanocomposites of interest in packaging food applications. J Plast Film Sheet 23:133–148
Sanchez-Garcia MD, Lopez-Rubio A, Lagaron JM (2010a) Natural micro and nanobiocomposites with enhanced barrier properties and novel functionalities for food biopackaging applications. Trends Food Sci Tech 21:528–536
Sanchez-Garcia MD, Lagaron JM, Hoa SV (2010b) Effect of addition of carbon nanofibers and carbon nanotubes on properties of thermoplastic biopolymers. Compos Sci Technol 70(7):1095–1105
Sandros MG, Shete V, Benson DE (2006) Selective, reversible, reagentless maltose biosensing with core-shell semiconducting nanoparticles. Analyst 131:229–235
Sanguansri P, Augustin MA (2006) Nanoscale materials development – a food industry perspective. Trends Food Sci Tech 17:547–556
Sanna V, Lubinu G, Madau P, Pala N, Nurra S, Mariani A, Sechi M (2015) Polymeric nanoparticles encapsulating white tea extract for nutraceutical application. J Agric Food Chem 63(7):2026–2032
Santos DT, Martín Á, Meireles MAA, Cocero MJ (2012) Production of stabilized sub-micrometric particles of carotenoids using supercritical fluid extraction of emulsions. J Supercrit Fluids 61:167–174
Sasmal A, Nayak PL, Sasmal S (2009) Degradability studies of green nanocomposites derived from soy protein isolate (SPI)-furfural modified with organoclay. Polym-Plast Technol Eng 48:905–909

Sharmila RJ, Premkumar S, Alagar M (2010) Preparation and characterization of organoclay-filled, vinyl ester-modified unsaturated polyester nanocomposites. High Perform Polym 22:16–27
Shimoni G, Levi CS, Tal SL, Lesmes U (2013) Emulsions stabilization by lactoferrin nano-particles under in vitro digestion conditions. Food Hydrocoll 33:264–272
Shlyapnikov YM, Shlyapnikova EA, Simonova MA, Shepelyakovskaya AO, Brovko FA, Komaleva RL, Grishin EV, Morozov VN (2012) Rapid simultaneous ultrasensitive immunodetection of five bacterial toxins. Anal Chem 84(13):5596–5603
Siegrist M, Stampfli N, Kastenholz H, Keller C (2008) Perceived risks and perceived benefits of different nanotechnology foods and nanotechnology food packaging. Appetite 51:283–290
Simoniana AL, Good TA, Wang S-S, Wild JR (2005) Nanoparticle-based optical biosensors for the direct detection of organophosphate chemical warfare agents and pesticides. Anal Chim Acta 534:69–77
Singh BN, Kim KH (2005) Effects of divalent cations on drug encapsulation efficiency of deacylated gellan gum. J Microencapsul 22:761–771
Singh AK, Senapati D, Wang S, Griffin J, Neely A, Candice P, Naylor KM, Varisli B, Kalluri JR, Ray PC (2009) Gold nanorod based selective identification of Escherichia coli bacteria using two-photon rayleigh scattering spectroscopy. ACS Nano 3(7):1906–1912
Sinha Ray S, Okamoto M (2003) New polylactide/layered silicate nanocomposites: open a new dimension for plastics and composites. Macromol Rapid Commun 24:815–840
Sinha Ray S, Yamada K, Okamoto M, Ueda K (2003) Biodegradable polylactide/montmorillonite nanocomposites. J Nanosci Nanotechnol 3:503–510
Sliwinska M, Wisniewska P, Dymerski T, Namiesnik J, Wardencki W (2014) Food analysis using artificial senses. J Agric Food Chem 62(7):1423–1448
Smolkova B, Yamani NE, Collins AR, Gutleb AC, Dusinska M (2015) Nanoparticles in food. Epigenetic changes induced by nanomaterials and possible impact on health. Food Chem Toxicol 77:64–73
Sonkaew P, Sane A, Suppakul P (2012) Antioxidant activities of curcumin and ascorbyl dipalmitate nanoparticles and their activities after incorporation into cellulose-based packaging films. J Agric Food Chem 60:5388–5399
Sorrentino A, Gorrasi G, Tortora M, Vittoria V, Costantino U, Marmottini F, Pasella F (2005) Incorporation of Mg–Al hydrotalcite into a biodegradable poly (epsilon, caprolactone) by high energy ball milling. Polymer 46:1601–1608
Sorrentino A, Gorrasi G, Vittoria V (2007) Potential perspectives of bio-nanocomposites for food packaging applications. Trends Food Sci Tech 18:84–95
Sothornvit R, Rhim JW, Hong SI (2009) Effect of nano-clay type on the physical and antimicrobial properties of whey protein isolate/clay composite film. J Food Eng 91:468–473
Sowasod N, Charinpanitkul ST, Tanthapanichakoon W (2008) Nanoencapsulation of curcumin in biodegradable chitosan via multiple emulsion/solvent evaporation. Int J Pharm 347:93–101
Speranskaya ES, Beloglazova NV, Lenain P, De Saeger S, Wang Z, Zhang S, Hens Z, Knopp D, Niessner R, Potapkin DV, Goryacheva IY (2014) Polymer-coated fluorescent CdSe-based quantum dots for application in immunoassay. Biosens Bioelectron 53:225–231
Sriupayo J, Supaphol P, Blackwell J, Rujiravanit R (2005) Preparation and characterization of α-chitin whisker-reinforced chitosan nanocomposite films with or without heat treatment. Carbohydr Polym 62:130–136
Staiano M, Matveeva EG, Rossi M, Crescenzo R, Gryczynski Z, Gryczynski I, Iozzino L, Akopova I, D'Auria S (2009) Nanostructured silver-based surfaces: new emergent methodologies for an easy detection of analytes. ACS Appl Mater Interfaces 1(12):2909–2916
Su XL, Li Y (2004) Quantum dot biolabeling coupled with immunomagnetic separation for detection of *Escherichia coli* O157:H7. Anal Chem 76(16):4806–4810
Sun L-M, C-l Z, Li P (2011) Characterization, antimicrobial activity, and mechanism of a high-performance—epigallocatechin-3-gallate (EGCG)—Cu^{2+}/Polyvinyl Alcohol (PVA) nanofibrous membrane. J Agric Food Chem 59:5087–5092

Sun X, Jia M, Guan L, Ji J, Zhang Y, Tang L, Li Z (2015) Multilayer graphene–gold nanocomposite modified stem-loop DNA biosensor for peanut allergen-Ara h1 detection. Food Chem 172:335–342

Sung S-Y, Sin LT, Tee TT, Bee S-T, Rahmat AR, Rahman WAWA, Tan A-C, Vikhraman M (2013a) Antimicrobial agents for food packaging applications. Trends Food Sci Tech 33:110–123

Sung YJ, Suk HJ, Sung HY, Li T, Poo H, Kim MG (2013b) Novel antibody/gold nanoparticle/magnetic nanoparticle nanocomposites for immunomagnetic separation and rapid colorimetric detection of *Staphylococcus aureus* in milk. Biosens Bioelectron 43:432–439

Suntres ZE, Shek PN (1996) Alleviation of paraquat-induced lung injury by pre-treatment with bifunctional liposomes containing alpha-tocopherol and glutathione. Biochem Pharmacol 52:1515–1520

Tamjidi F, Shahedi M, Varshosaz J, Nasirpour A (2013) Nanostructured lipid carriers (NLC): a potential delivery system for bioactive food molecules. Innov Food Sci Emerg Technol 19:29–43

Tang S, Zou P, Xiong H, Tang H (2008) Effect of nano-SiO_2 on the performance of starch/polyvinyl alcohol blend films. Carbohydr Polym 72:521–526

Tang D, Sauceda JC, Lin Z, Ott S, Basova E, Goryacheva I, Biselli S, Lin J, Niessner R, Knopp D (2009) Magnetic nanogold microspheres-based lateral-flow immunodipstick for rapid detection of aflatoxin B2 in food. Biosens Bioelectron 25:514–518

Teeranachaideekul V, Müller RH, Junyaprasert VB (2007) Encapsulation of ascorbyl palmitate in nanostructured lipid carriers (NLC)—effects of formulation parameters on physicochemical stability. Int J Pharm 340:198–206

Tominaga M, Nomura S, Taniguchi I (2009) D-Fructose detection based on the direct heterogeneous electron transfer reaction of fructose dehydrogenase adsorbed onto multi-walled carbon nanotubes synthesized on platinum electrode. Biosens Bioelectron 24:1184–1188

Tsai P-J, Chen Y-S, Sheu C-H, Chen C-Y (2011) Effect of nanogrinding on the pigment and bioactivity of djulis (*Chenopodium formosanum* Koidz.). J Agric Food Chem 59:1814–1820

Tunc S, Angellier H, Cahyana Y, Chalier P, Gontard N, Gastaldi E (2007) Functional properties of wheat gluten/montmorillonite nanocomposite films processed by casting. J Membr Sci 289:159–168

Valodkar M, Bhadoria A, Pohnerkar J, Mohan M, Thakore S (2010) Morphology and antibacterial activity of carbohydrate-stabilized silver nanoparticles. Carbohydr Res 345:1767–1773

Vanderroost M, Ragaert P, Devlieghere F, Meulenaer BD (2014) Intelligent food packaging: the next generation. Trends Food Sci Tech 39:47–62

Varshney M, Yang L, Su XL, Li Y (2005) Magnetic nanoparticle-antibody conjugates for the separation of Escherichia coli O157: H7 in ground beef. J Food Protect 68(9):1804–1811

Varshney M, Li Y, Srinivasan B, Tung S (2007) A label-free, microfluidics and interdigitated array microelectrode-based impedance biosensor in combination with nanoparticles immunoseparation for detection of Escherichia coli O157: H7 in food samples. Sensor Actuat B-Chem 128(1):99–107

Vartiainen J, Tuominen M, Nättinen K (2010) Bio-hybrid nanocomposite coatings from sonicated chitosan and nanoclay. J Appl Polym Sci 116:3638–3647

Veleirinho B, Lopes-da-Silva JA (2009) Application of electrospun poly (ethylene terephthalate) nanofiber mat to apple juice clarification. Process Biochem 44(3):353–356

Vermeiren L, Devlieghere F, van Beest M, de Kruijf N, Debevere J (1999) Developments in the active packaging of foods. Trends Food Sci Tech 10:77–86

Verran J, Packer A, Kelly P, Whitehead KA (2010) Titanium coating of stainless steel as an aid to improved cleanability. Int J Food Microbiol 141:134–139

Vinayaka AC, Basheer S, Thakur MS (2009) Bioconjugation of CdTe quantum dot for the detection of 2,4-dichlorophenoxyacetic acid by competitive fluoroimmunoassay based biosensor. Biosens Bioelectron 24:1615–1620

Viswanathan S, Wu L-C, Huang M-R, Ho JA (2006) Electrochemical immunosensor for cholera toxin using liposomes and poly(3,4-ethylenedioxythiophene)-coated carbon nanotubes. Anal Chem 78(4):1115–1121

Viswanathan S, Radecka H, Radecki J (2009) Electrochemical biosensor for pesticides based on acetylcholinesterase immobilized on polyaniline deposited on vertically assembled carbon nanotubes wrapped with ssDNA. Biosens Bioelectron 24:2772–2777

Vladimiriov V, Betchev C, Vassiliou A, Papageorgiou G, Bikiaris D (2006) Dynamic mechanical and morphological studies of isotactic polypropylene/fumed silica nanocomposites with enhanced gas barrier properties. Compos Sci Technol 66:2935–2944

von Bultzingslöwen C, McEvoy AK, McDonagh C, MacCraith BD, Klimant I, Krause C, Wolfbeis OS (2002) Sol–gel based optical carbon dioxide sensor employing dual luminophore referencing for application in food packaging technology. Analyst 127(11):1478–1483

Vrignaud S, Hureaux J, Wack SV, Benoit J-P, Saulnier P (2012) Design, optimization and in vitro evaluation of reverse micelle-loaded lipid nanocarriers containing erlotinib hydrochloride. Int J Pharm 436:194–200

Wang M, Li Z (2008) Nano-composite ZrO_2/Au film electrode for voltammetric detection of parathion. Sensor Actuat B Chem 133(2):607–612

Wang R, Dong W, Ruan C, Kanayeva D, Tian R, Lassiter K, Li Y (2008) TiO_2 nanowire bundle microelectrode based impedance immunosensor for rapid and sensitive detection of *Listeria monocytogenes*. Nano Lett 8(9):2625–2631

Wang L, Chen W, Xu D, Shim BS, Zhu Y, Sun F, Liu L, Peng C, Jin Z, Xu C, Kotov NA (2009) Simple, rapid, sensitive, and versatile SWNT-paper sensor for environmental toxin detection competitive with ELISA. Nano Lett 9(12):4147–4152

Wang Z, Miu T, Xu H, Duan N, Ding X, Li S (2010a) Sensitive immunoassay of *Listeria monocytogenes* with highly fluorescent bioconjugated silica nanoparticles probe. J Microbiol Methods 83:179–184

Wang P, Hu X, Cheng Q, Zhao X, Fu X, Wu K (2010b) Electrochemical detection of amaranth in food based on the enhancement effect of carbon nanotube film. J Agric Food Chem 58:12112–12116

Wang Z, Xu H, Wu J, Ye J, Yang Z (2011a) Sensitive detection of *Salmonella* with fluorescent bioconjugated nanoparticles probe. Food Chem 125:779–784

Wang H, Li Y, Wang A, Slavik M (2011b) Rapid, sensitive, and simultaneous detection of three foodborne pathogens using magnetic nanobead–based immunoseparation and quantum dot–based multiplex immunoassay. J Food Protect 74(12):2039–2047

Wang H, Zhang Y, Li H, Du B, Ma H, Wu D, Wei Q (2013a) A silver-palladium alloy nanoparticle-based electrochemical biosensor for simultaneous detection of ractopamine, clenbuterol and salbutamol. Biosens Bioelectron 49:14–19

Wang X, Li L, Wang Y, Xu C, Zhao B, Yang X (2013b) Application of reduced graphene oxide and carbon nanotube modified electrodes for measuring the enzymatic activity of alcohol dehydrogenase. Food Chem 138:2195–2200

Wang Y, Yuan L, Yao C, Ding L, Li C, Fang J, Sui K, Liu Y, Wu M (2014) A combined toxicity study of zinc oxide nanoparticles and vitamin C in food additives. Nanoscale 6:15333–15342

Wang M, Gao Y, Sun Q, Zhao J (2015) Ultrasensitive and simultaneous determination of the isomers of Amaranth and Ponceau 4R in foods based on new carbon nanotube/polypyrrole composites. Food Chem 172:873–879

Warner MG, Grate JW, Tyler A, Ozanich RM, Miller KD, Lou J, Marks JD, Bruckner-Lea CJ (2009) Quantum dot immunoassays in renewable surface column and 96-well plate formats for the fluorescence detection of botulinum neurotoxin using high-affinity antibodies. Biosens Bioelectron 25(1):179–184

Warriner K, Reddy SM, Namvar A, Neethirajan S (2014) Developments in nanoparticles for use in biosensors to assess food safety and quality. Trends Food Sci Tech 40:183–199

Wei H, Wang E (2008) Fe3O4 magnetic nanoparticles as peroxidase mimetics and their applications in H_2O_2 and glucose detection. Anal Chem 80:2250–2254

Weir A, Westerhoff P, Fabricius L, Hristovski K, von Goetz N (2012) Titanium dioxide nanoparticles in food and personal care products. Environ Sci Technol 46:2242–2250

Wongsasulak S, Pathumban S, Yoovidhya T (2014) Effect of entrapped α-tocopherol on mucoadhesivity and evaluation of the release, degradation, and swelling characteristics of zein-chitosan composite electrospun fibers. J Food Eng 120:110–117

Wu CL, Zhang MQ, Rong MZ, Friedrick K (2002) Tensile performance improvement of low nanoparticles filled-polypropylene composites. Compos Sci Technol 62:1327–1340

Wu CL, Zhang MQ, Rong MZ, Friedrich K (2005a) Silica nanoparticles filled polypropylene: effects of particle surface treatment, matrix ductility and particle species on mechanical performance of the composites. Compos Sci Technol 65:635–645

Wu L, Yuan X, Sheng J (2005b) Immobilization of cellulase in nanofibrous PVA membranes by electrospinning. J Membr Sci 250(1):167–173

Wu Q, Henriksson M, Liu X, Berglund LA (2007) A high strength nanocomposite based on microcrystalline cellulose and polyurethane. Biomacromolecules 8:3687–3692

Wu Y, Luo Y, Wang Q (2012) Antioxidant and antimicrobial properties of essential oils encapsulated in zein nanoparticles prepared by liquid–liquid dispersion method. LWT Food Sci Technol 48:283–290

Xie J, Hsieh YL (2003) Ultra-high surface fibrous membranes from electrospinning of natural proteins: casein and lipase enzyme. J Mater Sci 38(10):2125–2133

Xie L, Lv X-Y, Han Z-J, Ci J-H, Fang C-Q, Ren P-G (2012) Preparation and performance of high-barrier low density polyethylene/organic montmorillonite nanocomposite. Polym-Plast Technol Eng 51:1251–1257

Xiong HG, Tang SW, Tang HL, Zou P (2008) The structure and properties of a starch-based biodegradable film. Carbohydr Polym 71:263–268

Xu G, Chi Y, Li L, Liu S, Kan X (2015) Imprinted propyl gallate electrochemical sensor based on graphene/single walled carbon nanotubes/sol–gel film. Food Chem 177:37–42

Yakes BJ, Lipert RJ, Bannantine JP, Porter MD (2008) Detection of *Mycobacterium avium* subsp. Paratuberculosis by a sonicate immunoassay based on surface-enhanced Raman scattering. Clin Vaccine Immunol 15(2):227–234

Yang L, Li Y (2005) Quantum dots as fluorescent labels for quantitative detection of *Salmonella typhimurium* in chicken carcass wash water. J Food Protect 68(6):1241–1245

Yang H, Qu L, Wimbrow AN, Jiang X, Sun Y (2007) Rapid detection of *Listeria monocytogenes* by nanoparticle-based immunomagnetic separation and real-time PCR. Int J Food Microbiol 118(2):132–138

Yang J, Wang P, Zhang X, Wu K (2009) Electrochemical sensor for rapid detection of triclosan using a multiwall carbon nanotube film. J Agric Food Chem 57:9403–9407

Yano K, Usuki A, Okad A (1997) Synthesis and properties of polyimide–clay hybrid films. J Polym Sci A Polym Chem 35:2289–2294

Yebo NA, Sree SP, Levrau E, Detavernier C, Hens Z, Martens JA, Baets R (2012) Selective and reversible ammonia gas detection with nanoporous film functionalized silicon photonic micro-ring resonator. Opt Express 20(11):11855–11862

Yoksan R, Chirachanchai S (2010) Silver nanoparticle-loaded chitosan starch based films: fabrication and evaluation of tensile, barrier and antimicrobial properties. Mater Sci Eng C 30:891–897

Yu C-C, Lee C-L, Pan T-M (2006) A novel formulation approach for preparation of nanoparticulate red mold rice. J Agric Food Chem 54:6845–6851

Yun M, Choe JE, You J-M, Ahmed MS, Lee K, Ustundag Z, Jeon S (2015) High catalytic activity of electrochemically reduced graphene composite toward electrochemical sensing of Orange II. Food Chem 169:114–119

Zaru M, Manca ML, Fadda AM, Antimisiaris SG (2009) Chitosan-coated liposomes for delivery to lungs by nebulisation. Colloids Surf B 71:88–95

Zeng H, Gao C, Wang Y, Watts PCP, Kong H, Cui X, Yan D (2006) *In situ* polymerization approach to multiwalled carbon nanotubes-reinforced nylon 1010 composites: mechanical properties and crystallization behavior. Polymer 47:113–122

Zhang C, Ding Y, Ping Q, Yu L (2006) Novel chitosan-derived nanomaterials and their micelle-forming properties. J Agric Food Chem 54:8409–8416

Zhang X, Geng P, Liu H, Teng Y, Liu Y, Wang Q, Zhang W, Jin L, Jiang L (2009) Development of an electrochemical immunoassay for rapid detection of E. coli using anodic stripping voltammetry based on Cu@Au nanoparticles as antibody labels. Biosens Bioelectron 24:2155–2159

Zhang Y, Zhang X, Lu X, Yang J, Wu K (2010) Multi-wall carbon nanotube film-based electrochemical sensor for rapid detection of Ponceau 4R and Allura Red. Food Chem 122:909–913

Zhang X, He S, Chen Z, Huang Y (2013a) $CoFe_2O_4$ nanoparticles as oxidase mimic-mediated chemiluminescence of aqueous luminol for sulfite in white wines. J Agric Food Chem 61:840–847

Zhang J, Nie S, Wang S (2013b) Nanoencapsulation enhances epigallocatechin-3-gallate stability and its antiatherogenic bioactivities in macrophages. J Agric Food Chem 61:9200–9209

Zhang J, Wang M, Shentu C, Wang W, Chen Z (2014) Simultaneous determination of the isomers of Ponceau 4R and Amaranth using an expanded graphite paste electrode. Food Chem 160:11–15

Zhao XJ, Hilliard LR, Mechery SJ, Wang YP, Bagwe RP, Jin SG, Tan W (2004) A rapid bioassay for single bacterial cell quantitation using bioconjugated nanoparticles. Proc Natl Acad Sci U S A 101(42):15027–15032

Zhao ZW, Chen XJ, Tay BK, Chen JS, Han ZJ, Khor KA (2007) A novel amperometric biosensor based on ZnO:Co nanoclusters for biosensing glucose. Biosens Bioelectron 23:135–139

Zhao Q, Wei F, Luo Y-B, Ding J, Xiao N, Feng Y-Q (2011) Rapid magnetic solid-phase extraction based on magnetic multiwalled carbon nanotubes for the determination of polycyclic aromatic hydrocarbons in edible oils. J Agric Food Chem 59:12794–12800

Zhou LT, Li RY, Li ZJ, Xia QF, Fang YJ, Liu JK (2012) An immunosensor for ultrasensitive detection of aflatoxin B-1 with an enhanced electrochemical performance based on graphene/conducting polymer/gold nanoparticles/the ionic liquid composite film on modified gold electrode with electrodeposition. Sensor Actuat B-Chem 174:359–365

Zhu S, Chen J, Li H, Cao Y, Yang Y, Feng Z (2014) Preparation and properties of montmorillonite/poly (ethylene glycol) grafted polypropylene/polypropylene nanocomposites. Appl Clay Sci 87:303–310

Zimet P, Rosenberg D, Livney YD (2011) Re-assembled casein micelles and casein nanoparticles as nano-vehicles for omega-3 polyunsaturated fatty acids. Food Hydrocoll 25(5):1270–1276

Chapter 4
Nanoencapsulation of Bioactive Compounds for Nutraceutical Food

Pooja J. Rao and Madhav M. Naidu

Abstract Nanoscience is gaining a rapid attention in food research. An application of nanosciecne in food science is the encapsulation of bioactive compounds at the nano size for incorporation in food products. Encapsulation is known to enhance the water solubility of bioactive compounds, including essential oils. Encapsulation thus provide a germ free product while retaining the taste and aroma of essential oils. Nanoencapsulation enhances solubility, bioavailability and health beneficial activities. This effect is due to the nanosize that increases surface area, and, in turn, activity. For example, curcumin in turmeric, omega-3-fatty acid in fish oil, vitamins from fruits show antimicrobial, antioxidant and anticarcinogenic activities. After consumption of the food the bio-actives are released at the targeted organ of the body and utilized for its nutritional property. These nutraceutical foods will influence the tastes, health and habits of the population and nature of the food products available in the market. The present chapter describes various bioactive compounds used for nanoencapsualtion with novel techniques.

Keywords Nanoencapsulation • Food bioactives • Omega-3 fatty acid • Vitamin

4.1 Introduction

'Nutraceutical' is a combination of two different terms: Nutrition and Pharmaceutical. This term was first coined by Dr. Stephen De Felice, founder of the 'Foundation for Innovation in Medicine' in Cranford, New Jersy, in 1979 (De Felice 1995). He elaborated "A Nutraceutical is any substance that may be considered food or part of food and provides medical or health benefits, including the prevention and treatment of disease, such products may range from isolated nutrients, dietary supplements and diets of genetically engineered "designer" foods, herbal products and foods such as

---

P.J. Rao (✉) • M.M. Naidu
Department of Spice & Flavour Science, Council of Scientific Research Institute- Central Food Technological Research Institute (CSIR-CFTRI), 570020 Mysuru, Karnataka, India
e-mail: poojarao@cftri.res.in

S. Ranjan et al. (eds.), *Nanoscience in Food and Agriculture 2*, Sustainable Agriculture Reviews 21, DOI 10.1007/978-3-319-39306-3_4

cereals, soups and beverages." This term could cover all aspects of other terms like dietary supplements, fortified foods, functional foods and medicinal foods under its broad meaningful definition. Although a universally accepted definition is still not available, standards have been set by United States – Food and Drug Administration (FDA), European Union – European Food Safety Authority (EFSA) and Canada – Food Directorate of Health Canada (FDHC) on the nutrient labeling, health claims and marketing of the nutraceutical products (Alberta and Boye 2015). The health ministry of Canada released their version of nutraceutical as "a product isolated or purified from the food, generally sold in medicinal form not associated with food and demonstrated to have a physiological benefit. It also provides benefit against chronic disease" (Bull Esther 2000). However, in recent years several products have been developed under the name of functional foods or nutraceutical products/foods while, there is a slight difference between two commonly used terms *viz 'nutraceutical'* and *'functional foods'*. The term *functional food* was first used in Japan in 1980s for foods fortified with special constituents possessing beneficial physiological effects (Siró et al. 2008). According to FDHC (Anon 2002), a functional food is similar in appearance to, or may be, a conventional food, is consumed as part of a usual diet, and is demonstrated to have physiological benefits and/or reduce the risk of chronic disease beyond basic nutritional functions. The Ministry of Agriculture, Fisheries and Food, Britain has defined functional food as "a food that has a component incorporated into it to give it a specific medical or physiological benefit, other than purely nutritional benefit" (Cockbill 1994). Whereas, nutraceutical products or nutraceutical incorporated foods are considered to provide the basic nutritional physiological benefits along with prevention/treatment of chronic diseases other than anaemia. Examples of nutraceuticals include fortified dairy products (e.g. milk) and citrus fruits (Kalra 2003). It is also expected that a scientifically proven nutraceutical food will negate the chances of adverse effects on human body, which are regular features of pharmaceutical drugs. The European Nutraceutical Association (ENA) has a simple yet precise way to explain nutraceutical. It mentions "according to conventional view, foodstuffs primarily provide nutrition in the form of substances which the human body needs for normal development and maintenance of bodily functions. Medicines, on the other hand, are always seen as therapeutic agents in the context of disease and health. The more knowledge is obtained about the health-promoting effect of foods, the more blurred this seemingly strict separating line becomes between these two product groups. This is reflected in the word nutraceutical combining both the ideas of nutrition and the concept of the pharmaceutical. And it is now a well established term." (Koch et al. 2014)

In last few years, awareness has grown substantially in managing health, wellness and reducing risks of common or chronic health problems through food. In this context, the targeted role of nutraceutical foods towards several diseases (Fig. 4.1) is crucial in shifting consumer attention from pharmaceutically derived supplements and drugs to disease preventive foods. The transformation of traditional and convenience food to nutraceutical food is achieved by incorporating bioactive compounds of various spices, vegetables, fruits and plant/animal products. The bioactive compounds like curcumin, lycopene, omega-3 fatty acid, lutein, vitamin A, vitamin D, probiotic bacteria, etc. have numerous health-beneficial activity targeted to specific body parts (Table 4.1).

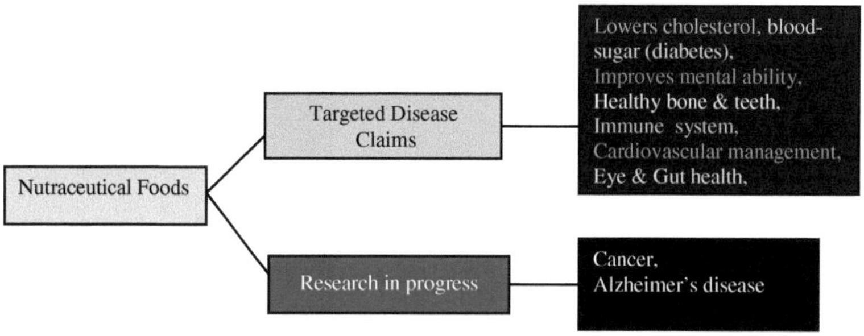

Fig. 4.1 Target diseases to be treated by nutraceutical foods as an alternative to pharmaceutical drugs

Table 4.1 List of major nutraceutical bioactives and their relationship with biological activity

Nutraceutical bioactives	Biological/health activity
Vitamins	Cardiovascular, brain, gut, immune, bone, dental, antioxidants, metabolism
Vit. C/A/E/K/D/B1/B2/B6/B12/Folic acid/Niacin/Biotin, Panthothenic acid	
Carotenoids	
β-carotene/Lycopene/Lutein/Zexanthin/Astaxanthin	
Minerals	
Ca/Mg/Zn/Fe/Na/K/Se	
Proteins/peptides	Muscle, bone, energy
Immunoglobulins/Bioactive peptides/Lactoferrin/Digestive enzymes	
Carbohydrates	
Poly-dextrose/Starch/Pectins/Guar gums	
Fatty acids	Brain, cardiovascular, vision
DHA/EPA/ARA/GLA/CLA/Structured lipids	
Fibers	Gut, cardiovascular
Inulin/Psyllium/Insoluble fibers	
Prebiotic	Metabolism, gut, antioxidants
FOS/inulin/GOS/Poly-dextrose/Oligosachharides	
Probiotic	
Lactobacilli/Bifidobacteria/other cultures	
Phytochemicals	Antioxidant, cardiovascular, dental, weight management
Phytosterols/Isoflavanones/Lignans/Polyphenols (curcumin)/Tocotrienols/Isothiocyanates	

Source: Arnoldi 2010 (http://www.vub.ac.be/MICH/eafp/eafp_pages/archive/pdfpres/catania/Arnoldi.pdf)
DHA Docosahexaenoic acid, *EPA* Eicosapentaenoic acid, *ARA* Arachidonic acid, *GLA* Gamma-linolenic acid, *CLA* Conjugated linoleic acid, *FOS* Fructooloigosachharide, *GOS* Galactooligosachharide

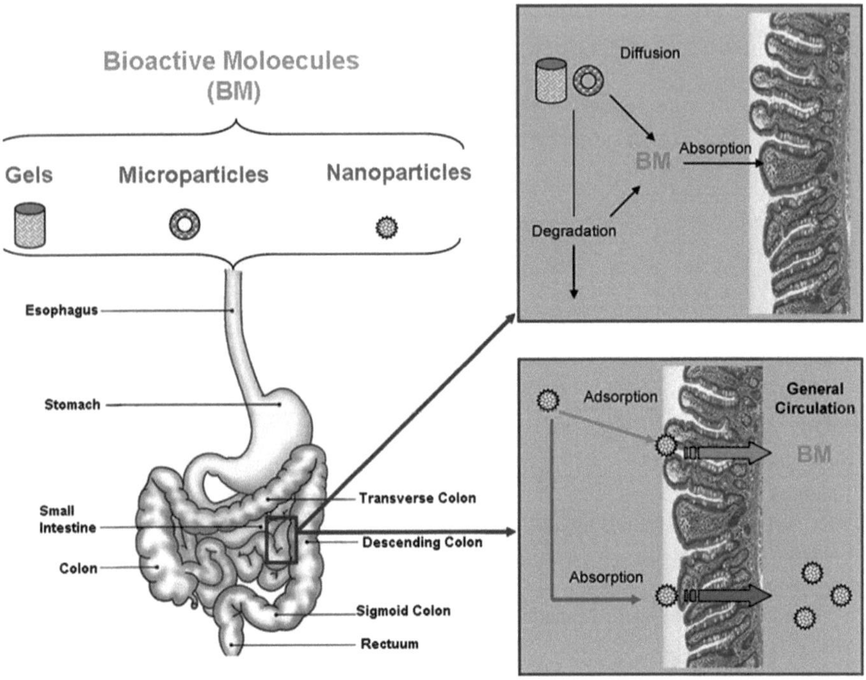

Fig. 4.2 Schematic representation of different absorption mechanisms of bioactive molecules (Source: Chen et al. 2006. Reproduced with permission of Elsevier)

However, their stability, water solubility and bioavailability are very less. The encapsulation of bioactive compounds in a suitable carrier material enhances water solubility, prevents degradation due to environmental and biological conditions and masks their taste or flavor as per the requirement. The size and nature of encapsulated particle are the decisive factors to attain aforementioned objectives. Microencapsulation is a known technique since 1980s to enhance the stability, bioactivity, applicability in powder form and reduce the volatility, toxicity of bioactive compounds (Desai & Park 2005). Nonetheless the lumps or bigger particle size and high concentration of wall materials were constraints. These would finally influence its absorption in the gastrointestinal tract (Fig. 4.2).

The encapsulation of bioactive compounds in the scale less than 100–500 nanometer (nm; mainly lower than 100 nm) is called as 'nanoencapulation' and the material is known as 'nanomaterial'. In the last decade, nanoencapsulation has established itself as a promising discipline for the delivery of many nanoencapsulated lipophilic, hydrophilic bioactive compounds and living cells that cannot be taken in pure form and needed to be incorporated in a delivery system. Further, a delivery system must fulfill following criteria:

- A vehicle for carrying the functional ingredient to the desired site of action.
- Protect the functional ingredient from chemical or biological degradation (for example, oxidation) during processing, storage, and utilization; this maintains the functional ingredient in its active state.

- Capable of controlling the release of the functional ingredient, such as the release rate or the specific environmental conditions that trigger release (for example, pH, ionic strength, or temperature).
- The delivery system has to be compatible with other components in the system, as well as being compatible with the physicochemical and qualitative attributes (that is, appearance, texture, taste, and shelf-life) of the final product (Weiss et al. 2006).

The nanoencapsulated bioactives or functional ingredients are reported to have controlled release property, enhanced water solubility and bioavailability. The reason behind higher water solubility and slow release of nanoencapsulated food materials compared to their native counterparts are (Kawashim 2001; Peppas 1997; Chen et al. 2006; Zhong et al. 2007; Bilia et al. 2014)–

1. reduction in size of the particles to 'nano' range leads to increase in surface area bringing maximum active sites on the surface thereby, augmenting their penetration and biological activities.
2. the tailor-made carrier material controls the bioactive compound release for a gradual and longer duration and thus allows the maximum absorption of the available concentration of the later in the GI tract and the residence time of the nanoparticle due to hydrophobic or electrostatic interaction with the biological system (Fig. 4.2).

The food nanomaterials are consisted of inner food bioactive compound called as *core material* and outer encapsulating carrier called as *shell material*. The nanomaterials may be in particle, capsule or fiber forms (Weiss et al. 2006). It is important to note that *shell materials* should be bio-compatible or natural and have generally regarded as safe status to facilitate its incorporation in food matrix without apprehension, contrary to pharmaceutical applications where choices are vast and selectivity is not that stringent. The alterations in shell materials influence the physico-chemical property of nanostructures.

The core materials are the functional or bioactive components to be encapsulated within the interfacial region. Encapsulating functional components slow down the chemical degradation process by engineering the properties of interfacial layer surrounding them (McClements and Decker 2000). Moreover, nanoencapsulated products are formed due to self-assembly of reactants that help to enhance stability of products; generally upto a pH of 2, allows pasteurization (thereby achieving thermodynamic stability) and provides better control of reaction rate. Apart from these, the nanosturctures have an advantage of being clear in visual and textural effects thus can be a part of clear beverages also (Garti 2008). Table 4.2 lists various nanostructures that have potential applications in nutrient delivery systems. Different techniques like high pressure homogenization, microfluidization, ultrasonication, etc., along with appropriate combination of raw materials may be used to synthesize nanoemulsions of oil-in-water (O/W), oil-water-oil (O/W/O) or water-oil-water (W/O/W), liposomes, solid lipid nanoparticles etc.

The present chapter covers the research activities on specific bioactives (vitamin A/D, omega-3 fatty acid and essential oils from spices) as they drew lot of attention

Table 4.2 List of nanostructures with suitable shell materials and applications for nutrient delivery

Nanostructure material	Probable raw material & other information	Potential applications
Nanoemulsions	Lipids/Polymers; droplet size 50–200 nm. Relatively stable	Delivery of both hydrophobic and hydrophilic compounds, multilayer system possible thus simultaneous or sequential delivery of single or multiple compounds
Solid lipid nanoparticles	Crystalline or semi-crystalline stabilized by a surfactant coating	Delivery of hydrophobic compound
Liposomes	Capsules consisting of lipid bi-layer with aqueous interior. Generally, phosphholipids such as phosphatidyl choline	Delivery of both hydrophobic or hydrophilic compound. Can be part of tailor made foods
Micelles	Droplets of surfactants (lipids/bio-polymers) in a liquid	Mainly drug-delivery of hydrophobic compounds
Casein	Milk protein that self-assembles into miceller structure	Delivery of mineral, protein, vitamins, spice bio-active compounds
Whey proteins	β-lactoglobuline and alpha-lactoalbumin used. Fibrils, hydrogels, nanoparticles formed depending on processing conditions. Resistant to stomach and enzymes	Delivery of hydrophobic component; can be added in food (affect mouth feel)
Chitosan	Carbohydrate isolated from crustaceous. Muco-adhesive, bio-compatible, non-toxic. Nanocapsules, hydrogels are formed	Delivery of various component to oral/intestinal mucosa, as part of multi-component and layered system
Silica	Bio-degradable, Bio-compatible. Highly nano porous	Delivery of various hydrophilic nutrients to stomach

Source: ObeservatoryNANO briefing 18, July 2011 (http://www.nanopinion.eu/sites/default/files/observatorynano_briefing_no_18_improving_delivery_of_essential_vitamins_minerals.Pdf)

in recent years as core materials because of their potentiality to prevent night blindness, bone problems, cardiovascular disease and antioxidant/anti-inflammatory/anti-cancerous properties. Nonetheless, the following section discusses their encapsulation in various shell materials where high encapsulation efficiency (>90 %), enhanced water solubility and significantly improved biological activity was reported by employing the nanoencapsulation process.

4.2 Vitamins, Surfactants, Proteins, Biocompatible Polymers, Lipids

Vitamins are group of organic substances, present in minute amounts in natural food stuffs that are essential to normal metabolism; insufficient amounts in the diet may cause deficiency diseases. Vitamins A, D, E and K are fat-soluble and vitamins C and all the B vitamins are water-soluble. Except Vitamin D and K these biochemicals cannot be synthesized *de novo* in animal kingdom, therefore, plants or natural food stuffs are the basic source (MedlinePlus). These are generally present in fruits, vegetables, milk products, legumes, meat, fish oil, etc. To prevent or cure the vitamin deficiency diseases a lot of attention has been given in recent times to develop nutraceutical foods incorporated with nanoencapsulated bioactive compounds of vitamins. In the following section, research trends focused on vitamin A and vitamin D are mentioned.

Vitamin A is a group of unsaturated organic compounds that includes all-trans retinol, retinal, retinoic acid many provitamin A carotenoids (particularly β-carotene). It is important for human body because of its antioxidant, anti-cancer, cardiovascular protective properties and to prevent or cure the night blindness (Albanes 1999; Edge et al. 1997; Erhatrt et al. 2003; Rock 1997). These trans-retinols form electron dense regions and are susceptible to all oxidative species or radicals produced by light with wavelength less than 415 nm. These compounds may undergo isomerization in aqueous solution (Carlotti et al. 2002; Loveday and Singh 2008). Moreover, their solubility in water is very less and in oil it is ~0.1 % (Borel et al. 1996). In view of this, nanoencapsulation in suitable carriers is a potential technique to overcome the aforementioned problems and provide stability to lipophilic compounds along with high water dispersibility. Table 4.3 lists the different techniques employed by researchers to nanoencapsulate vitamins and study their physico-chemical and biological properties.

4.2.1 β-Carotene-Surfactants

The nanoemulsion synthesized using oil-in-water (O/W) technique with high pressure homogenization has been used most widely to study the effect of various carriers on particle size of nanoparticle or nanocapsule, their stability and release profile of bioactive compound (Yuan et al. 2008; Silva et al. 2011). The bioactive compound β-carotene can be dispersed in organic solvent or medium chain triglyceride along with emulsifier like Tween 20/40/60/80 to form emulsion. However, the usage of organic solvents is not safe; therefore, it has to be removed from the final nanoemulsion. The physico-chemical parameters such as high speed homogenization time, high pressure homogenization rate, temperature, cycle, pressure and type/concentration of emulsifier greatly influence the particle size, storage stability of nanoemulsion and release profile of β-carotene. The response surface method and 2^3 factorial level designs indicate that with increase in temperature upto 65 °C and pressure 110–130 MPa the particle size of nanoemulsion decreases but a higher temperature degrades the stability.

Table 4.3 Nanoencapsulation of carotenoids, vitamins within different shell materials and physico-chemical, biological activities tested (MCT-medium chain triglycerides)

Nanoparticle synthesis process	Core-shell material	Physico-chemical parameters/Activity tested	References
Oil-in-water Homogenization	β-carotene-MCT/ Tween20/40/60/80	Optimization of core-shell concentration using RSM	Yuan et al. (2008)
High energy emulsification evaporation	β-carotene/hexane/ Tween20	Stability & release profile	Silva et al. (2011)
Maillard reaction	β-carotene-Casein-graft-dexter copolymer	Stability & release profile	Pan et al. (2014)
Nanaoencapsulation	Carotenoids- Cationic/ anionic polyelectrolyte		Gheith et al. (2014)
Multilamellar liposomes	Retinol- phospholipid/α--tocopherol	Stability under varying temperature, light and pH condition	Lee et al. (2002)
Homogenization evaporation	β-carotene-sodium caseinate, whey protein isolate, soybean protein isolate	Antioxidant activity on Caco-2 cells	Yi et al. (2015)
High pressure homogenization	β-carotene-barley protein nanoparticle matrix	Bioavailability- Simulated Intestinal fluid Toxicity -in vitro	Yang et al. (2014)
Hot melt homogenization, ultrasonication	Trans-retinol- Palm oil based solid lipid nanoparticle	Antioxidant	Jee et al. (2006)
Casein self-assembled micelles	Vitamin D_2- bovine caseins	Protection against UV-light-induced degradation bioavailablity of vitamin D2 (clinical study)	Semo et al. (2007); Livney and Douglas (2014) Haham et al. (2012)
Nanoprecipitation technique	Calcidiol/calcitriol- Poly-lactic acid	Growth inhibitory efficiency on human breast adenocarinoma cells *in vitro*	Almouazen et al. (2013)
Ionic gelation method	Fucoxanthin- chitosan/ sodium tripolyphosphate/ glycolipid nanogels	Antioxidant activity and stability	Hindupur and Vallikannan (2015)

4.2.2 *β-Carotene-Proteins & Starch*

Yi et al. (2015) studied the physico-chemical property of β-carotene-sodium caseinate nanoemulsion and used the Stokes settling velocity equation to predict the stability of the same. Later, they encapsulated β-carotene by sodium caseinate, whey protein isolate, and soybean protein isolate by the homogenization-evaporation method forming nanoparticles of 78, 90 and 370 nm diameter (Yi et al. 2015). The nanoencapsulated products were reported to have high *in vitro* antioxidant activity and cellular antioxidant activity compared to free β-carotene.

Chemically treated Caco-2 cells were used to ascertain cellular antioxidant activity and the data was used to calculate the medium effective dose values for free β-carotene, sodium caseinate/whey protein isolate/soybean protein isolate-encapsulated β-carotene nano-particles (14.40, 13.02, and 17.11 mg/mL, respectively). This indicated the cellular antioxidant activity of β-carotene was improved after encapsulation and advantageous release property was observed by using whey protein isolate as shell material.

Yang et al. (2014) encapsulated β-carotene in barley protein nanoparticle matrix and studied its bioavailability, toxicity in simulated intestinal fluid, *in vitro*, respectively. The solubility of β-carotene was also studied by incorporating it in (i) Casein-*graft*-dextran copolymer produced using the Maillard reaction (Pan et al. 2007) where, β-carotene and casein formed the core and dextran formed the shell, and (ii) nano-composites of dextrin of high amylose corn starch prepared through a blending process with immiscible phases of aqueous dextrin and organic β-carotene solutions (Kim et al. 2013). A United States patent mentioned nanoencapsulation of carotenoids in cationic/anionic polyelectrolyte resulting in formation of capsule matrix in the size range 10–1000 nm (Gheith et al. 2014).

4.2.3 All-Trans Retinol-Lipids Nanoparticles

Solid lipid nanoparticles consist of a core solid lipid with the bioactive being a part of lipid ingredient matrix. Surfactants are used to stabilize the particles and crystallized lipids increase the control of release (Weiss et al. 2008). Loading of all-trans retinol into the solid matrix palm oil based solid-lipid nanoparticles was reported to be an effective means to protect them against chemical degradation and enhance stability. Various physicochemical properties of all-trans retinol-loaded solid-lipid nanoparticles, including mean particle diameter and zeta potential, were modulated by changing the total amount of eggPhosphatidylcholine, Tween 80 and their mixing ratio surfactant. The mean particle diameter and zeta potential of the smallest solid-lipid nanoparticles were 96 nm and –28 mV, respectively. The co-loading of antioxidants like butylate hydroxyanisol, butylate hydroxytoluene, vitamin C, α-tocopherol improved the stability of all-trans retinol loaded solid-lipid nanoparticles while the physico-chemical parameters were not affected. The encapsulation efficiency of vitamin A was significantly enhanced from 74 % to 100 % with the usage of antioxidants where the maximum efficiency was achieved with the butylate hydroxyanisol-butylate hydroxytoluene (Jee et al. 2006). However, the reason behind higher stability and encapsulation efficiency was to be ascertained. Lee et al. (2002) incorporated retinol into multilamellar liposomes that were prepared from soybean phosphatidylcholine and encapsulation was achieved upto 98.14 +/– 0.93 % at pH 9.0 at a ratio of 0.01:1 (wt:wt) retinol:phospholipid. The time course of the retinol degradation in the aqueous solution in liposomes was compared to that of free retinol and free retinol with alpha-tocopherol under a variety of conditions of pH (3, 7, and 11), temperature (4, 25, 37, and 50 °C), and light exposure (dark, visible, and UV). The protective effect of the liposome incorporation was greater in solid gel phase at low temperatures, neutral, high pH and in the dark. This effect was attributed to lower molecular oxygen permeability in the solid gel phase compared to the fluid liquid crystalline phase of lipids.

4.2.4 *Vitamin D, Proteins, Biocompatible Polymer*

Casein micelles are generally regarded as safe materials and naturally created nanocapsules. To harness its potential fat-soluble vitamin D_2 (essential for calcium metabolism) was encapsulated in self-assembled bovine caseins (Semo et al. 2007; Livney and Douglas 2014). The vitamin was about 5.5 times more concentrated within the micelles than in the serum where it was present lightly bonded to residual soluble caseins. The study showed that re-assembled casein micelles could provide partial protection against ultraviolet light-induced degradation to vitamin D_2 and the product could be used as nutraceuticals within food products. A clinical study in a randomized double blinded placebo controlled with 87 human volunteers showed high bioavailablity of vitamin D_2 (Haham et al. 2012). Calcitriol (1,25-dihydroxyvitamin D_3) an active metabolite of Vitamin D_3 is a potential anticancer agent but hypercalcemia limits the achievement of effective serum concentration. Almouazen et al. (2013) nanoencapsulated the bioactive compound in biodegradable polymer (poly-lactic acid) and tested its growth inhibitory efficiency on human breast adenocarinoma cells (MCF-7) *in vitro*. Calcidiol/calcitriol loaded nanoparticles had good encapsulation efficiencies (around 90 %) associated with sustained releases over 7 days and enhanced stability. Moreover, loaded nanoparticles showed similar growth inhibition to nonencapsulated metabolites of vitamin D_3 on day 4 and higher activities on days 7 and 10 after treatment initiation. A US patent application (Mousa 2013) also mentions incorporation of vitamin D in modified chitosan (biocompatible polymer) nanoformulation and a method of using the nanoformulation to treat a disorder and improve efficacy of current therapies where resistance develops in a patient who is administered with a therapeutically effective amount of the nanoformulation for treating the disorder.

4.2.5 *Fucoxanthin-Biocompatible Polymer*

Apart from vitamin A, D and β-carotene there is a non-provitamin A, Fucoxanthin, which is also a carotenoid and known for antioxidant and other activities. The nanoencapsulation of fucoxanthin in different ratios of chitosan/sodium tripolyphosphate/glycolipid nanogels was done by ionic gelation method (Hindupur and

Vallikannan 2015). The glycolipid offered an enhanced stability ($t_{1/2}$, 45 h) to fucoxanthin nanogels with chitosan as compared to the one with no glycolipid ($t_{1/2}$, 15 h) and standard fucoxanthin ($t_{1/2}$, 5 h). The bioavailability of fucoxanthin encapsulated in chitosan nanogels with glycolipid was higher (~68 %) compared to nanogels without glycolipid (51 %), fucoxanthin with glycolipid (31.5 %) and native fucoxanthin as control (21.5 %).

In short, the stability, bioavailability and biological properties of hydrophobic fat soluble carotenoids and vitamins can be improved by nanoencapsulation in oil-surfactant/protein nanoemulsion, casein micelles, medium chain triglycerides surfactant based solid lipid nanoparticles, soybean phosphatidylcholine based multilamellar liposomes and modified biodegradable polymers. The choice of shell material governs the incorporation in food matrices, the release profile of the bioactive compound and the retention period of nanoparticle in the biological system.

4.3 Omega-3 Fatty Acid, Polysaccharides, Proteins, Biocompatible Polymer

Omega-3 fatty acids (also called ω-3 fatty acids or *n*-3 fatty acids are polyunsaturated fatty acids with a double bond (C=C) at the third carbon atom from the end of the carbon chain. The fatty acids have two ends, the carboxylic acid (–COOH) end, which is considered the beginning of the chain, i.e., "alpha", and the methyl (CH_3) end, which is considered the "tail" of the chain, i.e., "omega." It is named as omega as the position of first double bond is counted from the methyl or omega end. The three types of omega-3 fatty acid are α-Linolenic acid (found in plant oils like walnut oil, flaxseed oil), Eicosapentaenoic acid and docosahexaenoic acid both commonly found in marine oils, fish oils, egg oil, etc. Omega-3 fatty acids are important for normal metabolism (Anon 2005) and many health benefits have also been attributed to it (particularly docosahexaenoic acid) such as anti-thrombotic (Tamura et al. 1992), anti-inflammatory (Ren et al. 2007), reducing risk of cardiac/circulatory disorder (GISSI 1999), cognitive control and aids neural and brain development (Lauritzen et al. 2000; McCann and Ames 2005), however, contradictory reports are also available (Sala-Vila and Calder 2011; Rizos et al. 2012). Since the conversion of shorter-chain omega-3 fatty acid α-Linolenic acid (18 carbons and 3 double bonds), to eicosapentaenoic acid (20 carbons and 5 double bonds) and finally to docosahexaenoic acid (22 carbons and 6 double bonds) is limited in mammals due to low activity of rate limiting enzyme Δ-6desaturase (Carlier et al. 1991), it is important to fortify our food with this compound. Studies related with physicochemical parameter of nanoencapsulated with omega-3-fatty acid and fortified food are mentioned below.

4.3.1 Omega-3 Fatty Acid-Polysaccharide, Protein

In 2007, Borneo et al. reported incorporation of 400 mg of eicosapentaenoic acid and docosahexaenoic acid encapsulated in starch and pectin matrix, in cream-filled sandwich cookies and successfully tested the stability of fortified food in different atmospheric and packaging conditions without affecting its sensory properties. The maximum 5 % loss was observed after 28 days of storage. Later Zimet and Livney (2009) and Zimet et al. (2011) used β-lactoglobulin along with low methoxyl pectin and re-assembled casein micelles to encapsulate docosahexaenoic acid. The addition of pectin in higher amounts created negatively charged particles that contained ~166 times higher docosahexaenoic acid concentration than the surrounding serum. This enabled the formation of dilutable nanoparticle dispersions, which formed transparent solutions containing 0.05 % β- lactoglobulin and docosahexaenoic acid at a 1:2 (β- lactoglobulin:docosahexaenoic acid) molar ratio, with a very good colloidal stability and average particle size of ~100 nm. The entrapment by β-lactoglobulin and nanocomplexes with the pectin provided good protection against degradation of docosahexaenoic acid during an accelerated shelf-life stress test (about 5–10 % loss during 100 h at 40 °C compared to ~80 % loss of unprotected docosahexaenoic acid). The addition of calcium and phosphate decreased the particle size of docosahexaenoic acid loaded casein from 288 nm to 50 nm. The product thus formed could be used for enrichment of clear acid drinks.

Gökmen et al. (2011)) enriched the bread with functional molecules i.e., omega-3 fatty acids. High amylase corn starch was used to form nanosized complexes with flax seed oil that was converted to powder of microparticles by spray drying. The particles incorporated into bread formulation at different amounts to investigate their effects on bread quality characteristics showed significant decrease in lipid oxidation due to encapsulation. Increasing the amount of particles in dough significantly decreased the formation of acrylamide and hydroxymethyl furfural in breads. Scanning electron microscopic analysis of bread demonstrated that particles added to dough remained intact in the crumb, but partially destroyed in the crust. Comparing to its free form, addition of nanoencapsulated flax seed oil increased final product quality and safety by lowering lipid oxidation and formation of harmful compounds in breads during baking.

The high pressure homogenization is the widely used technique for 'nano' formation but other techniques have also been investigated. Jafari et al. (2008) compared the microfluidization and ultrasonication techniques to study the surface oil content and surface oil coverage of nanoencapsulated fish oil powder. Maltodextrin combined with surface-active biopolymer, modified starch or whey protein concentrate, at a ratio of 3:1 was used as shell material. The lowest uncapsulated surface oil was obtained with microfluidization method, mainly due to its capacity to produce emulsion in the nano-range (d_{43} of 210–280 nm).

4.3.2 Omega-3 Fatty Acid-Biocompatible Polymer

The biocompatible polymers (poly lactic acid) can also be used for the formation of nanocapsules of α-Linolenic acid (Habib et al. 2012). A modified emulsion-diffusion technique resulted in formation of nanocapsules with encapsulation efficiency of 93 %. Tween 20, Pluronic 60 and gelatin were tested as stabilizers; however, the method involved the usage of different organic solvents like acetone and ethyl acetate. Both particle size and zeta potential were influenced by the type of stabilizer, organic solvent and the organic to aqueous phase ratio. Acetone was superior to ethyl acetate, and Tween 20 was superior to each of Pluronic-60 and gelatin in obtaining smaller, less aggregated nanocapsules. An organic to aqueous phase ratio of 1:5 was shown to be more suitable for the formation of smaller nanocapsules, particularly when acetone was used as the organic solvent.

Although omega-3 fatty acids containing food products are widely available and useful, the sensory profile and dietary habits pose a limitation to its approval by a large section of society. Thus, development of a fortified food with nanoencapsulated omega-3 fatty acid helps to resolve aforementioned challenges and provide an opportunity to prevent or cure many health problems (Table 4.4) by using nutraceutical food to meet Food and Drug Administration recommendations.

4.4 Spice Bioactives, Essential Oils, Polysaccharides, Proteins, Biocompatible Polymer, Lipids

Spices are used worldwide in adequate amounts as a part of food for its seasoning and flavor properties. These special properties are, basically, attributed to the particular essential oil/oils present in the spices. Moreover, in countries like India and China the medicinal and preservative properties of spices are well documented in ancient literature and a lot of research and review articles have also been published world over (Aggarwal et al. 2006; Bharati et al. 2004; Suresh babu and Srinivasan 1998; Srinivasan 2014). It is well known that the spices in whole or powder form are susceptible to degradation caused by atmospheric conditions such as humidity, light, heat or microbes. The extraction of volatile/non-volatile oil can ameliorate the problems related with microbes, however, the stability of essential oils at room temperature and procedure to consume are major issues. Microencapsulation of essential oil has been extensively studied and reported; however, nanoencapsualtion of spice bioactives/essential oil is the current research trend for several advantageous reasons and to harness its potential for many pharmacokinetic applications (Ammon and Wahi 1991). The bioactive compound, curcumin, from the spice turmeric has gained lot of attention due to its anti-oxidant, antihepatoma activity (Yen et al. 2010), anti-inflammatory (Wang et al. 2008), hypocholesterolemic effect (Manjunatha & Srinivasan 2007), anti-diabetic (Suryanarayana et al. 2005), anti-arthritis (Liacini et al. 2002), chemopreventive (Surh 1999), anti-microbial

Table 4.4 Nanoencapsulation of omega-3 fatty acids with different shell materials and its effect on of fortified food

S.No	Synthesis process	Core-shell material	Studies of nanostructures/ fortified food	References
1	Conventional bakery process	EPA & DHA – starch and pectin matrix	Stability in different atmospheric and packaging conditions, sensory properties of sandwich cookies	Borneo et al. (2007)
2	Spontaneous binding method, casein micelle formation	DHA – β-lactoglobulin, low methoxyl pectin, casein micelles	Degradation studies during accelerated shelf-life stress	Zimet and Livney (2009), Zimet et al. (2011)
3	High pressurc homogenization	Omega-3 fatty acids-nanosized complexes of High amylase corn starch, flax seed oil	Formation of lipid oxidation products and thermal process contaminants in fortified bread	Gökmen et al. (2011)
4	Emulsion-diffusion technique	ALA – PLA, Tween 20, Pluronic 60, gelatin	Effect of organic solvents and stabilizer on particle size and stability	Habib et al. (2012)
5	Microfluidization, ultrasonication,	Fish oil – maltodextrin, biopolymer	Surface oil content and surface oil coverage of nanoencapsulated fish oil powder	Jafari et al. (2008)
	Spray drying			

EPA Eicosapentaenoic acid, *DHA* Docosahexaenoic acid, *ALA* α-Linolenic acid, *PLA* Poly lactic acid

(Bhawana et al. 2011), anti-cancer (Sahu et al. 2008) treating Alzheimer's disease (Yang et al. 2005) and wound healing properties (Phan et al. 2001) (Table 4.5).

4.4.1 *Curcumin-Polysaccharides/Biocompatible Polymer/ Lipids*

Garti et al. (2007) synthesized the nano-sized self-assembled concentrates, composed of an aqueous phase, an oil phase, a surfactant a co-solvent and co-surfactant, which could be diluted to any desired extent in either oil or water maintaining their structure and the active material. Later, curcumin was reported to be nanoencapsulated in pharmaceutically safe carriers (Neven et al. 2011; Sally and Greg 2007) and tested for anti-inflammatory property and Alzheimer's disease. The enhancement in bioavailability of nanoencapsulated curcumin was reported by nanoencapsulating in biocompatible polymers (Yallapu et al. 2010), organic materials such as dichloromethane (Bhawana et al. 2011) or Silica gel (Bitar et al. 2012) and different pharmacological activities of nanocurcumin were tested. Yen et al. 2010 assessed

Table 4.5 *In vitro* and *in vivo* Pharmacokinetic studies of curcumin encapsulated in different carriers

Encapsulation	Pharmacokinetic studies	Model	References
Curcumin-poly vinyl pyrollidene nanoemulsion	Antioxidant, antihepatoma	human hepatoma cell lines	Yen et al. (2010)
Curcumin-medium chain triglyceride, Tween 20 nanoemulsion	Anti-inflammatory	Mouse ear inflammation model	Wang et al. (2008)
Native curcumin	Hypocholesterolemic effect	Female wistar rats	Manjunatha et al. (2007b)
Native curcumin	Anti-diabetic	Rats	Suryanarayana et al. (2005)
Native curcumin	Anti-arthritis	*In vitro*	Liacini et al. (2002)
Native curcumin	Chemopreventive	*In vitro*	Surh (1999)
Curcumin-dichloromethane nanoemulsion	Anti-microbial	*In vitro*	Yamamoto et al. (2005)
Native curcumin	Alzheimer's disease	*In vivo*	Yang et al. (2005)
Curcumin-solid lipid nanoparticle		*In vitro, in vivo* (mice)	Sally and Greg (2007)
Native curcumin	Wound healing	*In vitro*	Phan et al. (2001)
Curcumin-casein micelles	Anti-cancer	*In vitro* HeLa cells	Sahu et al. (2008)

antioxidant and antihepatoma activity of nanocurcumin against human hepatoma cell lines. The particle size of nanocurcumin-in-polyvinyl pyrrolidone was 142 nm and the anticancer property of it was higher than dimethyl sulfoxide dissolved curcumin. The nanocurcumin-in-dichloromethane (Bhawana et al. 2011) exhibited higher antimicrobial activities against Gram-positive bacteria than Gram-negative bacteria. Furthermore, its antibacterial activity was much better than antifungal activity. The size of the nanocurcumin was in the range of 2–40 nm. The anticancer property of nanocurcumin-in-Si gel was much higher than its native counterpart. Gangwar et al. (2013) determined it against HeLa cell lines normal fibroblast. Xie et al. (2011) studied the release kinetics of nanocurcumin in poly-lactic glycolic acid *in vitro* and pharmacokinetics *in vivo*. After oral administration of nanocurcumin-polyglycolic acid, the relative bioavailability was 5.6-fold and had a longer half-life compared with that of native curcumin. The results showed that the effect in improving oral bioavailability of curcumin may be associated with improved water solubility, higher release rate in the intestinal juice, enhanced absorption by improved permeability, inhibition of P-glycoprotein-mediated efflux, and increased residence time in the intestinal cavity. The nanocurcumin encapsulated in organic/polymer based material could find potential in nutraceutical products as medicine but in food the regulations and choices are quite stringent. The use of appropriate food grade dispersing, emulsifying/stabilizing agent is crucial in deciding the particle size of nanocurcumin, its water solubility, release kinetics and therapeutic properties.

Yu and Huang (2012)) studied the metastable solubility of curcuminoids in medium chain triglycerides, coconut oil, canola oil and corn oil and observed that it was highest in corn oil. However, the lipolysis *in vitro* showed maximum bioavailability in medium chain triglycerides based nanocurcumin products. The monostearin, a generally regarded as safe material, was used as organo-gelator to form the food-grade organogel, in which the bioaccessibility of curcuminoids was not affected by the gel formation. It was concluded that this newly-developed organogel had the potential to be used for oral delivery of curcuminoids as well as incorporated in functional foods systems. Wang et al. (2008) used medium chain triglycerides and Tween 20 to prepare nanocurcumin products and studied the anti-inflammation properties. Apart from medium chain triglycerides as dispersing agent, the short chain triglycerides could be a better option because of presence of more number of polar groups on its surface. In such process, the usage of any organic solvents could totally be eliminated and a way to green chemistry approach could be given for food applications. The research work related to it is being carried out in our group and first part of it has been communicated (Rao and Khanum 2016).

4.4.2 Curcumin-Proteins

The nanoencapsulation of curcumin in casein (Sahu et al. 2008; Esmaili et al. 2011), caseinate (Pan et al. 2013), alginate materials (Song et al. 2012) and whey protein (Sneharani et al. 2010) are also gaining prominence because of their generally regarded as safe material status and being nanometer particle size range. The complex formation of nano–sized curcumin in bovine casein micelles was studied using fluorescence spectroscopy (Sahu et al. 2008) and it was found that the complexes were formed through hydrophobic interactions and the binding constant for the casein micelle-curcumin interaction was calculated to be $1.48 \times 10^4\,M^{-1}$. Cytotoxicity studies of HeLa cells revealed that the inhibitory concentration of free curcumin and the casein micelle-curcumin complex was 14.85 and 12.69 μM, respectively. Curcumin dissolved in permissible limits of ethanol was nanoencapsulated in β-casein (Esmaili et al. 2011) and sodium caseinate (Pan et al. 2013, 2014). The stability of curcumin-sodium caseinate nanoemulsion was studied with respect to variation in pH. The higher antioxidant and anticancer property of nanocurcumin was observed. Curcumin was encapsulated in camel β-casein also and critical micelle concentration of later was determined at 25, 30 and 37 °C using pyrene fluorescence and the solubility of curcumin was evaluated according to the solvent-evaporation technique. Presence of camel β-casein increased the solubility of curcumin at least 2500-fold (Esmaili et al. 2011). Marcolino et al. (2011) studied the complex formation of curcumin with cyclodextrin achieved by three methods *viz.* coprecipitation, kneading and mixing. It was observed that complexation of colorants with β-cyclodextrin promoted an intensification of color and increased water solubility. The sensorial profile of cheese and yogurt products incorporated with curcumin-β- cyclodextrin showed well accepted level. Therefore, the complexation of these natural colorants with β- cyclodextrin favors their use in low-fat foods, broadening the field of industrial application.

4.4.3 Curcumin-Lipids

Liposomes, small vesicles with phospholipid bilayer membrane, are potential vehicle because of their small size, biodegradability, hydrophobic and hydrophilic nature and low toxicity (Takeuchi et al. 2005). Chen et al. (2012) nanoencapsulated curcumin in liposomes prepared by using phospholipid, cholesterol and Tween 80. The particle size was in the range 82–92 nm and encapsulation efficiency 80–82 %. The antimelanoma activity (*in vitro*/*in vivo)* of different samples were studied, the curcumin loaded liposomes showed significant improvement against tumors.

4.4.4 Essential Oilss from Different Spices

Other essential oils like eugenol, thymol, carvacrol, p-cymene, *Nigella sativa* oil, garlic oil, cinnamon oil have also been reported to be nanoencapsulated and studied for their higher water solubility, bioavailability compared to their native counterparts. A review article by Bilia et al. (2014) presents a relevant literature about nanoencapsulation of essential oils, its limits and challenges for clinical purpose. Eugenol is an active component of clove essential oils and is used for its antibiotic, antiseptic and counter-irritation action in Indian medicine (Pruthi 1976). Choi et al. (2009) encapsulated eugenol as a volatile active substance by inclusion with β-cyclodextrin and 2-hydroxypropyl-b-cyclodextrin, and by an emulsion–diffusion method with polycaprolactone. Thermogravimetric analysis revealed the encapsulation efficiency of polycaprolactane, β-cyclodextrin eugenol and 2- hydroxypropyl-β-cyclodextrin eugenol inclusion complexes were 100 %, 90.9 % and 89.1 %, respectively. polycaprolactane encapsulated eugenol had higher oxidation stability compared to its other counterparts. Woranuch et al. (2013a, b) studied the high thermal stability and antimicrobial property of eugenol encapsulated in chitosan nanoparticles synthesized via an emulsion–ionic gelation crosslinking method and its application for food packaging applications. Chitosan nanoparticles were also used to encapsulate the oregano essential oils (Hosseini et al. 2013). The antimicrobial studies showed an initial burst effect followed by a slow drug release. The two phase release of essential oil, conataining either eugenol or transcinnamaldehyde, was also reported from polyglycolic acid nanoparticle (Gomes et al. 2011). The nanoparticles were prepared by ultrasoniacting organic phase, where poly vinyl alcohol was used as a surfactant. These nanoparticles are suitable for hydrophobic antimicrobial release carriers, with an entrapment efficiency ranging from 92 % to 98 % for both compounds and final mean size under 200 nm, for a 16 % (w/w) antimicrobial theoretical loading. These were expected to find applications in food packaging and direct food applications.

The poly-glycolic acid nanocapsules were also used to nanoencapsulate carvacrol (Iannitelli et al. 2011) with particle size ~200 nm. The encapsulation efficiency was 26 % and the loading capacity was 21 %. These were also studied for antimicrobial property. Zhang et al. (2014) encapsulated thymol in zein (corn prolamin protein) nanoparticle and studied the effect of soium caseinate, chitosan hydrochloride on the physico-chemical property of nanoparticles. The particle size of the nanoparticle increased from 118 nm to 200 nm with the addition of sodium caseinate. It also helped in stabilizing and water redispersibility of nanoparticle. The antimicrobial, antifungal activity of carvacrol and thymol against a food born pathogen (*Listeria monocytogenes)* were studied after nanoencapsulating in liposomes (Liolios et al. 2009). The bioavailability, endurance to humidity and ultra-violte light were also found to be enhanced (Coimbra et al. 2011).

The encapsulation of essential oils has also been done in solid lipid nanoparticles. Though the processing conditions are quite challenging, it imparts successful immobilization of active compound, chemical protection, less leakage and sustained release (Weiss et al. 2008). Alhaj et al. 2010 encapsulated *Nigella sativa* oil in solid lipid nanoparticles where, palm oil, softisan 154 and essential oils were used as lipid matrix and while sorbitol, water as surfactant and the particle size was in the range of 66–142 nm. Though they studied the dermal applications, it has potential in nutraceutical food industry due to antioxidant, anti-inflammatory, anticancer, analgesic, antimicrobial and medicinal properties.

4.5 Opportunities and Challenges

Opportunities are plenty for the research and development of clinically tailored nutraceutical foods that target particular diseases and different geographically based population. The growing awareness about the adverse effects of drugs and the health benefits obtained with consumption of nutraceutical or functional foods is well publicized and explicitly indicative of the demand/preferences for minerals/vitamins/essential bioactive-compounds incorporated foods rather than capsules or medicines. It was estimated in a report that the food encapsulation market might reach about €27.7 billion by the year 2015 (Anon 2010). The nutraceutical global market reached $ 142.1 billion in 2011 and expected to cross $ 250 billion by 2018 (ASA 2013). A worldwide surveys about the research activities, technology development and market assessments were carried out during the year 2009–2012, however, a statistical data to analyze the current status vis-à-vis earlier predictions are lacking. The products developed based on nanoencapsulation technologies are currently available in different parts of the world and some are listed in Table 4.6. The Nanoclusters™ system, such as Slim Shake Chocolate, contains silica nanoparticles coated with cocoa to enhance the chocolate flavor. NanoCluster™, from RBC Life

Table 4.6 Examples of functional/nutraceutical foods containing nanoencapsulated bioactive ingredients

Product name	Bioactive ingredients	Health claims	Manufacturer/country	References
Brook bond –Red label tea	Ursolic acid, Withaferin A, Glycyrrhetinic acid, 6-Gingerol	Vitality enhancement (Improvement in NK cell activity)	Unilever/India	Narayana (2008)
Becel pro-activ® Margarine	Plant sterol	Lowering cholesterol	Unilever	Narayana (2008)
Largo® Fortified juice	Inulin, l-carnitine, vitamins, Ca, Mg	–	Estonia	Alberta and Boye (2015)
Blue Band Goede Start Enriched bread	fibers, vitamins, B_1, B_3, B_6, Fe, Zn, Inulin	–	Unilever	Benkouider (2005)
Essentra® (Bars, candies, beverages)	Extract of ayurvedic botanical Aswagandha (Withania somnifera)	Improves energy, mental cognition, weight management, lowers stress, cardiovascular disease	Nutragenesis, USA	www.nutragenesis.com
OptiNutrin® (Candies, beverages)	Extract of Mailake mushrooms	Immune booster	Nutragenesis, USA	www.nutragenesis.com
Nano-curcumin 25 % Ingredient	Curcumin	Various pharmacological application	Konark herbal & healthcare, India	http://www.konarkherbal.com/curcumin.html
NUTRASORB™ PROCES Ingredient	Polyphenols, gingerols, beneficial alkaloids	Health promotion	Nutrasórb	Nutrasorb.com/process/
Yakult Fortified milk	*Lactobacillus casei Shirota*	Healthy gut	Yakult, Japan	Siró et al. (2008)

Sciences® Inc. (Irving, TX, USA), is another such delivery system for food products (Ranjan et al. 2014).

The US and Europe, currently, are the largest markets for nutraceutical products accounting for 36 % and 25 %, respectively, while country like India make just 2 % of global market. The nutraceutical market in India was estimated to grow to USD 2731 million in 2016 at a compound annual growth rate of 13 % (Biospectrum Asia Edition, 16 March 2012). Sastry et al. (2010) summarized the percentage of patents filed and various articles published in different food sectors. The nutraceutical nanofood witnessed 6 % of patents and 4 % of articles published in nano-food sector. This indicated that the initial research results were protected even before the product development and commercialization (Dasgupta et al. 2015). The venture capital investments for research and product commercialization in both developed and developing countries are quite promising where governments are tackling problems related with high-calorie and low nutritional food, malnutrition in children and pregnant/lactating females, aging population, life threatening diseases etc. The examples of currently available functional or nutraceutical foods include yogurts enriched with prebiotics/probiotics, breakfast foods/cereals/snacks fortified with vitamins/minerals/fibers/proteins, fiber rich sports drinks for energy restoration, dairy products like cheese, spreads containing phytosterols for lowering cholesterol and docosohexanoic acid/eicosapentaenoic acid fortified baked goods for vegans (Alberta and Boye 2015; Siró et al. 2008). Major companies like Nestlé, Unilever, pepsico, Coca-cola, DANONE, Nutragenesis and Amway are making significant investments in nutraceutical foods and beverages. Several beverages fortified with probiotics and foods enriched with vitamins, minerals, dietary fibers are commercially available in the market as one stop solution, owing to their multi-functional health benefits. Becel pro-active® was the first margarine enriched with plant sterol developed by Unilever in 1980s. In late 2003, Unilever introduced white bread enriched with fibers, vitamins, B_1, B_3, B_6, Fe, Zn, Inulin, the nutritione normally available in brown bread (Benkouider 2005). Today, Unilever India is providing tea fortified with bioactives of herbs under the product name Red Label Tea (Narayana 2008). The Freshlay Foods are providing VITA eggs by enriching with ω-3 fatty acid, Vit. D, E, $B_{12,}$ Folic acid and Selenium. An example of fortified beverage is Yakult, a milk product fortified with *Lactobacillus casei* Shirota. Various nano-encapsulated bioactive compounds are also commercially available and can be used in food products.

Challenges are profound as far as the incorporation of "nanoencapsulated natural material to food" is concerned. From research perspectives, it is important to have-

- A protocol to test the efficacy of the claimed product,
- A mandatory use of generally regarded as safe materials as encapsulating agents to provide controlled release through oral route,
- A complete study to highlight the residency period of nanofood in gastro-intestinal tract, penetration or absorption of bioactive compound in the body and its influence on biological process such as activation of antioxidant defenses,

Table 4.7 Recommended intake quantity of some of the bioactive compounds

Bioactive compound	Recommendation	Organization	Year	References
EPA and DHA	250 mg/d EPA and DHA for adults	European Food Safety Authority	2010	Gebauer et al. (2006)
	250 EPA and DHA mg/d plus 100–200 mg DHA for pregnant/lactating women	European Society of Cardiology	2003	Gebauer et al. (2006)
	1000 mg/d of ω-3 EAA/ DHA for patients with CHD			
Dietary or supplemntal Vitamin A as preformed Vitamin A (micrograms of RAE/day)	400–600 mcg/day (Infants/children) 700–900 mcg/day	Food & Drug Administration, USA	2001	Higdone (2003)
	(adults)			
Vitamin D	400 mcg (Infants)	Food & Nutrition Board of Institute of Medicine	2010	Higdone (2014)
	600 mcg (children/adults)			
Curcumin	3 mg/kg body weight/day	European Food Safety Authority	2010	GRAS Notice No. 460, 2013

EPA Eicosapentaenoic acid, *DHA* docosahexaenoic acid, *RAE* Retinol Activity Equivalents, *CHD* Coronary heart disease

signal transduction pathways, cell survival-associated gene expression, cell proliferation and differentiation and preservation of mitochondrial integrity (Gupta et al. 2013).

- A comprehensive knowledge about most suitable core-shell material combination for foods, beverages targeting the general well being or the chronic diseases.
- Sufficient clinical trials regarding intake concentration based on age, sex and health status, toxicity involved and internal or external applications of product are also essential.
- Apart from targeted-delivery of bioactives, such foods should be acceptable to consumer in terms of appearance, taste and texture.

Table 4.7 mentions the recommended intake quantity for different bioactive compounds; however, prescribed consumption per serving/day of nanoencapsulated nutraceutical food is scant and needs to be given a priority attention by the global food regulatory authority. Friends for the Earth have raised genuine concerns over potential health risk and have called for a moratorium on the commercial release of food products that contain nanomaterials until nano-specific safety laws are established and the public is involved in decision making (www.foeeurope.org). It will be a challenge to spread awareness in distant villages, to educate the people about

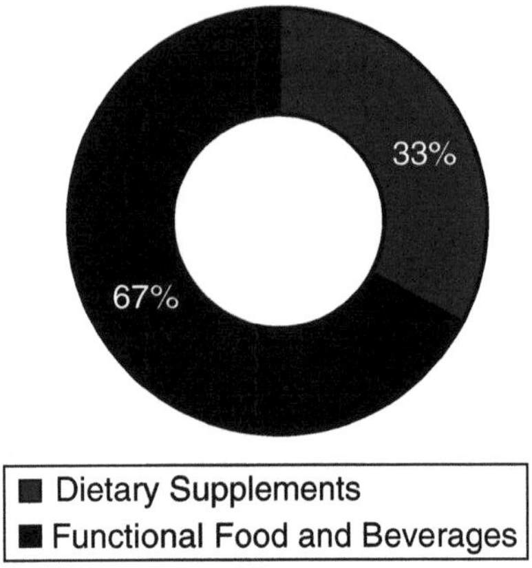

Fig. 4.3 Functional food market share growth in India (Source: http://www.frost.com/prod/servlet/cio/236145272)

frequent use of nanomaterial incorporated food as the associated risk factor cannot be ignored to avoid/manage potential hazard to surface water or soil after disposal or excretion of nano-based foods.

4.6 Future Trends

Future prospect of nutraceutical food development lies in a reasonable merger of food and pharmaceutical companies, particularly in developing countries in an organized way to target different genre of society. The pharmaceutical companies' interest in nutraceutical market would provide necessary research base to food companies to substantiate the claims and fulfill the quality parameter norms as set by food and drug administration or other organizations and translate the product from basic food to special disease preventive/curative food.

In the long run, nutraceutical market could also influence the funds granted to health sector for expensive medicines. The progress is expected to lead to a cheaper and less time consuming research for development of health beneficial nutraceutical products. The global scenario is expected to have highest market for functional beverages approximately USD 87 billion, followed by food based products approximately USD 67 billion and around USD 51 billion for supplement sectors at compound annual growth rate of 8.8 %, 6.4 % and 4.8 %, respectively, during the period from 2011 to 2016 (BCC Research 2011). The Indian market is expected to have functional food and beverages as major share compared to supplements (Fig. 4.3).

4.7 Conclusion

Novel nano-encapsulation technologies for food have been studied in the last decade for enhanced bioavailability and efficacy of bioactive compound in treating diseases or maintaining the well being of human population, increased stability and storage of food products due to anti-oxidant and anti-microbiological nature of bioactives, masking the unpleasant taste/odor of some pungent compounds, etc. Of them some products have already entered the market and are poised to make a big impact in the coming years. The economics and market growth will be governed by clinical trials, substantiated health claims, and targeted population as per geographical conditions and the cost and affordability of the nutraceutical products. The potential health benefits or risk of the nutraceutical products are the key factors to be addressed by following systematic and enormous scientific studies.

References

Aggarwal BB, Bhatt ID, Ichikawa H, Ann KS, Sethi G, Sandur SK, Natarajan C, Seeram N, Shishodia S (2006) Curcumin- biological and medicinal properties. p 297. www.indsaff.com/10%20curcumin%20biological.pdf. Retrieved 17 December 2014

Albanes D (1999) Beta-carotene and lung cancer study. Am J Clin Nutr 69:1345S–1350S

Alberta NA, Boye JI (2015) Current and emerging trends in the formulation and manufacture of nutraceuticals and functional food products. In: Boye JC (ed) Nutraceutical and functional food processing technology, Ith edn. WILEY Blackwell, Chichester, pp 1–52

Alhaj NA, Shamsudin MN, Alipiah NM, Zamri HF, Bustamam A, Ibrahim S, Abdullah R (2010) Characterization of *Nigella sativa* L. essential oil-loaded solid lipid nanoparticles. Am J Pharmacol Toxicol 5(1):52–57

Almouazen E, Bourgeois S, Jordheim LP, Fessi H, Briançon S (2013) Nano-encapsulation of Vitamin D_3 active metabolites for application in chemotherapy: formulation study and in vitro evaluation. Pharm Res 30:1137–1146

Ammon HPT, Wahi MA (1991) Pharmacology of Curcuma longa. Planta Med 57:1–7

Anon (2002) Sami Labs – pioneer in nutraceuticals. The Hindu Newspaper, Aug 05, 2002. http://www.hinduonnet.com/thehindu/biz/2002/08/05/stories/2002080500040200.htm. Retrieved 29-04-2015

Anon (2005) Omega-3 fatty acids and health: fact sheet for health professionals, US National Institutes of Health, Office of Dietary Supplements. http://ods.od.nih.gov/factsheets/Omega3FattyAcidsandHealth-HealthProfessional/. Retrieved 12 April 2014

Anon (2010) Food encapsulation: a global strategic business report. Global Industry Analysts, Inc., San Jose

Arnoldi A (2010) Nutritional pharmacy. http://www.vub.ac.be/MICH/eafp/eafp_pages/archive/pdfpres/catania/Arnoldi.pdf

ASA & Associates (2013) http://www.asa.in/pdfs/surveys-reports/Nutraceutical-Products-in-India.pdf

BCC Research (2011) FOD013D nutraceuticals: global markets and processing technologies. http://www.bccresearch.com/market-research/food-and-beverage/nutraceuticals-markets-processing-technologies-fod013d.html, Published July 2011. Retrieved 15 May 2015

Benkouider C (2005) Dining with the Dutch. Functional foods and nutraceuticals. http://www.ffnmag.com/ASP/articleDisplay.asp?strArticleId=753&strSite=FFNSITE&Screen=CURRENTISSUE. Retrieved 15 May 2015

Bharati AC, Shishodia S, Reuben JM, Weber D, Alexanian R, Raj-Vadhan S, Estrov Z, Talpaz M, Aggarwal BB (2004) Nuclear factor-kappaB and STAT3 are constitutively active in CD138+ cells derived from multiple myelom patients and suppression of these transcription factors leads to apoptosis. Blood 103(8):3175–3184

Bhawana, Bansiwal RK, Buttar HS, Jain VK, Jain N (2011) Curcumin nanoparticles: prepration, characterization and antimicrobial study. J Agric Food Chem 59:2056–2061

Bilia AR, Guccione C, Isacchi B, Righeschi C, Firenzuoli F, Bergonzi MC (2014) Essential oils loaded in nanosystems: a developing strategy for a successful therapeutic approach. Evid Based Complement Alternat Med 2014:1–14, article ID 651593

Bitar A, Ahmad NM, Fessi H, Elasissari A (2012) Silica based nanoparticles for biomedical applications. Drug Discov Today 17:1147–1154

Borel P, Grolier P, Armand M, Partier A, Lafont H, Lairon D, Azais-Braesco V (1996) Carotenoids in biological emulsions: solubility, surface-to-core distribution, and release from lipid droplets. J Lipid Res 37(2):250–261

Borneo R, Kocer D, Ghai G, Tepper BJ, Karwe MV (2007) Stability and consumer acceptance of long-chain omega-3 fatty acids (Eicosapentaenoic Acid, 20:5, n-3 and Docosahexaenoic Acid, 22:6, n-3) in cream-filled sandwich cookies. J Food Sci 72(1):S49–S54

Carlier H, Bernard A, Caselli C (1991) Digestion and absorption of polyunsaturated fatty acids. Reprod Nutr Dev 31(5):475–500

Carlotti ME, Rosatto V, Gallarate M (2002) Vitamin A and vitamin A palmitate stability over time and under UVA and UVB radiation. Int J Pharm 240(1–2):85–94

Chen H, Weiss J, Shahidi F (2006) Nanotechnology in nutraceutical and functional foods. Food Technol 60(3):30–36

Chen Y, Wu Q, Zhang Z, Yuan L, Liu X, Zhou L (2012) Preparation of curcumin-loaded liposomes and evaluation of their skin permeation and pharmacodynamics. Molecules 17:5972–5987

Choi MJ, Soottitantawat A, Nuchuchua O, Min SG, Ruktanonchai U (2009) Physical and light oxidative properties of eugenol encapsulated by molecular inclusion and emulsion diffusion method. Food Res Int 42(1):148–156

Cockbill CA (1994) Food law and functional foods. Br Food J 96:3–4

Coimbra M, Isacchi B, Bloois LV, Torano JS, Ket A, Wu X, Broere F, Metselaar JM, Rijcken CJF, Storm G, Bilia R, Schiffelers RM (2011) Improving solubility and chemical stability of natural compounds for medicinal use by incorporation into liposomes. Int J Pharm 416(2):433–442

Dasgupta N, Ranjan S, Mundekkad D, Ramalingam C, Shanker R, Kumar A (2015) Nanotechnology in agro-food: from field to plate. Food Res Int 69:381–400

De Felice LS (1995) The nutraceutical revolution, its impact on food industry. Trends Food Sci Technol 6:59–61

Desai KGH, Park HJ (2005) Recent developments in microencapsulation of food ingredients. Dry Technol 23:1361–1394

Edge R, McGarvey DJ, Truscott TG (1997) The carotenoids as antioxidants - a review. J Photochem Photobiol B Biol 41:194–197

Erhatrt JG, Meisner C, Bode JC, Bode C (2003) Lycopene, beta-carotene and colorectal adenomas. Am J Clin Nutr 78:1219–1224

Esmaili M, Ghaffari M, Moosavi-Movahedi Z, Atri MS, Sharifizadeh A, Farhadi M, Yousefi R, Chobert JM, Haertle T, Moosavi-Movahedi AA (2011) Beta casein-micelle as a nano vehicle for solubility enhancement of curcumin; food industry application. LWT Food Sci Technol 44:2166–2172

Esther B (2000) What is nutraceutical? Pharml J 265:57–58

European Nutraceutical Association (ENA). The term nutraceutical. http://www.enaonline.org/index.php?lang=en&path=about. Retrieved 1 June 2015

Gangwar RK, Tomar GB, Dhumale VA, Zinjarde S, Sharma RB, Datar S (2013) Curcumin conjugated silica nanoparticles for improving bioavailability and its anticancer applications. J Agric Food Chem 61:9632–9637

Garti N (2008) Starch as encapsulation material to control digestion rate in the delivery of active food component. In: Garti N (ed) Delivery and controlled release of bioactives in foods and nutraceutical. Woodhead Publishing Limited, Cambridge, MA, pp 279–291

Garti N, Aserin A, Spernath A, Amar KS (2007) Nano-sized self-assembled liquid dilutable vehicle. US patent – US 7,182,950 B2

Gebauer SK, Psota TL, Harris WS, Kris-Etherton PM (2006) ω-3 fatty acid dietary recommensation and food sources to achieve essentiality and cardiovascular benefits. Am J Clin Nutr 83(6):S1526–S1535

Gheith MK, Fang YL, Wong MS (2014) Method for nanoencapsulation. US patent – US 8,663,690 B2

GISSI (1999) Dietary supplementation with ω-3 polyunsaturated fatty acids and vitamin E after myocardial infarction: results of the GISSI-Prevenzione trial. Gruppo Italiano per lo Studio della Sopravvivenza nell'Infarto miocardico. Lancet 354(9177):447–455

Gökmen V, Mogol BA, Lumaga RB, Fogliano V, Kaplun Z, Shimoni E (2011) Development of functional bread containing nanoencapsulated omega-3 fatty acids. J Food Eng 105:585–591

Gomes C, Moreira RG, Castell-Perez E (2011) Poly (DL lactide-co-glycolide) (PLGA) nanoparticles with entrapped trans-cinnamaldehyde and eugenol for antimicrobial delivery applications. J Food Sci 76(2):N16–N24

GRAS Notice (GRN) No. 460. (2013) http://www.fda.gov/Food/IngredientsPackagingLabeling/GRAS/NoticeInventory/default.htm, http://www.fda.gov/ucm/groups/fdagov-public/@fdagov-foods-gen/documents/document/ucm346902.pdf. Retrieved 15 May 2015

Gupta SK, Yadav SK, Patil SMM (2013) Nutraceutical–a bright scope and opportunity of Indian healthcare market. Int J Res Dev Pharm Life Sci 2(4):478–481

Guri A, Gülseren I, Corredig M (2013) Utilization of solid lipid nanoparticles for enhanced delivery of curcumin in cocultures of HT29-MTX and Caco-2 cells. Food Funct 4:1410–1419

Habib SM, Amr AS, Hamadneh IM (2012) Nanoencapsulation of Alpha-linolenic acid with modified emulsion diffusion method. J Am Oil Chem Soc 89:695–703

Haham M, Ish-Shalom S, Nodelman M, Duek I, Segal E, Kustanovich M, Livney YD (2012) Stability and bioavailability of vitamin D nanoencapsulated in casein micelles. Food Funct 3:737–744

Higdone J (2003) lpi.oregonstate.edu/mic/vitamins/vitamin-A#RDA

Higdone J (2014) lpi.oregonstate.edu/mic/vitamins/vitamin-D

Hindupur R, Vallikannan B (2015) Biodegradable chitosan-glycolipid hybrid nanogels: a novel approach to encapsulate fucoxanthin for improved stability and bioavailability. Food Hydrocoll 43:717–725

Hosseini SF, Zandi M, Rezaei M, Farahmandghavi F (2013) Two-step method for encapsulation of oregano essential oil in chitosan nanoparticles: preparation, characterization and in vitro release study. Carbohydr Polym 95(1):50–56. http://www.foeeurope.org/activities/nanotechnology/Documents/Nano_food_report

Iannitelli A, Grande R, Stefano AD, Giulio MD, Sozio P, Bessa LJ, Laserra S, Paolini C, Protasi F, Cellini L (2011) Potential antibacterial activity of carvacrol-loaded poly(DL-lactide-co-glycolide) (PLGA) nanoparticles against microbial biofilm. Int J Mol Sci 12(8):5039–5051

Jafari SM, Assadpoor E, Bhandari B, He Y (2008) Nano-particle encapsulation of fish oil by spray drying. Food Res Int 41(2):172–183

Jee JP, Lim SJ, Park JS, Kim CK (2006) Stabilization of all-trans retinol by loading lipophilic antioxidants in solid lipid nanoparticles. Eur J Pharm Biopharm 63(2):134–139

Kalra EK (2003) Nutraceutical-definition and introduction. Am Assoc Pharm Sci J 5(3):27–28

Kawashim Y (2001) Nanoparticulate systems for improved drug delivery. Adv Drug Deliv Rev 47:1–2

Kim JY, Seo TR, Lim ST (2013) Preparation of aqueous dispersion of beta-carotene nano-composites through complex formation with starch dextrin. Food Hydrocoll 33:256–263

Koch A, Brandenburger S, Türpe S, Birringer M (2014) The need for a legal distinction of nutraceuticals. Food Nutr Sci 5:905–913

Konark Herbal and Health Care (2004) http://www.konarkherbal.com/curcumin.html

Lauritzen I, Blondeau N, Heurteaux C, Widmann C, Romey G, Lazdunski M (2000) Polyunsaturated fatty acids are potent neuroprotectors. EMBO J 19(8):1784–1793

Lee SC, Yuk HG, Lee DH, Lee KE, Hwang YI, Ludescher RD (2002) Stabilization of retinol through incorporation into liposomes. J Biochem Mol Biol 35(4):358–363

Liacini A, Sylvester J, Li WQ, Zafarullah M (2002) Inhibition of interleukin-1-stimulated MAP kinases, activating protein-1 (AP-1) and nuclear factor kappa B (NF-kappaB) transcription factors downregulates matrix metalloproteinase gene expression in articular chondrocytes. Matrix Biol 21(3):251–262

Liolios CC, Gortzi O, Lalas S, Tsaknis J, Chinou I (2009) Liposomal incorporation of carvacrol and thymol isolated from the essential oil of *Origanum dictamnus* L. and in vitro antimicrobial activity. Food Chem 112(1):77–83

Livney YD, Douglas D (2014) Casein micelles for nanoencapsulation of hydrophobic compounds. European patent – EP 2 007 212 B1

Loveday SM, Singh H (2008) Recent advances in technologies for vitamin A protection in food. Trends Food Sci Technol 19(12):657–668

Manjunatha H, Srinivasan K (2007) Hypocholesterolemic and antioxidant effect of curcumin and capsaicin in induced hyper-cholesterolemic rats. Lipids 42:1133–1142

Marcolino VA, Zanin GM, Durrant LR, Benassi MDT, Matioli G (2011) Interaction of curcumin and bixin with β-cyclodextrin: complexation methods, stability, and applications in food. J Agric Food Chem 59:3348–3357

McCann JC, Ames BN (2005) Is docosahexaenoic acid, an ω-3 long chain polyunsaturated fatty acid, required for development of normal brain function? An overview of evidence from cognitive and behavioral tests in humans and animals. Am J Clin Nutr 82(2):281–295

McClements DJ, Decker EA (2000) Lipid oxidation in oil-in-water emulsions: impact of molecular environment on chemical reactions in heterogenous food system. J Food Sci 65(8):1270–1282

Mousa SA (2013) Nanoformulation of vitamin D derivatives and/or vitamin D metabolites. Patent application – US 2013/0149385 A1

Narayana ADB (2008) Introduction of functional foods & nutraceuticals in the diet. 2nd international food regulatory summit. http://newsletters.cii.in/newsletters/foodsafety_qualityyear08/images/dba_narayana.pdf

Neven P, Serteyn D, Delarge J, Scheer E (2011) Water soluble curcumin compositions for use in anti-cancer and anti-inflammatory therapy. Patent application US 2011/0257126 A1

O'Rourke Eleanor (2011) AGRIFOOD. http://www.nanopinion.eu/sites/default/files/observatorynano_briefing_no_18_improving_delivery_of_essential_vitamins_minerals.pdf

Pan X, Yao P, Jiang M (2007) Simultaneous nanoparticle formation and encapsulation driven by hydrophobic interaction of casein-*graft*-dextran and *β*-carotene. J Colloid Interface Sci 315:456–463

Pan K, Zhong Q, Baek J (2013) Enhanced dispersibility and bioactivity of curcumin by encapsulation in casein nanocapsules. J Agric Food Chem 61:6036–6043

Pan K, Luo Y, Gan Y, Baek SJ, Zhong Q (2014) pH-driven encapsulation of curcumin in self assembled casein nanoparticles for enhanced dispersibility and bioactivity. Soft Matter 10:6820

Pandey M, Verma RK, Saraf SA (2010) Nutraceuticals: new era of medicine and health. Asian J Pharm Clin Res 3(1):11–15

Peppas NA (1997) Hydrogels and drug delivery. Curr Opin Colloid Interface Sci 2:531–537

Phan TT, See P, Lee ST, Chan SY (2001) Protective effects of curcumin against oxidative damage on skin cells in vitro: its implication for wound healing. J Trauma 51:927–931

Pruthi JS (1976) Spices and condiments, Ith edn. National Book Trust of India, New Delhi, p 91

Ranjan S, Dasgupta N, Chakraborty AR, Samuel SM, Chidambaram R, Kumar A, Rishi S (2014) Nanoscience and nanotechnologies in food industries: opportunities and research trends. J Nanopart Res 16(6):2464

Rao PJ, Khanum H (2016) A green chemistry approach for nanoencapsulation of bioactive compound – Curcumin. LWT Food Sci Technol 65:695–702

Ren J, Han EJ, Chung SH (2007) *In vivo* and *in vitro* anti-inflammatory activities of alpha-linolenic acid isolated from *Actinidia polygama* fruits. Arch Pharm Res 30(6):708–714

Rizos EC, Ntzani EE, Bika E, Kostapanos MS, Elisaf MS (2012) Association between Omega-3 fatty acid supplementation and risk of major cardiovascular disease events: a systematic review and meta-analysis. J Am Med Assoc 308(10):1024–1033

Rock CL (1997) Carotenoids: biology and treatment. Pharmacol Ther 75:185–197
Sahu A, Kasoju N, Bora U (2008) Fluorescence study of the curcumin-casein micelle complexation and its application as a drug nanocarrier to cancer cells. Biomacromolecules 9:2905–2912
Sala-Vila A, Calder PC (2011) Update on the relationship of fish intake with prostate, breast, and colorectal cancers. Crit Rev Food Sci Nutr 51(9):855–871
Sally AF, Greg MC (2007) Bioavailable curcuminoid formulations for treating alzheimer's disease and other age-related disorders. Patent application WO/2007/103435, PCT/US2007/005829
Sastry RK, Rashmi HB, Rao NH, Ilyas SM (2010) Integrating nanotechnology into agri-food systems research in India: a conceptual framework. Technol Forecast Soc Chang 77:639–648
Semo E, Kesselman E, Danino D, Livney YD (2007) Casein micelle as a natural nano-capsular vehicle for nutraceuticals. Food Hydrocoll 21:936–942
Silva DH, Cerqueira MA, Souza BWS, Ribeiro C, Avides MC, Quintas MAC, Coimbra JSR, Carneiro-da-Cunha MG, Vicente AA (2011) Nanoemulsions of beta-carotene using a high-energy emulsification–evaporation technique. J Food Eng 102:130–135
Siró I, Kápolna E, Kápolna B, Lugasi A (2008) Functional food: product development, marketing and consumer acceptance- a review. Appetite 51(3):456–467
Sneharani AH, Karakkat JV, Singh SA, Appu Rao AG (2010) Interaction of curcumin with β-lactoglobulin;stability, spectroscopic analysis, and molecular modeling of the complex. J Agric Food Chem 58:11130–11139
Song S, Wang Z, Qian Y, Zhang L, Luo E (2012) The release rate of curcumin from calcium alginate beads regulated by food emulsifiers. J Agric Food Chem 60:4388–4395
Srinivasan K (2014) Antioxidant potential of spices and their active constituents. Crit Rev Food Sci Nutr 54:352–372
Suresh babu P, Srinivasan K (1998) Ameliorationo of renal leasions associated with diabetes by dietary curcumin in streptozotocin diabetic rats. Mol Cell Biochem 181(1–2):87–96
Surh YJ (1999) Molecular mechanisms of chemopreventive effects of selected dietary and medicinal phenolic substances. Mutat Res 428:305–327
Suryanarayana P, Saraswat M, Mrudula T, Krishna TP, Krishnaswamy K, Reddy GB (2005) Curcumin and turmeric delay streptozotocin induced diabetic cataract in rats. Invest Ophthalmol Vis Sci 46:2092–2099
Takeuchi H, Matsui Y, Sugihara H, Yamamoto H, Kawashima Y (2005) Effectiveness of submicron-sized, chitosan-coated liposomes in oral administration of peptide drugs. Int J Pharm 303:160–170
Tamura Y, Hirai A, Terano T, Saitoh H (1992) Anti-thrombotic and anti-atherogenic action of eicosapentaenoic acid. Nippon Rinsho 50(2):403–407
The Bureau of Nutritional Sciences, of the Food Directorate of Health Canada (2002) Policy paper—nutraceuticals/functional foods and health claims on foods. http://www.hc-sc.gc.ca/fn-an/label-etiquet/claims-reclam/nutra-funct_foods-nutra-fonct_aliment-eng.php
Vitamins, MedlinePlus. Trusted health information for you, US National Institutes of Health. http://www.nlm.nih.gov/medlineplus/vitamins.html. Retrived 29 April 2015
Wadhwa M, Charles A (2011) Nutraceutical FICCI whitepaper, Frost & Sullivan. http://www.frost.com/prod/servlet/cio/236145272. Retrieved 29 April 2015
Wang X, Jiang Y, Wang YW, Huang MT, Ho CT, Huang Q (2008) Enhancing anti-inflammation activity of curcumin through O/W nanoemulsions. Food Chem 108:419–424
Weiss J, Takhistov P, McClements J (2006) Functional materials in food nanotechnology. J Food Sci 71(9):r107–r116
Weiss J, Decker EA, McClements DJ, Kristbergsson K, Helgason T, Awad T (2008) Solid lipid nanoparticles as delivery systems for bioactive food components. Food Biophys 3:146–154
Woranuch S, Yoksan R (2013a) Eugenol-loaded chitosan nanoparticles: I. Thermal stability improvement of eugenol through encapsulation. Carbohydr Polym 96:578–585
Woranuch S, Yoksan R (2013b) Eugenol-loaded chitosan nanoparticles: II. Application in bio-based plastics for active packaging. Carbohydr Polym 96:586–592

Xie X, Tao Q, Zou Y, Zhang F, Guo M, Wang Y, Wang H, Zhou Q, Yu S (2011) PLGA nanoparticles improve the oral bioavailability of curcumin in rats: characterizations and mechanisms. J Agric Food Chem 59:9280–9289
Yallapu MM, Gupta BK, Jaggi M, Chauhan SC (2010) Fabrication of curcumin encapsulated PLGA nanoparticles for improved therapeutic effects in metastatic cancer cells. J Colloid Interface Sci 351:19–29
Yamamoto Y, Aoyama S, Hamaguchi N, Rhi GS (2005) Antioxidative and antihypertensive effects of Welsh onion on rats fed with a high-fat highsucrose diet. Biosci Biotechnol Biochem 69:1311–1317
Yang F, Lim GP, Begum AN, Ubeda OJ, Simmons MR, Ambegaokar SS, Chen PP, Kayed R, Glabe CG, Frautschy SA, Cole GM (2005) Curcumin inhibits formation of amyloid beta oligomers and fibrils, binds plaques, and reduces amyloid *in vivo*. J Biol Chem 280(7):5892–5901
Yang J, Zhou Y, Chen L (2014) Elaboration and characterization of barley protein nanoparticles as an oral delivery system for lipophilic bioactive compounds. Food Funct 5:92–101
Yen FL, Wu TH, Tzeng CW, Lin LT, Lin CC (2010) Curcumin nanoparticles improve the physicochemical properties of curcumin and effectively enhance its antioxidant and antihepatoma activities. J Agric Food Chem 58:7376–7382
Yi J, Lam TI, Yokoyama W, Cheng WLW, Zhong F (2015) Beta-carotene encapsulated in food protein nanoparticles reduces peroxyl radical oxidation in Caco-2 cells. Food Hydrocoll 43:31–40
Yu H, Huang Q (2012) Improving the oral bioavailability of curcumin using novel organogel-based nanoemulsions. J Agric Food Chem 60:5373–5379
Yuan Y, Gao Y, Maoa L, Zhao J (2008) Optimisation of conditions for the preparation of β-carotene nanoemulsions using response surface methodology. Food Chem 107:1300–1306
Zhang Y, Niu Y, Luo Y, Ge M, Yang T, Yu L, Wang Q (2014) Fabrication, characterization and antimicrobial activities of thymolloaded zein nanoparticles stabilized by sodium caseinate-chitosan hydrochloride double layers. Food Chem 142:269–275
Zhong Y, Madhujit T, Mahfouz N, Shahidi F (2007) Compositional characteristics of muscle and visceral oil from steelhead trout and their oxidative stability. Food Chem 104:602–608
Zimet P, Livney YD (2009) Beta-lactoglobulin and its nanocomplexes with pectin as vehicles for ω-3 polyunsaturated fatty acids. Food Hydrocoll 23:1120–1126
Zimet P, Rosenberg D, Livney YD (2011) Re-assembled casein micelles and casein nanoparticles as nano-vehicles for ω -3 polyunsaturated fatty acids. Food Hydrocoll 25:1270–1276

Chapter 5
Biofortification of Food with Minerals and Vitamins Encapsulated in Silica

Sneha Singh, Nikita Rathi, Ashwini Angal, Prabhat Parida, and Debabrata Rautaray

Abstract Micronutrients such as minerals and vitamins are a class of food compounds that are essential for human health. Minerals and vitamins are indeed important for biological processes. The micronutrients actually occurring in dietary sources are not stable and get degraded during food processing and cooking. A major part of the population is suffering from vitamin deficiency due to change in the life style, food affordability, food habits, and food accessibility. As a consequence supplements have been made available in the market. Supplements are prepared by expensive formulations to increase vitamin bioavailability. Many additives are added to stabilize and enhance the delivery of active compounds. Last, people become reluctant to consume tablets because they do not trust what looks like pharmaceuticals. Hence, the alternative is the fortification of food, that is adding micronutrients to food. Examples include vitamin D fortified milk and iodized salt. For vitamins, biocompatible carrier molecules, such as silica, are needed because vitamins degrade easily in the presence of in light, pH, oxidation and reducing agents. Here we review the causes and consequences of micronutrient deficiencies, the bioavailability of nutrients, and vitamins, minerals and silica for food. We also studied of the release of nutrients entrapped in silica. Results show that the release of nutrients from silica in simulated intestinal fluid is better than in simulated gastric fluid.

Keywords Silica • Nutrition • Micronutrients • Vitamins • Minerals • Nutrient carrier • Nutrient release • Encapsulation • Gastro-intestinal tract

---

S. Singh • N. Rathi • A. Angal • P. Parida • D. Rautaray (✉)
TATA Chemicals Innovation Centre,
Survey No 315, Hissa No 1-14, Ambedveth (V), Paud road, Mulshi, 412111 Pune, Maharasthra, India
e-mail: snehi009@gmail.com; nikitarathi1989@gmail.com; ashwini.angal@tatachemicals.com; pkparida@tatachemicals.com; drautaray@tatachemicals.com

S. Ranjan et al. (eds.), *Nanoscience in Food and Agriculture 2*, Sustainable Agriculture Reviews 21, DOI 10.1007/978-3-319-39306-3_5

5.1 Introduction

The understanding that micronutrients are essential for human growth and that they have to be part of the diet was a major stimulus for nutrition science. The importance of vitamins led to the development of processes for their large scale production, formulation and incorporation in food fortification so that the product could be ultimately consumed by humans. This inspired pharmaceutical and nutritional companies to develop different food formulations. Industrial production was not the complete solution yet, and new challenges arose with the incorporation of micronutrient and nutraceuticals in end-user applications such as tablets, vitamin water, beverages, yoghurts and other foodstuffs, as the nutrients in them are often chemically unstable. The nutrients could be easily degraded in natural form via oxidation, hydrolysis, light and heat (Gharsallaoui et al. 2007). New technologies had to be explored to maintain the stability of the micronutrients. Therefore, new approaches have been invented for the delivery of micronutrient by using carrier molecules to protect them from natural environmental condition. Thereby the stability and shelf-life of the compounds are prolonged and they can be released in a controlled and tailored manner. Further benefits of formulation and encapsulation processes include easier handling, improved sensory properties with respect to appearance and taste and uniform dispersion of low-concentrated actives (Gouin 2004). The food administrations already setup the dietary standards and nutrient requirements with recommended dietary allowance for the optimal intake of vitamins and minerals depending on age, gender and risk groups. In order to secure a sufficient intake of nutrients in diet, a number of countries implemented fortification programs of staple food. Examples are the fortification of flour with folic acid in the US, Canada and Latin American countries, the fortification of margarine with vitamin A and D or the fortification of milk and juices with vitamin D and fortification of flour or sugar with vitamin A especially in low-income countries.

5.2 Micronutrient Deficiencies in Developing Countries

Malnutrition in developing countries among women, children and adolescents is an emergency, needing immediate attention if these countries have to have inclusive growth and development. Without good nutrition, neither communicable nor non-communicable diseases can be controlled. Malnutrition is the worst form of non-communicable disease and is important risk factor for chronic diseases at a later stage. Nutrition is one of the bases of judging a national development. Besides deficiencies of calories and protein, deficiency of micronutrients like vitamins and minerals are also widespread (Indian national nutrition monitoring bureau 2003, 2006). Micronutrients deficiency is referred to as the hidden hunger since often at times it is a hidden killer or crippler, and extracts heavy human and economic cost. These contribute significantly in biological functions, because they are needed for utilization of proteins and calories, they act as co-factors in functions of many enzymes (Bender 2003) and to fight infections from a young age. In addition to serious detrimental effects on bodily functions, the existence of malnutrition has profound implication on economic development and productivity in terms of huge public health costs. For example, Indian foods that are mostly cereal based are also deficient in micronutrient such as iron, folic acid, calcium, vitamin A, riboflavin. Among the micronutrient deficiencies, iron deficiency anaemia is the most serious public health problem (Indian national nutrition monitoring bureau, 2003, 2006). Estimates of iron deficiency anaemia in women and children have varied from 50 to 60 %; pregnant women being particularly susceptible. Iodine deficiency disease is another worrisome public health problem (National nutrition monitoring bureau 2006). Though, its magnitude has declined in recent years after the introduction of iodised salt, but the problem still persists, and not confined to the sub Himalayan regions as earlier thought. Fortunately, some of the severe vitamins- deficiency diseases such as beri-beri: thiamine-vitamin B1 deficiency, pellagra –niacin deficiency and scurvy –vitamin C deficiency has disappeared. Blindness due to vitamin A deficiency (Aykroyd 1930) and rickets due to vitamin D deficiency remain as clinical rather than public health problems. However, milder clinical manifestations and biochemical evidence of these deficiencies is rampant. Also osteoporosis in adults, particularly women after menopause due to calcium and vitamin D deficiency is common (Ross et al. 2011a, b).

5.3 Consequences of Micronutrients Deficiencies

The micronutrient deficiency results into severe depression related problem and mental illness. The affected person will develop poor appetite, anxiety. Nutrient such as amino acids, minerals, vitamin B, omega-3 fatty acids are known to be the precursors of neurotransmitters and various biological pathway and deficiency of these can lowers the secretion of the neurotransmitter with serious consequences of mental disorder (Sathyanarayana Rao et al. 2008). Micronutrient malnutrition is a major obstruction to socioeconomic development of the nation, which eventually has a deleterious effect over the underprivileged groups. Micronutrient malnutrition leads to high economic loss, reduction in the capability and potential of manpower, high rates of illness and mortality due to deficiency disease. It adversely affects human health, learning ability and productivity.

Apart from human suffering due to morbidity and mortality, malnutrition in general and micronutrient deficiencies in particular have high economic cost. Cost of treating malnutrition is 27 times more than investment required for its prevention (Leadership agenda 2010). According to a panel of Nobel laureates, the top 10 priorities selected for advancing global welfare using methodologies based on the theories of welfare economics, in Copenhagen Consensus 2008, five were in the area of nutrition – micronutrient supplements, micronutrients fortification, bio-fortifications, de-worming and other nutrient programs at school and community level (Copenhagen

Consensus 2008). These were also needed to achieve the millennium development goals.

5.4 Causes of Micronutrient Deficiency

The over exploitation of current agricultural resources has led to a development of nutrient deficient crops. The agricultural produce which traditionally had the necessary nutrients such as vitamins and minerals has progressively shown deficiency due to the inherently low nutrients in the soil or to the low application by farmer and low uptake by plants. The application of toxic chemical pesticides to the crops also affects the quality of soil and hinders the absorption of micronutrient by the plants and the crops remain nutrient deficient (Hornick 1992). Not only inherently low nutritional value of crops but also the lack of awareness of dietary rich food, malabsorption, & affordability are some of the major factors which responsible for the micronutrient deficiency.

5.5 Dietary Reference Intakes

Recommended dietary allowance is the average daily dietary nutrient intake level sufficient to meet the nutrient requirement of nearly 97–98 % healthy individuals in a particular life stage and gender group (Dietary Reference Intakes 2001). Adequate Intake is a recommended average daily intake level based on observed or experimentally determined approximations or estimates of nutrient intake by a group(s) of apparently healthy people, that are assumed to be adequate used when an recommended dietary allowance cannot be determined. Tolerable upper intake level is the highest average daily nutrient intake level that is likely to pose no risk of adverse health effects for almost all individuals in the general population. As intake increases above the upper intake level, the potential risk of adverse effect increases. Estimated average requirement is the average daily nutrient intake level estimated to meet the requirement of half of the healthy individuals in a particular life stage and gender group.

The recommended dietary allowance is derived from (i) the individual variability, and (ii) the nutrient bio-availability from the habitual diet. Individual variability: Definition of recommended dietary allowance takes into account the variability that exists in the requirement of a given nutrient between individuals in a given population group. The distribution of nutrient requirement in a population group is considered normal and the recommended dietary allowance corresponds to a requirement, which covers most of the individuals (97.5 %) in a given population. This is termed as a safe level of intake of a nutrient, that is, the chances of individuals having requirements above the recommended dietary allowance is only 2.5 %. This principle is used in case of all nutrients except energy, since in the case of energy, intakes either the excess or below the actual requirement of energy are not safe. In case of other nutrients the recommended dietary allowance is 25 % (+2SD) higher than the mean requirement, 12.5 % being considered as the extent of individual variability in the requirements of all those nutrients.

5.6 Bioavailability of Nutrients

The release of nutrient from food, its absorption in the intestine and bio response is the bioavailability of a given nutrient from a diet. This bioavailability factor is quite important in case of calcium and protein and trace elements like iron and zinc. In case of iron, the amount to be present in the diet is 20–30 times higher than the actual iron requirement to account for the low bio-availability of iron from a given diet, particularly a cereal-based diet. Recommended dietary allowance represents the level of the nutrient to be consumed daily to meet all the requirement of most of the individuals in a given population. However, it must be recognized that recommended dietary allowance is not meant to be used as standard to determine whether or not a given individual requirement has been met, since it is a level above the requirement of most individuals in a given population. Recommended dietary allowance value of a nutrient is valid only when all other dietary nutrient intakes are satisfactory.

5.7 Essential Nutrients and Importance of Vitamins and Minerals

Essential nutrients are nutrients that cannot be synthesized by the body, however it has an eminent role in normal body function and metabolism, therefore it must be obtained from dietary source. Categories of essential nutrients include vitamins, fats, essential amino acids and dietary minerals. Some essential nutrients may also be toxic when intake in high doses and can cause hyperactive diseases symptoms (Lichtenstein and Russell 2005).

Vitamins and minerals are the major constituent, which has the significant role in the metabolic and biological process of the body and also responsible for the reproductive function Hurley and Doane (1989). As vitamins are organic nutrient without caloric content and are essential for health, hence though minerals also perform diverse and critical roles in human health and disease. Vitamins can be obtained by dietary sources in adequate amount with the exception of vitamin D which are synthesized in the presence of sunlight. Dietary sources of vitamins, such as fruits and vegetables are preferable for vitamin and minerals supplements; perhaps this is because fruits and vegetables contain vitamins in useful proportions, as well as fiber and minerals, which together play a role in good nutrition and disease prevention

(Fragakis 2003). Vitamins come under two categories, fat soluble vitamins and water soluble vitamin. Both classes of vitamins are equally important for proper functioning of body.

What are vitamins? Vitamins are organic compounds which occur in very small quantities in food but are very important to life for specific regulatory functions and the maintenance of life and normal growth (Lieberman and Bruning 1990). All vitamins contain carbon, hydrogen, and oxygen but some of them contain nitrogen, sulfur and cobalt. Vitamins are measured also in terms of micrograms equivalent to one millionth (1/1,000,000) of a gram or in milligrams equivalent to one thousandth (1/1000) of a gram. Vitamins are divided into two groups: water-soluble (B-complex vitamins and C vitamins) and fat-soluble vitamins (A, D, E and K). Unlike water-soluble vitamins that need regular replacement in the body (Fukuwatari and Shibata 2008) fat-soluble vitamins are stored in the liver and fatty tissues, and are eliminated much more slowly than water-soluble vitamins (Maqbool and Stallings 2008). Complete classification and details of the vitamins are shown in Table 5.1.

Few vitamins in plants and animals are preformed, meaning that they are already in active form and ready for their biological function in the body. Some vitamins are present in plants as precursors, meaning that they have to be changed to active form to be ready for biological function. An example of this is carotene or pro-vitamin A. Carotenoid called a precursor of vitamin A (Wald 1935). There are also man-made vitamins which are synthesized in the laboratory. Hence, these are called synthetic vitamins. These synthetic vitamins are mostly used for therapeutic purposes. An individual who eats a well-balanced meal does not need synthetic vitamins because he is assured of the normal intake of vitamins from food sources. The condition that results from a deficient intake of vitamins is referred to as *avitaminosis.* On the other hand, too much intake results to excessive accumulation in the body resulting to *hypervitaminosis.* This is especially true with fat-soluble vitamins.

5.8 Fat-Soluble Vitamins

They constitute of vitamins that show solubility in fats like oil or organic solvents. They comprise mainly vitamin A, vitamin D, vitamin E and vitamin K. Vitamin A is an unsaturated hydrocarbon, which consist of retinol, retinal, retinoic acid and several provitamin A caroteniods and beta-carotene (Fennema 2008). Vitamin D is mainly a cholesterol derived vitamin. Vitamin E comprises of groups mainly: tocopherols and tocotrienols; each comprising of four forms (α) alpha, (β) beta, (γ) gamma and (δ) delta. Though the deficiency of fat-soluble vitamins were rare earlier, but now days due to the adaptation of a healthy life-style by many people their deficiencies are increasing. People avoid the intake of fat or fat containing compounds in any form resulting in the mal-absorption of the vitamins in the body.

Vitamin A All forms of vitamin A have a beta-ionone ring to which an isoprenoid chain is attached, called a retinyl group. Both structural features are essential for

Table 5.1 Classification of vitamins (Kutsky 1973)

Generic name	Chemical name	Solubility	Deficiency disease	Overdose disease	Food source
Vitamin A	Retinol, retinal	Fat	Night-blindness, Hyperkeratosis and Keratomalacia	Hypervitamin osis A	Orange, Ripe yellow fruits, carrot, liver, leafy vegetables, soya milk
Vitamin B_1	Thiamine	Water	Beriberi, Wernicke-Korsakoff syndrome	Drowsiness or muscle relaxation with large doses	Pork, oatmeal, brown rice, liver, egg
Vitamin B_2	Riboflavin	Water	Ariboflavinosis	–	Dairy products, beans
Vitamin B_3	Niacin, niacinamide	Water	Pellagra	Liver damage	Meat, Fish, eggs, mushrooms
Vitamin B_5	Pantothenic acid	Water	Paresthesia	Diarrhea, nausea, heartburn	Meat, broccoli, avocados
Vitamin B_6	Pyridoxine	Water	Anemia peripheral neuropathy	Impairment of proprioceptio n nerve damage	Meat, Nuts, banana
Vitamin B_7	Biotin	Water	Dermatitis, enteritis	–	Egg yolk, lover, peanuts
Vitamin B_9	Folic acid	Water	Megaloblast	–	Leafy vegetables, pasta, liver
Vitamin B_{12}	Cyanocobalami ne	Water	Megaloblastic anemia	Acne-like rash	Meat products
Vitamin C	Ascorbic acid	Water	Scurvy	Vitamin C megadose	Citrus fruits, liver
Vitamin D	Cholecalciferol	Fat	Rickets and Osteomalacia	Hypervitamin osis D	Fish, eggs, liver
Vitamin E	Tocopherols Tocotrienols	Fat	Hemolytic anemia	Heart failure	Fruits, vegetables, nuts, seeds
Vitamin K	Phylloquinone	Fat	Bleeding diathesis	Increased coagulation	Spinach, egg yolk, liver

vitamin activity. Many different geometric isomers of retinol, retinal and retinoic acid are possible as a result of either a trans or cis configuration of four of the five double bonds found in the polyene chain. The cis isomers are less stable and can readily convert to the all-trans configuration. Nevertheless, some cis isomers are found naturally and carry out essential functions (Nasab et al. 2013) (Fig. 5.1).

Fig. 5.1 Chemical structure of vitamin A (US National library of medicine, NIH)

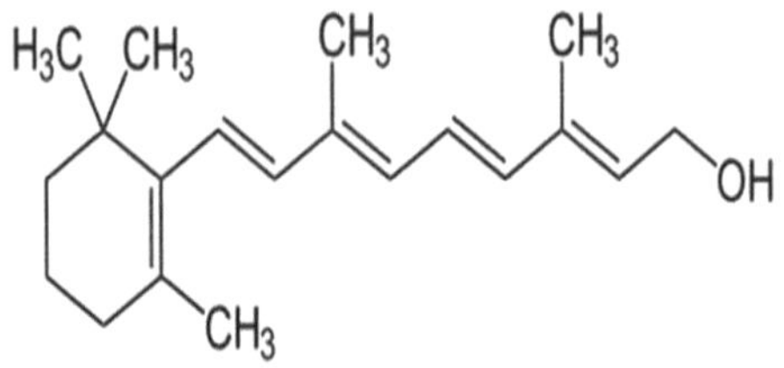

Significance of Vitamin A It is essential for normal vision, for maintaining the integrity of epithelial tissues and for a wide variety of metabolic functions (Tanumjhardjo 2011). Considerable amounts of vitamin A can be stored in the liver and made available for use as the need arises. Although the role of vitamin A in the visual process is well known, its mechanism of action in other metabolic processes is, as yet, not well understood. A series of studies (Sivakumar et al. 1997) on pregnant women were carried out at national institute of nutrition, correlating the duration of vitamin A supplementation, that was required to prevent the fall of plasma vitamin A, invariably occurring in the last few weeks of pregnancy, compared to those who did not receive any supplement. It was found that a daily dose of 780 RE was able to not only prevent the fall in plasma level but also increase neonatal status. Such amounts of intake were associated with improved fetus-placental function reversing the changes in protein catabolism and oxidative stress (Sivakumar et al. 1997).

Dietary Sources of Vitamin A Foods provide vitamin A either in the preformed state (directly as retinol or retinyl esters of fatty acids) from animal sources such as milk (28 μg 3 %), butter (684 μg 76 %), egg (140 μg 16 %) and carrot (835 μg 93 %) or its precursor carotenoids, especially carotene, derived from leafy vegetables and yellow- and orange-coloured fruits and vegetables. β-carotene and other carotenoids are variably converted into vitamin A in the body (Tang et al. 2005). Since β-carotene forms a major source of dietary vitamin A in many developing countries including India, the efficiency with which it is absorbed and utilized becomes important in translating β-carotene values into retinal equivalents. Although the enzymatic conversion of one molecule of β-carotene to retinol should theoretically result in two molecules of retinol, because of physiological inefficiency, the maximum conversion that has been shown in experimental animals is around 50 % (Olsan 1989). In addition, the efficiency of β-carotene absorption is variable, depending upon the food source. Red palm oil contains easily absorbable carotenes. Based upon available evidence, the food and agriculture organization –united nations/world health organization expert group (WHO Tech. Rep, 1967) assumed that about one third of dietary β-carotene is absorbed. As a result, a factor of 0.16 was used to convert β-carotene to retinol. In two studies carried out in India, one in adults (Rao and Rao 1970) and the other in children (Lala and Vinodini 1970) absorption of β-carotene from green leafy vegetables was found to be much higher, ranging from 50 to 99 %. Lower figures were obtained for the absorption of β-carotene from other sources

Table 5.2 Recommended intake of vitamin A (Retinol equivalents, RE, μg/d)

Group	Sub group	FAO/WHO[a] (2005) Retinol RE, μg/d	FAO/WHO (2011) Retinol RE, μg/d	FAO/WHO (2011) B-carotene[b] RE, μg/d
Adult men		600	600	4800
Adult women		500		
Pregnant women		800	800	6400
Lactating women		850	950	7600
Infants	0–6 months	375	350	–
	6–12 months	400		2800
Children	1–6 years	400/450	400	3200
	7–9 years	500	600	4800
Adolescents	10–17 years	600	600	4800

[a]Food and agriculture organization –United Nations/World health organization, [b]A conversion ratio of 1:8 is used

such as carrots and papayas. There is epidemiological evidence to support this contention (Lala and Vinodini 1970; Pereira and Begum 1969) (Table 5.2).

Deficiency of Vitamin A One of the earliest manifestations of vitamin A deficiency is night blindness and more severe deficiencies include ocular changes leading to blindness, particularly in young children. Though there has been a significant reduction in severe clinical forms of Vitamin A deficiency over the last 3 decades, vitamin A deficiency is still a public health problem with 40–60 % child population in developing countries showing inadequate biochemical status of vitamin A (World Health Organization global database, 1995–2005). Very low intakes, poor bio-availability of pro-vitamin A from the predominantly vegetarian diet and recurrent infections are thought to be the main reasons for widespread prevalence of vitamin A deficiency.

Role of Vitamin A A generalized scheme for retinoid metabolism is that dietary retinyl esters, retinol, and provitamin A carotenoids (such as β-carotene) are taken into the body. Vitamin A (by definition all-trans-retinol) may be esterified into retinyl esters and stored. In times of dietary retinoid insufficiency, retinyl ester stores are hydrolyzed to retinol for delivery to peripheral tissues. Both all-trans-retinol and β-carotene may be converted enzymatically to all-trans-retinal. Retinal either can be enzymatically oxidized to retinoic acid, which regulates transcription of over 500 retinoid-responsive genes, or reduced enzymatically to retinol. When retinoic acid is no longer needed, it is catabolized and eliminated from the body (Li et al. 2014).

Vitamin D It refers to a group of fat-soluble seco-steroid compounds. Two nutritionally significant compounds are vitamin D2 (ergocalciferol) and vitamin D3 (cholecalciferol). Vitamin D3 is metabolised to the active steroid hormone 1,25-dihydroxyvitamin D3 by successive hydroxylations in the liver and kidney. Vitamin D2

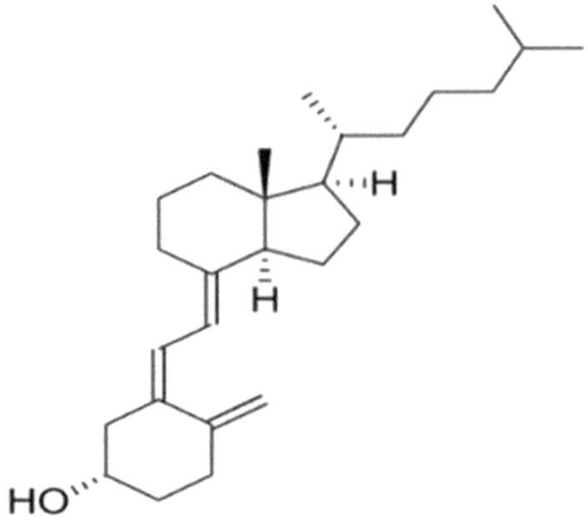

Fig. 5.2 Chemical structure of vitamin D (US National library of medicine, NIH)

is metabolised to 1,25-dihydroxyvitamin D2 by the same enzyme systems. One milligram of vitamin D is equivalent to 40,000 international units (Khandare et al. 2005) (Fig. 5.2).

Significance of Vitamin D It is a unique vitamin and its availability in the body largely depends on its synthesis in the skin when exposed to sunlight and hence, its dietary requirement is usually very small especially in the Indian context. In humans, the most important related compounds of vitamin D are vitamin D2 and vitamin D3 (Holick 2006). Cholecalciferol (vitamin D_3) and ergocalciferol (vitamin D_2) are unique as they constitute the vitamin D and can be ingested from the diet and/or supplements (Calvo et al. 2005; Norman 2008). Apart from its classical calcitropic etiological role in the development of rickets (in children) and osteomalacia (in adults) (Wolf 2004), leading to weak bones and physical deformation, vitamin D has been found to have many non-clacitropic health promoting effect (Pittas et al. 2010). Diseases such as osteoporosis, diabetes, psoriasis, hypertension, arthritis, multiple sclerosis and cardio vascular diseases are known to have the involvement of vitamin D leading to the suggestion that the above spectrum of deficiency diseases be called vitamin D deficiency diseases as in the case of other micronutrients (Chung et al. 2009; McLaren 2006). The best evidence of benefit is for bone health (Ross et al. 2011a, b) and a decrease in mortality in elderly women (Bjelakovic et al. 2011).

Dietary Sources of Vitamin D It can be photolytically formed from 7-dehydrocholesterol (as provitamin D3) in the skin when exposed to sunlight (UV-B rays) (Houghton and Vieth 2006). It can be also obtained from the diet both through plant sources as provitamin D2 (ergosterol) and get converted to D2 (ergocalciferol) in the skin or through animal foods as preformed vitamin D3 or its provitamin (Holick 2007). It is estimated that about one tenth of the body's requirement is derived from dietary sources in the Indian context; mainly the exposure to sunlight being the limiting factor (McLaren 2006). Vitamin D3 is also found in oily fish and cod liver oil (Holick 2007) (Table 5.3).

Table 5.3 Recommended intake of vitamin D

Age	Male	Female	Pregnancy	Lactation
0–12 months	400 IU (10 mcg)	400 IU (10 mcg)	–	–
1–13 years	600 IU (15 mcg)	600 IU (15 mcg)	–	–
14–18 years	600 IU (15 mcg)	600 IU (15 mcg)	600 IU (15 mcg)	600 IU (15 mcg)
19–50 years	600 IU (15 mcg)	600 IU (15 mcg)	600 IU (15 mcg)	600 IU (15 mcg)
51–70 years	600 IU (15 mcg)	600 IU (15 mcg)	–	–
>70 years	800 IU (20 mcg)	800 IU (20 mcg)	–	–

IU international units, *mcg* micrograms

Deficiency of Vitamin D Its deficiency lead to abnormal calcium homeostasis resulting in defective mineralization of the growing long bones (rickets in children) or a decrease in the mineral content of the matrix of the bones (osteomalacia in adults) ending with weakened bones. A poor calcium deposition during growth phase and faulty economy in adults lead to osteoporosis and vitamin D and calcium supplementation is shown to be beneficial. While the rabid forms of bone disease are no longer prevalent in children now, growing urbanization, reduced physical activity and low exposure to sunlight are believed to contribute to a spurt of vitamin deficiency diseases as evidenced from low circulating vitamin D levels. The emergence of sub-clinical vitamin D deficiency in the country has been reviewed by Goswami (Goswami et al. 2008). In recent years, in endemic areas of fluorosis, excess of fluoride in drinking water was found to be associated with rachitic bone deformities and vitamin deficiency diseases in young children (Khandare et al. 2005)

Role of Vitamin D It is activated in two series of hydroxylation reactions in liver and kidney to active hormones endogenously which mediate increased intestinal absorption and renal tubular resorption of calcium and contribute to an increase in the availability of calcium (Cranney et al. 2007; Bischoff-Ferrari et al. 2012). The first hydroxylated compound, 25 hydroxy cholecalciferol and the next 1, 25 dihydroxy cholecalciferol are the well-known metabolites of vitamin D. The first one is the circulating and storage form and represents the status of vitamin D. 1, 25 dihydroxy cholecalciferol is the active hormone metabolite which induces the formation of calcium transport protein involved in calcium absorption (Cranney et al. 2008). Blood calcium is maintained in a narrow range by the concerted actions of parathyroid hormone, thyrocalcitonin and vitamin D. A lowering in calcium intake triggers an elevation in Parathyroid hormone levels which enhances the release of 1, 25 dihydroxy cholecalciferol so as to increase calcium absorption and thus normalize blood calcium. A lack of vitamin D will reflect in alterations in the calcitropic hormones and not always in serum calcium. Vitamin D is now considered more as a pro-hormone, than as a vitamin. It can be synthesized in the body in adequate amounts by simple exposure to bright sunlight even for 5 min per day (Crissey et al. 2003). Habitual Indian diets do not provide even 10 % of the requirement. Based on healing response in rickets, increased calcium absorption or circulating levels of 25

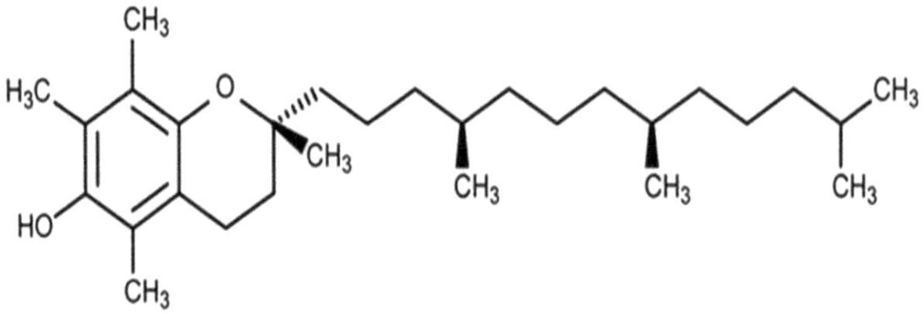

Fig. 5.3 Chemical structure of vitamin E (US National library of medicine, NIH)

hydroxy cholecalciferol and also turnover of isotopic vitamin D, the requirements of vitamin D have been determined over the years. The World Health Organization expert committees recommended 100 Units (2.5 μg)/d for adult males in 1988 (World Health Organization 1988) and increased them later in 2005 to 200 Units (5 μg)/d (Food and agriculture organization/World health organization 2004). This is obviously due to progressive decrease in the exposure to sunlight and the need to obtain the requirements from dietary and supplement sources. Hence, many foods such as milk and vegetable oils are subjected to mandatory fortification with vitamin D in the developed countries.

Vitamin E Vitamin E is not a single molecule, but a family of 8 related molecules called tocopherols and tocotrienols. Moreover, each of the different tocopherols exists in eight stereoisomers. However, having described this molecular complexity, it turns out that dietary vitamin E is predominantly alpha and gamma-tocopherol. The structure of alpha-tocopherol is depicted below; note the long hydrocarbon chain similar to the tail of a fatty acid (Fig. 5.3).

Significance of Vitamin E It is a form of fat-soluble compound that include both tocopherols and tocotrienols (Brigelius-Flohe and Traber 1999). It helps to stop the production of reactive oxygen species formed when fat undergoes oxidation (Herrera and Barbas 2001; Packer et al. 2001). Vitamin E has many biological functions, the antioxidant function being the most important one (Bell 1987). Other functions comprises of enzymatic activites, gene expression and neurobiological functions. It plays a very important role in cell signaling (Azzi 2007). Vitamin E mainly acts as a peroxyl radical scavenger, preventing the propagation of free radicals in cells. As it is fat-soluble it is incorporated into the cell membrane, which protects them from oxidative damage. Vitamin E plays an important role in gene expression too (Devaraj et al. 2001). It has an important role in neurological functions (Muller 2010) and inhibition of platelet aggregation (Atkinson et al. 2008). Vitamin E protects lipids and prevents the oxidation of polyunsaturated fatty acids too (Whitney et al. 2011).

Dietary Sources of Vitamin E Vegetable oils and invisible fat of cereals and other foods like nuts and vegetables contribute to adequate tocopherol content in Indian

Table 5.4 Recommended Intake of Vitamin E (National Institute of Health 2009)

Group	Sub-group	mg/day
Infants	0–6 months	4
	7–12 months	5
Children	1–3 years	6
	4–8 years	7
	9–13 years	11
Adolescents	14 and above	15

diets. About 60 % of vitamin E in the diet comes from vegetable oil (soybean, corn, cottonseed, and safflower). This also includes products made with vegetable oil (margarine and salad dressing). Vitamin E sources also include fruits and vegetables, grains, nuts (almonds and hazelnuts), seeds (sunflower) and fortified cereals (National Institute of Health 2009) (Table 5.4).

Deficiency of Vitamin E Its deficiency is rare. Cases of vitamin E deficiency usually only occur in premature infants and in those unable to absorb fats. Since vegetable oils are good sources of vitamin E, people who excessively reduce their total dietary fat may not get enough vitamins E. The deficiency is rare but if occurs leads to various problems like myopathies, peripheral neuropathy, ataxia impairment of the immune system, and red blood cell destruction (Kowdley et al. 1992). Vitamin E obtained from food usually does not pose a risk for toxicity. Supplemental vitamin E is not recommended due to lack of evidence supporting any added health benefits. Megadoses of supplemental vitamin E may pose a hazard to people taking blood-thinning medications such as Coumadin (also known as warfarin) and those on statin drugs.

Role of Vitamin E It is hydrophobic and is absorbed similarly to other dietary lipids. After getting soluble by bile acids, it is absorbed into small intestinal epithelial cells, incorporated into chylomicrons, and transported into blood via lymphatic. Once in the circulation, vitamin E is liberated from chylomicrons and much is taken up by the liver, where it is repackaged into very low density lipoproteins and secreted again into blood. Ultimately, vitamin E is transported in blood bound to a variety of lipoproteins, from which it is taken up by tissues throughout the body. Vitamin E is stored within the fat droplets of adipose tissue cells (Gagne et al. 2008).

5.9 Water-Soluble Vitamins

These are dissolved by the water and not stored by the body, henceforth required in adequate level in diet on daily basis. As excess of water-soluble vitamins gets eliminated in urine so therefore generally it's not toxic and harmful. The major dietary sources of water-soluble vitamins are fruit, vegetables and grains. Water-soluble vitamins are categorized into vitamin C, vitamin B and folic acid (Emerson and folkers 1951). Although vitamins and dietary minerals are present in the foods however it is still found inadequate in major population and people become vulnerable to deficiency diseases. Preparation of food can also result in the loss of vitamins from vitamin enriched food because water soluble vitamins are heat labile and light sensitive so it can easily be degraded at high temperature or by exposure in light and air (Kreutler and Czajka-Narins 1987). In a survey low blood, folate level was found in most of the school going children (6–16), (Sivakumar et al. 2006) though their diet contained a mean intake of 155 % recommended dietary allowances for folic acid still they were found folate deficient. It was obvious that may be there is loss of dietary folate during cooking of food or in the processing of fortified food. Folate is essential in biological methylation reactions and nucleotide synthesis and impairment of these processes are thought to be involved in cancer development (Kim 2004)

Folic Acid Chemically folates are known as pteroyl-L-glutamic acid, its structure composed of three distinct moieties, heterocyclic pteridine ring, para-aminobenzoic acid and glutamic acid. Iwater-soluble vitamin and is member of vitamin B family (Blakley 1969) (Fig. 5.4)

It is an essential micronutrient in diet and is required for the healthy bodily function and metabolic process (Carmel 2005). Naturally, folate is present in very low concentration in food. Fortification can enrich the folic acid content by commercially adding the folic acid formulation in the dietary food (Cuskelly et al. 1996). Folic acid fortification is a process in which the stable formulation of folic acid is added to food so as to increase the blood folate levels. In USA, and other countries folic acid fortification is made mandatary so as to reduce the deficiency disease develops due to inadequate intake of folate (Bailey et al. 2010a, b). The Recommended dietary allowance for folic acid is age dependent like for infants it is 60 mcg/day, children 80–140 mcg/day, adults 200 mcg/day and for pregnant women 400 mcg/day (Subar et al. 1989).

Sources of Folic Acid Folate rich food includes leafy vegetables such as spinach, asparagus, lettuce, (Houlihan et al. 2011). Legumes such as dried fresh beans, Brussels sprouts peas and lentils, egg yolk, baker's yeast, fortified grain products, breakfast cereals fortified with folate, sunflower seeds (Table 5.5).

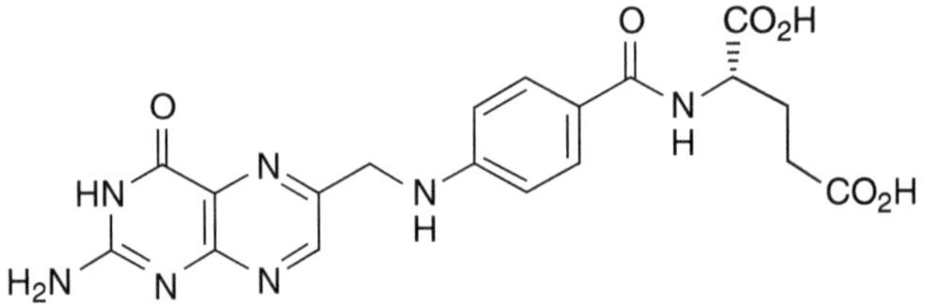

Fig. 5.4 Chemical structure of folic acid (US National library of medicine, NIH)

Table 5.5 Recommended dietary allowances for folic acid [Food and nutrition board, institute of medicine (1998)]

Age	Recommended dietary allowance (dietary folate equivalent-DFE)	Upper intake level
0–6 months	65 mcg DFE	–
7–12 months	80 mcg DFE	–
1–3 years	150 mcg DFE	300 mcg
4–8 years	200 mcg DFE	400 mcg
9–13 years	300 mcg DFE	600 mcg
14–18 years	400 mcg DFE	800 mcg
19+years	400 mcg DFE	1000 mcg
Pregnant women	600 mcg DFE	1000 mcg

**mcg* microgram
1 mcg DFE=1 mcg food folate, 1 mcg DFE=0.6 micrograms folic acid from fortified foods or dietary supplements consumed with foods, 1 mcg DFE=0.5 micrograms folic acid from dietary supplements taken on an empty stomach

Functions of Folic Acid Folic acid is important in DNA synthesis and replication process. It has a biochemical role in mediating the transfer of single carbon moieties and act as cofactors for biosynthesis of metabolites (Wagner 1995). Human cannot synthesize folate, therefore folate has to be supplied through the diet to meet their daily requirement as folic acid is essential for its various biological functions. In aids to this folate is also important during the period of frequent cell division and growth, in infancy and pregnancy. Adequate folate intake during pregnancy would help to protect against neural tube defect (Shaw et al. 1995). Folic acid consumption also found in reducing the risk of colorectal cancer (Sanjoaquin et al. 2005). Human needs folate for the production of red blood cells and white blood cells as it helps in maintenance and production of new cells.

Absorption Mechanism of Folic Acid In the diet, folates are present in the conjugated form of polyglutamates with variable number of glutamate residue (Bernstein et al. 1970) Polyglutamate is converted into monoglutamate by folate reductase in the intestine before absorption. Folic acid is more stable than the natural food folates; however it loses its activity during the processing of food. Natural food folate/folic acid is not biologically active, but is converted into dihydrofolate by dihydrofolate synthase in the liver and tetrahydrofolate by tetrahydrofolate reductase and become active for biological function (Crider et al. 2011).

Deficiency of Folic Acid Folic acid deficiency is rare and uncommon. Chronic and severe forms of folic acid deficiency lead to hemopoiesis and megaloblastic anemia characterized by large, abnormally, nucleated erythrocytes. Both vitamin B12 and folic acid were found to be associated with elevated blood homocysteine level, which was reported as a risk factor for cardiovascular diseases (Brattstrom 1996). Since folate deficiency limits cell division, erythropoiesis, production of red blood cells is hindered and leads to megaloblastic anemia. In addition, inadequate intake of folic acid in pregnant women causes fetal neural tube defect and other birth defects (Crider et al. 2011) neural tube defect occurs when the neural tube fails to close early in embryonic stage. Folate deficiency may also lead to diarrhea, soreness and ulceration of tongue, depression. (Bailey and Gregory 2006). The deficiency in dietary folate have also been reported to increase the risk of Colorectal cancer (Kim 2004).

5.10 Minerals

Minerals are nutrient ions present in the soil and cannot be synthesized by living being, such as plants and animals, however it is absorb by plants from soil and aided to the food web Minerals are one of the most important constituent require for metabolism and all biochemical functioning of body (Underwood and Suttle 1999). Main dietary source of minerals includes fruits and vegetables. Minerals are also present in drinking water depending upon the geographical location.

Significance of Minerals They are minute amounts of inorganic nutrient that are essential for bone and tooth formation. They also help in biological activity as enzyme action, muscle contraction, nerve reaction, and blood clotting. Minerals play an important role in regulation of intracellular and extracellular body fluid and Generation of energy from the food (Soetan et al. 2010). Minerals enriched food includes meat, cereals, fish, milk and dairy foods, vegetables, fruit (especially dried fruit) and nuts. Essential minerals include calcium and iron, although there are also many other types of minerals that are an important part of a healthy diet.

Minerals are broadly classified into macro-minerals and trace minerals. Body requires macro-minerals in large quantity than trace mineral. The macro-mineral group includes calcium, phosphorus, magnesium, sodium, potassium, chloride, and sulfur (Eruvbetine 2003). A trace mineral are those, which required by the body in small quantity. So even though your body needs trace minerals, it needs just a tiny bit of each one. Trace minerals includes iron, manganese, copper, iodine, zinc, cobalt, fluoride, and selenium.

Calcium is a mineral nutrient that plays an important role in bone health, muscle contraction, blood clotting, nerve conduction, enzyme regulation, (Wosje and Specker 2000) and required in substantial amounts, but many diets are deficient in calcium making supplementation necessary or desirable. Similarly, iron deficiency is one of the leading risk factors for death world- wide, affecting an estimated 2 billion people (Zimmermann and Hurrell 2007). The high prevalence of iron deficiency in the developing world has substantial health and economic costs, including poor pregnancy outcome, impaired school performance, and decreased productivity. Knowing the benefits of mineral nutrients and the fact that there is a low intake via food, novel strategies have been aimed to supplement these nutrients to meet the dietary recommendations.

Calcium It is an essential nutrient for human metabolism. It is a major element in the human body. The key role of calcium is to maintain the health and strength of bones. It also has other important functions like muscle contraction, nerve induction, blood coagulation, membrane permeability (Berivan and Nuray 2005). The recommended dietary allowance of calcium can be met by the consumption of healthy calcium rich diet. In some individual particularly in, elderly people calcium supplements required to achieve their recommended dietary calcium intake (McCabe et al. 2004). Calcium fortified food and beverages are the other option for the people who are lactose intolerant and those who does not consume dairy products to meet their calcium levels by these means. People of different age groups require different level of calcium, it is especially important for children of growing age, older women, pregnant women (Heyman 2006). The recommended dietary allowance for calcium varies, for children it is 600–800 mg/day, adults 1000 mg/day, pregnant women 1200 mg/day (Bailey et al. 2010a, b). The deficiency leads to severe disorder like osteoporosis. Calcium plays a key role in wide range of biological function either in the free form or bound complexes. Approximately 99 % of the calcium is present in the skeleton system as calcium phosphate composition and the remaining one percent is present in soft tissue and blood.

Calcium Sources Calcium enriched food includes dairy based products such as milk, yoghurt, cottage cheese. These are the major contributor of calcium. Other natural calcium rich food sources are vegetables such as broccoli, cabbage, spinach but the bioavailability is poor. Fortified foods with good calcium content are Tofu, cereals, fruit juices. The two main forms of calcium precursors are calcium citrate and calcium carbonates, which are used as calcium supplements (Straub 2007). The bioavailability of calcium carbonate is high when consumed with food in comparison to calcium citrate. Other calcium supplements include gluconate, lactate, and phosphate. In fortified juices calcium malate is well absorbed (Andon et al. 1996) (Table 5.6).

Functions of Calcium Calcium is required for the development and maintenance of skeleton (Heaney 2002). Calcium helps in construction, formation of bone and teeth. Apart from these, calcium plays a crucial role in regulation of cardiovascular

Table 5.6 Recommended dietary allowance for calcium (Bailey et al. 2010a, b)

Age	Recommended dietary allowance (RDA)	Upper intake level (UL)
0–6 months	200 mg	1000 mg
7–12 months	260 mg	1500 mg
Children (1–3 years)	700 mg	2500 mg
Children (4–6 years)	1000 mg	2500 mg
Children (9–13 years)	1300 mg	3000 mg
Adolescents (10–18 years)	1300 mg	3000 mg
Adults (19–50 years)	1000 mg	2500 mg
Pregnant women	1300 mg	3000 mg

and neuromuscular function. All muscles use Ca^{2+} as their main regulatory and signaling molecule (Berchtold et al. 2000). It acts as an important co-factor in the blood clotting mechanism and it also help in wound healing. Calcium is an important component in the production of enzymes and hormones, which are essential for metabolic process. Calcium regulates normal functioning of kidney. Extracellular calcium is essential for maintaining the potential difference across the cell membrane. Calcium is also important in the functioning of parathyroid, thyroid gland and regulation of hormones (Brown 1991). Vitamin D3 (calcitriol) plays a pivotal role in calcium absorption.

Absorption Mechanism of Calcium Parathyroid hormone and calcitonin are the two hormones, which regulates calcium metabolism cascade. The Parathormone is secreted by parathyroid gland; it increases the calcium concentration in blood whereas calcitonin secreted by thyroid gland decreases the calcium concentration in the blood. The fall in the calcium level increases the secretion of PTH from parathyroid gland. PTH acts directly on bone to stimulate resorption and release of Ca^{2+} into the extracellular space and acts on kidney to increase calcium reabsorption and phosphate excretion (Potts and Gardella 2011). It enhances the absorption of calcium in intestine by increasing the production of vitamin D by up regulating the enzymatic process of conversion of vitamin D2 to Vitamin D3 (Peacock 2010).

Calcium Deficiency Calcium is very vital in strengthening of bones and proper functioning of body. Its deficiency diseases may lead to skeleton abnormalities, osteoporosis in adults and rickets in children. Apart from loss in bone density, muscle cramp, hypertension (McCarron, 1989) and impairment of nervous system are the symptoms of calcium deficiency. Major part of the calcium is deposited as bone mineral; its inadequate intake may lead to accelerated bone loss and osteoporosis (Parfitt et al. 1995). In mechanism of muscle contraction, calcium is essential and lack of which may results into painful muscle cramp. Deficiency of calcium also causes neuromuscular irritability as it maintains the ionic channels within the nervous system.

Iron It is an essential micronutrient, which is responsible for various metabolic processes in human body. Iron is essential for many complex biological processes that occur at molecular level like transportation of oxygen to red blood cells and muscles (Anderson and Fitzgerald 2010). Iron is a major part of hemoglobin structure, which carries oxygen to the red blood cells and myoglobin, which carries oxygen to the muscles. Iron also act as important co-factor for various enzymes in the food iron is present in two form heme iron and non-heme iron. Almost two third of the body iron is present as hemoglobin as red blood cells. In addition, iron is involved in energy producing reaction in the body. Excess iron is stored in the body in the form of ferretin protein in the liver (Ponka et al. 1998). The rich source of iron includes meat, fish, dried beans, spinach, dried fruits, and spinach. Adequate iron intake is important, particularly in childhood and pregnancy, where the developing baby is dependent upon its mother's iron supplies. To meet the recommended iron level adults need to absorb 1–1.5 mg of iron per day. The recommended dietary allowance for iron varies considerably based on the age, for children its 7–8 mg, adults 15–18 mg, pregnant women 27 mg per day (Institute of Medicine, Food and Nutrition Board 2001). Iron deficiency anemia occurs when there is inadequate amount of iron in erythrocytes due to blood loss, pregnancy, poor absorption. Iron absorption is inhibited by many factors and food sources. Phosvitin protein present in eggs binds to iron and prevents the body from absorbing from food. Daily consumption of calcium in large quantity reduces the iron absorption (Halberg et al. 1992) Oxalate present in tea, coffee, and cocoa impairs the absorption of iron.

Sources of Iron The food sources rich in iron occur in two form heme form and non-heme form. Heme iron is rich in food that contains natural hemoglobin such as in red meat, fish (Hurrell 1997). Vegetables such as lentils, beans contain iron in non-heme form. Iron in heme form has high bioavailability than those in non-heme form. Fortified iron based food contain non-heme form of iron (Miret et al. 2003) (Table 5.7).

Functions of Iron The main role of iron in the body is in the production of red blood cells, a process known as haematopoiesis. It combines with globin protein to form hemoglobin, five membered ring structures which is the pigment of erythrocytes (Webb 1992). Iron regulates the proper functioning of immune system and helps in maintenance of the physical growth of cells (Beard 2001). Iron act as a catalyst and regulates the oxidation and reduction reaction in the body.

Absorption Mechanism of Iron Iron metabolism is a set of chemical reactions in body, which helps in the maintaining of human homeostasis. The rate of dietary iron absorption in human body is a set of many interdependent factors and is a dynamic process (Herndon et al. 1957). The efficiency of dietary iron is dependent on food source, as Iron absorption is decreased with the food rich in oxalate, phytate, polyphenol as these binds with it and reduces its absorption. Absorption of iron takes place in duodenum of intestine (Anderson et al. 2009). The iron in ferrous state is in the form of heme iron and absorbed more efficiently than non-heme iron, which is in ferric state. The enzyme Duodenal cytochrome B present in the enterocytes cells

Table 5.7 Recommended dietary allowance for iron (Food and Nutrition Board, Institute of Medicine 2001a)

Age	Recommended dietary allowance (RDA)	Upper intake level (UL)
7–12 months	11 mg	40 mg
Children (1–3 years)	7 mg	40 mg
Children (4–8 years)	10 mg	40 mg
Children (9–13 years)	8 mg	40 mg
Adolescents (14–18 years)	11 mg	45 mg
Adults (19–50 years)	8 mg	45 mg
Pregnant women	27 mg	45 mg

of intestinal lining plays an essential role in the reduction of iron from its ferric to ferrous state. The non-heme iron binds to apoferritin and stored as ferretin protein (Dennis et al. 1999).

Iron Deficiency Iron deficiency anemia is one of the most severe and important nutritional deficiency, affecting all the age groups, especially pregnant women, infants, adolescent boys and girls Pollitt et al. (1976). Anemia is a blood disorder occurs when there is insufficient hemoglobin in erythrocytes. Inadequate intake of dietary iron will eventually results in the depletion of stored iron in the body and the person develops anemia. Apart from anemia iron deficiency also impairs the cognitive development in children (Seshadri and Gopaldas 1989). It affects the immune system mechanism (Beard 2001). Iron deficiency in pregnant women causes problems of hemorrhage, birth defects and both maternal and prenatal are affected. Iron deficiency occurs due to abnormal blood loss, bleeding can be caused due to ulcers, hemorrhoids, cancer. In a survey conducted by the National Institute of Nutrition it was reported that more than 70 % of the children in India are suffering from anemia. (Gomber et al. 1998). In infants, iron deficiency also results in poor brain development (Grantham McGregor and Ani 1999).

5.11 Silica as a Nutrient

In 1939, the Nobel Prize winner chemistry, Professor Adolf Butenant, proved that life cannot exist without Silica. Silica has been long suspected to be important in maintaining health in humans (Becker et al. 1983). Silica plays an important role in many body functions and has direct effect to mineral absorption. The average human body holds approximately seven grams of Silica, quantity more than important mineral such as iron. As we age, scientific measurements have shown that the human body retains less and less silica. Silica does not occur in sufficient amounts in a wide variety of foodstuffs due to food processing in modern food. Hence silica supplementation is essential. Silica is needed for carrying out ongoing metabolic work that is vital to life. Many studies that prove the favorable influence of vegetal silica on the development of animals have been under taken. It is essential to the development of the skeleton and mineralization. Silica's absence results in skeletal deformities. Average daily intakes of silicon have been suggested to range from about 20–50 mg/day (Pennington 1991). Ample evidence exists to indicate that silicon can be accepted as an essential nutrient for higher animals, including humans. Findings from animals indicate that silicon nurtures apparently affects macromolecules, such as glycosaminoglycan, collagen and elastin, and thus is needed for healthy bones, brains and blood vessels. Although more should be known about the physiologic functions and requirement for silica before doing so, it is seductive to speculate about specific disorders that can be augmented or caused by inadequate silicon nutrition; those that have been proposed are atherosclerosis, osteoarthritis and hypertension. Even if these speculations are not found to be true, because the silicon content in human diets can be easily be lower than that inducing changes in animals especially those containing refined and animal product foods, and because the response of animals to silicon deprivation can be enhanced by stressors commonly found with humans, such as low dietary calcium, high dietary aluminum and low estrogen status, finding pathologic conditions caused by silicon deprivation would not be surprising. Thus, silicon probably should be considered a nutrient of concern for humans

5.12 Silica as a Carrier Molecule

Till date a lot of search has been done in encapsulating the vitamins or minerals on materials like chitosan (Britto et al. 2012) gelatin or polylactidecoglycolide (Ignjatovic et al. 2013). Silica has attracted considerable attention due to their ability to enhance solubility of poorly water-soluble compounds, as well as to achieve sustained release, leading to improved bioavailability (Kapoor et al. 2010; Zhang et al. 2010). The advantages of silica particles as a carrier molecule is that it provides large surface area, high pore volume, tunable pore size, and well defined pore structure, as well as mechanical and chemical stability and protection of the integrity of the molecular structure of the compound (Kilpelainen et al. 2009; Slowing et al. 2008; Salonen et al. 2005). Recently porous silica particles were developed for intracellular controlled drug/gene delivery (Vivero-Escoto et al. 2010; Hom et al. 2011) and targeted drug delivery to cancer cells (Mamaeva et al. 2011; Rosenholm et al. 2010). Moreover the safety of silica as delivery systems has been highlighted in many reports, which makes it desirable as perfect delivery system (Chervenkov et al. 2007). Studies have even found that oral administration of silicon dioxide did not result in significant changes in the biochemical parameters characterizing liver functions and gastric mucosal damage (Chervenkov et al. 2007; Tanaka et al. 2010).

Silica surface chemistry is very well studied, thus suitable surface modifications enables the loading of diverse nutrients and minerals, through the non-covalent electrostatic, hydrogen bonding and van der walls force of interaction. Characteristic property of silica having high porosity enables the nutrients to be loaded within the pores, canals or adsorbed on the surface. The advantage of silica particle is that it itself is an inherent nutrient which is essentially required for our body. Probably one of the most challenging tasks in the formulation and loading of micronutrients is to control the release rate while assuring protection of the active compound during processing, transport and storage. Generally, the release of an entrapped compound from a dense particle is slower than from a highly porous structure, which also inversely affects the stability of the active compound (Matalanis et al. 2011). In this report, silica is used as a carrier molecule for the loading of nutrients, increase the efficiency and stability of nutrients and study this formulation through sustained release profile in in-vitro gastro-intestinal conditions.

5.13 Silica as a Carrier for Delivery of Nutrients

Though nutrients are available in natural food sources but still there is a huge percentage of population suffering from their deficiency. A number of supplements are available in the market but these are mainly pharmaceutical supplements (Yuan et al. 2010). To increase the bioavailability of the vitamins in these supplements they are prepared by expensive formulations like micro encapsulations. There are many additives added to them to stabilize the molecules and enhance their delivery. These additives need not have any biological importance and the body needs to spend extra energy in their excretion. And people are usually reluctant to consume these tablets. Hence a need of a carrier molecule which serves the purpose of both carrier and nutrient was needed. The carrier would not only deliver the vitamins in a stable form but also show biological functions in the body, and its preparation should be cost effective.

The main concern though becomes providing the right nutrition at the affordable cost. We propose herein the use of nutrient loaded silica particles to provide the much needed nutrition for the society. The choice of the material comes from the facts that (i) Silicon in silica is inherently a nutrient that is essential for our bodies for bone and cartilage development (Jugdaohsingh 2007; Vasanthi et al. 2012) (ii) Nutrient loaded silica nanoparticles serve dual purpose: inherent nutrient as well as carrier –a novel way of providing quick nutrition in the form of food additives or health drinks, (iii) Highly porous food grade silica provides high surface area suitable to load large quantities of nutrients both water soluble or insoluble (Ciriminna et al. 2011). The concept of our work is centered on using these silica particles which by themselves act as inherent nutrient, as carriers of other nutrients such as vitamin A, D3 and E and calcium, iron, vitamin B9 (folic acid) –a novel way of providing quick nutrition in the form of food additives or health drinks.

5.14 Experimental Section

5.14.1 Loading of Fat Soluble Nutrients onto Silica

Fat soluble vitamins such as retinyl acetate (vitamin A), cholecalciferol (vitamin D_3) and α-tocopherol acetate (vitamin E) were obtained from sigma Aldrich. The commercial silica used was of food grade quality. It was amorphous in nature and did not contain any heavy metal. Other reagents used for various experiments & analysis were of analytical grade.

***Ex-Situ* Loading of the Vitamin E** 6.39 g of vitamin E was weighed and dissolved in 100 ml of absolute Ethanol. To this homogenous solution 10 g of commercial food grade Silica was weighed and added slowly with continuous stirring. The mixture was kept in rotatory shaker for overnight stirring at 120 rpm. The samples were then centrifuged at 4000 rpm for 5 min and supernatant was collected for analysis. The solids which settled down were washed again with ethanol to remove any unbound vitamin E by centrifuging at 4000 rpm for 5 min. Again the supernatant was collected for analysis and the powder was dried in a desiccator for 24 h. Once the samples dried it was grinded and used for further analysis by particle size analyzer, Fourier transform infrared spectroscopy, UV-vis spectroscopy, thermo gravimetric analysis, BET surface area and high performance liquid chromatography. 0.1 mg of samples was taken and dissolved in absolute ethanol. This dispersed sample was used to measure the particle size and the zeta potential of the silica before and after loading of vitamin. IR spectra of the samples were recorded in the range of 400–4000 cm^{-1}. The UV-vis absorbance of the supernatant collected was measured using a UV-vis spectrometer and the amount of vitamin present was calculated by plotting against a suitable standard. High performance liquid chromatography in a mobile phase used for quantification of vitamin loaded onto silica particles. Quantification studies resulted in 20 % loading of vitamin E onto silica particles.

Ex-Situ **Loading of Vitamin A and D onto Silica Matrix** 2 g of vitamin A and 5 g of vitamin D were weighed separately and dissolved separately in 2 conical flasks containing 100 ml of absolute ethanol (100 %) each. To each of these homogenous solution 10 g of silica powder was added slowly with continuous stirring. The mixtures were kept in rotatory shaker for overnight stirring at 120 rpm. The samples were then centrifuged at 4000 rpm for 5 min and supernatant was collected separately for analysis. The powder which settled down was washed again with ethanol to remove any unbound vitamin A and D by centrifuging at 4000 rpm for 5 min. Again the supernatants were collected for analysis and the powders were dried at 40 °C under vacuum. Once the samples were dried they were grinded using a mortar and pestle, and were further characterized. Quantification studies resulted in 13–15 % loading of vitamin A and 7–8 % loading on vitamin D3 onto silica particles.2

5.14.2 In-Vitro *Release Studies of Fat Soluble Vitamins*

The *in-vitro* release profile of the vitamins A, D and E loaded onto silica were determined by gastric and intestinal release processes. Samples were subjected to simulated environments of the gastro-intestinal tract by preparation of simulated fluids, maintenance of pH and temperature and time estimation.

Release at Gastric Conditions Accurately weighed 0.1 mg nutrient loaded silica was transferred to a beaker containing 15 ml of simulated gastric juice fluid. The simulated gastric fluid was prepared by dissolving 2 g of sodium chloride in 100 ml of distilled water. 7 ml of concentrated hydrochloric acid was added drop wise. 3.2 g of pepsin was added to the solution and the volume is made up to 1000 ml and checked for the pH. The gastric fluid is acidic so the pH was maintained between 1 and 1.3. The test solution had a pH of about 1.3. Initially the mixture was mixed slightly and placed in a shaker incubator rotating at 50 rpm and maintained at 37 °C. The supernatant was removed at a time interval of 30 min for next 2 h. The supernatant was filtered through 0.45 μm Millipore membrane filter and then was analyzed for the amount of vitamin present by high performance liquid chromatography method.

Release at Intestinal Conditions For intestinal release study the pH of the solution obtained after gastric digestion was adjusted to neutral pH by addition of simulated intestinal fluids and the above gastric juice treated sample was added to it after filtration through 0.45 μm Millipore membrane. The intestinal simulated fluid was prepared by dissolving 6.8 g of monobasic potassium phosphate in 250 ml of water, mix properly and 77 ml of 0.2 N sodium hydroxide and 500 ml of water was added. 10 g of pancreatin was added. Mixture was mixed thoroughly and pH was adjusted to 6.8 with either 0.2 N sodium hydroxide or 0.2 N hydrochloric acid. The volume was made up to 1000 ml with distilled water. The bottles were again placed in shaker incubator rotating at 50 rpm and maintained at 37 °C. The supernatant was removed at a time interval of 30 min for next 2 h. The supernatant was filtered through 0.45 μm Millipore membrane filter and then was analyzed for the amount of vitamin present by high performance liquid chromatography method.

5.14.3 Loading of Water Soluble Nutrients onto Silica

Folic acid, calcium chloride, ferrous sulfate was obtained from Sigma Aldrich. The commercial silica used was of food grade quality. It was amorphous in nature and did not contain any heavy metal.

***Ex-Situ* Loading of Folic Acid** 2 g of folic acid is dissolved in aqueous solution of 6 % sodium hydrogen carbonate and to this homogenous solution 10 g of food grade silica was added while stirring. The mixture was kept in rotatory shaker for overnight stirring at 120 rpm. The samples were then centrifuged at 4000 rpm for 5 min and supernatant was collected for analysis. The solids which settled down were dried under vacuum at 30 °C. Once the samples dried it was grinded and used for further analysis by Fourier Transfer Infrared Spectroscopy, UV-Spectroscopy, Thermo gravimetric analysis, BET surface area and high performance liquid chromatography 0.1 mg of samples was taken and dissolved in water. This dispersed sample was used to measure the zeta potential of the Silica before and after loading of the folic acid. IR spectra of the samples were recorded in the range of 400–4000 cm^{-1}. The UV-vis absorbance of the supernatant collected was measured using UV-vis spectrometer and the amount of folate present was calculated by plotting against a suitable standard. High performance liquid chromatography in a mobile phase used for quantification of folates loaded onto silica particles. Quantification studies resulted in 19 % loading of folates onto silica particles.

***In-Situ* Loading of Folic Acid** 2 g of folic acid was added to 100 ml of 14 % commercial food grade sodium silicate and stirred for 5 min. In 100 ml of 2.5 M of sulfuric acid, 2 % of gelatin was dissolved and stirred for 10 min. Thereafter, 54 ml of folic acid dissolved in sodium silicate solution is added drop wise to sulfuric acid-gelatin solution until neutralization (pH ~ 6.5–7). The samples obtained were centrifuged at 4000 rpm for 5 min and was dried under vacuum at 30 °C. Once the samples dried it was grinded and used for further analysis. Quantification studies resulted in ~18 % loading of folates onto silica particles. The samples were analyzed by Fourier transfer infrared spectroscopy, UV-vis spectroscopy, Thermo gravimetric analysis, BET surface area and high performance liquid chromatography.

Both in-situ and ex-situ loading methods were employed for the loading of nutrients. Water soluble folic acid was loaded on to the silica carrier molecule by both *in-situ* and *ex-situ* techniques. Iron and calcium were loaded using *in-situ* method. As loading efficiency was very poor using *ex-situ* method for minerals.

5.15 Loading of Calcium onto Silica

Calcium Chloride was used as precursor for the loading of calcium. Calcium chloride (20 g) was dissolved in 100 ml of 2.5 M Sulfuric acid and 2 % gelatin solution. 10 g of silica synthesized from sodium silicate was added to the solution and left for overnight stirring. The samples were centrifuged at 4000 rpm for 5 min and were dried under hot air oven. The dried samples were grinded into fine powder. The powder generated was analyzed by different characterization technique to ensure the loading efficiency.

5.16 Loading of Iron onto Silica

Method 1 Ferrous sulfate was used as precursor for the loading of iron. To 100 ml of 1.5 M of sulfuric acid 2 % gelatin was dissolved and stirred for 10 min. 20 g of ferrous sulfate was added to the above solution. Ascorbic acid was used as capping agent for iron to maintain it in its ferrous state. It was taken in a ratio of ferrous sulfate (1): ascorbic acid (0.5). The reaction mixture was left for stirring. Sodium Silicate solution was added drop wise and checked for neutralization (pH = 6.5–7). The samples were centrifuged at 4000 rpm for 5 min and were dried under hot air oven. The samples were dried and grinded into fine powder. The powder generated was analyzed by different characterization technique. 68 ml sodium silicate was consumed in reaction.

Method 2 100 ml of 0.5 M ortho-phosphoric acid (H_3PO_4) was taken in a 1 liter beaker and 20 g of ferrous sulfate heptahydrate ($FeSO_4.7H_2O$) was added slowly while stirring using magnetic stirrer. Added sodium silicate (7 % SiO_2 with pH value of ~13) to the above solution until pH 7 was achieved. The above solution was stirred for 1 h to facilitate precipitation. The precipitated solution was centrifuged at 4000 rpm for 5 min. The supernatant was separated and solid cake collected after centrifugation. The cake was washed twice with distilled water while centrifuging at 4000 rpm for 5 min each time, to remove any free uncoordinated iron. The pellet from the centrifuge was dried in an oven at 60 C for 12 h. Ground the dry powder obtained using a kitchen grinder. The silica-iron complex powder obtained was characterized by inductively coupled plasma (ICP-MS) method.

5.16.1 In-Vitro *Release Studies of Water Soluble Vitamins*

The *in-vitro* study was conducted to quantify the amount of folic acid, Iron and calcium leached out from the matrix in the fluid during the gastric and intestinal digestion. The recommended amount of folic acid, iron and calcium was added to the fluid. The release amount of folic acid was measured by high performance liquid chromatography (high performance liquid chromatography) and iron, calcium by inductively coupled plasma spectroscopy.

Gastric Digestion Gastric fluid was prepared to conduct the digestion process. 2 g of sodium chloride was added to 100 ml of water. 7 ml of concentrated hydrochloric acid was added drop wise. 3.2 g of Pepsin was added to the solution and the volume is made up to 1000 ml and checked for pH. The gastric fluid is acidic so the pH was maintained around 1–1.3. Pepsin was added prior to the acidification of the samples to pH2 to pH4. Acidification of the samples to pH2 or 4 is important, because pepsin begins to denature itself and thus will 1.lose its activity at pH $\geq$5.

Digestion Process The procedure used for in-vitro gastric digestion. Weighed amount of samples was added in 50 ml of gastric juices. The mixture was stirred for 2 min. The beaker was placed in a shaker incubator at 50 rpm. The temperature of shaker incubator was maintained at room temperature 37 °C. The reaction mixture was stirred continuously for 2 h and the supernatant was collected in the interval of half an hour and kept for analysis. The nutrient components present in the supernatant represent the soluble components were quantified by inductively coupled plasma spectroscopy and High performance liquid chromatography.

Intestinal Digestion Intestinal fluid was prepared to setup an experimental digestion process. 6.8 g of monobasic potassium phosphate was dissolved in 250 ml of water. To this 77 ml of 0.2 N sodium Hydroxide dissolved in 500 ml of water was mixed and stirred. 10 g of pancreatin was added and the volume was made up to 1000 ml with distil water. The pH was adjusted to 6.8–7 with 0.1 N sodium hydroxide solutions. Before the start of the intestinal digestion, the samples were neutralized to pH5.5–6 prior to the addition of pancreatin which includes pancreatic enzymes such as pancreatic amylase, lipase, ribonuclease, and proteases such as trypsin and bile salts. The final pH is adjusted between 6.5 and 7.

Digestion Process Weighed amount of mineral were added to 50 ml of intestinal juices in a beaker. The mixture was stirred for 2 min. The beaker was placed in a shaker incubator which was set at 50 rpm. The temperature of shaker incubator was maintained at room temperature 37 °C. The reaction mixture was stirred continuously for 2 h and the supernatant was collected in the interval of half an hour and kept for analysis. The nutrient components present in the supernatant represent the soluble components and were analysed by inductively coupled plasma spectroscopy (ICP), high performance liquid chromatography (high performance liquid chromatography).

5.17 Results and Discussion

5.17.1 Characterization and Analysis of Fat Soluble Nutrients on Silica Matrix

Thermo gravimetric analysis of silica loaded with vitamin E is studied. As shown in Table 5.8 in control sample which consisted of only bare food grade silica sample 1 (CS-1) the weight loss was observed in one step from 30 °C–192 °C which is due to loss of water. For bare food grade silica sample 2 (CS-2) the weight losses were seen between 30 °C–172 °C which is also due to loss of water; as shown below in Fig. 5.5a, b. After loading of the vitamins onto the silica the weight loss was seen in three steps, addition to water loss two more steps of weight loss was noticed. For CS-1 loaded with vitamin E loss was seen in between 201.82 °C–362.73 °C and 362.73 °C–515.45 °C as shown in Fig. 5.6a. For CS-2 loaded with Vitamin E the weight loss was seen in between 202.67 °C–346.93 °C and 346.93 °C–451.42 °C as shown in Fig. 5.6b.

From the thermo gravimetric measurement data one can clearly see that after loading of vitamin E onto silica matrix there is a significant weight loss which was not seen in bare silica, and the weight loss is around the temperature corresponding to the boiling point of vitamin E i.e. 201 °C. Hence the percentage loss in weight can be due to the loss of vitamin E from the silica matrix, as silica itself is very stable at high temperature. Hence from the TGA analysis it was found that loading percentage of vitamin E on CS-1 was 19.937 % and on CS-2 was 18.994 %. TGA could not be conducted for vitamin A and vitamin D due to its low melting point.

UV-Visible (UV–vis) Spectroscopy Once vitamins are loaded onto silica matrix, it is washed as per procedure explained in the experimental section. The absorbance of the supernatant obtained after centrifugation and after washing was taken.

Table 5.8 Thermo gravimetric analysis

Sample code	Temperature (°C)	Percentage loss (%)	Percentage remaining (%)
Bare food grade commercial silica (CS-1)	30.00	–	99.791
	30–192.66	6.506	93.285
	217.63–573.68	1.607	91.582
Bare food grade commercial silica (CS-2)	30.00	–	99.637
	30–172.76	6.373	93.264
	316.21–713.26	2.259	90.124
	30.7–179.64	7.592	91.628
	274.22–725.78	3.753	86.644
Vitamin E loaded on CS-1 Silica (*ex-situ*)	30.2	–	98.611
	30.2–149.09	3.007	95.604
	201.82–362.73	15.296	79.876
	362.73–515.45	4.641	75.235
Vitamin E loaded on CS-2 Silica (*ex-situ*)	30.00	–	100.026
	30–135.17	2.892	97.133
	202.67–346.93	13.604	82.818
	346.93–451.42	4.188	78.63
	561.47–742.71	1.202	75.162

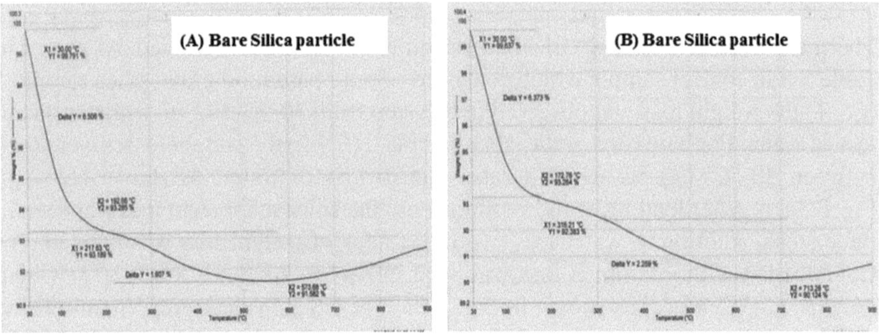

Fig. 5.5 (**a**) Thermo gravimetric measurement of commercial food grade silica particles (CS-1); (**b**) Thermo gravimetric measurement of commercial food grade silica particles (CS-2)

Vitamin A showed absorbance at 325 nm, vitamin D at 264 nm and vitamin E at 285 nm. The absorbance was quantified against a standard. The absorbance and standards of the vitamins are shown below in Figs. 5.7, 5.8, and 5.9. From the UV-vis analysis the percent of vitamins in the solution was calculated, and from that the percent remaining in the powder was estimated. Table 5.9 lists the percent of vitamins loaded onto silica matrix.

High Performance Liquid Chromatography The high performance liquid chromatography analysis of the vitamins loaded onto silica powder was carried out by extracting the vitamins from the powder by standard extraction technique.

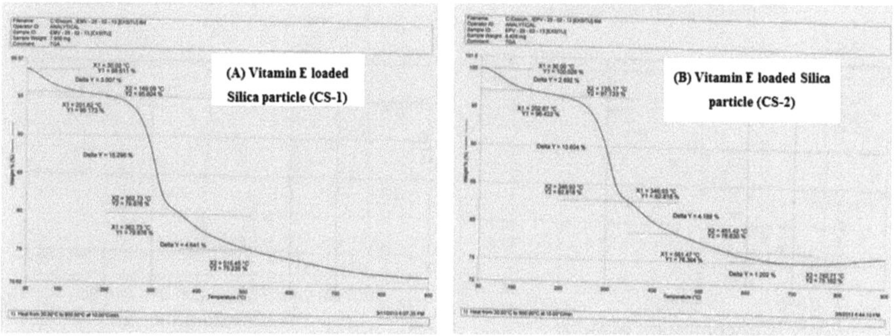

Fig. 5.6 (**a**) Thermo gravimetric measurement of silica particles (CS-1) loaded with vitamin E; (**b**) Thermo gravimetric measurement of silica particles (CS-2) loaded with vitamin E

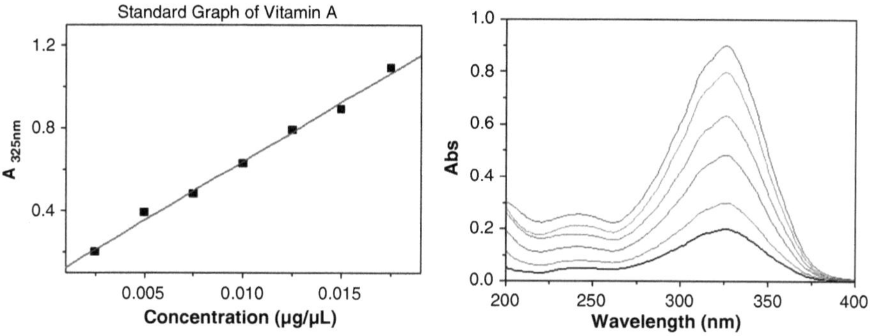

Fig. 5.7 The absorbance of the supernatant and washed solution of vitamin A was quantified using a standard plotted absorbance against concentration as shown above

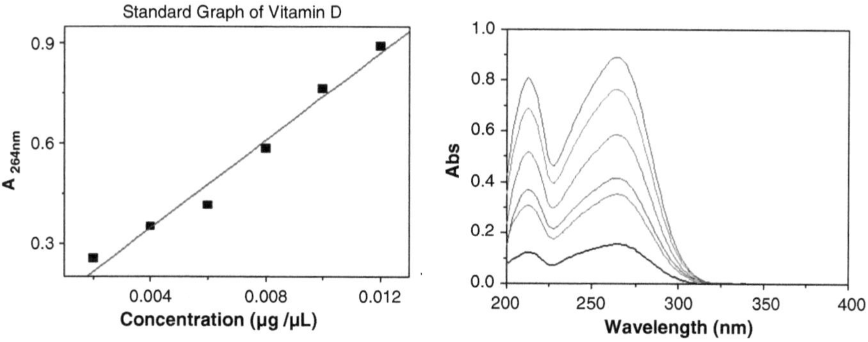

Fig. 5.8 The absorbance of the supernatant and washed solution of vitamin D was quantified using a standard plotted absorbance against concentration as shown above

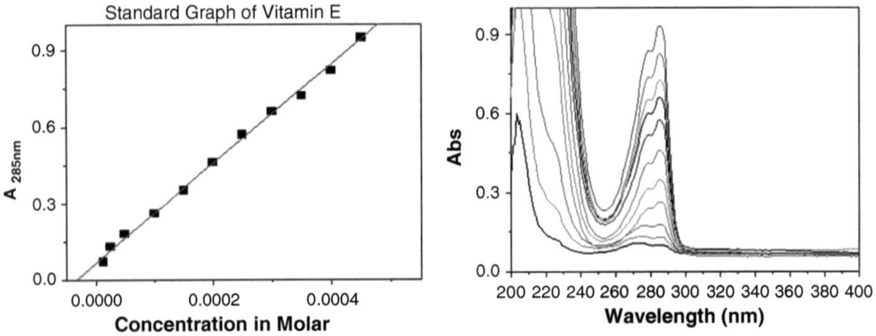

Fig. 5.9 The absorbance of the supernatant and washed solution of Vitamin E was quantified using a standard plotted absorbance against concentration as shown above

Table 5.9 Estimation of vitamins loaded onto silica matrix through UV–vis spectroscopy analysis

Sample	Silica	Loaded percent (%)
Vitamin A	CS-1	13.7
	CS-2	15.2
Vitamin D	CS-1	8.3
	CS-2	7.2
Vitamin E	CS-1	21.38
	CS-2	20.49

Table 5.10 Loading percentage estimated by high performance liquid chromatography (HPLC)

Samples		Percentage estimated by HPLC method (%)
Vitamin A	CS-1	9.87
	CS-2	8.78
Vitamin D	CS-1	6.33
	CS-2	5.21
Vitamin E	CS-1	17.79
	CS-2	15.27

Calibration curve was obtained analyzing five times each, five different solutions of known concentrations of analytes which included 10 ppm, 25 ppm, 50 ppm, 75 ppm and 100 ppm of concentration. The curve equation $y = bx + m$ calculated with linear regression method was used to determine samples concentration. Results are reported in Table 5.10.

Fourier Transform Infrared Spectroscopy (FTIR) In order to understand the characteristic of vitamins loaded onto silica, FTIR spectroscopy was used. FTIR spectroscopy will help to understand whether there is any interaction between the loaded vitamin molecules with the silica matrix. The FTIR spectra of different vitamins loaded onto silica are shown below in the figures (Figs. 5.10, 5.11, 5.12, and 5.13).

5.17.2 Stability Studies of Fat Soluble Vitamins–Silica Complex

FTIR spectrum of the same sample was taken after an interval of time to check the stability of the samples. It was noticed that there was no change in peaks in the IR spectrum for any of the samples of silica loaded with vitamins. As shown in Figs. 5.14, 5.15, and 5.16, all the significant peaks of the vitamins observed in the loaded silica are retained by the silica even after ~2 months under storage at the room temperature.

5.18 Zeta Potential Measurements

The results of zeta potential for silica particles without and with different vitamins are shown in Table 5.11. Zeta potential was reported as the average and standard deviation of measurements, with five reading taken per sample. From the table below we can see that zeta potential of silica particles after vitamin loading clearly

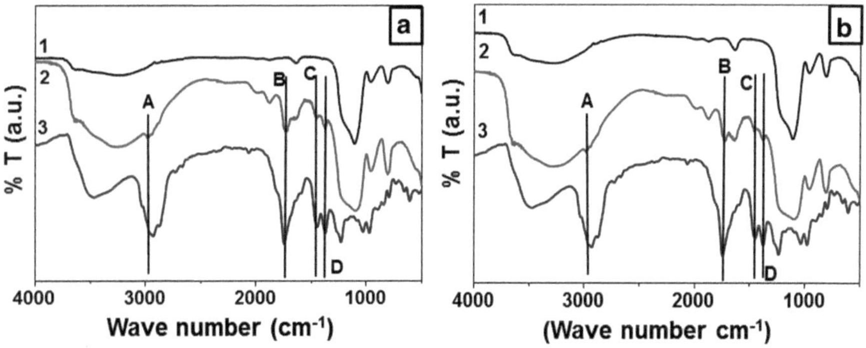

Fig. 5.10 (**a**): CS-1 loaded with vitamin A-acetate via *ex-situ* method. Here *curve 1* belongs to bare silica, *Curve 2* belongs to vitamin A loaded onto CS-1 and *curve 3* is for standard vitamin A. ***A*** CH_2/CH_3 asymmetrical stretching was seen at 2978 cm^{-1} which slows a slight shift which may be due to interaction between the CH/CH and the silica molecules. ***B*** C=O stretching was seen at 1735 cm^{-1}. ***C*** C-CH bending at 1439 cm^{-1}. D) OH bending at 1372 cm^{-1}
(**b**): CS-2 loaded with vitamin A-Acetate via ex-situ method. Here also *curve 1* belongs to bare silica, *curve 2* belongs to vitamin A loaded onto CS-2 and *curve 3* is for standard vitamin A. ***A*** CH2/CH3 asymmetrical stretching was seen at 2978 cm^{-1} which slows a slight shift which may be due to interaction between the CH2/CH3 and the silica molecules. ***B*** C=O stretching was seen at 1735 cm^{-1}. ***C*** C-CH bending at 1439 cm^{-1}. ***D*** OH bending at 1372 cm^{-1}

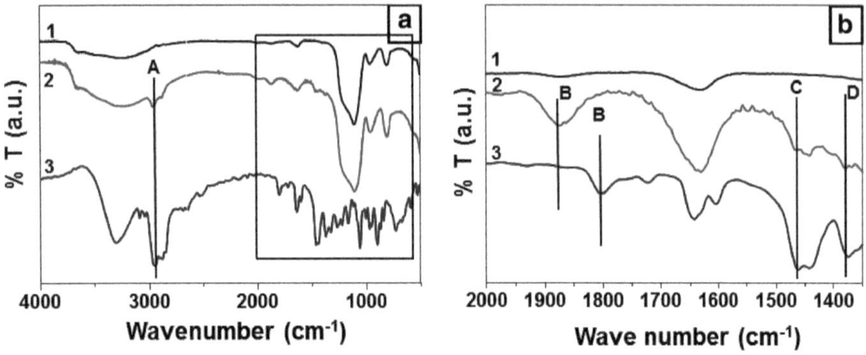

Fig. 5.11 (**a**): CS-1 loaded with vitamin D via *ex-situ* method. Here *curve 1* belongs to bare silica, *Curve 2* belongs to vitamin D loaded onto CS-1 and *curve 3* is for standard vitamin D. ***A*** CH_2/CH_3 asymmetrical stretching at 2948 cm^{-1}. Squared region expanded in Fig. 5.8b
(**b**): ***B*** C=C stretching shifts from 1800 to 1870 cm^{-1}; ***C*** CH bending at 1450 cm^{-1}. ***D*** OH bending at 1370 cm^{-1}

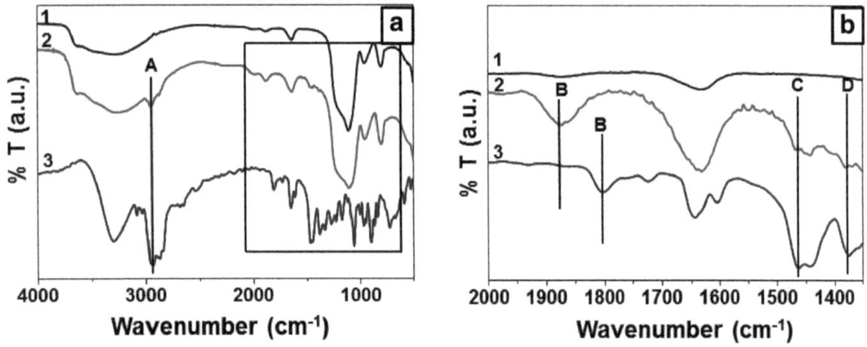

Fig. 5.12 (**a**): CS-2 loaded with vitamin D via *ex-situ* method. Here *curve 1* belongs to bare silica, *curve 2* belongs to Vitamin D loaded onto CS-1 and the *curve 3* is for standard vitamin D. ***A*** CH_2/CH_3 asymmetrical stretching at 2948 cm^{-1}. Squared region expanded in Fig. 5.9b
(**b**): C=C stretching shifts from 1800 cm^{-1} to 1870 cm^{-1}. ***C*** CH bending at 1450 cm^{-1}. ***D*** OH bending at 1370 cm^{-1}

shows significance change. This change is clearly due to the attachment of the vitamins onto silica surface as all the vitamins itself bear a negative zeta potential.

5.18.1 Release Profile Studies of Fat Soluble Vitamins from Silica Matrix

Bioavailability is defined as the amount of an ingested nutrient that is absorbed and available for physiological functions, is dependent on digestion, release from the food matrix, absorption by intestinal cells, and transport to body cells. *In-vitro*

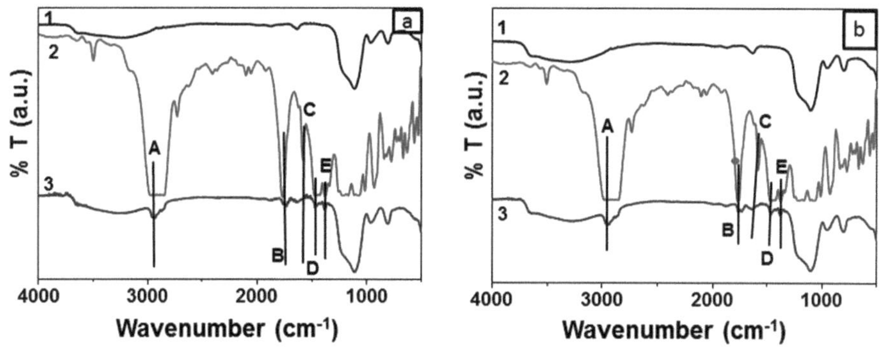

Fig. 5.13 (**a**) CS-1 loaded with vitamin E-Acetate via *ex-situ* method. Here *curve 1* belongs to bare silica, *curve 2* belongs to vitamin E loaded onto CS-1 and *curve 3* is for standard vitamin E. ***A*** CH_2/CH_3 asymmetrical stretching at 2946 cm^{-1}. ***B*** C=O stretching with a slight shift from 1763 cm^{-1} to 1735 cm^{-1}. ***C*** C-C=C stretching shifts from 1577 cm^{-1} to 1633 cm^{-1}. ***D*** Phenyl skeletal bending at 1465 cm^{-1}. ***E*** Methyl symmetrical bending at 1372 cm^{-1}
(**b**) CS-2 loaded with vitamin E-Acetate via *ex-situ* method. Here *curve 1* belongs to bare silica, *curve 2* belongs to vitamin E loaded onto CS-1 and *curve 3* is for standard vitamin E. ***A*** CH_2/CH_3 asymmetrical stretching at 2946 cm^{-1}. ***B*** C=O stretching with a slight shift from 1763 cm^{-1} to 1735 cm^{-1}. ***C*** C-C=C stretching shifts from 1577 cm^{-1} to 1633 cm^{-1}. ***D*** Phenyl skeletal bending at 1465 cm^{-1}. ***E*** Methyl symmetrical bending at 1372 cm^{-1}

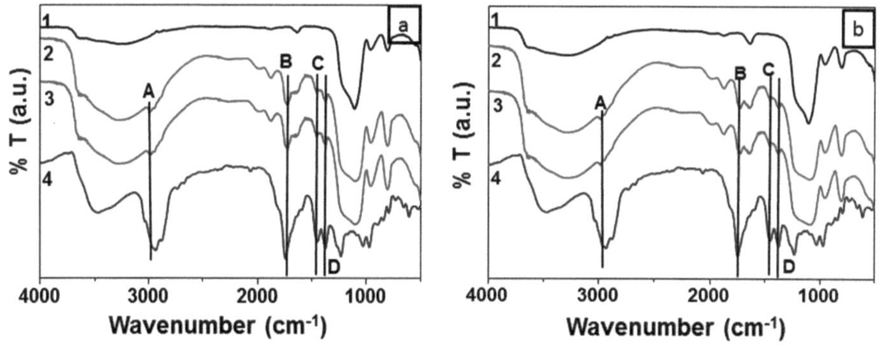

Fig. 5.14 (**a**): CS-1 loaded with vitamin A-Acetate via *ex-situ* method. Here *curve 1* belongs to bare silica, *curve 2* belongs to vitamin A loaded onto CS-1 on Day 1, *curve 3* is vitamin A loaded onto CS-1 on 60th day and *curve 4* is for standard vitamin A. As compare to previous results no loss of peaks are observed
(**b**): CS-2 loaded with vitamin A-Acetate via ex-situ method. Here *curve 1* belongs to bare silica, *curve 2* belongs to vitamin A loaded onto CS-2 on Day 1, *curve 3* belongs to vitamin A loaded onto CS-2 on 60th day and *curve 4* is for standard vitamin A. As compare to previous results no loss of peaks are observed

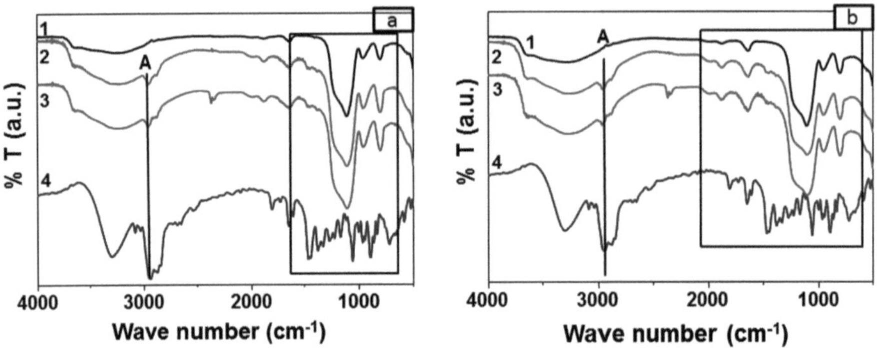

Fig. 5.15 (**a**): CS-1 loaded with vitamin D via ex-situ method. Here *curve 1* belongs to bare silica, *curve 2* belongs to vitamin D loaded onto CS-1 on Day 1, *curve 3* belongs to vitamin D loaded onto CS-1 on 60th day and *curve 4* is for standard vitamin D. As compare to previous results no loss of peaks are observed
(**b**): CS-2 loaded with vitamin D via ex-situ method. Here *curve 1* belongs to bare silica, *curve 2* belongs to vitamin D loaded onto CS-2 on Day 1, *curve 3* belongs to vitamin D loaded onto CS-2 on 60th day and *curve 4* is for standard vitamin D. As compare to previous results no loss of peaks are observed

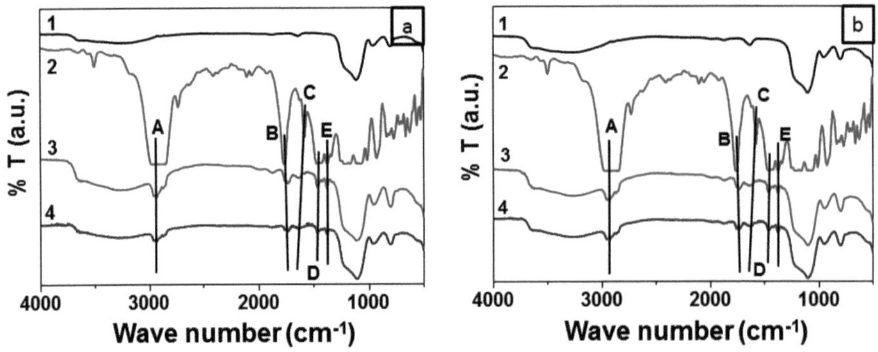

Fig. 5.16 (**a**): CS-1 loaded with vitamin E via ex-situ method. Here *curve 1* belongs to bare silica, *curve 2* belongs to vitamin E loaded onto CS-1 on Day 1, *curve 3* belongs to vitamin E loaded onto CS-1 on 60th day and *curve 4* is for standard vitamin E. As compare to previous results no loss of peaks are observed
(**b**): CS-2 loaded with vitamin E via ex-situ method. Here *curve 1* belongs to bare silica, *curve 2* belongs to vitamin E loaded onto CS-2 on Day 1, *curve 3* belongs to vitamin E loaded onto CS-2 on 60th day and *curve 4* is for standard vitamin E. As compare to previous results no loss of peaks are observed

bioaccessibility/bioavailability methods are useful to understand interactions between nutrients and food components. The effects of luminal factors such as pH and enzymes, nature of the matrix, processing protocols on micronutrient absorbability can be well understand by in-vitro bioavailability methods. *In vitro* methods are less expensive, faster, and offer better controls of experimental variables than

Table 5.11 Zeta potential of silica with and without vitamins

Sample	Zeta potential
CS-1	−40.2 ± 0.6
CS-2	−30.6 ± 0.7
Vitamin A	−7.36 ± 0.1
Vitamin A + CS-1	−23.2 ± 0.3
Vitamin A + CS-2	−25.1 ± 0.2
Vitamin D3	−12.5 ± 0.28
Vitamin D3+ CS-1	−4.32 ± 0.96
Vitamin D3 + CS-2	−8.44 ± 0.56
Vitamin E	–
Vitamin E + CS-1	−2.51 ± .036
Vitamin E + CS-2	−8.44 ± 0.98

human or animal studies (Sandberg 2005). However, *in-vitro* studies cannot be substituted for *in vivo* studies, and should be therefore regarded as a screening, or categorizing method. These are methods that can provide knowledge, when vast number of factors is considered that can affect nutrient absorption. The release of nutrient from matrix of the carrier molecule silica is conducted through in in-vitro studies. An *in-vitro* digestion includes the experimental setup of human digestive system through two-step digestion process that includes a gastric and intestinal digestion. The nutrients were first dissolved in gastric fluid for 2 h. The mixture was centrifuged and the pellet was added to the intestinal fluid.

A release profile of the vitamins D was studied under gastric and intestinal simulated conditions. The values obtained by high performance liquid chromatography analysis are plotted as time versus concentrations as shown in Fig. 5.17a, b.

From the graph one can clearly understand that the release of the vitamin occurs mostly in the intestinal fluids and not in the gastric fluids. A continuous release profile of the vitamins E was studied under gastric and intestinal simulated conditions. The values obtained by high performance liquid chromatography analysis are plotted as time versus concentrations as shown in Fig. 5.18a, b.

5.18.2 Discussion

Simple homogenization process of vitamins solutions stirred with silica was used for loading of the silica particles. The characterization of vitamin samples by various techniques showed that the vitamins are loaded onto silica particles. Though

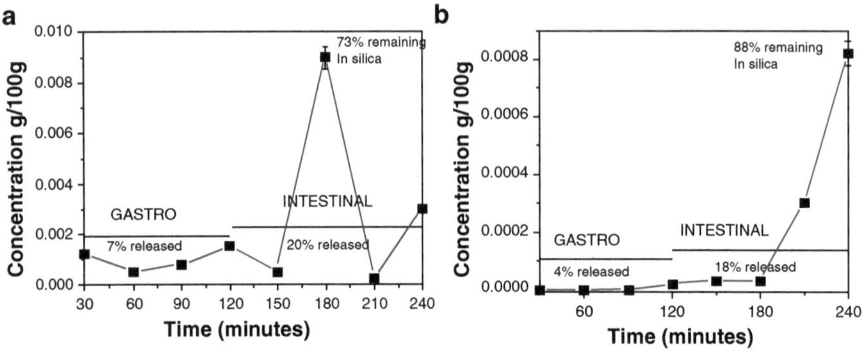

Fig. 5.17 (**a**) shows the release profile of vitamin D from CS-1; (**b**) shows the release profile of vitamin D from CS-2

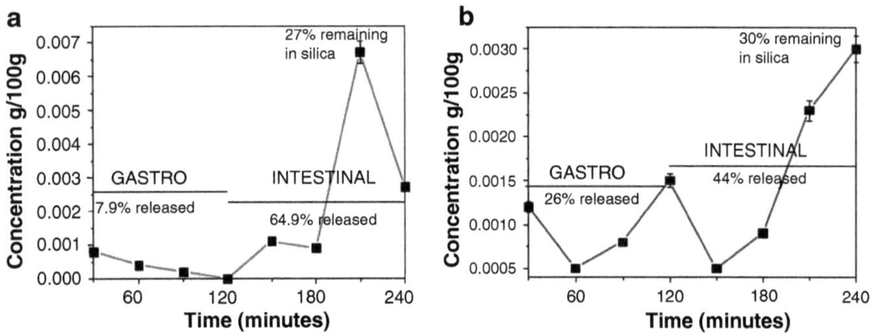

Fig. 5.18 (**a**) Shows the release profile of vitamin E from CS-1; (**b**) Shows the release profile of vitamin E from CS-2

structural features of the loaded silica does not show any visible change before and after loading, however FTIR and Zeta potential results show distinct changes. This clearly indicates the silica particles retain their structural functionality even after loading and there is surface interaction between the silica and vitamins. The stability studies of vitamins loaded silica matrix show that natural vitamins which are easily destroyed in natural environment due to light or oxidation, are found to be stable up to 60 days even after storage at room temperature. These samples were then subjected to various simulated environments to test for their release from silica matrix. The release profile studies were carried out in simulated gastric and intestinal environments. All the vitamins showed negligible release in the simulated gastric juice whereas considerable amount was released in the intestinal simulated fluids. This characteristic is an important feature as all the fat-soluble vitamins are absorbed in the intestinal region of the human gastro-intestinal tract. Bile forms a micelle and helps in the transport of these vitamins into the blood vessels via the villi of small intestinal region.

5.18.3 Analysis of Water Soluble Nutrients on Silica Matrix

Loading of nutrient on to silica was performed using both *in-situ* as well as *ex-situ* process. Calcium, folic acid loading was done by *ex-situ* method while iron loading was done using *in-situ* method. The loading percentage of calcium and iron was estimated by inductively coupled plasma spectroscopy (ICP). Folic acid loading percentage was determined by high performance liquid chromatography (Tables 5.12 and 5.13).

UV- Visible Spectrophotometry The supernatant obtained from silica/folic acid synthesis was analyzed by UV–vis spectrophotometry to determine the amount of folic acid loaded within the silica matrix. Folic acid showed absorbance at 280 nm. The absorbance was quantified against a standard. The absorbance and standard of the folic are shown below in Fig. 5.19. From the UV–vis analysis the percent of folic acid in the solution was calculated, and from that the percent remaining in the powder was estimated. The percentage loading result is shown in Table 5.14.

The graph of the linear relationship between folic acid concentration (mg/10 mL) and absorbance at 280 nm as obtained from the samples with known folic acid concentration is shown in the Fig. 5.19a and UV spectra of the solution with a known concentration of folic acid is in Fig. 5.19b.

Fourier Transformation Infrared Spectroscopy The structural characteristics of silica particles with folic acid loaded onto it and the qualitative composition of the sample is confirmed by IR spectroscopy. The IR spectra of folic acid loaded on CS-1 silica particles are presented in Fig. 5.20a. The IR spectra of iron and calcium loaded on silica are presented in Fig. 5.20b, c.

Besides the characteristic groups for silica, the spectra also shows characteristic groups for folic acid loaded on Silica (CS-1). The significant peaks of folic acid are found at 3545 cm^{-1} belongs to the hydroxyl (O-H) stretching, while the peaks at 3328 cm^{-1} are N-H stretching vibration. C-O bond stretching vibration of carboxyl group appears at 1695 cm^{-1}, while the band at 1644 cm^{-1} belongs to C-O bond

Table 5.12 Loading of iron and calcium percentage onto silica matrix as estimated by inductively coupled plasma spectroscopy

Sample	Loading percentage (%)
Iron loading on to food grade silica	21
Calcium loading on to food grade silica	10

Table 5.13 Loading folic acid percentage onto silica matrix as estimated by high performance liquid chromatography

Sample	Loading percentage (%)
Folic acid loading on to commercial food grade silica (CS 1)	10

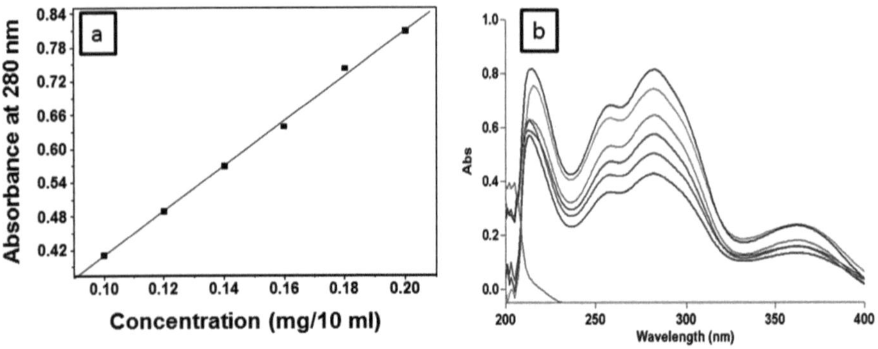

Fig. 5.19 (**a**) Graph of the linear relationship between absorbance and known folic acid concentration; (**b**) UV spectra of the solution with a known concentration of folic acid

Table 5.14 Folic acid loading percentage estimated by UV–vis spectrophotometer

Sample	Percentage loading (%)
Folic acid loading on to commercial food grade silica (CS 1)	10.6

stretching vibration of -$CONH_2$ group. The band at 1602 cm^{-1} relates to the bending mode of N H vibration and at 1482 cm^{-1} demonstrates the characteristic absorption band of phenyl ring. All these peaks are visible and shown in Fig. 5.20a.

In Fig. 5.20b, the band at 3545 cm^{-1} which belongs to the hydroxyl (O-H) stretching, 3328 cm^{-1} belongs to N-H stretching vibration peak, 1695 cm^{-1} belongs to C-O bond stretching vibration of carboxyl group, 1644 cm^{-1} belongs to C-O bond stretching vibration of -$CONH_2$ group and 1602 cm^{-1} band relates to the bending mode of N-H vibration. There is band shift of Si-O-Si asymmetrical bond at 1100 cm^{-1} in the (calcium loaded silica sample) and there is a loss of peak at 800 cm^{-1} belongs to Si-O-Si asymmetrical bond.

Iron does not have its characteristic peaks until it is not present in its complex form. The Fig. 5.20c shows that there is no major change in the significant peaks of silica and hence its stability is not affected by the minerals loaded onto it.

Zeta Potential Measurement The determinations of zeta potential for silica particles without and with loaded nutrient are shown in Table 5.15. Zeta potential was reported as the average and standard deviation of measurements, with five readings taken per sample. Zeta potential is an important property of the particle in dispersion as it has exerts a significant influence on their stability based on surface charge.

From Table 5.15 we can observe that there is fluctuation in the charge distribution of nutrient loaded silica when compared with their respective bare silica. Folic acid is negatively charged particles so it develops a net charge on the surface of silica. It affects the ion distribution and increases the charge of silica loaded with folic acid. Both iron and calcium are divalent cations and so it affects the charge

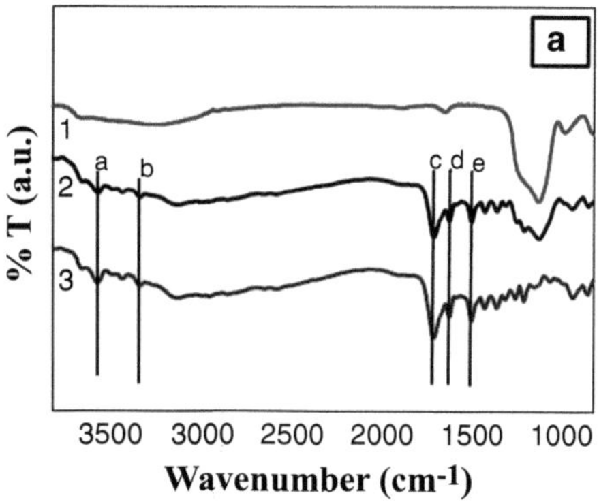

Fig. 5.20a (**a**) Spectra 1- FTIR spectra of commercial silica (CS-1); Spectra 2- FTIR spectra of folic acid loaded onto commercial Silica; Spectra 3- FTIR spectra of pure folic acid

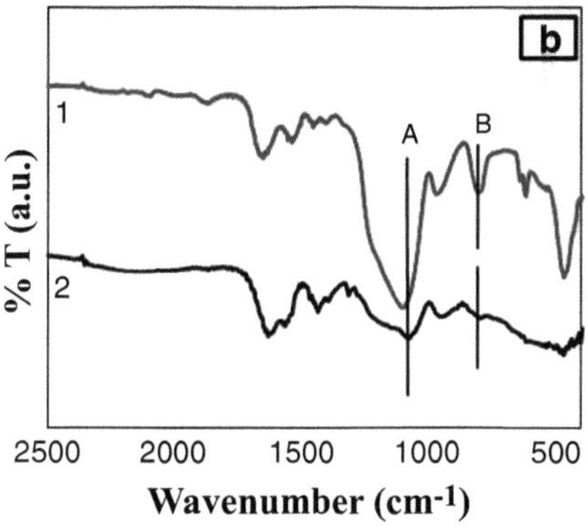

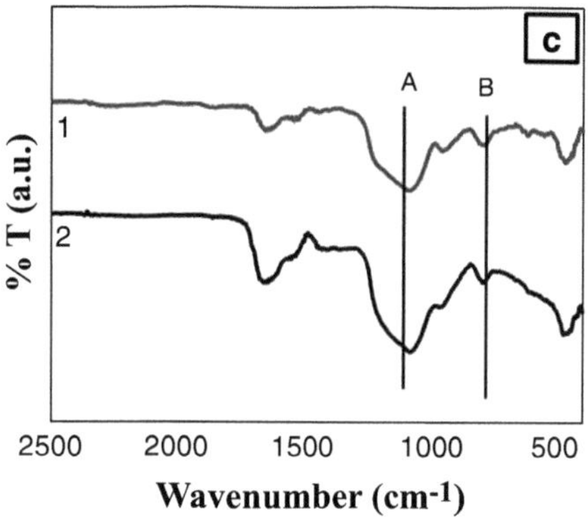

Fig. 5.20b–c (**b**) FTIR spectra of bare silica (*Curve 1*) and FTIR spectra of calcium loaded onto silica matrix (*Curve 2*). (**c**) FTIR spectra of bare silica (*Curve 1*) and FTIR spectra of iron loaded onto silica matrix (*Curve 2*)

Table 5.15 Zeta potential measurement of silica and silica loaded with folic acid, iron and calcium

Sample	Zeta potential
Silica (CS1)	(–) 38.8 ± 0.1
Folic acid + Silica	(–) 49.5 ± 0.1
Iron + Silica	(–) 24 ± 0.1
Calcium + Silica	(–) 14.9 ± 0.1

distribution of silica. The sample calcium loaded silica showing reduction in the charge as calcium itself is positively charge and stabilizes with negatively charged silica.

Energy-Dispersion Analysis of x-ray (EDAX) EDAX is a qualitative analysis technique for chemical composition of a sample. A chemical mapping of the nutrient loaded onto silica matrix was done using EDAX and the presence of element in each sample was analyzed (Figs. 5.21a–c, 5.21d–e, and 5.21f–g.

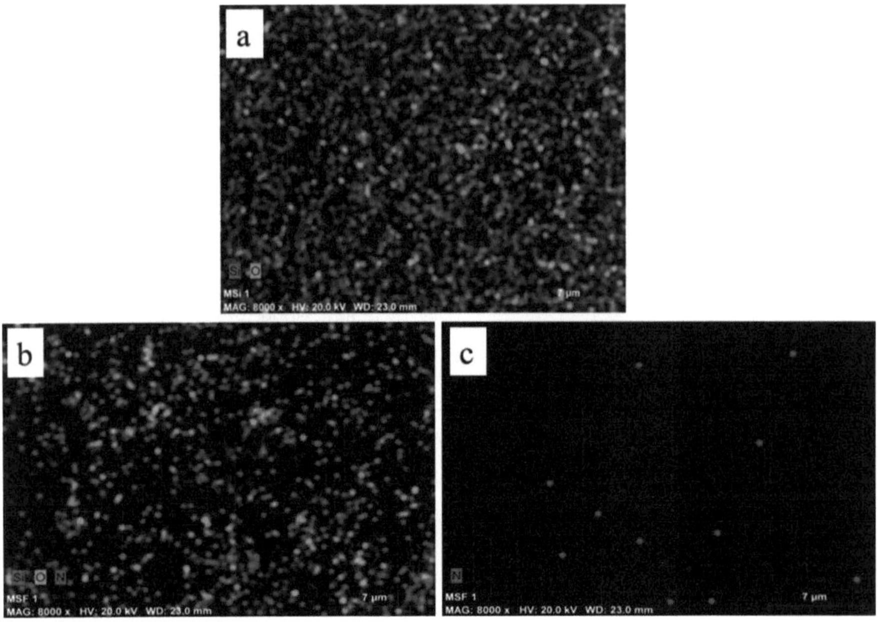

Fig. 5.21a–c (**a**) Chemical mapping (EDAX) of silica powder depicts presence of Si and O elements; (**b**) Chemical mapping of silica loaded with folic acid depicting presence of 3 elements such as Si, Oand N; (**c**) Chemical mapping of only nitrogen in the silica loaded folic acid sample (by masking Si and O elements)

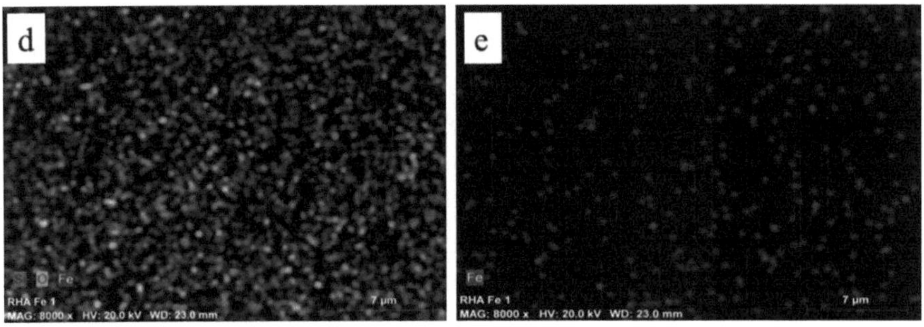

Fig. 5.21d–e (**d**) Chemical mapping (EDAX) of silica loaded with iron depicting presence of all 3 elements in the sample such as Si, O and Fe; (**e**) Chemical mapping of iron present in the iron loaded silica sample by masking Si and O in the image

From the micrographs (Fig. 5.21), it is clearly seen that the nutrient loaded samples shows the presence of individual signature elements such as N (nitrogen signal coming from amide group of folic acid), iron, and calcium when compared with the standard silica. The EDAX spectra recorded ~7 % percentage of nitrogen present in the folic acid loaded silica sample, ~21 % iron recorded in the iron loaded silica sample, and ~12 % calcium recorded calcium loaded silica sample.

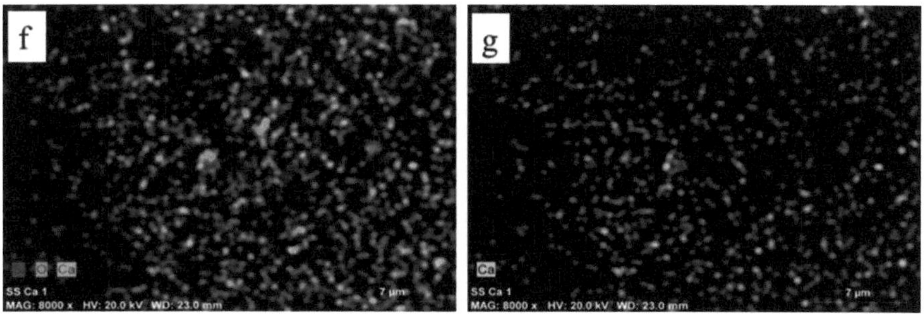

Fig. 5.21f–g (**f**) Chemical mapping (EDAX) of silica loaded with calcium depicting presence of Si, O and Ca elements. (**g**) Chemical mapping of calcium present in the calcium loaded silica sample by masking Si and O in the image

5.18.4 Release Profile Studies of Water Soluble Vitamins from Silica Matrix

The experimental study was conducted to quantify the release of loaded nutrients from the silica matrix. The protocol includes two steps process, gastric digestion and intestinal digestion. The sample is first dissolved in gastric fluid and then in intestinal fluid. After gastric digestion the sample were centrifuged and supernatant is removed and sent for analysis. The pellet obtained is then dissolved in intestinal fluid. The nutrients or compounds present in the supernatant represent the soluble components and are measured by inductively coupled plasma mass spectroscopy, high performance liquid chromatography. It is a kinetic study done over 4 h and is represented as graph in Fig. 5.22.

In Figs. 5.22, 5.23 and 5.24, it is observed that the release of nutrients from the silica CS-1 is constant till 2 h, after which a sudden increase in the release was observed with time interval. The first 2 h study was under in-vitro gastric condition between pH 1 and 2. The nutrient released in the gastric condition is negligible amount. However, in the next 2 h of study under *in-vitro* intestinal condition, the release of the micronutrient was increased. The release of nutrients in the intestinal region is more important that in the gastric region. The uptake of iron, folic acid and calcium, was of particular interest due to the way it is absorbed in the human body and processed through the intestines.

In Fig. 5.22, folic acid loaded on silica (CS-1) found to be constant in first half study then there is increase in release with time interval. The total input of the sample was 0.05 g which contains 4.8 mg of folic acid according to 9.6 % loading. The total amount released 1.07 mg. The percentage released in the leaching study was found 22.29 % which meets typical recommended dietary allowance values for folic acid.

Figure 5.23 shows the release profile of calcium from silica matrix. The release is seen to be constant in the first half study then there is increase in release with time. The total input of the sample was 1 g which contains 0.1 g of calcium accord-

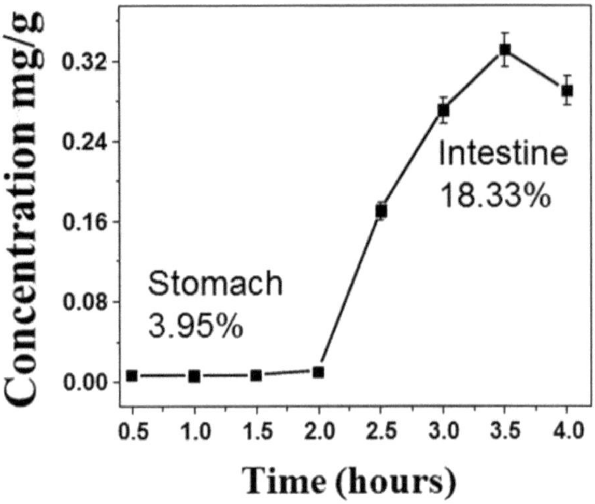

Fig. 5.22 Shows release profile of folic acid from silica matrix

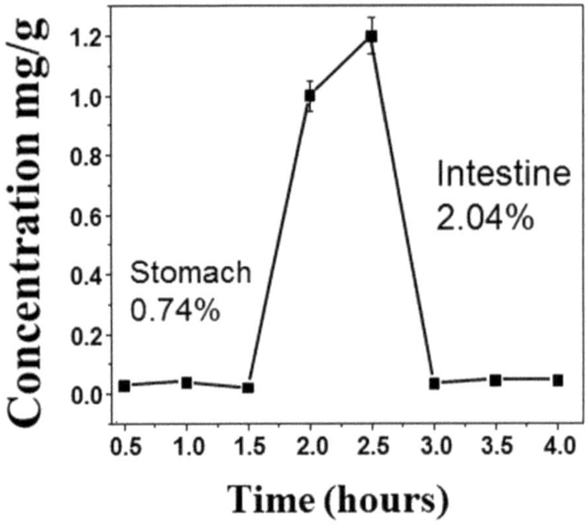

Fig. 5.23 Shows release profile of calcium from the silica matrix

ing to 10 % loading. The total amount released 2.78 mg. The percentage released in the leaching study was found 2.78 % which is found to be lower than specified recommended dietary allowance limits for calcium. Theoretically, if 10 g of sample is added to meet the recommended dietary allowance for calcium i.e. 1 g, the release will be 27.8 mg.

Figure 5.24 shows release profile of iron from silica matrix. The release is seen to be constant in the first half study and we observed variation in the release with time. The total input of the sample was 0.1 g which contains 0.01 g of iron according to 10 % loading. The total amount released 5.1 mg. The percentage released in the leaching study was found 5.1 %.

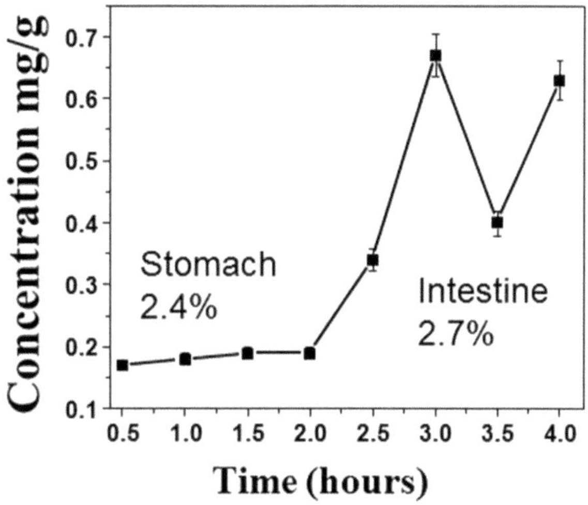

Fig. 5.24 Shows release profile of iron from the silica matrix

5.18.5 *Discussion –Water Soluble Vitamin Silica Complex*

The loading efficiency and percentage of folic acid was found to be increased when attempted with ex-situ process. Folic acid is very sensitive to light, so the samples were dried in room temperature in the dark. In this work the folic acid and silica ratio was taken as 2:10 and the percentage loading was analyzed to be 9.6 %. Calcium loading was carried out using ex-situ process. In most of the published data focuses on the use of calcium carbonate as a precursor or source for loading or encapsulation Han et al. (2007) because it is the most bioavailable form. In the present work three salts calcium carbonate, calcium chloride and calcium gluconate were attempted for the loading of calcium onto the silica matrix out of which calcium chloride shows the highest percentage loading i.e. 10 %. Though calcium chloride has low bioavailability, in the release profile study it was observed that the 2.78 % of calcium was released from the matrix. Iron loading was carried out by using both in-situ as well as ex-situ methods with and without surface modifications of silica matrix. In the ex-situ process the percentage loading was found to be very low, therefore in-situ process of loading iron onto silica was developed. Ascorbic acid was added as surface modifying agent so as to prevent the oxidation of iron from its ferrous form to ferric form. Ferrous sulfate was used as the precursors for iron, a 10 % loading of iron onto silica matrix was observed.

Silica due to its characteristic features like high porosity, negative surface charge, stability and biocompatibility forms a good substrate that can be used for carrier applications. Silica itself is a nutrient and believed to be required by body for maintenance of bone health. Two commercial food grade silica samples were used are amorphous in nature with surface area of 185 m^2/g (CS-1) and 133 m^2/g (CS-2). Based on the results presented above we have made an attempt to load both fat soluble and water soluble nutrients onto silica matrix and also demonstrated their *in-vitro* release profile under simulated gastro-intestinal tract fluids. The future aspect of the work would include the bioavailability study of these loaded silica particles and its application as food fortifications.

5.19 Conclusion

An attempt was made for the development of a nutrition loaded silica additive where the silica matrix provides chemical and physical stabilization of the entrapped/coated nutrients. Simple methods were employed for loading of nutritional molecules onto silica matrix. Detailed analysis of the nutrient loaded silica particles show stability of nutrient even after storage under ambient conditions for many weeks. The release of nutrients from silica particles in simulated gut-intestine fluids was studied. The idea of combining the inherent nutrition properties of silica with its ability to load essential nutrients could wide open the possibilities of potential use in food as an additive or in the form of a nutritional drink.

Acknowledgements NR and SS would like to acknowledge VIT University; Vellore to allow a 6 month internship training at the Tata Chemicals Ltd. Pune, India. All the authors would like to acknowledge the useful technical discussions and support of their colleagues Dr. Anand Gole, Mr. Shivkant Verma, Mr. Mangesh Kokate, Ms. Mayura Lolage of Tata Chemicals Ltd. The authors would also gratefully like to acknowledge support for this research from Tata Chemicals Ltd through its President –Innovation Centre, Dr. Arup Basu, and Head –Innovation Centre, Dr. Anil Kumar.

References

Anderson GJ, Fitzgerald C (2010) Iron: an essential nutrient. Colorado State University Extension. A division of the office of Engagement No. 9:356

Anderson GJ, Frazera DM, McLarenb GD (2009) Iron absorption and metabolism. Curr Opin Gastroenterol 25(2):129–135. doi:10.1097/MOG.0b013e32831ef1f7

Andon MB, Peacock M, Kanerva RL, De Castro JAS (1996) Calcium absorption from apple and orange juice fortified with calcium citrate malate. J Am Coll Nutr 15:313–316

Anonymous (2008) The Copenhagen consensus 2008-results. The Copenhagen Consensus cent, Frederiksberg, pp 1–6. Available at http://www.copenhagenconsensus.com/Home.aspx, Accessed 24 Apr 2015

Atkinson J, Epand RF, Epand RM (2008) Tocopherols and tocotrienols in membranes: a critical review. Free Radic Biol Med 44(5):739–764. doi:10.1016/j.freeradbiomed.2007.11.010

Aykroyd WR (1930) Functional night-blindness due to vitamin deficiency. Lancet 1:824

Azzi A (2007) Molecular mechanism of alpha-tocopherol action. Free Radic Biol Med 43(1):16–21. doi:10.1016/j.freeradbiomed.2007.03.013

Bailey LB, Gregory JFR (2006) Folate. In: Bowman B, Russell R (eds) Present knowledge in nutrition, vol 1. International Life Sciences Institute, Washington, pp 278–301

Bailey RL, Dodd KW, Gahche JJ et al (2010a) Total folate and folic acid intake from foods and dietary supplements in the United States. Am J Clin Nutr 91(1):231–237. doi:10.3945/ajcn.2009.28427

Bailey RL, Dodd KW, Goldman JA, Gahche JJ, Dwyer JT, Moshfegh AJ, Sempos CT, Picciano MF (2010b) Estimation of total usual calcium and vitamin D intakes in the United States. J Nutr 140(4):817–822. doi:10.3945/jn.109.118539

Beard JL (2001) Iron biology in immune function: muscle metabolism and neuronal functioning. J Nutr 131(2S-2):568S–580S

Becker CH, Matthias D, Wossmann H, Schwartz A, Engler E (1983) Investigations on a possible medical importance of silicon. In: Anke M, Baumann W, Briunlich H, Bruckner C (eds) Spurenelement-symposium, 4th edn. Friedrich-Schiller-Universitat, Jena, pp 142–148

Bell EF (1987) History of vitamin E in infant nutrition. Am J Clin Nutr 46:183–186

Bender DA (2003) Nutritional biochemistry of the vitamins. Cambridge University Press, Cambridge. ISBN 978-0-521-80388-5

Berchtold MW, Brinkmeier H, Muntener M (2000) Calcium ion in skeletal muscle: its crucial role for muscle function, plasticity, and disease. Am Physiol Soc 80(3):1215–1265. PMID: 10893434

Berivan T, Nuray N (2005) Importance of calcium. Turk J Med Sci 35:197–201

Bernstein LH, Gutsein S, Weiner S, Efron G (1970) The absorption and malabsorption of folic acid and its polyglutamates. Am J Med 48(5):570–579. doi:10.1016/0002-9343(70)90006-9

Bischoff-Ferrari HA, Willett WC, Orav EJ, Oray EJ, Lips P, Meunier PJ, Lyons RA, Flicker L, Wark J, Jackson RD, Cauley JA, Meyer HE, Pfeifer M, Sanders KM, Stahelin HB, Theiler R, Dawson-Hughes B (2012) A pooled analysis of vitamin D dose requirements for fracture prevention. N Engl J Med 367(1):40–49. doi:10.1056/NEJMoa1109617

Bjelakovic G, Gluud LL, Nikolova D, Whitfield K, Wetterslev J, Simonetti RG, Bjelakovic M, Gluud C (2011) Vitamin D supplementation for prevention of mortality in adults cochrane database of systematic reviews July. Cochrane Database Syst Rev. doi:10.1002/14651858.CD007470.pub2

Blakley RL (1969) The biochemistry of folic acid and related pteridines. North Holland research monographs. In: Newbergen H, Taton EL (eds) Frontiers of biology, vol 13. North Holland Publishing Company, Amsterdam, p 569

Brattstrom L (1996) Vitamins as homocysteine-lowering agents. J Nutr 126(4):1276S–1280S

Brigelius-Flohe B, Traber MG (1999) Vitamin E: function and metabolism. FASEB 13(10):1145–1155

Britto D, Moura MR, Aouada AF, Mattoso HCL, Assis OBG (2012) N, N, N-trimethyl chitson nanoparticles as a vitamin carrier system. Food Hydrocoll 27:487–493. doi:10.1016/j.foodhyd.2011.09.002

Brown EM (1991) Extracellular Ca2+ sensing, regulation of parathyroid cell function, and role of Ca2+ and other ions as extracellular (first) messengers. Am Physiol Soc 71(2):371–411. PMID: 2006218

Calvo MS, Whiting SJ, Barton CN (2005) Vitamin D intake: a global perspective of current status. J Nutr 135(2):310–316. PMID: 15671233

Carmel R (2005) Folic acid. In: Shils M, Shike M, Ross A, Caballero B, Cousins R (eds) Modern nutrition in health and disease. Lippincott Williams and Wilkins, Baltimore, pp 470–481

Chasteen ND, Harrison PM (1999) Mineralization in ferritin: an efficient means of iron storage. J Struct Biol 126(3):182–194. doi:10.1006/jsbi.1999.4118

Chervenkov T, Ivanova D, Galunska B, Gerova D, Yankova T (2007) Toxicity of polymeric nanoparticles with respect to their application as drug carriers. In: Simeonova P, Opopol N, Luster M (eds) Nanotechnology- toxicological issue and environmental safety. Springer, Dordrecht, pp 111–118

Chung M, Balk EM, Brendel M, Ip S, Lau J, Lee J, Lichtenstein A, Patel K, Raman G, Tatsioni A, Terasawa T, Trikalinos TA (2009) Vitamin D and calcium: a systematic review of health outcomes. Evid Rep/Technol Assess 183:1–420

Ciriminna R, Sciortino M, Alonzo G, Schrijver AD, Pagliaro M (2011) From molecules to systems: sol-gel microencapsulation in silica-based materials. Chem Rev 111:765–789

Cranney A, Horsley T, O'Donnell S, Weiler H, Puil L, Ooi D, Atkinson S, Ward L, Moher D, Hanley D, Fang M, Yazdi F, Garritty C, Sampson M, Barrowman N, Tsertsvadze A, Mamaladze V (2007) Effectiveness and safety of vitamin D in relation to bone health. Evidence report/ technology assessment August (158):1–235

Cranney A, Weiler HA, O'Donnell S, Puil L (2008) Summary of evidence-based review on vitamin D efficacy and safety in relation to bone health. Am J Clin Nutr August 88(2):513S–519S

Crider KS, Bailey LB, Robert J, Berry L (2011) Folic acid food fortification—its history, effect, concerns, and future directions. Nutrients 3(3):370–384. doi:10.3390/nu3030370

Crissey SD, Ange KD, Jacobsen KL, Slifka KA, Bowen PE, Stacewicz-Sapuntzakis M, Langman CB, Sadler W, Kahn S (2003) Serum concentrations of lipids, vitamin D metabolites, retinol, retinyl esters, tocopherols and selected carotenoids in twelve captive wild felid species at four zoos. J Nutr 133(1):160–166

Cuskelly GJ, McNulty H, Scott JM (1996) Effect of increasing dietary folate on red-cell folate: implications for prevention of neural tube defects. Lancet 347(9002):657–659. doi:10.1016/ S0140-6736(96)91205-2

Devaraj S, Hugou I, Jialal I (2001) Tocopherol decreases CD36 expression in human monocyte-derived macrophages. J Lipid Res 42:521–527

Emerson G, Folkers K (1951) Water soluble vitamins. Annu Rev Biochem 20:559–598. doi:10.1146/annurev.bi.20.070151.003015

Eruvbetine D (2003) Canine nutrition and health. A paper presented at the seminar organized by Kensington Pharmaceuticals Nig. Ltd., Lagos on August 21, 2003

Fennema O (2008) Fennema's food chemistry. CRC Press Taylor and Francis, Boca Raton, pp 454–455

Food and Agriculture Organization/World Health Organization (1967) Report of the Joint FAO/ WHO Expert Group. Requirements of vitamin A, thiamine, riboflavin and niacin, WHO technical report series no. 362. WHO, Geneva

Food and Agriculture Organization/World Health Organization (2004) Vitamin and mineral requirements in human nutrition, 2nd edn, WHO, Geneva. Annex 2

Food and Nutrition Board, Institute of Medicine (1998) Dietary reference intakes: thiamin, riboflavin, niacin, vitamin B6, folate, vitamin B12, pantothenic acid, biotin and choline. National Academy Press, Washington. doi:10.17226/6015

Food and Nutrition Board, Institute of Medicine (2001a) Dietary reference intakes for vitamin A, vitamin K, arsenic, boron, chromium, copper, iodine, iron, manganese, molybdenum, nickel, silicon, vanadium and zinc. National Academy Press, Washington. doi:10.17226/10026

Food and Nutrition Board, Institute of Medicine (2001b) Dietary reference intakes. National Academy Press, Washington. doi:10.17226/10026

Fragakis AS (2003) The health professional's guide to popular dietary supplements. The American Dietetic Association, Chicago

Fukuwatari T, Shibata K (2008) Urinary water-soluble vitamins and their metabolic contecnts as nutritional markers for evaluating vitamin intakes in young Japanese women. J Nutr Sci Vitaminol 54(3):223–229. doi:10.3177/jnsv.54.223

Gagne A, Wei SQ, Fraser WD, Julien P (2008) Absorption, transport, and bioavailability of vitamin E and its role in pregnant women. J Obstet Gynaecol Can 31(3):210–217

Gharsallaoui A, Roudaut G, Chambin O, Voilley A, Saurel R (2007) Applications of spray-drying in microencapsulation of food ingredients. Food Res Int 40:1107–1121. doi:10.1016/j. foodres.2007.07.004

Gomber S, Kumar S, Rusia U (1998) Prevalence and etiology of nutritional anemias in early childhood in an urban slum. Ind J Med Res 107:269–273. PMID: 9701895

Goswami R, Mishra SK, Kochupillai N (2008) Prevalence and potential significance of vitamin D deficiency in Asian Indians. Indian J Med Res 127:229–238. PMID: 18497436

Gouin S (2004) Microencapsulation: industrial appraisal of existing technologies and trends. Trends Food Sci Technol 15:330–347

Grantham McGregor SM, Ani CC (1999) The role of micronutrients in psychomotor and cognitive development. Br Med Bull 55(3):511–527. doi:10.1258/0007142991902583

Halberg L, Rossander- Hultein L, Brune M, Gleerup A (1992) Calcium and iron absorption: mechanism of action and nutritional importance. Eur J Clin Nutr 46(5):317–327. PMID: 1600930

Han MR, Kwon MC, Lee HY, Kim JC, Kim JD, Yoo SK, Sin IS, Kim SM (2007) pH-dependent release property of alginate beads containing calcium carbonate particles. J Microencapsul 24(8):787–796. doi:10.1080/02652040701657820

Heaney RP (2002) The importance of calcium intake for lifelong skeleton health. Calcif Tissue Int 70(2):70–73. doi:10.1007/s00223-001-0032-3

Herndon JF, Rice EG, Tucker RG, Van Loon EJ, Greenberg SM (1957) Iron absorption and metabolism. J Nutr 64(4):615–623. PMID: 13549994

Herrera E, Barbas C (2001) Vitamin E: action, metabolism and perspectives. J Physiol Biochem 57(2):43–56. PMID: 11579997

Heyman MB (2006) Lactose intolerance in infants, children, and adolescents. Am Acad Pediatr 118:1279–1286. doi:10.1542/peds.2006-1721

Holick MF (2006) High prevalence of vitamin D inadequacy and implications for health. Mayo Clin Proc March 81(3):353–373. doi:10.4065/81.3.353

Holick MF (2007) Vitamin D deficiency. N Engl J Med 357(3):266–281. doi:10.1056/NEJMra070553

Holick MF, Biancuzzo RM, Chen TC, Klein EK, Young A, Bibuld D, Reitz R, Salameh W (2007) Vitamin D2 is as effective as vitamin D3 in maintaining circulating concentrations of 25-hydroxyvitamin D. J Clin Endocrinol Metab 93(3):677–681. doi:10.1210/jc.2007-2308

Hom C, Lu J, Liong M et al (2011) Mesoporous silica nanoparticles facilitate delivery of siRNA to shutdown signaling pathway in mammalian cells. Small 6(11):1185–1190. doi:10.1002/smll.200901966

Hornick SB (1992) Factors affecting the nutritional quality of crops. Am J Altern Agric 7(1–2):63–68

Houghton LA, Vieth R (2006) The case against ergocalciferol (vitamin D2) as a vitamin supplement. Am J Clin Nutr 84(4):694–697

Houlihan A, Pyke M, Scheelings P, Graham G, Eaglesham G, Wong L, Puwastien P, Jongjitsin W (2011) Folate content of Asian vegetables, Rural Industries Research and Development Corporation, 10/16

Hurley WL, Doane RM (1989) Recent developments in the roles of vitamins and minerals in reproduction. J Dairy Sci 72(3):784–804. doi:10.3168/jds.S0022-0302(89)79170-0

Hurrell RF (1997) Preventing iron deficiency through food fortification. Nutr Rev 55(6):210–222. PMID: 9279057

Ignjatovic N, Uskokovic V, Ajdukovic Z, Uskokovic D (2013) Multifunctional hydroxyapatite and poly (D, L–lactide–co–glycolide) nanoparticles for the local delivery of cholecalciferol. Mater Sci Eng C 33(2):943–950. doi:10.1016/j.msec.2012.11.026

Institute of Medicine, Food and Nutrition Board (2001) Dietary reference intakes for Vitamin A, Vitamin K, Arsenic, Boron, Chromium, Copper, Iodine, Iron, Manganese, Molybdenum, Nickel, Silicon, Vanadium, and Zinc : a report of the panel on micronutrients. National Academy Press, Washington, DC

Jugdaohsingh R (2007) Silicon and bone health. J Nutr Health Aging 11(2):99–110. PMCID: PMC2658806

Kapoor MP, Vinu A, Fujii W et al (2010) Self-assembly of mesoporous silicas hollow microspheres via food grade emulsifiers for delivery system. Microporous Mesoporous Mater 128:187–193. doi:10.1016/j.micromeso.2009.08.019

Khandare AL, Harikumar R, Sivakumar B (2005) Severe bone deformities in young children from vitamin D deficiency in fluorosis in Bihar-India. Calcif Tissue Int 76:412–418. doi:10.1007/s00223-005-0233-2

Kilpelainen M, Riikonen J, Vlasova MA et al (2009) In vivo delivery of a peptide, ghrelin antagonist, with mesoporous silica nanoparticles. J Control Release 137(2):166–170. doi:10.1016/j.jconrel.2009.03.017

Kim YI (2004) Folate, colorectal carcinogenesis and DNA methylation: lessons from animal study. Environ Mol Mutagen 44(1):10–25. doi:10.1002/em.2002

Kowdley KV, Mason JB, Meydani SN, Cornwall S, Grand RJ (1992) Vitamin E deficiency and impaired cellular immunity related to intestinal fat malabsorption. Gastroenterology 102(6):2139–2142

Kreutler PA, Czajka-Narins DM (1987) Nutrition in perspective, 2nd edn. Prentice Hall, Englewood Cliffs

Kutsky RJ (1973) Vitamin B6.7n handbook of vitamins and hormones. Van Nostrand Reinhold Co, New York, pp 60–61

Lala VR, Vinodini R (1970) Absorption of β-carotene from green leafy vegetables in undernourished children. Am J Clin Nutr 23(1):110–113. PMID: 5412643

Leadership agenda (2010) Coalition for sustainable nutrition security for India. http://www.nutritioncoalition.in/publication.asp

Li Y, Wongsiriroj N, Blaner WS (2014) Evolving insight into vitamin a transport. Sight and Life 28(2):36–43

Lichtenstein AH, Russell RM (2005) Essential nutrients: food or supplements? Where should the emphasis be? JAMA 294(3):351–358. doi:10.1001/jama.294.3.351

Lieberman S, Bruning N (1990) The real vitamin and mineral book. New york, Avery Group, 3, ISBN 0-89529-769-8.

Mamaeva V, Rosenholm JM, Bate-Eya LT et al (2011) Mesoporous silica nanoparticles as drug delivery system for targeted inhibition of notch signaling in cancer. Mol Ther 19(8):1538–1546. doi:10.1038/mt.2011.105

Maqbool A, Stallings VA (2008) Update on fat-soluble vitamins in cystic fibrosis. Curr Opin Pulm Med 14(6):574–581. doi:10.1097/MCP.0b013e3283136787

Matalanis A, Jones OG, McClements DJ (2011) Structured biopolymer-based delivery systems for encapsulation, protection, and release of lipophilic compounds. Food Hydrocoll 25(8):1865–1880. doi:10.1016/j.foodhyd.2011.04.014

McCabe LD, Martin BR, McCabe GP, Johnston CC, Weaver CM, Peacock M (2004) Dairy intakes affect bone density in the elderly. Am J Clin Nutr 80(4):1066–1074. PMID: 15447921

McCarron DA (1995) Calcium metabolism in hypertension. Kieo J Med 44(4):105–114. doi:10.2302/kjm.44.105

McLaren DS (2006) Vitamin D deficiency disorders (VDDD): a global threat to health. Sight Life Mag 3:6–15

Miret S, Simpson RJ, McKie AT (2003) Physiology and molecular biology of dietary iron absorption. Annu Rev Nutr 23:283–301. doi:10.1146/annurev.nutr.23.011702.073139

Muller DP (2010) Vitamin E and neurological function: review. Mol Nutr Food Res 54(5):710–718. doi:10.1002/mnfr.200900460

Nasab MK, Davoudi SM, Hamedani MA, Eshagian M, EdalatiNasa M (2013) Metabolism and deficiency of retinol in ruminant. J Am Sci 9(7s):57–65. doi:10.7537/j.issn.1545-1003

National Nutrition Monitoring Bureau (NNMB) (2003) Prevalence of micronutrient deficiencies. Technical Report No 22, National Institute of Nutrition, ICMR, Hyderabad

National Nutrition Monitoring Bureau (NNMB) (2006) Diet and nutrition status of populations and prevalence of hypertension among adults in rural areas. Technical Report No 24, National Institute of Nutrition, ICMR, Hyderabad

Norman AW (2008) From vitamin D to hormone D: fundamentals of the vitamin D endocrine system essential for good health. Am J Clin Nutr August 88(2):491S–499S. PMID: 18689389

Olsan JA (1989) Provitamin A function of carotenoids: the conversion of beta-carotene into vitamin A. J Nutr 119(1):105–108. PMID:2643691

Otten JJ, Hellwig JP, Meyers LD (eds) (2001) Dietary reference intakes: the essential guide to nutrient requirements. Institute of the medicine of the national academies, The National Academies Press, Washington, DC

Packer L, Weber SU, Rimbach G (2001) Molecular aspects of α-tocotrienol antioxidant action and cell signaling. J Nutr 131(2):369S–373S. PMID: 11160563

Parfitt AM, Villanueva AR, Foldes J, Rao DS (1995) Relations between histologic indices of bone formation: implications for the pathogenesis of spinal osteoporosis. J Bone Miner Res 10(3):466–473. doi:10.1002/jbmr.5650100319

Peacock M (2010) Calcium metabolism in health and disease. Clin J Am Soc Nephrol 5:S23–S30. doi:10.2215/CJN.05910809

Pennington AT (1991) Silicon in foods and diets. Food Addit Contam 89(1):97–118. PMID: 2015936

Pereira SM, Begum A (1969) Studies in the prevention of vitamin A deficiency. Indian J Med Res 22(7):858–862, PMID: 5797053

Pittas AG, Chung M, Trikalinos T, Mitri J, Brendel M, Patel K, Lichtenstein AH, Lau J, Balk EM (2010) Vitamin D and cardiometabolic outcomes: a systematic review. Ann intern med, March 152(5):307–314. doi:10.1059/0003-4819-152-5-201003020-00009

Pollitt E, Rudolph MD, Leibel L (1976) Iron deficiency and behavior. J Pediatr 88(3):372–381. doi:10.1016/S0022-3476(76)80250-8

Ponka P, Beaumont C, Richardson DR (1998) Function and regulation of transferrin and ferritin. Euro Pubmed Cen 35(1):35–54. PMID:9460808

Potts JT, Gardella TJ (2011) Parathyroid hormone and calcium homeostasis. Pediatric bone, 2nd edn, Biology and diseases, science direct, pp 109–140. doi:10.1016/B978-0-12-382040-2.10006-1

Rao CN, Rao BS (1970) Absorption of dietary carotenes in human subjects. Am J Clin Nutr 23(1):105–109. PMID: 5412642

Rosenholm J, Sahlgren C, Linden M (2010) Cancer-cell targeting and cell-specific delivery by mesoporous silica nanoparticles. J Mater Chem 20:2707–2713. doi:10.1039/B920076B

Ross AC, Manson JE, Abrams SA (2011a) Report on dietary reference intakes for calcium and vitamin D from the Institute of Medicine: What clinicians need to know. J Clin Endocrinol Metab 96:53–58. doi:10.1210/jc.2010-2704

Ross AC, Taylor CL, Yaktine AL, Del VHB (2011b) Dietary reference intakes for calcium and vitamin D. National Academies Press, Washington

Salonen J, Laitinen L, Kaukonen AM et al (2005) Mesoporous silicon microparticles for oral drug delivery: loading and release of five model drugs. J Control Release 108(2–3):362–374. doi:10.1016/j.jconrel.2005.08.017

Sandberg AS (2005) Methods and options in vitro dialyzability; benefits and limitations. Int J Vitam Nutr Res 75(6):395–404. doi:10.1024/0300-9831.75.6.395

Sanjoaquin MA, Allen N, Couto E, Roddam A, Key TJ (2005) Folate intake and colorectal cancer risk: a meta-analytical approach. Int J Cancer 113(5):825–828. doi:10.1002/ijc.20648

Sathyanarayana Rao TS, Asha MR, Ramesh BN, Jagannatha Rao KS (2008) Understanding nutrition, depression and mental illnesses. Indian J Psychiatry 50(2):77–82

Seshadri S, Gopaldas T (1989) Impact of iron supplementation on cognitive functions in preschool and school-aged children. Am J Clin Nutr 50(3):675–686. PMID: 2773845

Shaw GM, Schaffer D, Velie EM, Morland K, Harris JA (1995) Periconceptional vitamin use, dietary folate, and the occurrence of neural tube defects. Epidemiology 6(3):219–226. PMID: 7619926

Sivakumar B, Panth M, Shatrugna V, Raman L (1997) Vitamin A requirements assessed by plasma response to supplementation during pregnancy. Int J Vitam Nutr Res 67(4):232–236. PMID:9285251

Sivakumar B, Nair KM, Sreeramulu D (2006) Effect of micronutrient supplement on health and nutrition status of school children: biochemical status. Nutrition 22(1):S15–S25. doi:10.1016/j.nut.2005.07.012

Slowing II, Vivero–Escoto JL, Wu CW, Lin VSY (2008) Mesoporous silica nanoparticles as controlled release drug delivery and gene tranfection carriers. Adv Drug Deliv Rev 60(11):1278–1288. doi:10.1016/j.addr.2008.03.012

Soetan KO, Olaiya CO, Oyewole OE (2010) The importance of mineral elements for humans, domestic animals and plants. Afr J Food Sci 4(5):200–222

Straub DA (2007) Calcium supplementation in clinical practice: a review of forms, doses, and indications. Nutr Clin Pract 22:286–296

Subar AF, Block G, James LD (1989) Folate intake and food sources in the US population. Am J Clin Nutr 50:508–516

Tanaka T, Godin B, Bhavane R et al (2010) In vivo evaluation of safety of nanoporous silicon carriers following single and multiple dose intravenous administration in mice. Int J Pharm 402(1–2):190–197. doi:10.1016/j.ijpharm.2010.09.015

Tang G, Qin J, Dolnikowski GG, Russell RM, Grusak MA (2005) Spinach or carrots can supply significant amounts of vitamin A as assessed by feeding with intrinsically deuterated vegetables. Am J Clin Nutr 82(4):821–828. PMID: 16210712

Tanumjhardjo SA (2011) Vitamin A: biomarkers of nutrition for development. Am J Clinl Nutr 94(2):658S–665S. doi:10.3945/ajcn.110.005777

U.S National Library of Medicine. National Institute of Health, The Brooklyn Hospital Center, Brooklyn, NY

Underwood EJ, Suttle NF (1999) Mineral nutrition of livestock, 4th edn. Moredun Research Institute, Cabi Publishing, Midlothian. doi:10.1079/9780851991283.0000

Vasanthi N, Saleena LM, Raj SA (2012) Silicon in day today life. World Appl Sci J 17(11):1425–1440

Vitamin E: fact sheet for health professionals (2009) National Institute of Health, office of dietary supplements, US department of health and services

Vivero-Escoto JL, Slowing II, Trewyn BG, Lin VSY (2010) Mesoporous slica nanoparticles for intracellular controlled drug delivery. Small 6(18):1952–1967. doi:10.1002/smll.200901789

Wagner C (1995) Biochemical role of folate in cellular metabolism. In: Bailey LB (ed) Folate in health and disease. Marcel Dekker, New York, pp 23–42

Wald G (1935) Caroteniod and the visual cycle. J Gen Physiol 19(2):351. doi:10.1085/jgp.19.2.351

Webb EC (1992) Anemia and physical work capacity. Enzyme Nomenclature Academic Press, San Diego

Whitney E, Sharon RR (2011) In: Peggy W (ed) Understanding normal and clinical Nutrition, 9th edn. Wadsworth, Cengage Learning, Belmont

Wolf G (2004) The discovery of vitamin D: the contribution of Adolf Windaus. J Nutr 134(6):1299–1302. PMID: 15173387

World Health Organization (1988) Calcium requirements, WHO technical report series no 230. WHO, Geneva

World Health Organization (1995–2005) Global prevalence of vitamin A deficiency in populations at risk, WHO global database on vitamin A deficiency, Geneva, WHO, 2009

Wosje K, Specker BL (2000) Role of calcium in bone health during childhood. Nutr Rev 58(9):253–268. doi:10.1111/j.1753-4887.2000.tb01879

Yuan H, Luo K, Lai Y, Pu Y, He B, Wang G, Wu Y, Gu Z (2010) A novel Poly(l-glutamic acid) dendrimer based drug delivery system with both pH-sensitive and targeting functions. Mol pharm 7(4):953–962. doi:10.1021/mp1000923

Zhang YN, Zhi ZZ, Jiang TY, Zhang JH, Wang ZY, Wang SL (2010) Spherical mesoporous silica nanoparticals for loading and release of the poorly water-soluble drug telmisartan. J Control Release 145(3):257–263. doi:10.1016/j.jconrel.2010.04.029

Zimmermann MB, Hurrell RF (2007) Nutritional iron deficiency. Lancet 370(9586):511–520. doi:10.1016/S0140-6736(07)61235-5

Chapter 6
Nanomaterials for Monitoring and Remediation of Water Pollution

Xing-yan Xue, Rong Cheng, Lei Shi, Zhong Ma, and Xiang Zheng

Abstract Water shortage and pollution are now serious challenges for many countries. Nanomaterials are promising new tools for water quality management due to their unique physicochemical properties, high economic benefit, high removal efficiency and environmental friendliness. Here we present four types of nanomaterials for water and wastewater treatment: nanofiltration membranes, nano-photocatalytic materials, nano-adsorption materials and nano-reducing materials. We discuss their properties, application scope and mechanism of pollutant removal. We also review nanomaterials used for water quality monitoring, especially for the detection of the extremely low concentration organic pollutants, inorganic pollutants and pathogens. Such nanomaterials include carbon nanotubes, magnetic nanoparticles, noble metal nanomaterials and quantum dots.

Keywords Nanomaterials • Water treatment • Wastewater treatment • Monitoring

6.1 Introduction

Water is the source of life. Rapid development of economy, heavy application of chemicals and imperfection of water management policy resulted in a series of water problems. Currently, Water shortage and pollution are the two major problems in both developing and developed countries. Water shortage and pollution are influenced by many factors such as human activities, population growth, users demand and global climate change (Savage and Diallo 2005). Polluted waters usually contain suspended matter, heavy metal, organic matter, bacteria, virus and many other complex compounds. Conventional water and wastewater treatment technologies such as adsorption, precipitation, coagulation and activated sludge process have many drawbacks such as low treatment efficiency, high cost and

---

X.-y. Xue • R. Cheng (✉) • L. Shi • Z. Ma • X. Zheng (✉)
School of Environment and Natural Resources, Renmin University of China, Beijing 100872, China
e-mail: chengrong@ruc.edu.cn; zhengxiang7825@163.com

S. Ranjan et al. (eds.), *Nanoscience in Food and Agriculture 2*, Sustainable Agriculture Reviews 21, DOI 10.1007/978-3-319-39306-3_6

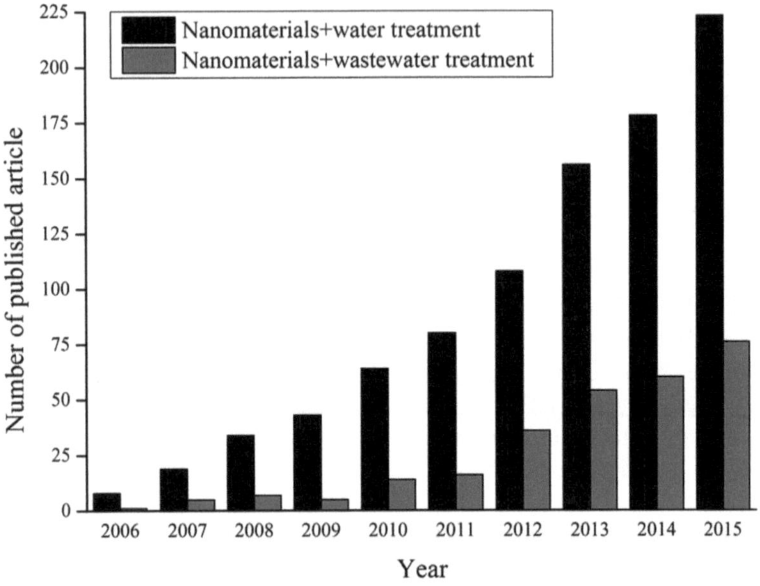

Fig. 6.1 Number of article published on the subjects of "nanomaterials + water treatment" and "nanomaterials + wastewater treatment" from 2006 to 2015

secondary pollution. Due to their unique physicochemical properties, economic benefit, efficiency and environmental friendliness, much attention has been paid to nanomaterials in water quality management recently (Fig. 6.1). Nanomaterials could remove many kinds of pollutants in water and wastewater treatment including refractory matters, organic matters and heavy metals.

Nanotechnology is related to the preparation of materials small than 100 nm in size. Several kinds of nanomaterials, such as iron-based nanomaterials, carbon-based nanomaterials and $nTiO_2$, have been widely studied in water quality management. These nanomaterials possess effective detection, adsorption and removal/degradation capacity to pollutants in water and wastewater. For example, nanoscale zero-valent iron has been proven to remove/degrade pollutants effectively including dyes (Fan et al. 2009), chlorinated solvents (Wang and Zhang 1997), pesticides (Elliott et al. 2009), lead (Ponder et al. 2001; Li and Zhang 2007) and cadmium (Li and Zhang 2007). Nano-Au can sensitively detected chlorpyrifos, malathion (Lisha et al. 2009), Hg^{2+} and CH_3Hg^+ (Lin and Tseng 2010).

So far, many relative reviews on nanomaterials in water monitoring and treatment have been reported. Andreescu et al. (2009) reviewed the application of advanced nanomaterials in environmental monitoring; Liu et al. (2014) reviewed the application of nanomaterials for monitoring emerging chemical pollutants; Qu et al. (2013) reviewed the performance of nanomaterials in water and wastewater treatment; Bystrzejewska-Piotrowska et al. (2009) reviewed the application of nanoparticles in environmental management. However, a comprehensive view on nanomaterials in water quality management is still lacking.

In this paper, an overview of recent advances in nanomaterials for water pollution remediation and monitoring were presented. Nanomaterials could be used in the treatment of surface water, ground water and wastewater. Nanofiltration membrane, nano-photocatalytic materials, nano-adsorption materials and nano-reducing materials could remove turbidity, hardness, heavy metal, organic matters and inorganic ions. Besides, nanomaterials including magnetic nanoparticles, carbon nanotubes and noble metal nanomaterials could be used in water quality monitoring. However, cost-effectiveness, technical hurdles and potential risk of nanomaterials are still challenges for their widespread application. Hence, more studies about nanomatrials in water pollution remediation and monitoring need to be done.

6.2 Application of Nanomaterials in Water and Wastewater Treatment

Nanomaterials are the materials which sized below 100 nm at least one dimension. Nanomaterials possess unique properties such as mechanical, electrical, optical, catalytic, magnetic and photonic properties in comparison to common materials. Based on the properties of nanomaterials and the removal principle of pollutants, nanomaterials in water and wastewater treatment can be classified as nanofiltration membrane, nano-photocatalytic materials, nano-adsorption materials and nano-reducing materials. Considering the kinds of pollutants, increasing production and usage of nanomaterials offer opportunities for the removal of various pollutants, which was listed in Table 6.1.

6.2.1 Nanofiltration Membranes

A nanofiltration membrane is a type of semipermeable membrane, which allows solvent molecules or some low molecular weight solutes or low ion permeation. Nanofiltration membranes possess pore size generally of 1–2 nm, molecular weight cut-off of 300–500 Da, water permeability of 5–50 L m^{-2} h^{-1} bar^{-1} and Operating pressure typically of 2–10 bar. Recently, the application of nanofiltration membrane has received wide attention due to its high flux, low investment, low operation pressure and cost. Cellulose acetate, polyamide, polyvinyl alcohol and sulfonated polysulfone can be used to form nanofiltration membrane (Savage and Diallo 2005; Cheng et al. 2011). Nanofiltration membrane could be used in the treatment of surface water, ground water and wastewater. Apart from the purpose to remove turbidity, hardness, fluorides, disinfection by-products and pesticides, recent studies have reported that nanofiltration has also being used for the removal of arsenic and emerging contaminants such as pharmaceuticals, hormones and personal care products (Table 6.2) (Mohammad et al. 2015). However, studies on the removal of the

Table 6.1 Common pollutants that can be removed by nanomaterials

Pollutants	Examples	References
Heavy metal ions	Pb^{2+}; Hg^{2+}; Ni^{2+}; Ag^{+}; Cd^{2+}; Cu^{2+}; Ni^{2+}; Zn^{2+}; Ca^{2+}; Mg^{2+}; As^{3+}; As^{5+}	Chang et al. (2014), Maher et al. (2014), Aarthi and Madras (2008)
Organic dyes	Orange II; Chrysoidine; Tropaeolin O; Acid Orange; Acid Red	Shu et al. (2010), Chen et al. (2011), Frost et al. (2010), Lin et al. (2008)
Chlorinated benzenes	Hexachlorobenzene; Pentachlorobenzene; Tetrachlorobenzenes; Trichlorobenzenes; Dichlorobenzenes; Chlorobenzene	Shih et al. (2011), Fan et al. (2011)
Trihalomethanes	Bromoform; Dibromochloromethane; Dichlorobromomethane	Lien (2005)
Chlorinated ethenes	Tetrachloroethene; Trichloroethene; *cis*-Dichloroethene; *trans*-Dichloroethene; 1,1-Dichloroethene; Vinyl chloride	Lien and Zhang (2001)
Inorganic anions	Dichromate; Arsenic; Perchlorate; Nitrate	Jiang et al. (2011), Zhang et al. (2011), Harisha et al. (2010)
Pesticides	Dichlorodiphenyltrichloroethane; Lindane; chlorpyriphos; pyrethroid; malathion	Joo and Zhao (2008), Shih et al. (2011), El-Sheikh et al. (2008), Parra et al. (2002)
Microorganism	Escherichia coli; Pseudomonas fluorescens; Aspergillus versicolor; poliovirus 1; hepatitis B virus; Herpes simplex virus; MS2 bacteriophage	Lee et al. (2008), Liga et al. (2011)
Others	Polychlorinated biphenyls; 2,4,6-Trinitrotoluene; Hormones	Mohammad et al. (2015)

pharmaceutical active compounds from drinking water and surface water by nanofiltration membranes are relatively scarce so far (Radjenović et al. 2008; Verliefde et al. 2007).

The most common application field of nanofiltration membrane is the softening of water. Fang et al. (2013) developed new composite nanofiltration hollow fiber membranes for surface water softening, the results showed that the new hollow fiber membranes rejections for Ca^{2+} and Mg^{2+} were 90 % while test for 3000 ppm total dissolved salt feed stream at 2 bar pressure, suggesting that the composite nanofiltration hollow fiber membranes could be effectively applied to surface water softening.

Numerous researches have proved that nanofiltration membrane showed an excellent performance on the remove of pesticides (Van der Bruggen et al. 2001; Van der Bruggen and Vandecasteele 2003). The removal efficiencies of pesticides with nanofiltration membrane were affected by pesticides, types of nanofiltration

Table 6.2 Pollutants that can be removed by nanofiltration membrane

Pollutants	Initial concentration	Transmembrane pressures	pH	Removal efficiency	References
As^{3+}	100 µg L^{-1}	0.55 MPa	9–10	40 %	Chang et al. (2014)
Ni^{2+}	1 ppm	6 bar	5.5	93 %	Maher et al. (2014)
Pb^{2+}	1 ppm	5.8 bar	5.5	86 %	Maher et al. (2014)
Mg^{2+}	200 ppm	5 bar	/	95.20 %	Zhong et al. (2012)
Mg^{2+}	51–128 ppm	2 bar	11	More than 90 %	Fang et al. (2013)
Ca^{2+}	128 ppm	2 bar	11	88 %	Fang et al. (2013)
Arsenic	0.0005 M	50 bar	8	98.98 %	Harisha et al. (2010)
Fluoride	20 ppm	14 kgf cm^{-2}	10.01	More than 98 %	Chakrabortty et al. (2013)
Orange II	50 ppm	5 bar	/	86.76 %	Zhong et al. (2012)
Safranin O	50 ppm	5 bar	/	99.98 %	Zhong et al. (2012)
Trichlorethylene	400 µg/L	15 bar	/	92 %	Ducom and Cabassud (1999)
Chlorpethylene	400 µg/L	15 bar	/	96 %	Ducom and Cabassud (1999)
Chloroform	200 µg/L	10 bar	/	79 %	Ducom and Cabassud (1999)
Phenol	$50–200 \times 10^{-3} \times kg\ m^{-3}$	$6*10^6\ N\ m^{-2}$	2–11	80 %	Bódalo et al. (2009)
Imidacloprid	0.5–1.5 mg L^{-1}	0.5; 1.0 MPa	/	2.92 % for nanofiltration membrane-4; 3.70 % for nanofiltration membrane-3; 54.6 % for nanofiltration membrane-2; 97.6 % for nanofiltration membrane-1	Kiso et al. (2000)
Dichlorvos	0.5–1.5 mg L^{-1}	0.5; 1.0 MPa	/	4.28 % for nanofiltration membrane-4; 13.0 % for nanofiltration membrane-3; 46.2 % for nanofiltration membrane-2; 86.7 % for nanofiltration membrane-1	Kiso et al. (2000)
Simazine	0.5–1.5 mg L^{-1}	0.5; 1.0 MPa	/	6.40 % for nanofiltration membrane-4; 9.15 % for nanofiltration membrane-3; 59.8 % for nanofiltration membrane-2; 96.7 % for nanofiltration membrane-1	Kiso et al. (2000)

(continued)

Table 6.2 (continued)

Pollutants	Initial concentration	Transmembrane pressures	pH	Removal efficiency	References
Simazine	100 μg L^{-1}	15 bar	3–10	More than 74 %	Musbah et al. (2013)
Atrazine	100 μg L^{-1}	15 bar	3–10	96 %	Musbah et al. (2013)
Diuron	100 μg L^{-1}	15 bar	3–10	More than 74 %	Musbah et al. (2013)
Desethylatrazine	100 μg L^{-1}	15 bar	3–10	More than 74 %	Musbah et al. (2013)
Simetryn	0.5–1.5 mg L^{-1}	0.5; 1.0 MPa	/	6.69 % for nanofiltration membrane-4; 6.95 % for nanofiltration membrane-3; 57.6 % for nanofiltration membrane-2; 98.6 for nanofiltration membrane-1	Kiso et al. (2000)
Atrazine	0.5–1.5 mg L^{-1}	0.5; 1.0 MPa	/	10.9 % for nanofiltration membrane-4; 14.9 % for nanofiltration membrane-3; 68.4 % for nanofiltration membrane-2; 97.5 % for nanofiltration membrane-1	Kiso et al. (2000)
Thiram	0.5–1.5 mg L^{-1}	0.5; 1.0 MPa	/	8.42 % for nanofiltration membrane-4; 18.7 % for nanofiltration membrane-3; 56.4 % for nanofiltration membrane-2; 97.7 % for nanofiltration membrane-1	Kiso et al. (2000)
Malathion	0.5–1.5 mg L^{-1}	0.5; 1.0 MPa	/	41.4 % for nanofiltration membrane-4; 42.0 % for nanofiltration membrane-3; 88.1 % for nanofiltration membrane-2; 99.64 % for nanofiltration membrane-1	Kiso et al. (2000)
Molinate	0.5–1.5 mg L^{-1}	0.5; 1.0 MPa	/	20.0 % for nanofiltration membrane-4; 20.4 % for nanofiltration membrane-3; 60.7 % for nanofiltration membrane-2; 98.5 % for nanofiltration membrane-1	Kiso et al. (2000)
Isoprothiolane	0.5–1.5 mg L^{-1}	0.5; 1.0 MPa	/	28.1 % for nanofiltration membrane-4; 36.3 % for nanofiltration membrane-3; 93.7 % for nanofiltration membrane-2; 99.76 % for nanofiltration membrane-1	Kiso et al. (2000)

Dizinon	0.5–1.5 mg L^{-1}	0.5; 1.0 MPa	/	44.6 % for nanofiltration membrane-4; 44.8 % for nanofiltration membrane-3; 95.1 % for nanofiltration membrane-2; 99.52 % for nanofiltration membrane-1	Kiso et al. (2000)
Chlorpyrifos	0.5–1.5 mg L^{-1}	0.5; 1.0 MPa	/	99.51 % for nanofiltration membrane-4; 99.32 % for nanofiltration membrane-3; More than 99.95 % for nanofiltration membrane-2; More than 99.95 % for nanofiltration membrane-1	Kiso et al. (2000)
Anilazine	0.5–1.5 mg L^{-1}	0.5; 1.0 MPa	/	21.8 % for nanofiltration membrane-4; 29.3 % for nanofiltration membrane-3; 72.8 % for nanofiltration membrane-2; 99.3 % for nanofiltration membrane-1	Kiso et al. (2000)
Pyridine	0.5–1.5 mg L^{-1}	0.5; 1.0 MPa	/	5.52 % for nanofiltration membrane-2; 18.5 % for nanofiltration membrane-1	Kiso et al. (2000)
2,3,5-trichloro-pyridine	0.5–1.5 mg L^{-1}	0.5; 1.0 MPa	/	95.6 % for nanofiltration membrane-4; 96.5 % for nanofiltration membrane-3; 88.9 % for nanofiltration membrane-2; 96.8 % for nanofiltration membrane-1	Kiso et al. (2000)

membrane, transmembrane pressures and solution pH value. Van der Bruggen et al. (2001) found the rejections for simazine, atrazine, diuron and isoproturon were all in the 90–95 % range with nanofiltration membrane. Kiso et al. (2000) studied the removal of 12 pesticides including imidacloprid, simetryn and dichlorvos with four nanofiltration membrane, the results show that the rejections of 12 pesticides decreased in an order from nanofiltration membrane-1 to nanofiltration membrane-4. The removal efficiencies of the pesticides except chlorpyrifos for each nanofiltration membrane were 86.7–99.95 % for nanofiltration membrane-1, 46.2–99.95 % for nanofiltration membrane-2, 3.7–99.32 % for nanofiltration membrane-3, and 2.9–99.51 % for nanofiltration membrane-4, respectively. The rejection of chlorpyrifos was more than 99 % for four types of nanofiltration membrane. Moreover, pesticides might transform and produce new transformation products during transporting to groundwater. The pesticide transformation products (PTPs) have been found in groundwater in recent years, which are different from pesticides (Mohammad et al. 2015). Madsen et al. (Madsen and Søgaard 2014) compared the performance of nanofiltration/low pressure reverse osmosis/reverse osmosis membrane for treatment of pesticides and pesticide transformation products in groundwater, the results show that low pressure reverse osmosis membrane could effectively reject the pesticides and pesticide transformation products, but nanofiltration membrane could not effectively reject pesticide transformation products. Therefore, some measures should be taken to reject pesticide transformation products, such as low pressure reverse osmosis membrane.

Sentana et al. (2010) studied the remove of the disinfection by-products by three commercial nanofiltration (nanofiltration 27, nanofiltration 90 and Desal-HL-51) membranes, the results show that the nanofiltration 90 membrane performed better in reducing in the formation of trihalomethane and haloacetic acid formation potential compared with nanofiltration 90 and Desal-HL-51 membranes.

The problem of arsenic (As) has received wide attention. Nanofiltration membrane was capable of converting As^{3+} into As^{5+} and removing As (Sen et al. 2010). However, most studies of the removal of arsenic from groundwater by nanofiltration membrane still at the laboratorial scale.

Hormones stem from agricultural, industrial, medical and domestic activities, which could enter aquatic environment and cause adverse effects to human health. Different studies investigated the feasibility of using nanofiltration to remove hormones, demonstrating that these compounds could be largely rejected according to the adsorption effect (Semião and Schäfer 2011; Schäfer et al. 2011; Sanches et al. 2012).

Although nanofiltration membrane has some great advantages such as high membrane flux, low investment and low operation cost, membrane fouling is still a great limitation. Membrane fouling is occurred when a particle or solute molecule is deposited on a membrane surface or in membrane pores, and then the membrane's performance is degraded. And it is determined by some factors including concentration polarization, membrane pore blocking and surface deposition. Membrane fouling can lead to some adverse effects such as flux decline, cost increase and membrane degradation. Hence, some measures should be taken to control membrane fouling such as feed pretreatment, membrane surface modification, physical cleaning and chemical cleaning (Mohammad et al. 2015; Hilal et al. 2004). Generally, physical cleaning such as washing, backwashing and immerse, gas-liquid mixing flushing is carried out regularly during the membrane operation. The water permeability of the membrane can achieve a certain degree of recovery in the short time, but the membrane flux will fall again. So physical cleaning can only be used to inhibit the growth of membrane fouling, and can't make the membrane flux completely recovered. The fouled materials can be washed with chemical agents such as acid, alkali. Acid could make insoluble substances convert into soluble substances. Alkali mainly remove protein, algae and other biological pollutants, colloid pollutants and organic pollutants (Lim and Bai 2003; Madaeni et al. 2001).

6.2.2 *Nanophotocatalytic Materials*

Photocatalysis is a promising process for the removal of trace contaminants and microbial pathogens (Aarthi and Madras 2007). But the limited photocatalytic activity limits its wide application. Compared to common photocatalytic materials, nano-photocatalytic materials have attracted more attention (Han and Ba 2009; Gupta and Tripathi 2011). Because of their large specific surfaces, nano-photocatalytic materials can enhance photocatalytic activity effectively (Lan et al. 2013). Up to now, $nTiO_2$, nZnO, nWO_3, $nBiVO_4$ and nAl_2O_3 are widely used nano-photocatalytic materials. Among them, $nTiO_2$ is the most commonly used nanomaterial in water and wastewater treatment due to its high reactivity, low toxicity, thermal stability, and abundance as raw material (Table 6.3) (Gupta and Tripathi 2011; Chen and Mao 2007).

In aqueous environment, $nTiO_2$ photocatalyst can generate an electron/hole pair when it is irradiated with energy greater than the band gap. Then, electron/hole pair can migrate to the surface of $nTiO_2$ and form reactive oxygen species (·OH, H_2O_2, etc.). The positive holes react with H_2O and form hydroxyl radical, which promote the oxidation of organics (Kwon et al. 2008). $nTiO_2$ has been successfully applied for the organic wastewater treatment, including dye wastewater (Aarthi and Madras 2007; Nagaveni et al. 2004), chemical industry wastewater, pesticide wastewater, oily wastewater (Yang et al. 2012); inorganic wastewater treatment and microbial control (Kwon et al. 2008; Qu et al. 2013; García et al. 2012). Although $nTiO_2$ has some great advantages in the field of water and wastewater treatment, especially for dye wastewater and paper mill wastewater, it has some drawbacks such as low absorb efficiency of visible light, low recycle rate and high cost. In order to overcome above-mentioned drawbacks, several approaches have been studied including dye sensitization, doping, coupling and capping of $nTiO_2$ (Gupta and Tripathi 2011). In particular, ion doping has received wide attention due to easy operation, high efficiency and more rapid reaction rate. Choi et al. (1994) reported that $nTiO_2$ doped with Fe^{3+}, Mo^{5+}, Ru^{3+}, Os^{3+}, Re^{5+}, V^{4+} and Rh^{3+} significantly increased the photochemical reactivity of $nTiO_2$ for the oxidation of trichloromethane and reduction of

Table 6.3 Pollutants that can be removed by $nTiO_2$

Pollutants	Initial concentration	$nTiO_2$ dose	pH	Removal efficiency	References
Cu^{2+}	145 μM	1g L^{-1}	10.8	100 %	Aarthi and Madras (2008)
Cr^{6+}	215 μM	1 g L^{-1}	7	63 %	Aarthi and Madras (2008)
Cu^{2+}	250–515 μM	1 g L^{-1}	/	About 16 %–40 % in 120 min	Vinu and Madras (2007)
Cr^{6+}	120–250 mM	1g L^{-1}	/	100 % in 5 min	Vinu and Madras (2007)
Hg^{2+}	150 ppm	5.0×10^{-3} M	natural	100 % in 130 min (253.7 nm light)	Skubal and Meshkov (2002)
Methylene blue	25–75 ppm	0.25–2 kg m^{-3}	/	Nearly 100 % in 150 min (The average solar intensity in a day between 10 a.m. and 2.30 p.m. was 0.753 kW m^{-2})	Nagaveni et al. (2004)
Orange G	25 ppm	1 kg m^{-3}	/	More than 90 % in 100 min	Nagaveni et al. (2004)
Orange G	40^{-1}	1 g L^{-1}	/	Nearly 100 % in 90 min (US + UV power intensity of 1.6 W cm^{-2})	Vinu and Madras (2008)
Remazol brill blue R	100 ppm	1 kg m^{-3}	/	About 70 % in 150 min	Nagaveni et al. (2004)
Remazol Brilliant Blue R	50 mg L^{-1}	1 g L^{-1}	/	About 80 % in 90 min	Vinu and Madras (2008)
Alizarin Red S	50 mg L^{-1}	1 g L^{-1}	/	the complete degradation in 60 min	Vinu and Madras (2008)
Indigo Carmine	50 mg L^{-1}	1 g L^{-1}	/	the complete degradation in 30 min	Vinu and Madras (2008)
4-chlorophenol	250 mg L^{-1}	2 g L^{-1}	5	The photocatalytic activity in the degradation of 4-chlorophenol was found to be higher for Mg^{2+} and Ba^{2+} doped nano TiO_2 than both pure nano TiO_2 and commercial TiO_2	Venkatachalam et al. (2007)
4-nitrophenol	30 mg L^{-1}	1 g L^{-1}	/	100 % in 30 min	Vinu and Madras (2007)
Phenol	30 mg L^{-1}	1 g L^{-1}	/	100 % in 30 min	Vinu and Madras (2007)
Cinosulfuron	15.5 μmol L^{-1}	2.5 g L^{-1}	/	More than 90 % in 60 min (The irradiation spectrum was cut-off below 340 nm)	Vulliet et al. (2002)
Triasulfuron	16 μmol L^{-1}	2.5 g L^{-1}	/	Nearly 100 % in 30 min	Vulliet et al. (2002)

Methamidophos	100 pm	0.5 g L^{-1}	10	100 % in 5 h (UV irradiation with light energy greater than 3.2 eV)	Dai et al. (2008)
Chlorbromuron	0.1 mmol L^{-1}	1 g L^{-1}	5–6	Nearly 100 % photodegradation in 2 h (UV40 mW cm^{-2})	Parra et al. (2002)
Chlortoluron	0.1 mmol L^{-1}	1 g L^{-1}	5–6	100 % in 1.5 h (UV40 mW cm^{-2})	Parra et al. (2002)
Metobromuron	0.1 mmol L^{-1}	1g L^{-1}	5–6	Nearly 100 % in 2 h (UV40 mW cm^{-2})	Parra et al. (2002)
Isoproturon	0.1 mmol L^{-1}	1 g L^{-1}	5–6	100 % in 1.5 h (UV 40 mW cm^{-2})	Parra et al. (2002)
Malathion	10 mg L^{-1}	10 mg L^{-1}	/	98.2 % (the UV luminous intensity was 0.4 mW cm^{-2})	Yu et al. (2010)
MS2 bacteriophage	~7*10^9 PFU mL^{-1}	100 mg L^{-1}	/	1.6 log inactivation of MS2 in 2 min (The total light intensity was 2.5 mW/cm^2)	Liga et al. (2011)

carbon tetrachloride. The dopant content influences the rate of electron/hole recombination and photocatalytic activity. As the dopant content increases, the electron/hole pairs within the space-charge region are separated by the large electric field before recombination. However, when the dopant content is high, the rate of electron/hole pairs recombination in the $nTiO_2$ increases. Therefore, there is an optimum content of dopant ion. Xin et al. (2007) studied the effect of the different doping ratio Fe^{3+}–$nTiO_2$, the results revealed that the $nTiO_2$ with a low doping concentration of Fe^{3+} (Fe/Ti lower than 0.03 mol) enhanced the photocatalytic activity of $nTiO_2$. However, the $nTiO_2$ containing a high doping concentration of Fe^{3+} (Fe/Ti higher than 0.03 mol) is unfavorable to photocatalytic reactions.

6.2.3 Nanoadsorption Materials

Adsorption is commonly employed as pretreatment or advanced treatment to remove organic pollutants, heavy metals and residual chlorine in water and wastewater. The efficiency of nano-adsorption materials is higher than that of conventional absorbents due to high specific surface area, associated sorption sites and surface chemistry.

6.2.3.1 Carbon Based Nanoadsorbents

Carbon nanotubes, a new type of nanomaterials, have received wide attention due to their unique properties, such as large specific surface area, high thermal stability and high chemical stability. Carbon nanotubes can be divided into single-walled carbon nanotubes and multi-walled carbon nanotubes. Carbon nanotubes are good adsorption materials for the remove of organic matter, heavy metal (Table 6.4).

Numerous studies have shown that carbon nanotubes were effective adsorbents and their efficiency was superior to activated carbon on adsorption of organic chemicals in water and wastewater treatment (Pan and Xing 2008; Su and Lu 2007; Wang et al. 2007). Su and Lu (2007) reported that the adsorption capacities of carbon nanotubes on natural dissolved organic matter (11.61 mg g^{-1}) is higher than that of granular activated carbon (3.55 mg g^{-1}), and the average weight losses of the carbon nanotubes (2.65 %) is lower than that of granular activated carbon (6.40 %). El-Sheikh et al. (2008) reported that the absorption capacity of multi-walled carbon nanotubes is three times that of activated carbon towards the pesticides. Long and Yang (2001) found that carbon nanotubes were better than activated carbon for dioxin removal. Its high adsorption capacity is mainly because of the large specific surface area and the pollutants-carbon nanotube interactions. In aqueous environment, carbon nanotubes strongly adsorb low molecular weight polar organic compounds due to the organic compounds-carbon nanotube interactions including hydrophobic effect, π–π interactions, hydrogen bonding, and electrostatic interactions (Pan and Xing 2008; Qu et al. 2013). Different adsorption mechanisms might

Table 6.4 Pollutants that can be removed by carbon nanotubes

Pollutants	Initial concentration	Carbon nanotubes dose	pH	Removal efficiency	References
Pb^{2+}	10–80 mg L^{-1}	0.5g L^{-1}	5	97.08 mg g^{-1}	Li et al. (2003a)
Pb^{2+}	2–14 mg L^{-1}	0.5 g L^{-1}	7	1 mg g^{-1}	Li et al. (2002)
Cu^{2+}	10 mg L^{-1}	5 g L^{-1}	9	Nearly 100 %	Pyrzyńska and Bystrzejewski (2010)
Cu^{2+}	5–30 mg L^{-1}	0.5 g L^{-1}	5	28.49 mg g^{-1}	Li et al. (2003a)
Cd^{2+}	2–15 mg L^{-1}	0.5g L^{-1}	5	10.86 mg g^{-1}	Li et al. (2003a)
Cd^{2+}	9.5 mg L^{-1}	0.5 g L^{-1}	7	1.1 mg g^{-1}	Li et al. (2003b)
Zn^{2+}	10–80 mg L^{-1}	0.5 g L^{-1}	7	10.21–11.23 mg g^{-1}	Lu et al. (2006)
Zn^{2+}	60 mg L^{-1}	0.5 g L^{-1}	1–5	37.03–46.94 mg $g^{-1}$and 30.3–34.36 mg g^{-1} from 5 to 45 °C	Lu and Chiu (2006)
Co^{2+}	10 mg L^{-1}	5 g L^{-1}	9	More than 90 %	Pyrzyńska and Bystrzejewski (2010)
Ni^{2+}	6–20 mg L^{-1}	0.3 g L^{-1}	6.55	9.8 mg g^{-1}	Chen and Wang (2006)
U(VI)	10^{-7}–10^{-4} M	1 g L^{-1}	4	5.0 mmol g^{-1}	Schierz and Zänker (2009)
Th(IV)	32.32 μmol L^{-1}	0.2 g L^{-1}	1.9	65.8 μmol g^{-1}	Chen et al. (2007)
Fluoride	12 mg L^{-1}	2 g L^{-1}	5–9	14.4 mg g^{-1}	Li et al. (2001)
$CHCl_3$	2 mg L^{-1}	0.33 g L^{-1}	3–11	2.41 mg g^{-1}	Lu et al. (2005)
$CHCl_2Br$	2 mg L^{-1}	0.33 g L^{-1}	3–11	1.23 mg g^{-1}	Lu et al. (2005)
$CHClBr_2$	2 mg L^{-1}	0.33 g L^{-1}	3–11	1.08 mg g^{-1}	Lu et al. (2005)
$CHBr_3$	2 mg L^{-1}	0.33 g L^{-1}	3–11	0.92 mg g^{-1}	Lu et al. (2005)
Phenol	0.1–100,000 mg L^{-1}	/	7	64.6 mg g^{-1}	Yang et al. (2008)
Aniline	0.1–100,000 mg L^{-1}	/	7	114.8 mg g^{-1}	Yang et al. (2008)
1,2-dichlorobenzene	20 mg L^{-1}	0.5 g L^{-1}	3–10	30.8 and 28.7 mg g^{-1} for as grown and graphitized CNT in 40 min	Peng et al. (2003)
Atrazine	5 mg L^{-1}	4 g L^{-1}	5	0.956 mg g^{-1}	El-Sheikh et al. (2008)
Propoxur	5 mg L^{-1}	4 g L^{-1}	5	0.625 mg g^{-1}	El-Sheikh et al. (2008)
Methidathion	5 mg L^{-1}	4 g L^{-1}	5	1.11 mg g^{-1}	El-Sheikh et al. (2008)

act simultaneously. The dominant adsorption mechanism might be affected by carbon nanotubes, organic chemicals and environmental conditions (Pan and Xing 2008).

Carbon nanotubes could adsorb heavy metals including Pb^{2+}, Cd^{2+}, Cu^{2+}, Co^{2+}, Ni^{2+} and Zn^{2+} effectively (Rao et al. 2007; Pyrzyńska and Bystrzejewski 2010; Li et al. 2005). Pyrzyńska and Bystrzejewski (2010) reported that carbon nanotubes have higher adsorption efficiency towards Co^{2+} and Cu^{2+} compared with activated carbons.

Regeneration is a key factor determining the cost-effectiveness of carbon nanotubes. Lu et al. (2006) reported that the adsorption of Zn^{2+} on single-walled carbon nanotubes and multi-walled carbon nanotubes can be reversed by 0.1 mol L^{-1} nitric acid solution and the adsorption capacity was maintained after ten cycles of the regeneration and reuse. This suggested that carbon nanotubes could be regenerated by reducing solution pH value. The adsorption capacity of carbon nanotubes was not much fluctuant after several cycles of adsorption/desorption reaction.

6.2.3.2 Metal Based Nanoadsorbents

Iron oxide nanomaterials as adsorbent has received wide attention due to its high surface area, low toxicity and easy synthesis (Deliyanni et al. 2004; Huang et al. 2007; Xu et al. 2012). Iron oxide nanomaterials Iron oxide is a general designation of a large class of substance including many types. nFe_3O_4, n-γFe_2O_3 and n-αFe_2O_3 are the most common three kinds of iron oxides nanomaterials in water and wastewater treatment.

Iron oxide nanomaterials could adsorb a variety of heavy metals including Pb^{2+}, Cd^{2+}, Cu^{2+}, Hg^{2+}, As^{3+} and Zn^{2+} (Li and Zhang 2006; Huang and Chen 2009; White et al. 2009); Organic pollutants (e.g., red dye, 1-naphthylamine, polycyclic aromatic hydrocarbons) (Iram et al. 2010; Hu et al. 2011; Zhang et al. 2010a) and radionuclides (Qu et al. 2013) (Table 6.5). Nassar (2010) reported that the maximum adsorption capacity of Pb^{2+} onto nFe_3O_4 was much higher than that of reported adsorbents. However, the adsorption of heavy metal onto iron oxide nanomaterials is still at the lab scale (Xu et al. 2012). Practical application is limited.

Other than iron oxide nanoparticle, $nTiO_2$, nZnO and nAl_2O_3 were also effective adsorbents for the removal of heavy metals, metallic pollutants and radionuclides (Hua et al. 2012). Similar to carbon nanotubes, metal oxide nano-adsorbents could also be regenerated by changing solution pH (Sharma et al. 2009), and then the adsorption capacity remained relatively stable (Hu et al. 2006). However, opposite results were also reported. Deliyanni et al. (2003) reported that adsorption of As^{5+} on akaganéite-type nanocrystals can be reversed, but the adsorption capacity would reduce about 25–30 % after each cycles of the regeneration and reuse. So akaganéite-type nanocrystals must be replaced after 2–4 regenerations.

Table 6.5 Pollutants that can be removed by iron oxide nanomaterials

Pollutants	Initial concentration	Nanomaterial Dose	pH	Removal efficiency	Reference
Cu^{2+}	0–20 mmol L^{-1}	5 g L^{-1}	4	0.5 mmol g^{-1}	Kim et al. (2003)
Cu^{2+}	200 mg L^{-1}	5 g L^{-1}	2–5.1	38.5 mg g^{-1}	Banerjee and Chen (2007)
Co^{2+}	0–100mg L^{-1}	20 g L^{-1}	5.5	27.5 mg g^{-1}	Chang et al. (2006)
Cu(II)	18 ppm	100 g L^{-1}	8	30 mg g^{-1}	Mahdavian and Mirrahimi (2010)
Cu(II)	100 mg L^{-1}	20 g L^{-1}	2 and 5	12.43 mg g^{-1}	Huang and Chen (2009)
Cr(VI)	10–50 mg L^{-1}	3 or 6 g L^{-1}	5	30 mg g^{-1}	Wei et al. (2009)
Cr(VI)	0–140 mg L^{-1}	5 g L^{-1}	2.5	15.3 mg g^{-1}	Yuan et al. (2009)
Cr(VI)	10–200 mg L^{-1}	1 g L^{-1}	2.5	25.8 mg g^{-1}	Hu et al. (2007)
Cr(VI)	100 mg L^{-1}	20 g L^{-1}	2 and 5	11.24 mg g^{-1}	Huang and Chen (2009)
Pb(II)	18 ppm	100 g L^{-1}	8	40 mg g^{-1}	Mahdavian and Mirrahimi (2010)
Pb(II)	100 mg L^{-1}	10 g L^{-1}	7	97.34 %	Ahmad et al. (2009)
Hg(II)	100 mg L^{-1}	10 g L^{-1}	7	90 %	Ahmad et al. (2009)
Cd(II)	1.8 ppm	100 g L^{-1}	8	5 mg g^{-1}	Mahdavian and Mirrahimi (2010)
Red dye	200 mg L^{-1}	1 g L^{-1}	6	90 %	Iram et al. (2010)
Fluoride	20 mg L^{-1}	1 g L^{-1}	6.5	91.74 mg g^{-1}	Zhao et al. (2010)
Phosphate	50 mg P L^{-1}	1 g L^{-1}	4	13.65 mg g^{-1}	Long et al. (2011)
1-naphthylamine	60 mg L^{-1}	0.4 g L^{-1}	6.5	200 mg g^{-1}	Hu et al. (2011)
Phenanthrene	0.4 ng mL^{-1}	0.2 g L^{-1}	3–10	more than 90 %	Zhang et al. (2010a)
Fluoranthene	0.4 ng mL^{-1}	0.2 g L^{-1}	3–10	more than 90 %	Zhang et al. (2010a)
Pyrene	0.4 ng mL^{-1}	0.2 g L^{-1}	3–10	more than 90 %	Zhang et al. (2010a)
Benzo(a) anthracene	0.4 ng mL^{-1}	0.2 g L^{-1}	3–10	more than 90 %	Zhang et al. (2010a)
Benzo[b] fluoranthene	0.4 ng mL^{-1}	0.2 g L^{-1}	3–10	more than 90 %	Zhang et al. (2010a)

6.2.3.3 Polymeric Nanoadsorbents

Recently, polymeric nano-adsorbents have emerged as a novel type of adsorbent materials for the removal of heavy metals and organic pollutants in water and wastewater. Regarding the environmental concerns, these adsorbents are typically made up of polystyrene or polyacrylic ester matrix (Ray and Shipley 2015). Although polymeric nano-adsorbents have excellent properties such as pore size distribution, large surface area, tunable surface chemistry and excellent mechanical rigidity, they have some drawbacks such as low adsorption capacities and high cost (Ray and Shipley 2015; Pan et al. 2009).

6.2.4 Nanoreducing Materials

As a kind of effective reductant for pollutants removal in water, nano zero-valent metals have attracted much attention science 1980s. Iron is a metal with standard redox potential ($E^0 = -0.44$ V). It is thus an effective reductant when reacting with oxidized contaminants in water. Nanoscale zero-valent iron is the particle size of zero-valent iron between 1 and 100 nm. The use of nanoscale zero-valent iron for the removal of contaminants in water and wastewater has received wide attention due to its high reduction performance, large specific surface area and high reactivity. Other types of nano-reducing materials have also been tested for water and wastewater treatment, such as nZn and nNi. In particular, nanoscale zero-valent iron is the most commonly used nanomaterial for groundwater remediation (Mueller et al. 2012). Nanoscale zero-valent iron has been successfully applied for the treatment of water and wastewater contaminated with chlorinated organic compounds (Arnold et al. 2002), arsenic (Kanel et al. 2005), heavy metals, including chromium (Scott et al. 2011), cadmium (Scott et al. 2011), copper (Li and Zhang 2007; Karabelli et al. 2008), silver (Li and Zhang 2007), zinc (Li and Zhang 2007; Klimkova et al. 2011), dyes (Lin et al. 2008; Fan et al. 2009) and phenol (Liu et al. 2005; Elliott et al. 2009; Crane and Scott TB 2012) (Table 6.6). Due to the significant variation in contaminant chemistry, numerous possible contaminant removal pathways have been performed, including sorption, complexation, (co)precipitation and surface mediated chemical reduction.

The pollutants removal by nanoscale zero-valent iron is affected by many factors such as type of pollutants, nanoscale zero-valent iron concentration, temperature and solution pH value. Lin et al. (2008) reported that the removal of AB24 dye by nanoscale zero-valent iron, the results showed that the degradation efficiency of AB24 dye increased with increasing nanoscale zero-valent iron concentration (0–4 g L^{-1}) and temperature (10–50 °C). The reaction rate is highly pH-dependent, the rate constants decreased as the pH increased from 3 to 6 or above 9 and the rate increased as pH increased from 6 to pH 9. The reduction of AB24 dye by ZVI dominated the surface reaction at pH lower than 6; whereas at pH higher than 6, the removal of AB24 dye was mainly due to an adsorption reaction. Li and Zhang (2007) reported that the removal efficiency of eight metal ions including Cd^{2+}, Ni^{2+}, Zn^{2+}, Cr^{6+}, Cu^{2+}, Pb^{2+} and Ag^{+} with nanoscale zero-valent iron is 36.5 %, 71.0 %, 92.5 %, 97.5 %, 99.7 %, 99.7 % and 99.8 %. In the study performed by Li and Zhang (2007), as for metals with standard potential E0 very close to or more negative than that of iron (E^0, −0.41 V), such as Zn^{2+} and Cd^{2+}, the removal mechanism is sorption and surface complexation. As for metals with E^0 greatly more positive than iron

Table 6.6 Pollutants that can be removed by nanoscale zero-valent iron

Pollutants	Initial concentration	Nanomaterial dose	pH	Removal efficiency	References
As^{3+}	5 mg L^{-1}	0.50 and 1.04 g L^{-1}	6.5	121 mg g^{-1}	Du et al. (2013)
As^{3+}	1 mg L^{-1}	1 g L^{-1}	4–10	88.6–99.9 % in 12 h	Kanel et al. (2005)
As^{5+}	5 mg L^{-1}	0.50 and 1.04 g L^{-1}	6.5	125 mg g^{-1}	Du et al. (2013)
As^{5+}	1 mg L^{-1}	1 g L^{-1}	7	Nearly 100 % in 10 min	Kanel et al. (2006)
Arsenic	2 mg L^{-1}	1 g L^{-1}	6.5	18.2 mg g^{-1}	Zhu et al. (2009)
Co^{2+}	1000 mg L^{-1}	2.5 g L^{-1}	10	More than 99.9 %	Üzüm et al. (2008)
Cu^{2+}	500 mg L^{-1}	1–3 g L^{-1}	4.67	43.3 %–99.9 %	Karabelli et al. (2008)
Cu^{2+}	10 mg L^{-1}	10 g L^{-1}	>5.7	96 % after 7 days	Scott et al. (2011)
Pb^{2+}	100 mg L^{-1}	1 g L^{-1}	4	More than 96 % in 140 min	Kim et al. (2013)
U(VI)	10 mg L^{-1}	10 g L^{-1}	>5.7	99 % after 7 days	Scott et al. (2011)
Cr(VI)	10 mg L^{-1}	10 g L^{-1}	>5.7	99 % after 7 days	Scott et al. (2011)
Cr(VI)	20 mg L^{-1}	2.0 g L^{-1}	3.0	>90 %	Liu et al. (2012)
Cr(VI)	34 mg L^{-1}	0.04–0.12 g L^{-1}	9	24 %–90 %	Xu and Zhao (2007)
Cr(VI)	20 mg L^{-1}	3 g L^{-1}	4.0	Nearly 100 % in 20 min	Shi et al. (2011)
Nitrate	50 mg N L^{-1}	2.4 g L^{-1}	5	97.2 %	Jiang et al. (2011)
Nitrate	50 mg L^{-1}	0.5 g L^{-1}	7.0	Nearly 100 % in 120 min	Zhang et al. (2011)
Perchlorate	200 mg L^{-1}	20 g L^{-1}	10 g L^{-1}	90 %,54.58 %,19.79 %,3.27 % at 75 °C,60 °C,40 °C,25 °C in 24 h	Cao et al. (2005)
Orange II	70 mg L^{-1}	2 g L^{-1}	natural	100 %	Luo et al. (2013)
Acid Blue 113 dye	100 mg L^{-1}	20 g L^{-1}	5.6	Almost 100 % after 30 min	Shu et al. (2010)
Methyl orange dye	100 mg L^{-1}	0.5 g L^{-1}	6.5	Nearly 100 % after 30 min	Chen et al. (2011)
Methylene blue dye	94 mg L^{-1}	1 g L^{-1}	5.9	89.36 % in 30–60 min	Frost et al. (2010)
AB24 dye	50 mg L^{-1}	0.3 g L^{-1}	5.6	completely removed in 20 min	Lin et al. (2008)
Azo dye methyl orange	100 mg L^{-1}	0.2–0.5 g L^{-1}	6	70.3 %–100 % in 60 min	Fan et al. (2009)

(continued)

Table 6.6 (continued)

Pollutants	Initial concentration	Nanomaterial dose	pH	Removal efficiency	References
Tetrachloroethene	20 mg L^{-1}	5 g L^{-1}	/	100 % in 90 min	Lien and Zhang (2001)
Trichloroethylene	10 mg L^{-1}	5 g L^{-1}	/	90 % in 30 min	Choe et al. (2001)
Trichloroethene	20 mg L^{-1}	5 g L^{-1}	/	100 % in 90 min	Lien and Zhang (2001)
Vinyl chloride	20 mg L^{-1}	5 g L^{-1}	/	100 % in 80 min	Lien and Zhang (2001)
Chloroform	10 mg L^{-1}	5 g L^{-1}	/	Completely dechlorinated within 5 min	Choe et al. (2001)
Nitrobenzene	10 mg L^{-1}	5 g L^{-1}	/	100 %	Choe et al. (2001)
Nitrotoluene	10 mg L^{-1}	5 g L^{-1}	/	85 %	Choe et al. (2001)
Dinitrobenzene	10 mg L^{-1}	5 g L^{-1}	/	80 %	Choe et al. (2001)
Dinitrotoluene	10 mg L^{-1}	5 g L^{-1}	/	70 %	Choe et al. (2001)
1,1-dichloroethene	20 mg L^{-1}	5 g L^{-1}	/	100 % in 90 min	Lien and Zhang (2001)
cis-1,2-dichloroethene	20 mg L^{-1}	5 g L^{-1}	/	100 % in 80 min	Lien and Zhang (2001)
trans-1,2-dichloroethene	20 mg L_{-1}	5 g L^{-1}	/	100 % in 90 min	Lien and Zhang (2001)
2,2-bis(4-Chlorophenyl)-1,1,1-trichloroethane	1.09 mg L^{-1}	0.05 g L^{-1}	7	over 90 % in 4 h	Tian et al. (2009)
p-chlorophenol	20–100 mg L^{-1}	0.3 g L^{-1}	/	100 % (4 days)	Cheng et al. (2007)
2,4,6-trinitrotoluene	80 mg L^{-1}	5 g L^{-1}	4	More than 99 % in 3 h	Zhang et al. (2010b)
Metronidazole	80 mg L^{-1}	0.1 g L^{-1}	5.6	Rapidly removed in 5 min	Fang et al. (2011)
Lindane	1 mg L^{-1}	0.02–0.5 g L^{-1}	7.9–8.3	45 % to nearly 100 % within 300 min	Joo and Zhao (2008)
Atrazine	5 mg L^{-1}	0.1 g L^{-1}	7.0–8.2	More than 95 % under anaerobic condition; 35 % under aerobic condition	Joo and Zhao (2008)
Hexachlorobenzene	5 mg L^{-1}	2–20 g L^{-1}	/	About 25–70 % degradation in 24 h	Shih et al. (2011)
Escherichia coli	1×10^{6}–2×10^{6} CFU mL^{-1}	1.2–110 mg L^{-1}	8	0.82 log inactivation/mg L^{-1} nanoFe-h	Lee et al. (2008)

such as Cu^{2+}, Ag^{+}, and Hg^{2+}, the removal mechanism is predominantly reduction. As for metal ions with E^0 slightly more positive than iron such as Ni^{2+} and Pb^{2+}, they can be immobilized at the nanomaterial surface by both sorption and reduction.

6.3 Applications of Nanomaterials for Water Quality Monitoring

Water quality monitoring is of importance to pollution sources control, water quality management and public health. With the development of water quality monitoring technology, some novel technologies were applied in microbial water quality monitoring including phylochips, quantitative real-time polymerase chain reaction and pyrosequencing (Aw and Rose 2012). Besides, previous studies showed that nanomaterials could be used in organic pollutants, inorganic pollutants and pathogen detection, including magnetic nanoparticles, carbon nanotubes, noble metal nanomaterials and quantum dots.

Pathogens detection in water is vital for human health. Many pathogens including bacteria such as Legionella, Escherichia coli and Helicobacter, viruses such as Enteroviruses, Hepatitis viruses and Rotaviruses, and protozoan such as Cryptosporidium and Giardia associated with drinking water are closely related to human diseases. Nanomaterial enabled pathogens sensors are consist of recognition agents, nanomaterials and a signal transduction mechanism. Among three components, nanomaterials are used to improve the detection sensitivity and response of pathogens due to their unique properties such as optical, electrochemical and magnetic properties.

Magnetic nanomaterials and carbon nanotubes have been applied for sample concentration and purification. Magnetic nanocomposite can be used to develop pathogen detection kits. Although carbon nanotubes performed the excellent sensitivity, heterogeneity is a great challenge. The carbon nanotubes production and purification processes often introduce contaminants and impurities, and even the carbon nanotubes structure degradation. Hence, it is necessary to produce homogeneous carbon nanotubes.

Hahn et al. (2005) used functionalized quantum dots to detect single cells of Escherichia coli O157: H7 serotype, the results showed that quantum dots were superior to traditional fluorescent dyes in terms of sensitivity and stability. Quantum dots can overcome the limitations of traditional fluorescence dyes and simultaneously detect multiple emission peaks from a single wavelength of light. For example, quantum dots can detect Escherichia coli O157: H7 and salmonella typhimurium simultaneously and expected to detect 3–4 species simultaneously in the near future (Yang and Li 2006). Taking advantage of the optical properties of quantum dots, they will help in pathogen detection certainly (Yang and Li 2006; Hahn et al. 2005). Quantum dots were capable of differentiating minute amounts of pathogenic bacterial cells among larger populations of innocuous cells due to their sensitivity and larger absorption cross sections (Hahn et al. 2005).

Nanomaterials can also be used in the detection of organic and inorganic pollutants. Nano-Au could detect chlorpyrifos and malathion at per billion levels from surface water (Lisha et al. 2009). Lysozyme Type VI-Stabilized Gold Nanoclusters was used to detect Hg^{2+} and CH_3Hg^+ (Lin and Tseng 2010), the limits of detection for Hg^{2+} and CH_3Hg^+ were estimated to be 3 pM and 4 nM. Lysozyme Type VI-Stabilized Gold Nanoclusters provided an about 330-fold improvement in the detection of Hg^{2+} in comparison to bovine serum albumin-stabilized gold Nanoclusters. More immortally, this probe was successfully used in seawater. Vega et al. (2007) reported the use of the carbon nanotube-modified glassy carbon electrode for the detection of phenolic estrogens.

6.4 Challenges of Applying Nanomaterials in Water Quality Management

Although nanomaterials enabled water and wastewater treatment and monitoring have shown great potentials in the future, cost-effectiveness and technical hurdles are still challenges for their development and commercialization. The cost of nanomaterials is relatively high. Studies on some nanomaterials such as nanoscale zero-valent iron are mostly carried out at the lab scale. Many laboratory studies have evaluated the performance of nanoscale zero-valent iron for removing various pollutants. However, the research on the treatment performance and the long-term performance of real water and waste water with nanomaterials is limited due to the short time of laboratory studies and complication of real water and wastewater.

In addition, potential risk of nanomaterials is another challenge for their widespread application. An increasing number of nanomaterials will be released to the environment due to the increasing application of nanomaterials in water and wastewater treatment and monitoring, which have attracted more and more concern. The environmental behavior and possible environmental effects of nanomaterials are still unknown. Human health risk assessment and ecological risk assessment of nanomaterials are limited (Moore 2006). Relevant laws and regulations were still lacking. Hence, more studies about nanotoxicology and nanoecotoxicology need to be done.

6.5 Conclusion

The application of nanomaterials in water quality management has received wide attention due to their unique electrical, mechanical, catalytic, optical, magnetic and photonic properties. Nanofiltration membrane could be used in the production of potable water and the removal of metals, disinfection by-products, pesticides and emerging contaminants from contaminated water. However membrane fouling is still a great limitation. $nTiO_2$ has some excellent performance in the field of water and wastewater treatment, especially for dye wastewater and paper mill wastewater. The surface modification of $nTiO_2$ is being studied for optimization. However, it has some drawbacks including narrow light response range and low recycle rate. Carbon nanotubes, iron oxides nanomaterials, $nTiO_2$ and polymeric nano-adsorbents have shown high adsorption capacities. Enhancing the regeneration of absorbents must be explored to reduce the cost in water quality management. In addition, nanoscale zero-valent iron could be used to remove heavy metals and organic pollutants by reduction or oxidation and degree of removal could be enhanced through functionalization. However, the persistence of activity is limited. Nanomaterials-based sensors have the potential to detect heavy metals, organic pollutants and pathogen in water and wastewater. Nanomaterials are used to improve the detection sensitivity and response speed of pollutions. In a word, nanomaterials have received extensive research in water pollution remediation and monitoring. However, there are many problems for their practical application need to be study and solve including cost-effectiveness, technical hurdles and potential risk of nanomaterials.

Acknowledgement This work was supported by the Fundamental Research Funds for the Central Universities, and the Research Funds of Renmin University of China (Grant No. 11XNK016), which are greatly acknowledged.

References

Aarthi T, Madras G (2007) Photocatalytic degradation of rhodamine dyes with nano-TiO_2. Ind Eng Chem Res 46:7–14

Aarthi T, Madras G (2008) Photocatalytic reduction of metals in presence of combustion synthesized nano-TiO_2. Catal Commun 9:630–634

Ahmad A, Rafatullah M, Sulaiman O, Ibrahim MK, Chii YY, Siddique BM (2009) Removal of Cu(II) and Pb(II) ions from aqueous solutions by adsorption on sawdust of Meranti wood. Desalination 247:636–646

Andreescu S, Njagi J, Ispas C, Ravalli MT (2009) JEM spotlight: applications of advanced NMs for environmental monitoring. J Environ Monit 11:27–40

Arnold WA, Winget P, Cramer CJ (2002) Reductive dechlorination of 1,1,2,2-tetrachloroethane. Environ Sci Technol 36:3536–3541

Aw TG, Rose JB (2012) Detection of pathogens in water: from phylochips to qPCR to pyrosequencing. Curr Opin Biotechnol 23:422–430

Banerjee SS, Chen DH (2007) Fast removal of copper ions by gum Arabic modified magnetic nano-adsorbent. J Hazard Mater 147:792–799

Bódalo A, Gómez E, Hidalgo AM, Gómez MM, Murcia MD, López I (2009) Nanofiltration membranes to reduce phenol concentration in wastewater. Desalination 245:680–686

Bystrzejewska-Piotrowska G, Golimowski J, Urban PL (2009) Nanoparticles: their potential toxicity, waste and environmental management. Waste Manag 29:2587–2595

Cao J, Elliott D, Zhang W (2005) Perchlorate reduction by nanoscale iron particles. J Nanopart Res 7:499–506

Chakrabortty S, Roy M, Pal P (2013) Removal of fluoride from contaminated groundwater by cross flow nanofiltration: transport modeling and economic evaluation. Desalination 313:115–124

Chang YC, Chang SW, Chen DH (2006) Magnetic chitosan nanoparticles: studies on chitosan binding and adsorption of Co(II) ions. React Funct Polym 66:335–341

Chang F, Liu W, Wang X (2014) Comparison of polyamide nanofiltration and low-pressure reverse osmosis membranes on as(III) rejection under various operational conditions. Desalination 334:10–16

Chen X, Mao SS (2007) Titanium dioxide nanomaterials: synthesis, properties, modifications, and applications. Chem Rev 107:2891–2959

Chen C, Wang X (2006) Adsorption of Ni(II) from aqueous solution using oxidized multiwall carbon nanotubes. Ind Eng Chem Res 45:9144–9149

Chen C, Li X, Zhao D, Tan X, Wang X (2007) Adsorption kinetic, thermodynamic and desorption studies of Th(IV) on oxidized multi-wall carbon nanotubes. Colloid Surf A Physicochem Eng Asp 302:449–454

Chen Z, Jin X, Chen Z, Megharaj M, Naidu R (2011) Removal of methyl orange from aqueous solution using bentonite-supported nanoscale zero-valent iron. J Colloid Interf Sci 363:601–607

Cheng R, Wang J, Zhang W (2007) Comparison of reductive dechlorination of p-chlorophenol using Fe^0 and nanosized Fe^0. J Hazard Mater 144:334–339

Cheng S, Oatley DL, Williams PM, Wright CJ (2011) Positively charged nanofiltration membranes: review of current fabrication methods and introduction of a novel approach. Adv Colloid Interf 164:12–20

Choe S, Lee SH, Chang YY, Hwang K, Khim J (2001) Rapid reductive destruction of hazardous organic compounds by nanoscale Fe^0. Chemosphere 42:367–372

Choi W, Termin A, Hoffmann MR (1994) The role of metal ion dopants in quantum-sized TiO_2: correlation between photoreactivity and charge carrier recombination dynamics. J Phys Chem 98:13669–13679

Crane RA, Scott TB (2012) Nanoscale zero-valent iron: future prospects for an emerging water treatment technology. J Hazard Mater 211:112–125

Dai K, Peng T, Chen H, Zhang R, Zhang Y (2008) Photocatalytic degradation and mineralization of commercial methamidophos in aqueous titania suspension. Environ Sci Technol 42:1505–1510

Deliyanni EA, Bakoyannakis DN, Zouboulis AI, Matis KA (2003) Sorption of as (V) ions by akaganeite-type nanocrystals. Chemosphere 50:155–163

Deliyanni EA, Lazaridis NK, Peleka EN, Matis KA (2004) Metals removal from aqueous solution by iron-based bonding agents. Environ Sci Pollut Res 11:18–21

Du Q, Zhang S, Pan B, Lv L, Zhang W, Zhang Q (2013) Bifunctional resin-ZVI composites for effective removal of arsenite through simultaneous adsorption and oxidation. Water Res 47:6064–6074

Ducom G, Cabassud C (1999) Interests and limitations of nanofiltration for the removal of volatile organic compounds in drinking water production. Desalination 124:115–123

Elliott DW, Lien HL, Zhang WX (2009) Degradation of lindane by zero-valent iron nanoparticles. J Environ Eng 135:317–324

El-Sheikh AH, Sweileh JA, Al-Degs YS, Insisi AA, Al-Rabady N (2008) Critical evaluation and comparison of enrichment efficiency of multi-walled carbon nanotubes, C18 silica and activated carbon towards some pesticides from environmental waters. Talanta 74:1675–1680

Fan J, Guo Y, Wang J, Fan M (2009) Rapid decolorization of azo dye methyl orange in aqueous solution by nanoscale zerovalent iron particles. J Hazard Mater 166:904–910

Fan X, Yang H, Tian W, Nie A, Hou T, Qiu F, Zhang X (2011) Catalytic oxidation of chlorobenzene over $MnOx/Al_2O_3$-carbon nanotubes composites. Catal Lett 141:158–162

Fang Z, Chen J, Qiu X, Qiu X, Cheng W, Zhu L (2011) Effective removal of antibiotic metronidazole from water by nanoscale zero-valent iron particles. Desalination 268:60–67

Fang W, Shi L, Wang R (2013) Interfacially polymerized composite nanofiltration hollow fiber membranes for low-pressure water softening. J Membr Sci 430:129–139

Frost RL, Xi Y, He H (2010) Synthesis, characterization of palygorskite supported zero-valent iron and its application for methylene blue adsorption. J Colloid Interface Sci 341:153–161

García A, Delgado L, Torà JA, Casals E, González E, Puntes V, Font X, Carrera J, Sánchez A (2012) Effect of cerium dioxide, titanium dioxide, silver, and gold nanoparticles on the activity of microbial communities intended in wastewater treatment. J Hazard Mater 199:64–72
Gupta SM, Tripathi M (2011) A review of TiO_2 nanoparticles. Chin Sci Bull 56:1639–1657
Hahn MA, Tabb JS, Krauss TD (2005) Detection of single bacterial pathogens with semiconductor quantum dots. Anal Chem 77:4861–4869
Han H, Ba R (2009) Buoyant photocatalyst with greatly enhanced visible-light activity prepared through a low temperature hydrothermal method. Ind Eng Chem Res 48:2891–2898
Harisha RS, Hosamani KM, Keri RS, Nataraj SK, Aminabhavi TM (2010) Arsenic removal from drinking water using thin film composite nanofiltration membrane. Desalination 252:75–80
Hilal N, Al-Zoubi H, Darwish NA, Mohamma AW, Arabi MA (2004) A comprehensive review of nanofiltration membranes: treatment, pretreatment, modelling, and atomic force microscopy. Desalination 170:281–308
Hu J, Chen G, Lo IMC (2006) Selective removal of heavy metals from industrial wastewater using maghemite nanoparticle: performance and mechanisms. J Environ Eng 132:709–715
Hu J, Lo IMC, Chen G (2007) Performance and mechanism of chromate (VI) adsorption by δ-FeOOH-coated maghemite (γ-Fe_2O_3) nanoparticles. Sep Purif Technol 58:76–82
Hu J, Shao D, Chen C, Sheng G, Ren X, Wang X (2011) Removal of 1-naphthylamine from aqueous solution by multiwall carbon nanotubes/iron oxides/cyclodextrin composite. J Hazard Mater 185:463–471
Hua M, Zhang S, Pan B, Zhang W, Lv L, Zhang Q (2012) Heavy metal removal from water/wastewater by nanosized metal oxides: a review. J Hazard Mater 211:317–331
Huang SH, Chen DH (2009) Rapid removal of heavy metal cations and anions from aqueous solutions by an amino-functionalized magnetic nano-adsorbent. J Hazard Mater 163:174–179
Huang YH, Hsueh CL, Cheng HP, Su LC, Chen CY (2007) Thermodynamics and kinetics of adsorption of Cu(II) onto waste iron oxide. J Hazard Mater 144:406–411
Iram M, Guo C, Guan Y, Ishfaq A, Liu H (2010) Adsorption and magnetic removal of neutral red dye from aqueous solution using Fe_3O_4 hollow nanospheres. J Hazard Mater 181:1039–1050
Jiang Z, Lv L, Zhang W, Du Q, Pan B, Yang L, Zhang Q (2011) Nitrate reduction using nanosized zero-valent iron supported by polystyrene resins: role of surface functional groups. Water Res 45:2191–2198
Joo SH, Zhao D (2008) Destruction of lindane and atrazine using stabilized iron nanoparticles under aerobic and anaerobic conditions: effects of catalyst and stabilizer. Chemosphere 70:418–425
Kanel SR, Manning B, Charlet L, Choi H (2005) Removal of arsenic(III) from groundwater by nanoscale zero-valent iron. Environ Sci Technol 39:1291–1298
Kanel SR, Greneche JM, Choi H (2006) Arsenic(V) removal from groundwater using nano scale zero-valent iron as a colloidal reactive barrier material. Environ Sci Technol 40:2045–2050
Karabelli D, Üzüm C, Shahwan T, Eroğlu AE, Scott TB, Keith R, Hallam KR, Lieberwirth L (2008) Batch removal of aqueous Cu^{2+} ions using nanoparticles of zero-valent iron: a study of the capacity and mechanism of uptake. Ind Eng Chem Res 47:4758–4764
Kim Y, Lee B, Yi J (2003) Preparation of functionalized mesostructured silica containing magnetite(MSM) for the removal of copper ions in aqueous solutions and its magnetic separation. Sep Sci Technol 38:2533–2548
Kim SA, Kamala-Kannan S, Lee KJ, Parkb Y, Shea PJ, Lee W, Kim H, Oh B (2013) Removal of Pb(II) from aqueous solution by a zeolite–nanoscale zero-valent iron composite. Chem Eng J 217:54–60
Kiso Y, Nishimura Y, Kitao T, Nishimura K (2000) Rejection properties of non-phenylic pesticides with nanofiltration membranes. J Membr Sci 171:229–237
Klimkova S, Cernik M, Lacinova L, Filip J, Jancik D, Zboril R (2011) Zero-valent iron nanoparticles in treatment of acid mine water from in situ uranium leaching. Chemosphere 82:1178–1184

Kwon S, Fan M, Cooper AT, Yang H (2008) Photocatalytic applications of micro-and nano-TiO_2 in environmental engineering. Crit Rev Env Sci Technol 38:197–226
Lan Y, Lu Y, Ren Z (2013) Mini review on photocatalysis of titanium dioxide nanoparticles and their solar applications. Nano Energy 2:1031–1045
Lee C, Kim JY, Lee WI, Nelson KL, Yoon J, Sedlak DL (2008) Bactericidal effect of zero-valent iron nanoparticles on Escherichia coli. Environ Sci Technol 42:4927–4933
Li X, Zhang W (2006) Iron nanoparticles: the core-shell structure and unique properties for Ni(II) sequestration. Langmuir 22:4638–4642
Li X, Zhang W (2007) Sequestration of metal cations with zerovalent iron nanoparticles a study with high resolution X-ray photoelectron spectroscopy (HR-XPS). J Phys Chem C 111:6939–6946
Li YH, Wang S, Cao A, Zhao D, Zhang X, Xu C, Luan Z, Ruan D, Liang J, Wu D, Wei B (2001) Adsorption of fluoride from water by amorphous alumina supported on carbon nanotubes. Chem Phys Lett 350:412–416
Li YH, Wang S, Wei J, Zhang X, Xu C, Luan Z, Wu D, Wei B (2002) Lead adsorption on carbon nanotubes. Chem Phys Lett 357:263–266
Li YH, Ding J, Luan Z, Di Z, Zhu Y, Xu C, Wu D, Wei B (2003a) Competitive adsorption of Pb^{2+}, Cu^{2+} and Cd^{2+} ions from aqueous solutions by multiwalled carbon nanotubes. Carbon 41:2787–2792
Li YH, Wang S, Luan Z, Ding J, Xu C, Wu D (2003b) Adsorption of cadmium(II) from aqueous solution by surface oxidized carbon nanotubes. Carbon 41:1057–1062
Li YH, Di Z, Ding J, Wu D, Luan Z, Zhu Y (2005) Adsorption thermodynamic, kinetic and desorption studies of Pb^{2+} on carbon nanotubes. Water Res 39:605–609
Lien HL (2005) Application and implication of zero-valent iron for transformation of disinfection by-products: a comparison study between nanoscale palladized iron and microscale iron. J Chin Inst Environ Eng 15:213–221
Lien HL, Zhang W (2001) Nanoscale iron particles for complete reduction of chlorinated ethenes. Colloids Surf A Physicochem Eng Asp 191:97–105
Liga MV, Bryant EL, Colvin VL, Li Q (2011) Virus inactivation by silver doped titanium dioxide nanoparticles for drinking water treatment. Water Res 45:535–544
Lim AL, Bai R (2003) Membrane fouling and cleaning in microfiltration of activated sludge wastewater. J Membr Sci 216:279–290
Lin YH, Tseng WL (2010) Ultrasensitive sensing of Hg^{2+} and CH_3Hg^+ based on the fluorescence quenching of lysozyme type VI-stabilized gold nanoclusters. Anal Chem 82:9194–9200
Lin YT, Weng CH, Chen FY (2008) Effective removal of AB24 dye by nano/micro-size zero-valent iron. Sep Purif Technol 64:26–30
Lisha KP, Anshup, Pradeep T (2009) Enhanced visual detection of pesticides using gold nanoparticles. J Environ Sci Health B 44:697–705
Liu Y, Majetich SA, Tilton RD, Sholl DS, Lowry GV (2005) TCE dechlorination rates, pathways, and efficiency of nanoscale iron particles with different properties. Environ Sci Technol 39:1338–1345
Liu T, Wang ZL, Zhao L, Yang X (2012) Enhanced chitosan/Fe^0-nanoparticles beads for hexavalent chromium removal from wastewater. Chem Eng J 189:196–202
Liu Q, Zhou Q, Jiang G (2014) Nanomaterials for analysis and monitoring of emerging chemical pollutants. TrAC-Trend Anal Chem 58:10–22
Long RQ, Yang RT (2001) Carbon nanotubes as superior sorbent for dioxin removal. J Am Chem Soc 123:2058–2059
Long F, Gong J, Zeng G, Chen L, Wang X, Deng J, Niu Q, Zhang H, Zhang X (2011) Removal of phosphate from aqueous solution by magnetic Fe–Zr binary oxide. Chem Eng J 171:448–455
Lu C, Chiu H (2006) Adsorption of zinc(II) from water with purified carbon nanotubes. Chem Eng Sci 61:1138–1145
Lu C, Chung YL, Chang KF (2005) Adsorption of trihalomethanes from water with carbon nanotubes. Water Res 39:1183–1189

Lu C, Chiu H, Liu C (2006) Removal of zinc(II) from aqueous solution by purified carbon nanotubes: kinetics and equilibrium studies. Ind Eng Chem Res 45:2850–2855
Luo S, Qin P, Shao J, Peng L, Zeng Q, Gu J (2013) Synthesis of reactive nanoscale zero valent iron using rectorite supports and its application for orange II removal. Chem Eng J 223:1–7
Madaeni SS, Mohamamdi T, Moghadam MK (2001) Chemical cleaning of reverse osmosis membranes. Desalination 134:77–82
Madsen HT, Søgaard EG (2014) Applicability and modelling of nanofiltration and reverse osmosis for remediation of groundwater polluted with pesticides and pesticide transformation products. Sep Purif Technol 125:111–119
Mahdavian AR, Mirrahimi MAS (2010) Efficient separation of heavy metal cations by anchoring polyacrylic acid on superparamagnetic magnetite nanoparticles through surface modification. Chem Eng J 159:264–271
Maher A, Sadeghi M, Moheb A (2014) Heavy metal elimination from drinking water using nanofiltration membrane technology and process optimization using response surface methodology. Desalination 352:166–173
Mohammad AW, Teow YH, Ang WL, Chung YT, Oatley-Radcliffe DL, Hilal N (2015) Nanofiltration membranes review: recent advances and future prospects. Desalination 356:226–254
Moore MN (2006) Do nanoparticles present ecotoxicological risks for the health of the aquatic environment? Environ Int 32:967–976
Mueller NC, Braun J, Bruns J, Černík M, Rissing P, Rickerby D, Nowack B (2012) Application of nanoscale zero valent iron (NZVI) for groundwater remediation in Europe. Environ Sci Pollut Res 19:550–558
Musbah I, Ciceron D, Saboni A, Alexandrova S (2013) Retention of pesticides and metabolites by nanofiltration by effects of size and dipole moment. Desalination 313:51–56
Nagaveni K, Sivalingam G, Hegde MS, Madras G (2004) Solar photocatalytic degradation of dyes: high activity of combustion synthesized nano TiO_2. Appl Catal B Environ 48:83–93
Nassar NN (2010) Rapid removal and recovery of Pb(II) from wastewater by magnetic nanoadsorbents. J Hazard Mater 184:538–546
Pan B, Xing B (2008) Adsorption mechanisms of organic chemicals on carbon nanotubes. Environ Sci Technol 42:9005–9013
Pan B, Pan B, Zhang W, Lv L, Zhang Q, Zheng S (2009) Development of polymeric and polymer-based hybrid adsorbents for pollutants removal from waters. Chem Eng J 151:19–29
Parra S, Olivero J, Pulgarin C (2002) Relationships between physicochemical properties and photoreactivity of four biorecalcitrant phenylurea herbicides in aqueous TiO_2 suspension. Appl Catal B Environ 36:75–85
Peng X, Li Y, Luan Z, Di Z, Wang H, Tian B, Jia Z (2003) Adsorption of 1, 2-dichlorobenzene from water to carbon nanotubes. Chem Phys Lett 376:154–158
Ponder SM, Darab JG, Bucher J, Caulder D, Craig L, Davis L, Edelstein N, Lukens W, Nitsche H, Rao L, Shuh DK, Mallouk TE (2001) Surface chemistry and electrochemistry of supported zerovalent iron nanoparticles in the remediation of aqueous metal contaminants. Chem Mater 13:479–486
Pyrzyńska K, Bystrzejewski M (2010) Comparative study of heavy metal ions sorption onto activated carbon, carbon nanotubes, and carbon-encapsulated magnetic nanoparticles. Colloid Surf A 362:102–109
Qu X, Alvarez PJJ, Li Q (2013) Applications of nanotechnology in water and wastewater treatment. Water Res 47:3931–3946
Radjenović J, Petrović M, Ventura F, Barceló D (2008) Rejection of pharmaceuticals in nanofiltration and reverse osmosis membrane drinking water treatment. Water Res 42:3601–3610
Rao GP, Lu C, Su F (2007) Sorption of divalent metal ions from aqueous solution by carbon nanotubes: a review. Sep Purif Technol 58:224–231
Ray PZ, Shipley HJ (2015) Inorganic nano-adsorbents for the removal of heavy metals and arsenic: a review. RSC Adv 5:29885–29907

Sanches S, Penetra A, Rodrigues A, Ferreira E, Cardoso VV, Benoliel MJ, Crespo MTB, Pereira VJ, Crespo JG (2012) Nanofiltration of hormones and pesticides in different real drinking water sources. Sep Purif Technol 94:44–53

Savage N, Diallo MS (2005) Nanomaterials and water purification: opportunities and challenges. J Nanopart Res 7:331–342

Schäfer AI, Akanyeti I, Semião AJC (2011) Micropollutant sorption to membrane polymers: a review of mechanisms for estrogens. Adv Colloid Interf 164:100–117

Schierz A, Zänker H (2009) Aqueous suspensions of carbon nanotubes: surface oxidation, colloidal stability and uranium sorption. Environ Pollut 157:1088–1094

Scott TB, Popescu IC, Crane RA, Noubactep C (2011) Nano-scale metallic iron for the treatment of solutions containing multiple inorganic contaminants. J Hazard Mater 186:280–287

Semião AJC, Schäfer AI (2011) Estrogenic micropollutant adsorption dynamics onto nanofiltration membranes. J Membr Sci 381:132–141

Sen M, Manna A, Pal P (2010) Removal of arsenic from contaminated groundwater by membrane-integrated hybrid treatment system. J Membr Sci 354:108–113

Sentana I, Rodríguez M, Sentana E, M'Birek C, Prats D (2010) Reduction of disinfection by-products in natural waters using nanofiltration membranes. Desalination 250:702–706

Sharma YC, Srivastava V, Singh VK, Kaul SN, Weng CH (2009) Nano-adsorbents for the removal of metallic pollutants from water and wastewater. Environ Technol 30:583–609

Shi L, Zhang X, Chen Z (2011) Removal of chromium(VI) from wastewater using bentonite-supported nanoscale zero-valent iron. Water Res 45:886–892

Shih Y, Hsu C, Su Y (2011) Reduction of hexachlorobenzene by nanoscale zero-valent iron: kinetics, pH effect, and degradation mechanism. Sep Purif Technol 76:268–274

Shu HY, Chang MC, Chen CC, Chen P (2010) Using resin supported nano zero-valent iron particles for decoloration of acid blue 113 azo dye solution. J Hazard Mater 184:499–500

Skubal LR, Meshkov NK (2002) Reduction and removal of mercury from water using arginine-modified TiO_2. J Photoch Photobio A 148:211–214

Su F, Lu C (2007) Adsorption kinetics, thermodynamics and desorption of natural dissolved organic matter by multiwalled carbon nanotubes. J Environ Sci Health A 42:1543–1552

Tian H, Li J, Mu Z, Li L, Hao Z (2009) Effect of pH on DDT degradation in aqueous solution using bimetallic Ni/Fe nanoparticles. Sep Purif Technol 66:84–89

Üzüm Ç, Shahwan T, Eroğlu AE, Lieberwirth I, Scott TB, Hallam KR (2008) Application of zero-valent iron nanoparticles for the removal of aqueous Co^{2+} ions under various experimental conditions. Chem Eng J 144:213–220

Van der Bruggen B, Vandecasteele C (2003) Removal of pollutants from surface water and groundwater by nanofiltration: overview of possible applications in the drinking water industry. Environ Pollut 122:435–445

Van der Bruggen B, Everaert K, Wilms D, Vandecasteele C (2001) Application of nanofiltration for removal of pesticides, nitrate and hardness from ground water: rejection properties and economic evaluation. J Membr Sci 193:239–248

Vega D, Agüí L, González-Cortés A, Yáñez-Sedeño P, Pingarrón JM (2007) Electrochemical detection of phenolic estrogenic compounds at carbon nanotube-modified electrodes. Talanta 71:1031–1038

Venkatachalam N, Palanichamy M, Murugesan V (2007) Sol–gel preparation and characterization of alkaline earth metal doped nano TiO_2: efficient photocatalytic degradation of 4-chlorophenol. J Mol Catal A Chem 273:177–185

Verliefde A, Cornelissen E, Amy G, Van der Bruggen B, van Dijk H (2007) Priority organic micropollutants in water sources in Flanders and the Netherlands and assessment of removal possibilities with nanofiltration. Environ Pollut 146:281–289

Vinu R, Madras G (2007) Kinetics of simultaneous photocatalytic degradation of phenolic compounds and reduction of metal ions with nano-TiO_2. Environ Sci Technol 42:913–919

Vinu R, Madras G (2008) Kinetics of sonophotocatalytic degradation of anionic dyes with nano-TiO_2. Environ Sci Technol 43:473–479

Vulliet E, Emmelin C, Chovelon JM, Guillard C, Herrmann J (2002) Photocatalytic degradation of sulfonylurea herbicides in aqueous TiO_2. Appl Catal B Environ 38:127–137
Wang CB, Zhang WX (1997) Synthesizing nanoscale iron particles for rapid and complete dechlorination of TCE and PCBs. Environ Sci Technol 31:2154–2156
Wang SG, Liu XW, Gong WX, Nie W, Gao B, Yue Q (2007) Adsorption of fulvic acids from aqueous solutions by carbon nanotubes. J Chem Technol Biotechnol 82:698–704
Wei L, Yang G, Wang R, Ma W (2009) Selective adsorption and separation of chromium(VI) on the magnetic iron–nickel oxide from waste nickel liquid. J Hazard Mater 164:1159–1163
White BR, Stackhouse BT, Holcombe JA (2009) Magnetic γ-Fe_2O_3 nanoparticles coated with poly-l-cysteine for chelation of as(III), Cu(II), Cd(II), Ni(II), Pb(II) and Zn(II). J Hazard Mater 161:848–853
Xin B, Ren Z, Wang P, Liu J, Jing L, Fu H (2007) Study on the mechanisms of photoinduced carriers separation and recombination for Fe^{3+}–TiO_2 photocatalysts. Appl Surf Sci 253:4390–4395
Xu Y, Zhao D (2007) Reductive immobilization of chromate in water and soil using stabilized iron nanoparticles. Water Res 41:2101–2108
Xu P, Zeng GM, Huang DL, Feng CL, Hu S, Zhao MH, Lai C, Wei Z, Huang C, Xie XG, Liu ZF (2012) Use of iron oxide nanomaterials in wastewater treatment: a review. Sci Total Environ 424:1–10
Yang L, Li Y (2006) Simultaneous detection of Escherichia coli O157: H7 and salmonella typhimurium using quantum dots as fluorescence labels. Analyst 131:394–401
Yang K, Wu W, Jing Q, Zhu L (2008) Aqueous adsorption of aniline, phenol, and their substitutes by multi-walled carbon nanotubes. Environ Sci Technol 42:7931–7936
Yang Y, Wang H, Li J, He B, Wang T, Liao S (2012) Novel functionalized nano-TiO_2 loading electrocatalytic membrane for oily wastewater treatment. Environ Sci Technol 46:6815–6821
Yu H, Wang X, Sun H, Huo M (2010) Photocatalytic degradation of malathion in aqueous solution using an Au–Pd–TiO_2 nanotube film. J Hazard Mater 184:753–758
Yuan P, Fan M, Yang D, He H, Liu D, Yuan A, Zhu J, Chen T (2009) Montmorillonite-supported magnetite nanoparticles for the removal of hexavalent chromium [Cr(VI)] from aqueous solutions. J Hazard Mater 166:821–829
Zhang S, Niu H, Hu Z, Cai Y, Shi Y (2010a) Preparation of carbon coated Fe_3O_4 nanoparticles and their application for solid-phase extraction of polycyclic aromatic hydrocarbons from environmental water samples. J Chromatogr A 1217:4757–4764
Zhang X, Lin Y, Shan X, Chen Z (2010b) Degradation of 2, 4, 6-trinitrotoluene (TNT) from explosive wastewater using nanoscale zero-valent iron. Chem Eng J 158:566–570
Zhang Y, Li Y, Li J, Hu L, Zheng X (2011) Enhanced removal of nitrate by a novel composite: nanoscale zero valent iron supported on pillared clay. Chem Eng J 171:526–531
Zhao X, Wang J, Wu F, Wang T, Cai Y, Shi Y, Jiang G (2010) Removal of fluoride from aqueous media by Fe_3O_4@$Al(OH)_3$ magnetic nanoparticles. J Hazard Mater 173:102–109
Zhong PS, Widjojo N, Chung TS, Weber M, Maletzko C (2012) Positively charged nanofiltration (NF) membranes via UV grafting on sulfonated polyphenylenesulfone (sPPSU) for effective removal of textile dyes from wastewater. J Membr Sci 417:52–60
Zhu H, Jia Y, Wu X, Wang H (2009) Removal of arsenic from water by supported nano zero-valent iron on activated carbon. J Hazard Mater 172:1591–1596

Chapter 7
Phytochemically Functionalized Silver and Gold Nanoparticles to Treat Microbes, Viruses and Cancer

Brajesh Kumar and Kumari Smita

Abstract There is a growing interest for phytochemically-functionalized nanoparticles because such nanoparticles are easy to synthesize, cheap, ecofriendly, and they have medicinal properties. Indeed, conventional synthesis of nanoparticles often requires toxic chemicals and high energy, whereas phytochemical synthesis involves the participation of phenolics, proteins, polysaccharides and other begnin biomolecules. Here we review the green synthesis of silver and gold nanoparticles, using phytochemicals and alternative energy. We also review applications to treat microbes. Silver and gold nanoparticles, sized from 10 to 100 nm, showed a significant effect on microbes *such as Escherichia coli*, *Staphylococcus aureus*, *Pseudomonas aeruginosa*, *Micrococcus luteus*, *Klebsiella pneumonia*, *Puccinia graminis*, *Aspergillus flavus*, *Aspergillus niger* and *Candida albicans*. Nanoparticles also showed inhibitory activities against various viruses and cancer cells like M13 phage virus, bacteriophage UZ1, HIV-1, murine norovirus 1, HeLa, MCF-7, A549, and HEp 2.

Keywords Silver nanoparticles • Gold nanoparticles • Phytochemicals • Green synthesis • Antimicrobial • Antiviral • Anticancer

---

B. Kumar (✉)
Centro de Nanociencia y Nanotecnologia, Universidad de las Fuerzas Armadas ESPE,
Av. Gral., Rumiñahui s/n, P.O. BOX 171-5-231B, Sangolqui, Ecuador

Department of Chemistry, TATA College, Kolhan University,
833202 Chaibasa, Jharkhand, India
e-mail: krmbraj@gmail.com; bkumar@espe.edu.ec

K. Smita
Centro de Nanociencia y Nanotecnologia, Universidad de las Fuerzas Armadas ESPE,
Av. Gral., Rumiñahui s/n, P.O. BOX 171-5-231B, Sangolqui, Ecuador
e-mail: kumarismi@gmail.com

S. Ranjan et al. (eds.), *Nanoscience in Food and Agriculture 2*, Sustainable Agriculture Reviews 21, DOI 10.1007/978-3-319-39306-3_7

7.1 Introduction

Nanotechnology is providing new consumer products including electronics, cosmetics, household appliances, textiles and food production as well as in medical and engineering products. The word nano, derived from the Greek nanos, meaning dwarf, is used to describe atomic and molecular assemblies of any material or property which occurs with shapes, sizes and surface properties on the nanometer scale (1–100 nm) (Dreaden et al. 2011; Chernousova and Epple 2013). Nanoparticles with controlled size and composition are of fundamental and technological interest as they provide solutions to technological and environmental challenges in the areas of solar energy conversion, catalysis, medicine and water treatment. Metal nanoparticles have a much greater surface area to volume ratio, which can exhibit completely new or improved properties compared to their bulk counterparts and are found to be interesting candidates for various applications (Raveendran et al. 2003). Up to now, metal nanoparticles are mostly prepared from noble metals such as silver, gold, platinum, and palladium. But beyond their beauty, gold (Au) and silver (Ag) nanoparticles are the most widely studied metal nanoparticles in the field of biological system, living organisms and medicine because of its chemical stability and facile surface functionalization properties (Dahl et al. 2007; Hutchison 2008). There are many methods available for the synthesis of metal nanoparticles such as ultraviolet irradiation, aerosol technologies, lithography, laser ablation, ultrasonic fields and photochemical reduction. However, these methods are not environmentally friendly. To address this challenge, there is a growing need to develop environmentally friendly processes of nanoparticle synthesis that do not use toxic chemicals and minimize the production of dangerous wastes. Therefore, an exhaustive study on the biological synthesis of nanoparticles has been carried out using a wide range of microorganisms, algae and plant extracts as possible ecofriendly alternatives to chemical and physical methods (Narayanan and Sakthivel 2011; Mittal et al. 2013). Using plants for nanoparticle synthesis can be advantageous over other biological processes by eliminating the elaborate process of maintaining cell cultures low cost (Kumar and Yadav 2009). It can also be inexpensive and suitably scaled up for large-scale synthesis of nanoparticles (Luangpipat et al. 2011). These reasons make the biological synthesis of nanoparticles using plant extracts are more valuable and as possible ecofriendly alternatives to chemical and physical methods.

7.2 Phytochemical Synthesis

Plants have a wide range of active constituents which has potent medicinal properties. In general, phytochemicals are bioactive compounds in fruits, vegetables, grains, oils, resins, seeds and other plant components that have been linked to reducing the incidence of major diseases (Jeetah et al. 2014; Liu 2004). It is necessary to understand the chemical structure and mechanism of different plant bioactive compounds. But Ag and Au bound phytochemical applications of plants are still unknown and the number of publications on phytochemical synthesized Ag/Au nanoparticles is gradually being increased since 2009. The Ag and Au nanoparticles are being extensively synthesized using plant extracts, their characterization and biological applications are shown in Fig. 7.1.

Although the exact mechanism for this phytochemical synthesis still remains to be completely unknown. However, a few hypotheses have been suggested to give some insights on the mechanical aspects of nanoparticles biosynthesis. Recent studies have shown that environmentally benign phytochemicals such as flavonoids, polyphenols, protein, polysaccharides, vitamins and fatty acids present in the plant extract play an important role in the reduction of M^{n+} ions via complexation and capping of the nanoparticles (Ahmad et al. 2009; Rajan et al. 2015). In Fig. 7.2, Ag^{+}/Au^{3+} ions can interact with -OH/-CHO groups, which subsequently undergo oxidation to -CHO and -COOH forms with consequent reduction of Ag^{+}/Au^{3+} to Ag/Au nanoparticles. Finally, –COOH groups help in the stabilization of nanoparticles. Nowadays, green synthesis of different size and shape Ag/Au nanoparticles using plant materials, including leaf (Shankar et al. 2012), flowers (Philip 2010), oil (Kumar et al. 2014a), seeds (Otari et al. 2014), fruits (Kumar et al. 2015a), bark (Sathishkumar et al. 2009), biomass (Kumar et al. 2014b), stem (Gangula et al. 2011) and root (Sreekanth et al. 2014) has been already reported in literature.

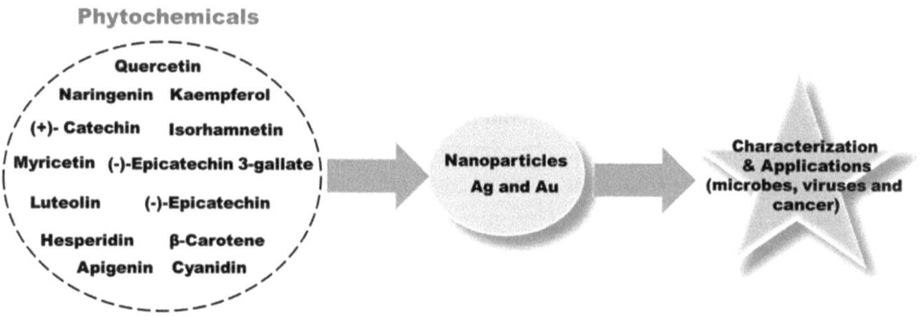

Fig. 7.1 Schematic scheme for the synthesis, characterization and application of Ag and Au nanoparticles

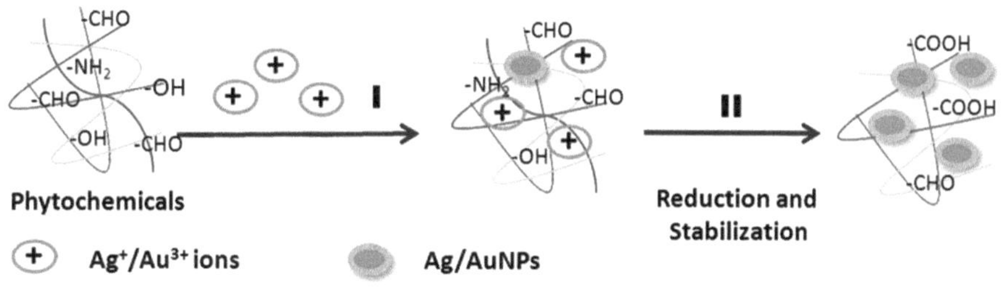

Fig. 7.2 Schematic diagram for the phytochemical synthesis of Ag and Au nanoparticles

Fig. 7.3 A schematic representation of plants (**a**–**e**) and insect (**f**) from Andean origin as a source of green synthesis of Ag nanoparticles

In continuation of our ongoing research on the phytochemical mediated synthesis of Ag nanoparticles, using plant material of Andean (Ecuador) origin, here in the current chapter will give a positive message that Ag/Au nanoparticles synthesized through phytochemical approach are much safer and further used for various human applications.

The most important example of such environmentally benign, cost-effective and renewable materials, or plants are Sacha inchi, Rambutan, Taxo, Capuli, and *Lantana camara* (Fig. 7.3a–e) as well as an insect, Cochineal (Fig. 7.3f). It is consumed mainly Ecuador, Peru and Colombia as fresh, cooked, jam, juice, frozen pulp or minor extent as wines.

Sacha inchi is a promising crop (*Plukentia volubilis* L.) which produces star shaped fruit, having a unique seed very rich in omega 3, 6 and 9 fatty acids and also known as the Inca peanut, native of the rain forest of the Andean region of South America. Its leaves contain terpenoids, saponins, phenolic compound and other components responsible for the antioxidant and antiproliferative activities. Ag nanoparticles are synthesized using Sacha inchi leaf extract under heating condition, showed surface plasmon resonance (SPR) band at $\lambda_{max}=430$ nm for spherical particle size ranging from 4 to 25 nm (Figs. 7.4a and 7.5a) (Kumar et al. 2014c). Whereas, the SPR peaks for Ag nanoparticles synthesized using Sacha inchi shell biomass appeared at $\lambda_{max}=420$ nm, predominantly spherical, with an average size of 7.2 nm (Figs. 7.4b and 7.5b). It could be as an effective photocatalyst for the decomposition of methylene orange dye (~60 %, pH 2.0) from aqueous solution (Kumar et al. 2014b). The Ag nanoparticles have been prepared through the reduction of silver ion by omega 3, 6, 9 fatty acids and amino acids of the Sacha inchi oil showed $\lambda_{max}=380$ and 480 nm (Fig. 7.4c). It reveals the formation of distorted cubic in

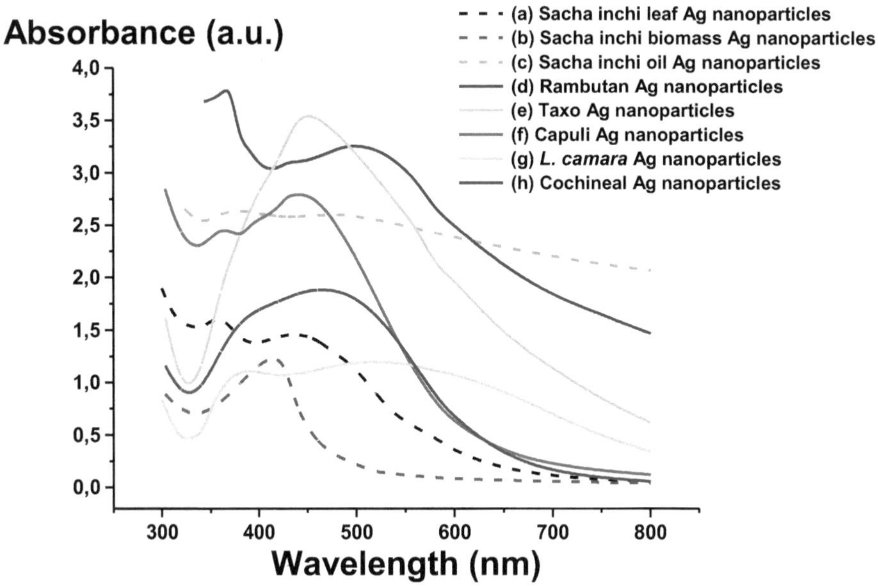

Fig. 7.4 A schematic representation of green synthesis of Ag nanoparticles

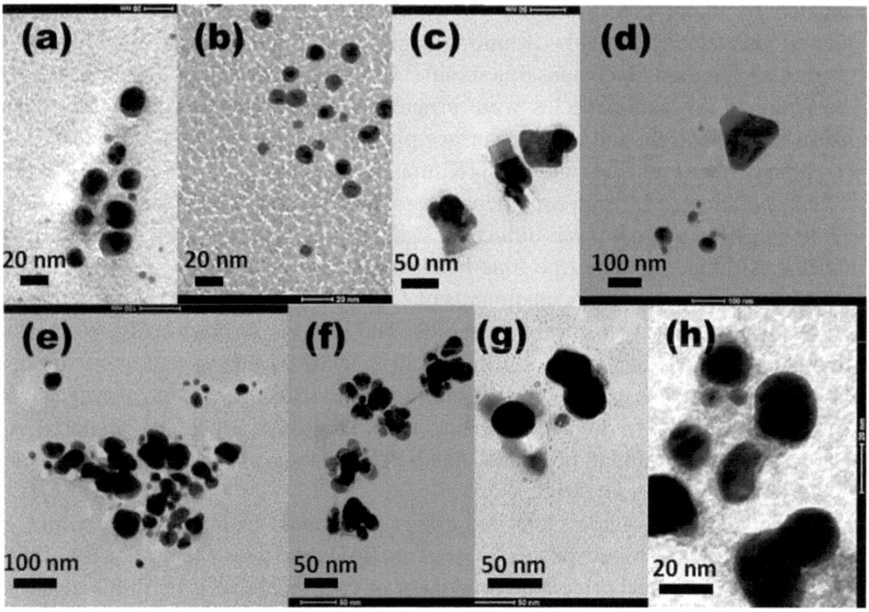

Fig. 7.5 Green synthesis of Ag nanoparticles using (**a**) Sacha inchi leaf, (**b**) Sacha inchi shell biomass, (**c**) Sacha inchi oil, (**d**) Rambutan peel, (**e**) Taxo fruit, (**f**) Capuli cherry, (**g**) *L. camara* berry and (**h**) Cochineal insect

shape nanostructure with 60 nm of average sizes (Fig. 7.5c) and can be used as an alternative agent for > 65 % removal of methylene blue from aqueous solution (Kumar et al. 2014a).

Rambutan (*Nephelium lappaceum* L.), a tropical fruit originated from Southeast Asia and its outer peel was rich in anthocyanins, ellagitannins, ellagic acid, corilagin geraniin, syringic acid and *p*-coumaric acid, a good source of natural antioxidants. The successful formation of triangular Ag nanoplates using Rambutan peel extract was confirmed by the appearance of two SPR bands at $\lambda_{max} = 370$ and 495 nm with average size 185 ± 79.3 nm (Figs. 7.4d and 7.5d). It also showed good scavenging efficacy (80 %, 0.05 mM) against 1,1-diphenyl-2- picrylhydrazyl (Kumar et al. 2015b).

The banana passion fruit, known in Ecuador as "taxo & tumbo" (*Passiflora tripartita* var. mollissima) belongs to the *Passifloraceae* plant family comprises around 530 species originated from temperate and tropical South America. The aromatic fruits consumed from prehispanic times, are very appreciated for the pleasant taste and acidic fruit juice due to the presence of luteolin and apigenin derivatives as isoorientin, orientin and isovitexin. The prepared Ag nanoparticles showed their $\lambda_{max} = 440$ nm and are found to be mostly spherical shapes with an average diameter of 49.7 ± 24.6 nm (Figs. 7.4e and 7.5e) (Kumar et al. 2015a).

Capuli (*Prunus serotina* Ehrh. var. Capuli) cherry are similar in shape and taste of black cherry and one of the most common fruits available on the highlands of Ecuador. Its major phytochemical composition is chlorogenic acid ((+)-catechin and (−)-epicatechin), proanthocyanidin (cyanidin-3-O-glucoside and cyanidin-3-O-rutinoside), flavonol glycosides (quercetin-3-xyloside, quercetin-3-arabinoside) etc. The obtained Ag nanoparticles were efficiently synthesized using Capuli cherry extract in visible light and showed surface plasmon resonance at $\lambda_{max} = 445$ nm with 40–100 nm size (Figs. 7.4f and 7.5f) (Kumar et al. 2016a).

Lantana camara L. is an ornamental garden weed plant and found various uses in folk medicine against fever, influenza, stomachache, chicken pox, measles, rheumatisms, asthma, etc. The ripe blue-black berries are also eaten in some tropical country and the major phytoconstituents of *L. camara* is Mono- and sesquiterpenes, Triterpenes, Iridoid glycosides, Flavanoids. The UV–vis and transmission electron microscopy (TEM) analysis confirmed that as synthesized Ag nanoparticles displayed SPR band at $\lambda_{max} = 390$ and 520 nm for spherical and 75.2 nm average sized (Figs. 7.4g and 7.5g). In addition, the antioxidant efficacy of as prepared Ag nanoparticles was found to be higher than berry extract against 1,1-diphenyl-2- picrylhydrazyl (Kumar et al. 2015c).

Cochineal is an organic violet-red colorant obtained from the dried bodies of females of the scale insect species, Coccus cacti (*Dactylopius coccus* Costa), which feed on wild cacti. The principal component of cochineal dye is carminic acid and has been used all over the world since ancient times as a colorant. The as-synthesized Ag nanoparticles showed the SPR band at $\lambda_{max} = 440$–460 nm and TEM indicated the formation of spherical nanoparticles of diameter ranging from 20 to 50 nm (Figs. 7.4h and 7.5h) (Kumar et al. 2016b).

Sathiskumar et al. (2010) reported the formation of spherical Ag nanoparticles using extracts of powder and the bark of *Curcuma longa* and reported that a bark extract can produce a large amount of Ag nanoparticles compared to the powder extract. Triangular, hexagonal, cubic, and circular Ag nanoparticles were synthesized from a *Pseudocydonia sinensis* fruit extract (Akhtar et al. 2013). The first scientific report describing the synthesis of colloidal Au nanoparticles came in 1857 when Michael Faraday found that the "fine particles" formed from the aqueous reduction of gold chloride by phosphorus could be stabilized by the addition of carbon disulfide, resulting in a "beautiful ruby fluid" (Faraday 1857). Whereas the first report for the synthesis of Ag/Au nanoparticles using living plants was claimed by Gardea-Torresdey et al. (2003). Narayanan and Sakthivel (2008) reported the extracellular synthesis of triangular, truncated and decahedral in Au nanoparticles with a mean size of 6.7–57.9 nm, using the leaf extract of *Coriandrum sativum*. Raghunandan et al. (2011) reported the stable poly-shaped Au nanoparticles with high purity using aqueous guava leaf extract under microwave- irradiation. Moreover, Huang et al. (2007) also reported the synthesis of Au nanoparticles using the sun-dried leaf extract of *Cinnamomum camphora* leaf extract. In addition, the synthesis of spherical Au nanoparticles promoted by the flower extract of *Nycanthes arbor-tristis* (Ratul Kumar et al. 2011) and the leaf extract of *Mangifera indica* (Daizy 2010) while, gold nanowires was reported from the pulp extract of *Beta vulgaris* (Laura Castro et al. 2010). Shankar et al., reported formation of Au nanotriangles by lemongrass leaf extract (Shankar et al. 2005).

7.3 Size, Shape and Surface Characterization of Nanoparticles

The nanoparticles are generally characterized by their size, shape, surface area, and disparity. UV–visible spectroscopy is a commonly used technique to determine surface plasmon resonance absorption band of synthesized nanoparticles in the range of 200–800 nm. The absorption measurements in the wavelength range of 380–480 and 515–550 nm is used to characterize Ag and Au nanoparticles, respectively. Dynamic light scattering is used to characterize the surface charge and hydrodynamic particle size distribution in a liquid. Fourier transform infrared spectroscopy is a useful technique for characterizing the surface chemistry and to hypothesize the associated functional groups/molecules from plant extracts involved in the synthesis of nanoparticles. Electrochemical measurements using cyclic voltamettry were carried out to know about the interaction of metal salts with biological components. Transmission electron microscopy and Scanning electron microscopy are used for morphological characterization on the nanometer to micrometer scale. The elemental composition of nanoparticles is commonly established using energy dispersive spectroscopy. Atomic force microscopy offers visualization in three dimensions and provides surface characterization on the atomic scale. X-ray diffraction studies were performed to identify the phase, purity and crystallite size of nanoparticles.

7.4 Effect of Morphology or Size and Shape on the Bioactivity of Nanoparticles

The SPR plays a major role in the determination of optical absorption spectra of nanoparticles, which shifts to a longer wavelength with an increase in particle size. The size and shape of the nanoparticles imply that it has a large surface area to come in contact with the bacterial cells and hence, it will have a higher percentage of interaction than bigger particles (Kreibig and Vollmer 1995; Mulvaney 1996). The nanoparticles smaller than 10 nm interact with bacteria and produce electronic effects, which enhance the bactericidal effect of Ag nanoparticles in a size dependent manner (Raimondi et al. 2005; Morones et al. 2005). Morones et al. (2005) confirmed the inhibition of microbial growth by differently shaped nanoparticles. According to Pal et al. (2007) the Ag nanoparticles with different shaped (truncated triangular =1 μg, spherical =12. 5 μg and rod=50–100 μg) have different effects on bacterial cell inhibition. The Au nanoparticles has recently been made in various biomedical nanotechnologies that have different morphologies (nanorods, nanoshells, nanocages, nanowires etc.) with the agreement between the optical properties of particles and the window of biotissue transparency (750–1000 nm). It preferentially accumulates at sites of tumor growth/inflammation and enter cells by mechanisms very different and much more rapid than those of small molecules (Khlebtsov and Dykman 2011). Pan et al. (2007) shown that the transition to the sizes of Au nanoparticles (1.2 nm, 1.8 nm and 15 nm) stabilized with triphenyl phosphine derivatives by using four cell lines (HeLa, Sk-Mel-28, L929, J774A1). According to the MTT assay data, 15 nm Au nanoparticles sharply decreases cytotoxicity, as compared to atomic clusters of approximately 1 to 2 nm, which can irreversibly bind to DNA and, possibly, to other key molecules as well (Pan et al. 2007). Recently, Dasgupta et al. (2016) studied the size dependent toxicological behavior of Ag nanoparticles prepared by thermal co-reduction approach. It was found that, both the 60 nm and 85 nm sized Ag nanoparticles were highly toxic to the bacterial as well as cancerous cell lines, but the microbial and cellular toxicological behavior for 60 nm was higher than and 85 nm (Dasgupta et al. 2015). Recently, Maddinedi et al. (2015) used Diastase, a natural enzyme for one pot aqueous synthesis of Au nanoparticles of tunable size and shape, including (a) 148 nm (hexagons and blunted triangular), (b) 23.2 nm (triangles, hexagons along with the more number of spherical particles) and (c) 9.7 nm (spherical). *In vitro* cytotoxicity study showed that the toxic effects of as-prepared Au nanocolloids are size and dose dependent towards HCT-116 and A-549 cancer cell lines. Smaller Au nanoparticles were found to be more toxic for both the cell lines compared to the larger. However, no significant toxicity was found with diastase, indicating the negligible role of capping agent on the toxicity of Au nanoparticles (Maddinedi et al. 2015). Thus, the bioefficacy of the nanoparticles depends on the size and shapes of the nanoparticles.

7.5 Antimicrobial Activity of Nanoparticles

The importance of Ag has been known since 1000 B.C. to make water potable and for centuries, it has been in use for the treatment of burns and chronic wounds (Rai et al. 2009; Castellano et al. 2007). In the 1940s, after penicillin was introduced the use of the Ag in the treatment of bacterial infections minimized (Chopra 2007). The antimicrobial property of Ag is related to the amount of Ag and the rate of Ag released. Ag in its metallic state is inert, but it reacts with the moisture in the skin and the fluid of the wound and gets ionized. The ionized silver is highly reactive, as it binds to tissue proteins and brings structural changes in the bacterial cell wall and nuclear membrane leading to cell distortion and death (Castellano et al. 2007). Ag also binds to bacterial DNA and RNA by denaturing and inhibits bacterial replication (Castellano et al. 2007). Among the non-organic antimicrobial agents, Ag ions or Ag nanoparticles are strong antimicrobial agents due to their extremely large surface area, which provides better contact with microorganisms. The bacterial membrane contains sulfur-containing proteins and the Ag/Au nanoparticles interact with these proteins in the cell as well as with the phosphorus containing compounds like vital enzymes or DNA. When Ag/Au nanoparticles enter the bacterial cell it forms a low molecular weight region in the center of the bacteria to which the bacteria conglomerates thus, protecting the DNA from the Ag^+ ions. The Au nanoparticles generate holes in the cell wall, resulting in leakage of the cell contents, formation of biofilm (Chwalibog et al. 2010) and finally cell death. In another way, it can bind to the DNA of bacteria and inhibit DNA transcription (Rai et al. 2010). The nanoparticles preferably attack the respiratory chain, cell division finally leading to cell death. The nanoparticles release Ag^+ ions in the bacterial cells, which enhance their bactericidal activity (Morones et al. 2005; Feng et al. 2000; Song et al. 2006). Small quantity of Ag is harmless to human cells, but is biocidal to microbial cells (Rai et al. 2009). Mechanism of action of Ag/Au nanoparticles is summarized in Table 7.1.

Table 7.1 Mechanism of action of Ag and Au nanoparticles

Nanoparticles	Mechanism of actions
Ag/Au	Inhibition of cell wall formation
	DNA intercalation
	Cell Signalling inhibition
	Cell membrane distruption
	Free radical formation
	RNA 30s interaction
	Enzymatic Inhibition
	Preventing biofilm formation

Silver nanoparticles synthesized through green method have been reported to have biomedical applications as well as in controlling the pathogenic microbes. So, Ag nanoparticles have been used most widely in the health industry, food storage, textile coatings and a number of environmental applications (Ahmed et al. 2016). Priya et al. (2014) reported that Ag nanoparticles synthesized from banana leaf extract exhibits inhibitory and lethal effects on both gram-positive and gram negative bacteria (*E. coli*, and *S. aureus*). The antibacterial activity of Au nanoparticles with Marigold flower extract exhibit a larger zone of inhibition compared to standard antibiotics (Krishnamurthy et al. 2012). The efficacy of the antibacterial activity of Au nanoparticles can be increased by the addition of antibiotics. Ghosh et al. (2012) have evaluated the *in vitro* synergistic antibacterial efficiency of *Dioscorea bulbifera* tuber synthesized Ag nanoparticles and antibiotics against *P. aeruginosa*, *E. coli*, and *A. baumannii*. Prakash et al. (2013) reported that *Mimusops elengi* leaf extract synthesized Ag nanoparticles showed maximum antibacterial activity against Gram-negative multi drug resistant pathogen, *Klebsiella pneumoniae* whereas moderate activity was observed with the Gram-positive pathogens *S. aureus and Micrococcus luteus*. Ag/Au nanoparticles synthesized from the leaf extract of *Mentha piperita* (Ali et al. 2011) and *Trianthema decandra* root (Geethalakshmi and Sarada 2012) demonstrated strong bactericidal activity against *E. coli* than *S. aureus* whereas Au nanoparticles showed antibacterial action only on *E. coli*. The coating of *Ananas comosus* fruit extract with Au nanoparticles fruit extract had a significant antibacterial activity against gram positive (*S. aureus*) and gram positive (*P. aeruginosa*) pathogen (Bindhu and Umadevi 2014), which are found in water. Recently, Saravanakumar et al. (2015) reported that, *Cassia tora* leaf extract synthesized Ag nanoparticles exhibited significant antibacterial activity against both gram positive (*S. aureus*, *B. subtilis*) and gram negative bacteria (*E.* coli and *P.* aeruginosa). The synthesized Au nanoparticles (45–75 nm) from aqueous seed extracts of *Abelmoschus esculentus* that were found to have significant antifungal activity tested against *Puccinia graminis tritci*, *Aspergillus flavus*, *Aspergillus niger* and *Candida albicans* using standard well diffusion method (Jayaseelan et al. 2013). Bankar et al. (2010) reported the banana peel (*Musa paradisiaca*) extract mediated Au nanoparticles displayed efficient antifungal and antibacterial activity towards the test pathogenic fungi, *C. albicans* (BX and BH), bacterial cultures, including *Citrobacter kosari*, *E. coli*, *Proteus valgaris*, *Pseudomonas aeruginosa*, *Enterobacter aerogenes* and *Klebsiella sp*. The synthesis of Au nanoparticles in aqueous medium using flower extracts of *Ixora coccinea* (Chetty flower) as reducing and stabilizing agent reported the antibacterial indicated that antibiotic with Au nanoparticles could be extracted from chetty flower which exhibited more zone of inhibition when compared to standard antibiotics and the zone of inhibition on *E.coli* and *Streptobacillus sp*. is more when compared to the zone of inhibition exhibited by *A. flavus* (Nagaraj et al. 2011). Basavegowda et al. (2014) reported that, as synthesized Au nanoparticles using *Hovenia dulcis* fruit exhibited moderate antibacterial activities against *E. coli* and *S. aureus* when compared with standard ciprofloxacin antibiotic. The above result indicates that phytochemical functionalized Ag/Au nanoparticles research is leading to product improvement, the quantity and quality enhancement in the field of pharmaceutical, agriculture, and food science & technology (Nandita et al. 2014; Shivendu et al. 2014).

7.6 Antiviral Activity of Nanoparticles

Viruses represent one of the leading causes of disease and death worldwide. The interaction of metal nanoparticles with viruses is another expanding field of research. Although, there have been relatively few reports on the antiviral activity of biosynthesized nanoparticles. Vijayakumar and Prasad (2009) described the effective antiviral activity of Ag nanoparticles prepared intracellularly using *Aspergillus ochraceus* against M13 phage virus using the plaque count method. In another study, *Lactobacillus fermentum* bacteria were used both as a reducing agent and as a scaffold for Ag nanoparticles. Their antiviral activity was determined for murine norovirus 1 and bacteriophage UZ1 (De Gusseme et al. 2010). It was noticed that Ag nanoparticles undergo a size-dependent interaction with HIV-1. Nanoparticles ranging in size from 1 to 10 nm readily interact with the HIV-1 virus via preferential binding to gp120 glycoprotein knobs. This specific interaction of Ag nanoparticles inhibits the virus from binding to host cells, demonstrated by *in vitro* study. Hence, Ag nanoparticles could find application in preventing as well as controlling HIV infection (Galdiero et al. 2011; Elechiguerra et al. 2005).

7.7 Anticancer Activity of Nanoparticles

Cancer is a major health problem and it arises from one single cell. The transformation of a normal cell into a tumour cell occurs in a multistage process and the changes are due to the interaction between genetic factors of a person and external agents of three categories like physical carcinogens, chemical carcinogens and biological carcinogens. Nowadays, Ag and Au nanoparticles are emerging as promising agents for cancer therapy. However, very few reports were available for the anticancer activities of plant extract synthesized Ag/Au nanoparticles against a variety of human cancer cell line such as HeLa cervical cancer cell line (Suman et al. 2013), MCF-7 breast cancer cell line (Vivek et al. 2012), A549 lung cancer cell line (Gengan et al. 2013), HEp 2 liver cancer cell line (Satyavani et al. 2011). MTT (3-(4, 5- dimethylthiazol-2yl)-2, 5- diphenyltetrazolium bromide) method was used to assess the biocompatibility and potential of nanoparticles for antiproliferative effect or other biomedical applications. It is based on the conversion of tetrazolium salts to formazan crystals by dehydrogenase activity in active mitochondria (Balachandran et al. 2015). Kumar et al. (2015a) evaluated the *In vitro* cytotoxicity of the Ag nanoparticles against human cancer cell lines A-549 and HeLa, from lung and cervix at different concentrations (0.01–20 μM). It showed that, there was no effect of Ag nanoparticles on cell proliferation of lung A-549 or HeLa cervical cells

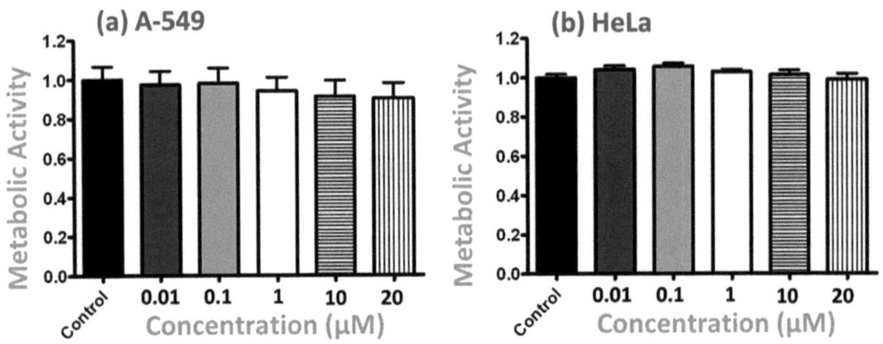

Fig. 7.6 Effect of Ag nanoparticles on cell proliferation of (**a**) A-549 lung cancer, and (**b**) HeLa cervical carcinoma cells

(Fig. 7.6) and these results suggest that the new Ag nanoparticles might be used as safe drug carriers.

Krishnaraj et al. (2014) reported the *in vitro* cytotoxic effect of *Acalypha indica* leaf extract synthesized Ag and Au nanoparticles against MDA-MB-231, human breast cancer cells. It showed that the cytotoxicity of nanoparticles was increased in a dose dependent manner with reduced viability of MDA-MB-231 cells (Krishnaraj et al. 2014). Ag nanoparticles synthesized using the calli extract of *Citrullus colocynthis* (Satyavani et al. 2011) showed better anticancer activity than the *Piper longum* leaf extract (Jacob et al. 2012) derived AgNPs against HEp 2 cell lines. Interestingly, the IC_{50} value was recorded at 500 nM concentration for *C. colocynthis* (Satyavani et al. 2011) whereas it was 31.25 μg/mL for *P. longum* (Jacob et al. 2012). Anand et al. (2015) reported the synthesis of Au nanoparticles using an Agroforestry waste as *Moringa oleifera* petals. Interestingly, the synthesized Au nanoparticles were significantly cytotoxic to A549 cells at both dilution 1:100, 1:50, 1:25, 1:10, and 1:3 with viabilities of 140 %, 139 %, 71 %, 49 %, and 12 %, respectively (Anand et al. 2014). Apoptosis is broadly considered as a distinctive mode of programmed cell death that eliminates genetically determined cells (Franco et al. 2009). The induction of apoptosis is confirmed by two factors, (a) reduced and shrunken cells and (b) DNA fragmentation (Sriram et al. 2010). Sukirtha et al. (2012) have also studied the *in-vivo* anticancer activity of *Melia azedarach* synthesized Ag nanoparticles and aqueous extract against Dalton's ascites lymphoma (DAL) induced mice. The Ag nanoparticles increased the life span of DAL induced mice in a dose dependent manner in comparison to extract The characteristic apoptotic features such as nuclear condensation, membrane blebbing, apoptotic bodies and cytoplasmic vacuoles were observed in Ag nanoparticles treated DAL cells at 700 μg/kg BW than the other experimental groups. The obtained results have clearly suggested that as synthesized Ag nanoparticles showed better cytotoxic activity than aqueous extract (Sukirtha et al. 2012). In the future, we might have to develop a new nanoparticles based antitumor drugs that do not induce an inflammatory response during apoptotic cell death.

7.8 Limitations

Although the continuous evolutions in the field of MNPs for medical imaging, drug delivery, therapeutics, diagnostics and engineering technology, there is a serious negative impacts on the environment and ecological systems should be carefully evaluated (Colvin 2003). Toxicity from Ag is observed in the form of argyria, due to consumption of silver-containing food supplement for a few months and Ag nanoparticles in most studies are suggested to be non-toxic (Leaper 2006). Small size and variable properties of Ag/Au nanoparticles are suggested to be hazardous to the environment due to of irreversibly binding to B-form DNA (Braydich-Stolle et al. 2005). Accumulation of Ag nanoparticles in various organs (lungs, kidneys, brain, liver, and testes) has been evidenced in animal studies (Kim et al. 2008). Most of the *in vitro* studies show dose dependence, in fact, higher doses of silver induce a stronger cellular toxicity. The generation of reactive oxygen species (ROS) by cells exposed to Ag nanoparticles has been shown in human lung fibroblast and human glioblastoma cells, and as a consequence DNA damage and cell cycle abnormalities have been observed (AshaRani 2009).

7.9 Conclusion

In conclusion, phytochemicals can be efficiently used in the synthesis of Ag and Au nanoparticles as a greener route as well as opening the way for exploring new plants for this purpose. Control over the shape and size of nanoparticles seems to be very easy with the use of plants and opened up a wide area of nanotechnology for human benefit. The mechanism of such nanoparticles synthesis due to the participation of phenolics, proteins, polysaccharides and other biomolecules is a very promising area of research. The phytochemical functionalized Ag and Au nanoparticles are preferred because they combine properties from both organic and inorganic nanomaterials, resulting in an increased biocompatibility and cellular adherence of the designed nanomaterials. They are found to be non-toxic to humans at low concentrations with their unique physio-chemical properties and proving as an alternative for the development of new antimicrobial, antiviral and anticancer agents. Hence, the development of green chemistry based technique for the production of high performance materials can provide an additional source of revenue for farmers, help in agro-industry diversification and best opportunities to add new vigor to this product with maximum benefits to the society and to the environment.

Acknowledgements This scientific work has been funded by the Prometeo Project of the Ministry of Higher Education, Science, Technology and Innovation (SENESCYT), Ecuador and TATA College, Kolhan University, Chaibasa, India.

References

Ahmad N, Alam MK, Singh VN, Sharma S (2009) Bio prospecting AgNPs from wild desmodium species. J Bionanosci 3:97–104

Ahmed S, Ahmad M, Swami BL, Ikram S (2015) A review on plants extract mediated synthesis of silver nanoparticles for antimicrobial applications: a green expertise. J Adv Res 7:17–28. http://dx.doi.org/10.1016/j.jare.2015.02.0072090–1232

Akhtar MS, Panwar J, Yun Y-S (2013) Biogenic synthesis of metallic nanoparticles by plant extracts. ACS Sustain Chem Eng 1:591–602

Ali M, Thajuddin D, Jeganathan N, Gunasekaran K (2011) Plant extract mediated synthesis of silver and gold nanoparticles and its antibacterial activity against clinically isolated pathogens. Colloids Surf B: Biointerfaces 85:360–365

Anand K, Gengan RM, Phulukdaree A, Chuturgoon A (2015) Agroforestry waste Moringa oleifera petals mediated green synthesis of gold nanoparticles and their anti-cancer and catalytic activity. J Ind Eng Chem 21:1105–1111

AshaRani PV, Low Kah Mun G, Hande MP, Valiyaveettil S (2009) Cytotoxicity and genotoxicity of silver nanoparticles in human cells. ACS Nano 3:279–290

Balachandran C, Duraipandiyan V, Emi N, Ignacimuthu S (2015) Antimicrobial and cytotoxic properties of Streptomyces sp. (ERINLG-51) isolated from Southern Western Ghats. South Indian J Biol Sci 1:7–14

Bankar A, Joshi B, Kumar AR, Zinjarde S (2010) Banana peel extract mediated synthesis of gold nanoparticles. Colloids Surf B: Biointerfaces 80:45–50

Basavegowda N, Idhayadhulla A, Lee YR (2014) Phyto-synthesis of gold nanoparticles using fruit extract of Hovenia dulcis and their biological activities. Ind Crop Prod 52:745–751

Bindhu MR, Umadevi M (2014) Antibacterial activities of green synthesized gold nanoparticles. Mater Lett 120:122–125

Braydich-Stolle L, Hussain S, Schlager J, Hofmann MC (2005) In vitro cytotoxicity of nanoparticles in mammalian germ line stem cells. Toxicol Sci 88:412–419

Castellano JJ, Shafii SM, Ko F, Donate G, Wright TE, Mannari RJ et al (2007) Comparative evaluation of silver-containing antimicrobial dressings and drugs. Int Wound J 4(2):114–122

Chernousova S, Epple M (2013) Silver as antibacterial agent: ion, nanoparticle, and metal. Angew Chem Int Ed 52:1636–1653

Chopra I (2007) The increasing use of silver-based products as antimicrobial agents: a useful development or a cause for concern? J Antimicrob Chemother 59:587–590

Chwalibog A, Sawosz E, Hotowy A, Szeliga J, Mitura S, Mitura K (2010) Visualization of interaction between inorganic nanoparticles and bacteria or fungi. Int J Nanomedicine 5:1085–1094

Colvin VL (2003) The potential environmental impact of engineered nanomaterials. Nat Biotechnol 21:1166–1170

Dahl JA, Maddux BLS, Hutchison JE (2007) Toward greener nanosynthesis. Chem Rev 107(6):2228–2269

Daizy P (2010) Rapid green synthesis of spherical gold nanoparticles using Mangifera indica leaf. Spectrochim Acta A 77:807–810

Dasgupta N, Ranjan S, Rajendran B, Manickam V, Ramalingam C, Avadhani GS, Kumar A (2016) Thermal co-reduction approach to vary size of silver nanoparticle: its microbial and cellular toxicology. Environ Sci Pollut Res 23:4149–4163. doi:10.1007/s11356-015-4570-z

Das RK, Gogoi N, Bora U (2011) Green synthesis of gold nanoparticles using Nyctanthes arbortristis flower extract. Bioprocess Biosyst Eng 34:615–619

De Gusseme B, Sintubin L, Baert L, Thibo E, Hennebel T, Vermeulen G et al (2010) Biogenic silver for disinfection of water contaminated with viruses. Appl Environ Microbiol 76:1082–1087

Dreaden EC, Mackey MA, Huang X, Kang B, El-Sayed MA (2011) Beating cancer in multiple ways using nanogold. Chem Soc Rev 40:3391–3404

Elechiguerra JL, Burt JL, Morones JR, Camacho-Bragado A, Gao X, Lara HH et al (2005) Interaction of silver nanoparticles with HIV-1. J Nanobiotechnol 3:6
Faraday M (1857) The Bakerian lecture: experimental relations of gold (and other metals) to light. Philos Trans R Soc Lond 147:145–181
Feng QL, Wu J, Chen GQ, Cui FZ, Kim TN, Kim JO (2000) A mechanistic study of the antibacterial effect of silver ions on Escherichia coli and Staphylococcus aureus. J Biomed Mater 52(4):662–668
Franco JL, Posser T, Dunkley PR (2009) Methyl mercury neurotoxicity is associated with inhibition of the antioxidant enzyme glutathione peroxidase. Free Radic Biol Med 7:449–457
Galdiero S, Falanga A, Vitiello M, Cantisani M, Marra V, Galdiero M (2011) Silver nanoparticles as potential antiviral agents. Molecules 16:8894–8918
Gangula A, Podila R, Ramakrishna M et al (2011) Catalytic reduction of 4-nitrophenol using biogenic gold and silver nanoparticles derived from Breynia rhamnoides. Langmuir 27:15268–15274
Gardea-Torresdey JL, Gomez E, Peralta-Videa J, Parsons JG, Troiani HE, Jose Yacaman M (2003) Alfalfa sprouts: a natural source for the synthesis of silver nanoparticles. Langmuir 19:1357–1361
Geethalakshmi R, Sarada DVSL (2012) Gold and silver nanoparticles from Trianthema decandra: synthesis, characterization, and antimicrobial properties. Int J Nanomedicine 7:5375–5384
Gengan RM, Anand K, Phulukdaree A, Chuturgoon A (2013) A549 lung cell line activity of biosynthesized silver nanoparticles using Albizia adianthifolia leaf. Colloids Surf B: Biointerfaces 105:87–91
Ghosh S, Patil S, Ahire M, Kitture R, Kale S et al (2012) Synthesis of silver nanoparticles using Dioscorea bulbifera tuber extract and evaluation of its synergistic potential in combination with antimicrobial agents. Int J Nanomedicine 7:483–496
Huang J, Li Q, Sun D, Lu Y, Su Y, Yang X, Wang H et al (2007) Biosynthesis of silver and gold nanoparticles by novel sundried Cinnamomum camphora leaf. Nanotechnology 18:105104–105114
Hutchison JE (2008) Greener nanoscience: a proactive approach to advancing applications and reducing implications of nanotechnology. ACS Nano 2(3):395–402
Jacob SJP, Finub JS, Narayanan A (2012) Synthesis of silver nanoparticles using Piper longum leaf extracts and its cytotoxic activity against Hep-2 cell line. Colloids Surf B: Biointerfaces 91:212–214
Jayaseelan C, Ramkumar R, Rahuman AA, Perumal P (2013) Green synthesis of gold nanoparticles using seed aqueous extract of Abelmoschus esculentus and its antifungal activity. Ind Crop Prod 45:423–429
Jeetah R, Bhaw-Luximon A, Jhurry D (2014) Nanopharmaceutics: phytochemical-based controlled or sustained drug-delivery systems for cancer treatment. J Biomed Nanotechnol 10:1–31
Khlebtsov N, Dykman L (2011) Biodistribution and toxicity of engineered gold nanoparticles: a review of in vitro and in vivo studies. Chem Soc Rev 40:1647–1671
Kim YS, Kim JS, Cho HS, Rha DS, Kim JM, Park JD et al (2008) Twenty-eight-day oral toxicity, genotoxicity, and genderrelated tissue distribution of silver nanoparticles in Sprague-Dawley rats. Inhal Toxicol 20:575–583
Kreibig U, Vollmer M (1995) Optical properties of metal clusters. Springer, Heidelberg
Krishnamurthy N, Nagaraj B, Malakar B, Liny P, Dinesh R (2012) Green synthesis of gold nanoparticles using Tagetes erecta L. (Mari Gold) flower extract & evaluation of their antimicrobial activities. Int J Pharm Bio Sci 3:212–221
Krishnaraj C, Muthukumaran P, Ramachandran R, Balakumaran MD, Kalaichelvan PT (2014) Acalypha indica Linn: biogenic synthesis of silver and gold nanoparticles and their cytotoxic effects against MDA-MB-23. Biotechnol Rep 4:42–49
Kumar V, Yadav SK (2009) Plant-mediated synthesis of silver and gold nanoparticles and their applications. J Chem Technol Biotechnol 84:151–157

Kumar B, Smita K, Cumbal L, Debut A (2014a) Sacha inchi (Plukenetia volubilis L.) oil for one pot synthesis of silver nanocatalyst: an ecofriendly approach. Ind Crop Prod 58:238–243
Kumar B, Smita K, Cumbal L, Debut A (2014b) Sacha inchi (Plukenetia volubilis L.) shell biomass for synthesis of silver nanocatalyst. J Saudi Chem Soc, article in press, http://dx.doi.org/10.1016/j.jscs.2014.03.005
Kumar B, Smita K, Cumbal L, Debut A (2014c) Synthesis of silver nanoparticles using Sacha inchi (Plukenetia volubilis L.) leaf extracts. Saudi J Biol Sci 21:605–609
Kumar B, Smita K, Cumbal L, Debut A, Camacho J et al (2015a) Pomosynthesis and biological activity of silver nanoparticles using Passiflora tripartita fruit extracts. Adv Mater Lett 6(2):127–132
Kumar B, Smita K, Angulo Y, Cumbal L (2016b) Green synthesis of silver nanoparticles using natural dyes of Cochineal. J Cluster Sci 27:703–713
Kumar B, Smita K, Cumbal L, Debut A (2015c) Lantana camara berry for the synthesis of silver nanoparticles. Asian Pacific J Trop Biomed 5(3):192–195
Kumar B, Angulo Y, Smita K, Cumbal L, Debut A (2016a) Capuli cherry-mediated green synthesis of silver nanoparticles under white solar and blue LED light. Particuology 24:123–128
Kumar B, Smita K, Cumbal L, Angulo Y (2015b) Fabrication of silver nanoplates using Nephelium lappaceum (Rambutan) peel: a sustainable approach. J Mol Liq 211:476–480
Laura Castro M, Blázquez L, González F, Jesús A (2010) Extracellular biosynthesis of gold nanoparticles using sugar beet pulp. Chem Eng J 164:92–97
Leaper DL (2006) Silver dressings: their role in wound management. Int Wound J 3(4):282–294
Liu RH (2004) Potential synergy of phytochemicals in cancer prevention: mechanism of action. J Nutr 134:3479S
Luangpipat T, Beattie IR, Chisti Y, Haverkamp RG (2011) Gold nanoparticles produced in a microalga. J Nanopart Res 13:6439–6445
Maddinedi SB, Mandal BK, Ranjan S, Dasgupta N (2015) Diastase assisted green synthesis of size-controllable gold nanoparticles. RSC Adv 5:26727–26733. doi:10.1039/C5RA03117F
Mittal AK, Chisti Y, Banerjee UC (2013) Synthesis of metallic nanoparticles using plant extracts. Biotechnol Adv 31:346–356
Morones JR, Elechiguerra JL, Camacho A, Ramirez JT (2005) The bactericidal effect of silver nanoparticles. Nanotechnology 16:2346–2353
Mulvaney P (1996) Surface plasmon spectroscopy of nanosized metal particles. Langmuir 12:788–800
Nagaraj B, Krishnamurthy NB, Liny P, Divya TK, Dinesh R (2011) Biosynthesis of gold nanoparticles of Ixora coccinea flower extract and their antimicrobial activities. Int J Pharm Bio Sci 2:557–565
Nandita D, Shivendu R, Deepa M, Chidambaram R, Ashutosh K, Rishi S (2014) Nanotechnology in agro-food: from field to plate. Food Res Int 69:381–400
Narayanan KB, Sakthivel N (2008) Coriander leaf mediated biosynthesis of gold nanoparticles. Mater Lett 62:4588–4590
Narayanan KB, Sakthivel N (2011) Green synthesis of biogenic metal nanoparticles by terrestrial and aquatic phototrophic and heterotrophic eukaryotes and biocompatible agents. Adv Colloid Interface Sci 169:59–79
Otari SV, Patil RM, Ghosh SJ, Pawar SH (2014) Green phytosynthesis of silver nanoparticles using aqueous extract of Manilkara zapota (L.) seeds and its inhibitory action against Candida species. Mater Lett 116:367–369
Pal S, Tak YK, Song JM (2007) Does the antibacterial activity of silver nanoparticles depend on the shape of the nanoparticle? A study of the gramnegative bacterium Escherichia coli. Appl Environ Microbiol 27(6):1712–1720
Pan Y, Neuss S, Leifert A, Fischler M, Wen F, Simon U, Schmid G, Brandau W, Jahnen-Dechent W (2007) Size-dependent cytotoxicity of gold nanoparticles. Small 3:1941–1949
Philip D (2010) Green synthesis of gold and silver nanoparticles using Hibiscus rosa sinensis. Phys E 42:1417–1424

Prakash P, Gnanaprakasam P, Emmanuel R, Arokiyaraj S, Saravanan M (2013) Green synthesis of silver nanoparticles from leaf extract of Mimusops elengi Linn. for enhanced antibacterial activity against multi drug resistant clinical isolates. Colloids Surf B: Biointerfaces 108:255–259

Priya B, Satapathy M, Mukhopahayay A, Das P (2014) Leaf extract mediated green synthesis of silver nanoparticles from widely available Indian plants: synthesis, characterization, antimicrobial property and toxicity analysis. Bioresour Bioprocess 1:1–10

Raghunandan D, Bedre D, Mahesh S, Basavaraja SD, Balaji SY, Manjunath Venkataraman A (2011) Microwave-assisted rapid extracellular synthesis of stable bio-functionalized silver nanoparticles from guava (Psidium guajava) leaf extract. J Nanopart Res 13:2021–2028

Rai M, Yadav A, Gade A (2009) Silver nanoparticles as a new generation of antimicrobials. Biotechnol Adv 27:76–83

Rai A, Prabhune A, Perry CC (2010) Antibiotic mediated synthesis of gold nanoparticles with potent antimicrobial activity and their application in antimicrobial coatings. J Mater Chem 20:6789–6798

Raimondi F, Scherer GG, Kotz R, Wokaun A (2005) Nanoparticles in energy technology: examples from electochemistry and catalysis. Angew Chem Int Ed 44:2190–2209

Rajan R, Chandran K, Harper SL, Yun SI, Kalaichelvan PT (2015) Plant extract synthesized silver nanoparticles: an ongoing source of novel biocompatible materials. Ind Crop Prod 70:356–373

Raveendran P, Fu J, Wallen SL (2003) Completely "Green" synthesis and stabilization of metal nanoparticles. J Am Chem Soc 125(46):13940–13941

Saravanakumar A, Ganesh M, Jayaprakash J, Jang HT (2015) Biosynthesis of silver nanoparticles using Cassia tora leaf extract and its antioxidant and antibacterial activities. J Ind Eng Chem 28(25):277–281

Sathishkumar M, Sneha K, Won SW, Cho C-W, Kim S et al (2009) Cinnamon zeylanicum bark extract and powder mediated green synthesis of nano-crystalline silver particles and its bactericidal activity. Colloids Surf B: Biointerfaces 73:332–338

Sathishkumar M, Krishnamurthy S, Yun YS (2010) Immobilization of silver nanoparticles synthesized using Curcuma longa tuber powder and extract on cotton cloth for bactericidal activity. Biores Technol 101:7958–7965

Satyavani K, Gurudeeban S, Ramanathan T, Balasubramanian T (2011) Biomedical potential of silver nanoparticles synthesized from calli cells of Citrullus colocynthis (L.) Schrad. J Nanobiotechnol 9:43

Shankar SS, Rai A, Ahmad A et al (2005) Controlling the optical properties of lemongrass extract synthesized gold nanotriangles and potential application in infrared-absorbing optical coatings. Chem Mater 17:566–572

Shankar SS, Rai A, Ahmad A et al (2012) Synthesis of metal nanoparticles in living plants. Ital J Agron 7:577–583

Shivendu R, Nandita D, Arkadyuti RC, Melvin SS, Chidambaram R, Rishi S, Ashutosh K (2014) Nanoscience and nanotechnologies in food industries: opportunities and research trends. J Nanopart Res 16(6):2464. doi:10.1007/s11051-014-2464-5

Song HY, Ko KK, Oh LH, Lee BT (2006) Fabrication of silver nanoparticles and their antimicrobial mechanisms. Eur Cell Mater 11:58

Sreekanth TVM, Ravikumar S, Eom I-Y (2014) Green synthesized silver nanoparticles using Nelumbo nucifera root extract for efficient protein binding, antioxidant and cytotoxicity activities. J Photochem Photobiol B Biol 141:100–105

Sriram MI, Kanth SB, Kalishwaralal K, Gurunathan S (2010) Antitumor activity of silver nanoparticles in Dalton's lymphoma ascites tumor model. Int J Nanomedicine 5:753–762

Sukirtha R, Priyanka KM, Antony JJ, Kamalakkannan S, Thangam R, Gunasekaran P (2012) Cytotoxic effect of green synthesized silver nanoparticles using Melia azedarach against in vitro HeLa cell lines and lymphoma mice model. Process Biochem 47:273–279

Suman TY, Radhika Rajasree SR, Kanchana A, Beena Elizabeth S (2013) Biosynthesis, characterization and cytotoxic effect of plant mediated silver nanoparticles using Morinda citrifolia root extract. Colloids Surf B: Biointerfaces 106:74–78

Vijayakumar PS, Prasad BLV (2009) Intracellular biogenic silver nanoparticles for the generation of carbon supported antiviral and sustained bactericidal agents o cytotoxic effect on MCF-7 cells. Langmuir 25:11741–11747

Vivek R, Thangam R, Muthuchelian K, Gunasekaran P, Kaveri K, Kannan S (2012) Green biosynthesis of silver nanoparticles from Annona squamosa leaf extract and its in vitro cytotoxic effect on MCF-7 cells. Process Biochem 47:2405–2410

Chapter 8
Nanotechnology for the Detection and Diagnosis of Plant Pathogens

Prem Lal Kashyap, Pallavi Rai, Shikha Sharma, Hillol Chakdar, Sudheer Kumar, K. Pandiyan, and Alok K. Srivastava

Abstract Rapid detection technologies with high sensitivity and selectivity for plant pathogens are essential to prevent disease spread with minimal loss to crop production and food quality assurance. Traditional laboratory techniques such as microscopic and cultural techniques are time-consuming and require complex sample handling. Immunological and molecular techniques are advanced but have some issues related to rapidity and signal strength. In this context, integration of immunological and molecular diagnostics with nanotechnology systems offers an alternative where all detection steps are done by a portable miniaturized device for rapid and accurate identification of plant pathogens. Further, nanomaterial synthesis by utilizing functionalized metal nanoparticles as a sensing component offer several desirable features required for pathogen detection. The sensitive nature of functionalized nanoparticles can be utilized to design phytopathogen detection devices with smart sensing capabilities for field use. This chapter provides an overview of the application of nanotechnology in the field of microbial diagnostics with special focus on plant pathogens.

Keywords Agriculture • Detection • Diagnosis • Nanosensor • Nanotechnology • Pathogen • Quantum dots

P.L. Kashyap (✉) • P. Rai • H. Chakdar • K. Pandiyan • A.K. Srivastava
ICAR-National Bureau of Agriculturally Important Microorganisms, Mau, Uttar Pradesh 275103, India

ICAR-Indian Institute of Wheat and Barley Research (IIWBR), Regional Station, Flowerdale, Shimla, Himachal Pradesh 171002, India
e-mail: plkashyap@gmail.com

S. Sharma
Department of Plant Pathology, Punjab Agricultural University (PAU), Ludhiana, Punjab 141004, India

S. Kumar
ICAR-Indian Institute of Wheat and Barley Research, Karnal, Haryana 132001, India

S. Ranjan et al. (eds.), *Nanoscience in Food and Agriculture 2*, Sustainable Agriculture Reviews 21, DOI 10.1007/978-3-319-39306-3_8

8.1 Introduction

Plant diseases are major limiting factor in sustainable crop production. It is estimated that about 20–30 % of the field crops are annually lost due to infection of diseases (Nezhad 2014; Sankaran et al. 2010; Mann et al. 2008). Although, combined infestation of pests and diseases in plants could result up to 82 % losses in attainable yield in case of cotton and over 50 % losses for other major crops (Pan et al. 2010; Thind 2012). Further, if we combine these losses with post-harvest spoilage and deterioration in quality, these losses become more critical particularly for resource poor countries like India. Usually, the bacterial, fungal, and viral infections, spread over larger area in crops, groves and plantations through accidental introduction of vectors or through infected seed or plant materials. Another route for the spread of pathogens is through ornamental plants that act as hosts. These plants are frequently sold through mass distribution before the infections are known. In this context, early detection of diseases is of key importance to prevent disease spread with minimal loss to crop production (Sankaran et al. 2010; Martinelli et al. 2014).

Traditional methods for identifying plant pathogens rely on the interpretation of visual symptoms and/or the isolation, culturing and laboratory identification of the pathogen. These techniques suffered from some major drawbacks such as lack of sensitivity, time-consuming and costly etc. Additionally, the accuracy and reliability of these assays depend largely on the experience and skill of the person making the diagnosis (Sankaran et al. 2010; Kashyap et al. 2011; Alvarez 2004). Table 8.1 provides comparative analysis of conventional (culture-based), immunological, nucleic acid based-assays and nanotechnological tools for detection and diagnosis of plant pathogens. From past two decades, several attempts have been devoted to the development of methods for detecting and identifying plant pathogens based on enzyme-linked immunosorbent assay (ELISA), biochemical assays based on specific protein and toxins, nucleic acid probe technology and polymerase chain reaction (PCR) amplification of nucleic acid sequences (McCartney et al. 2003; Sundelin et al. 2009; Kumar et al. 2013; Kumar and Kashyap 2013; Kashyap et al. 2013a; Singh et al. 2014).

Table 8.1 Comparison of diagnostic methods used for the detection of plant pathogens

Method	Assay duration (h)	Detection limit	Time before result	Specificity	On-field portability	Sensitivity
Plating technique	>72	1 cfu ml^{-1}	1–3 days	Good	Poor	Poor
Immunological technique	1–3	1 pg/mL	1–2 h	Moderate	Very Good	Moderate
Nucleic acid-based technique	1–3	10^{3} cfu ml^{-1}	6–12 h	Very good	Moderate	Good
Nanotechnology-based techniques	0.30–1.0	1 fmol l^{-1}	0.30–1.0 h	Excellent	Excellent	Very good

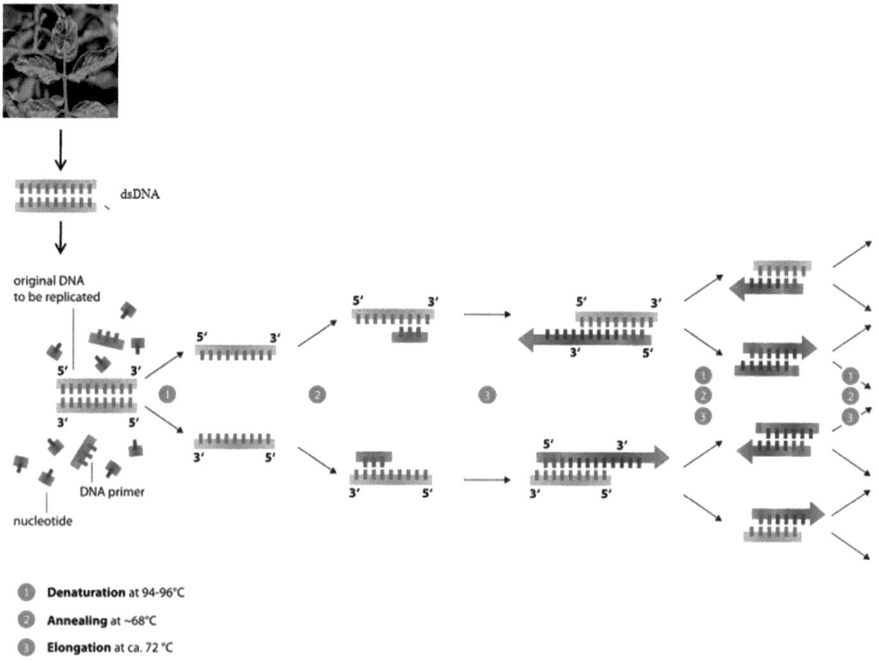

Fig. 8.1 Schematic illustration of Polymerase chain reaction (PCR). PCR is used to amplify, or create millions of identical copies of a particular DNA sequence within a tiny reaction tube. Prior to the initiation of each new round for DNA amplification, the DNA is denatured, two sets of DNA primers anneal to the denatured complementary strand. Then, primers lead DNA synthesis by the DNA polymerase. All reactions occur sequentially in template dependent manner

PCR has been widely used for the detection of plant diseases caused by fungi, bacteria, viruses and phytoplasma (Fang and Ramasamy 2015; Kashyap et al. 2011). It takes 5–24 h in detection that depends on the specific PCR variation used and this does not include any previous enrichment steps. Figure 8.1 illustrates the PCR method, the extracted and purified DNA followed by annealing of specific primer and an extension phase using a thermostable polymerization enzyme. Then each new double stranded DNA acts as target for a new cycle and exponential amplification is thus obtained. The presence of the amplified sequence is subsequently detected by gel electrophoresis. In addition to the basic PCR, variants of PCR methods such as reverse-transcription PCR (RT-PCR) has also been used for plant pathogen having RNA as a genetic material (López et al. 2003; Jeong et al. 2014). Multiplex PCR was proposed to enable simultaneous detection of several pathogens in a single reaction (López et al. 2003). Real-time PCR platforms, loop-mediated isothermal amplification (LAMP), isothermal and chimeric primer-initiated amplification of nucleic acids (ICAN) and microarray have also been used for on-site, rapid diagnosis of plant diseases based on the fungal, bacterial and viral nucleic acids (Jeong et al. 2014; DeBoer and Lopez 2012; Kashyap et al. 2013). Although these nucleic acid-based techniques and biochemical assays are very sensitive, accurate, and effective for confirming visual scouting, they are unreliable as screening tests to monitor plant health status before the appearance of symptoms. They require detailed sampling procedures, expensive infrastructure, and may feign the real status of pathogen infections. Unfortunately, these assays can be only effectively used for a restricted number of plants pathogens. Still, most of these methods cannot be applied for on-site pathogen detection in the agricultural fields. Furthermore, the high price and short shelf-life of molecular biology reagents, such as enzymes and primers, limit the application of molecular methods. As a result, developing low-cost methods to improve the accuracy and rapidity of plant pathogens diagnosis is needed. Recent advances have led to the development of functional nanoparticles (electronic, optical, magnetic, or structural) that can be covalently linked to biological molecules such as peptides, proteins, and nucleic acids.

One of the most promising nanomaterials is quantum dots (QD), which have been widely used in a broad range of bio-related applications including rapid detection of a particular biological marker with extreme accuracy (Kashyap et al. 2015). Biosensor, quantum dots, nanostructured platforms, nanoimaging and nanopore DNA sequencing tools have the potential to raise sensitivity, specificity and speed of the pathogen detection, facilitate high-throughput analysis, and can be used for high-quality monitoring and crop protection (Khiyami et al. 2014). Furthermore, nanodiagnostic kit equipments can easily and quickly detect potential plant pathogens, allowing experts to help farmers in the prevention of epidemic diseases. Currently, a vast library of nanostructures has been synthesized and documented, with different properties and applications (Savaliya et al. 2015; Khiyami et al. 2014). Figure 8.2 illustrates potential applications of nanotechnology in detection and diagnosis of plant pathogens. Briefly, the present article discusses the various applications of nanotechnology in plant pathogen detection for quicker, more cost-effective and precise diagnostic procedures of plant pathogens. Such an accurate technology may help to frame an effective integrated disease management system which may modify crop environments that adversely affect crop pathogens.

8.2 Nanotechnology-Based Diagnostic Systems

The integration of molecular diagnostics and nanotechnology is a promising technology for rapid and accurate identification of plant pathogens. Presently, several nanodevices and nanosystems have been used in diagnostics as well as sequencing single molecules of DNA. Assays with the use of nano-size devices to investigate DNA sequences and diagnose disease are becoming faster, more flexible and more sensitive. It is worth mentioning here that newly developed nanomaterials with special nanoscale characteristics offers tremendous breakthrough in plant pathogen detection and diagnosis technology (Khiyami et al. 2014). Table 8.2 provides an overview of major developments in nanotechnology-based systems for detection and diagnosis of plant pathogens. Besides this, nanotechnology is also driving the

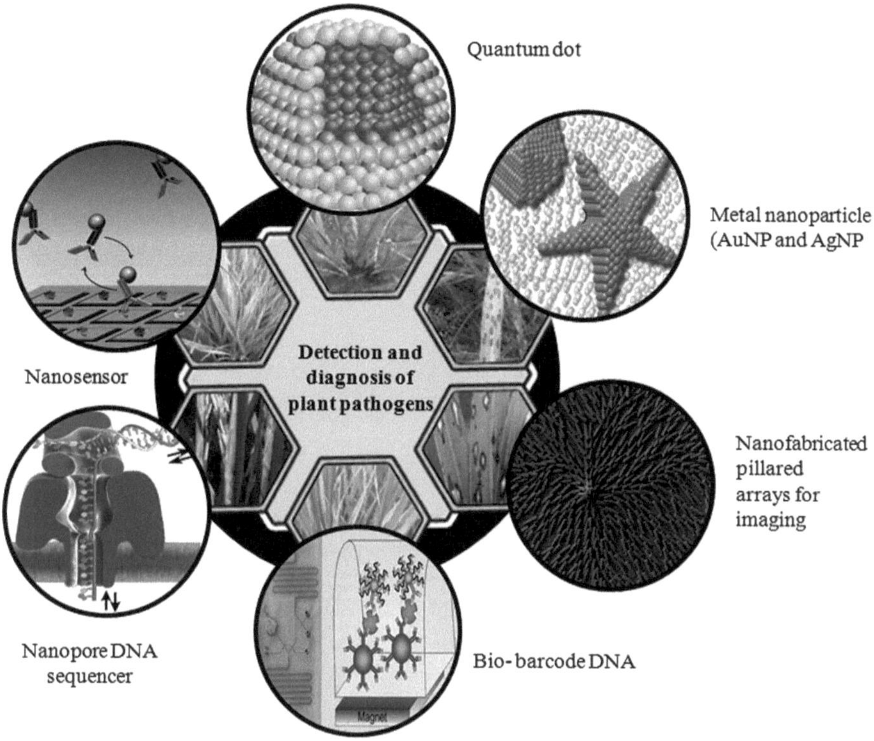

Fig. 8.2 Application of nanotechnology and currently synthesized nanomaterial for detection and diagnosis of plant pathogens. These include metallic, semiconductor and organic molecule nanomaterials of a variety of shapes, sizes and structures

development of lab-on- chip systems for detecting pathogens, toxicity in water, observing nutrients in irrigation water and controlling the quality in food products.

8.2.1 Functional Quantum Dot Nanoparticles Based Diagnostic System

Quantum dots (QD) are a class of luminescent semiconductor nanocrystals that emit light of specific wavelengths, in which the size of the nanoparticle determines the wavelength; the larger the size, higher the wavelength of the infra-red light emitted (Edmundson et al. 2014). They offer several advantages over organic dyes based broad excitation spectra. The quantum dots have narrow defined tunable emission peak, longer fluorescence lifetime, resistance to photobleaching and 10–100 times higher molar extinction coefficient. These properties of quantum dots allow multicolor quantum dots to be excited from one source by common fluorescent dyes without emission signal overlap and results in brighter probes comparing to

Table 8.2 Major breakthroughs in the development of nanotechnology-based systems for detection of plant pathogens

Year	Breakthrough(s)	Nanomaterial used	References
2009	Fluorescence silica nanoprobe as a biomarker for rapid detection of *Xanthomonas axonopodis* pv. *vesicatoria*, responsible for bacterial spot disease in tomatoes and peppers	Fluorescent silica nanoparticles (FSNP) combined with antibody molecules	Yao et al. (2009)
2010	Electrocatalytic oxidation of phytohormone salicylic acid at copper nanoparticles-modified gold electrode and its detection in oilseed rape infected with fungal pathogen *Sclerotinia sclerotiorum*	Copper nanoparticles-modified gold electrode	Wang et al. (2010)
2010	Surface plasmon resonance based immunosensor for Karnal bunt (*Tilletia indica*) diagnosis based on the experience of nano-gold based lateral flow immune-dipstick test	Nano-gold particles	Singh et al. (2010)
2012	Development of a quantum dots FRET-based biosensor for efficient detection of *Polymyxa betae*, a vector of beet necrotic yellow vein virus (BNYVV) responsible for *Rhizomania* disease in sugar beet	Tioglicolic acid-modified Cadmium-Telluride QD	Safarpour et al. (2012)
2012	Detection of *Candidatus Phytoplasma aurantifolia* with a quantum Dots FRET-based biosensor	Tioglicolic acid-modified cadmium-telluride quantum dots (CdTe-QD)	Rad et al. (2012)
2013	Synthesis of CuO nanoparticles and fabrication of nanostructural layer biosensors for detecting *Aspergillus niger* fungi	CuO nanoparticles and nanostructural layer biosensors	Etefagh et al. (2013)
2013	Development of a fluorescence resonance energy transfer (FRET)-based DNA biosensor for detection of synthetic oligonucleotide of *Ganoderma boninense*, an oil palm pathogen	Modified QD that contained carboxylic groups	Bakhori et al. (2013)
2013	Polypyrrole nanoribbon based chemiresistive immunosensors for viral plant pathogen detection	Nanoribbon	James (2013)
2014	Direct detection of orchid viruses using nanorod-based fiber optic particle plasmon resonance immunosensor	Nanorod	Lin et al. (2014)
2014	Electrochemical detection of p-ethylguaiacol, a fungi infected fruit volatile using metal oxide nanoparticles	TiO_2 or SnO_2 nanoparticles on screen-printed carbon electrodes	Fang et al. (2014)

(continued)

Table 8.2 (continued)

Year	Breakthrough(s)	Nanomaterial used	References
2014	Plant diseases detection using nanowire as biosensor transducer	Nanowire	Ariffin et al. (2014)
2015	Development of a helicase-dependent isothermal amplification (HDA) in combination with on-chip hybridization for the detection of selected *Phytophthora* species	Silver nanoparticle	Schwenkbier et al. (2015)

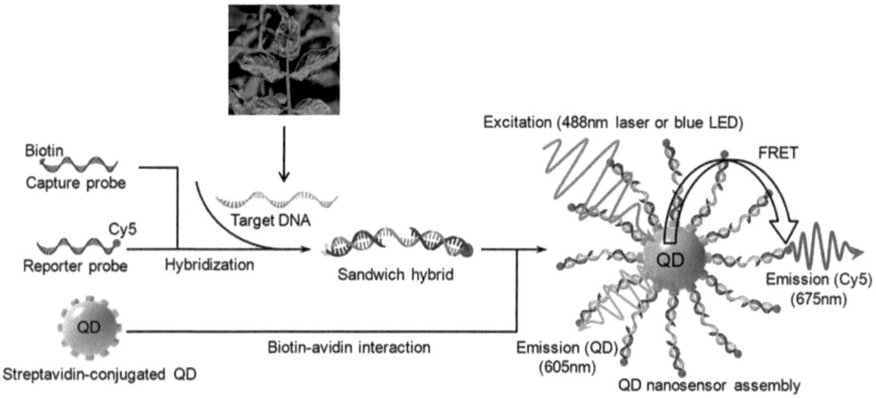

Fig. 8.3 Schematic illustration of quantum dot fluorescence resonance energy transfer (QD-FRET) sensors. Two probes labeled with biotin and Cy5 respectively hybridize to the target DNA (pathogen infected plant sample) and form a sandwich hybrid. The hybrids self assemble onto the quantum dots surface to form a QD-FRET nanosensor (Source: Chen et al. 2013)

conventional fluorophores (Zhao and Zeng 2015). Due to these advantages, QD-FRET-based nanosensors gained a wide spread popularity in agriculture and allied sectors. Figure 8.3 describes the schematic illustration of QD-FRET nanosensor. These sensors are most frequently applied in the domain of nucleic acid and enzyme activity detection (Stanisavljevic et al. 2015).

The mycosynthesis of semiconductor nanomaterials was first reported in unicellular yeast, which was capable of producing cadmium sulphide (CdS) crystallites in response to cadmium salt stress (Dameron et al. 1989). Different microbes have also been used for the biosynthesis of CdS (Yadav et al. 2015), however, limited studies have focused on its fluorescent properties. A proficient myco-mediated synthesis of highly fluorescent CdTe quantum dots was accomplished by *Fusarium oxysporum* when reacted with a mixture of $CdCl_2$ and $TeCl_2$ (Jain 2003; Kashyap et al. 2013b; Alghuthaymi et al. 2015). Knudsen et al. (2013) have shown that QD-based nanosensors are capable of probing multiple enzyme activities simultaneously. Recently, CdTe quantum dots has been used as biosensors by coating them with specific antibodies against *Polymyxa betae* specific glutathione-S-transferase (GST) protein (Safarpour et al. 2012). The mutual affinity of antigen and antibody brought the

CdTe quantum dots and rhodamine together close enough to allow the resonance dipole-dipole coupling required for fluorescence resonance energy transfer (FRET) to occur. The constructed immunosensor showed a high sensitivity, specificity and was successfully used for high-throughput screening of plant samples with consistent results within 30 min. On parallel lines, Rad et al. (2012) also developed a quantum dot (QD)-based nano-biosensor for highly sensitive detection of phytoplasma (*Candidatus Phytoplasma aurantifolia*) in infected lime trees. The developed immunosensor showed 100 % specificity with a detection limit of 5 *Ca. P. aurantifolia* μl^{-1}. Recently, an optical DNA biosensor based on fluorescence resonance energy transfer (FRET) utilizing synthesized quantum dot (QD) has been developed for the detection of specific-sequence of DNA for *Ganoderma boninense* (Bakhori et al. 2013). Modified quantum dots (5–8 nm) that contained carboxylic groups was conjugated with a single-stranded DNA probe (ssDNA) via amide-linkage.

Hybridization of the target DNA with conjugated QD-ssDNA and reporter probe labeled with Cy5 allows the detection of related synthetic DNA sequence of *Ganoderma boninense* gene based on FRET signals. The developed biosensor has shown high sensitivity with detection limit of 3.55×10^{-9} M. This approach is also capable in providing simple, rapid and sensitive method for detection of plant pathogens. Moreover, quantum dots can be excited using UV light and fluorescence can be visualized with the naked eye, this technology can be transferred into the field for immediate use. Research is just beginning for the use of quantum dots for detecting plant pathogens and toxins on and in foods and plants. Work must continue for the optimization of assays to obtain an accurate signal for low levels of pathogens in complex systems, whether they are food, plants or insects. The opportunities are endless for the applications of functional quantum dot nanoparticles based diagnostic system in agriculture and allied sectors, the field is ever-expanding and scientists are trying to keep up with the latest technologies that can be used to protect agricultural crops and food commodities from plant pathogens.

8.2.2 Metal Nanoparticles Based Diagnostic System

Metal nanoparticles have been applied in biosensors as marker tags to replace enzymes as the label. Striping voltammetry as an electrochemical technique can be applied to detect the metal nanoparticles directly making the assay simple to perform. Gold (AuNP) and silver nanoparticles (AgNP) can be used in these methods including different inorganic nanocrystals (ZnS, PbS and CdS) for analyte detection (Upadhyayula 2012). The unique physical and chemical properties of nanoparticles such as colloidal gold can provide excellent application in a wide range of biosensing techniques (Rosi and Mirkin 2005; Khan and Rizvi 2014). AuNPs have high surface-to-volume ratios and can be functionalized to detect specific pathogen targets, offering lower detection limits and higher selectivity than conventional strategies. AuNP system is composed of two types of AuNP. Each type of AuNP is coated with different thiol-oligonucleotides. The last 15 nucleotides are complementary to one-half of a target DNA sequence. When the target DNA is introduced into the system, the two types of AuNP both bind to the DNA and aggregate. This results in a color change from red to blue, which is a well-known behavior of AuNPs. Further, there is still a wide scope to improve the signal amplification efficiency by exploring the effects of the loading of detector probes and the number of gold nanoparticles in colour development. Figure 8.4 illustrates the schematic diagram of oligonucleotide-gold nanoparticle aggregation assay and detection strategy. This AuNP based technology is an excellent example of exploitation of the tunability of AuNP surface chemistry to optimize performance under real field conditions. Several products are available in the market such as Oxanica (UK) Quantum dots and MultiPlxBeads™ from Crystalplex Corporation, USA (Tothill 2011).

Nanoparticles can also be exploited in conductivity based sensors where they can induce a change in the signal upon the attachment of the nanoparticles tagged antibody with the antigen captured on the sensor surface (Servin et al. 2015). Various strategies such as antibody-antigen, adhesion receptor, antibiotic and complementary DNA sequence recognitions have been developed for a specific detection between target phytopathogenic cells and bio-functionalized nanomaterials (Conde et al. 2014). Gold nanoparticles are excellent markers to be used in biosensors due to ease in alternation of their optical or electrochemical procedures to identify pathogens. A number of nanoparticle-based experiments have been performed to develop biomolecular detection with DNA- or protein functionalized gold nanoparticles, which are used as the target-specific probes (Thaxton et al. 2006). These detection methods include conductive polymer nanowires (Pal et al. 2008), carbon nanotubes (Poonam and Deo 2008), nanoporous silicon (Yang et al. 2008) and gold nanoparticles (Wang et al. 2010). Singh et al. (2010) used nano-gold based immunosensors that could detect Karnal bunt disease in wheat (*Tilletia indica*) using surface plasmon resonance (SPR).

Wang et al. (2010) exploited indirect stimulus to develop a sensitive electrochemical sensor, using modified gold electrode with copper nanoparticles, to monitor the levels of salicylic acid in oil seeds to detect the pathogenic fungus, *Sclerotinia sclerotiorum*. They successfully and accurately measured salicylic acid using this sensor. Research on similar sensors and sensing techniques needs to be expanded for detecting pathogens, their by products or monitor physiological changes in plants due to infections. Dubertret et al. (2001) noticed the ability of gold nanoparticles to act as fluorescence quenchers and, therefore, it could be used to solve major drawbacks in molecular biology experiments. For instance, a DNA oligonucleotide could be synthesized, fluorescently labelled at its 5′ end and conjugated at the 3′ end with gold nanoparticles. The successful application of these oligonucleotides has been reported in the diagnosis of the phytoplasma associated with the flavescence dorée (FD) of grapevine (Firrao et al. 2005). Fan et al. (2003) reported that the gold nanoparticles efficiently quench the fluorescence of light harvester polymers, such as polyfluorene, and will open new perspectives in the development of optical performances of nanobiotransducers for diagnostic purposes. Furthermore, diagnostic probe made of a specific oligonucleotide bearing a fluorescein at its 5′ terminal and

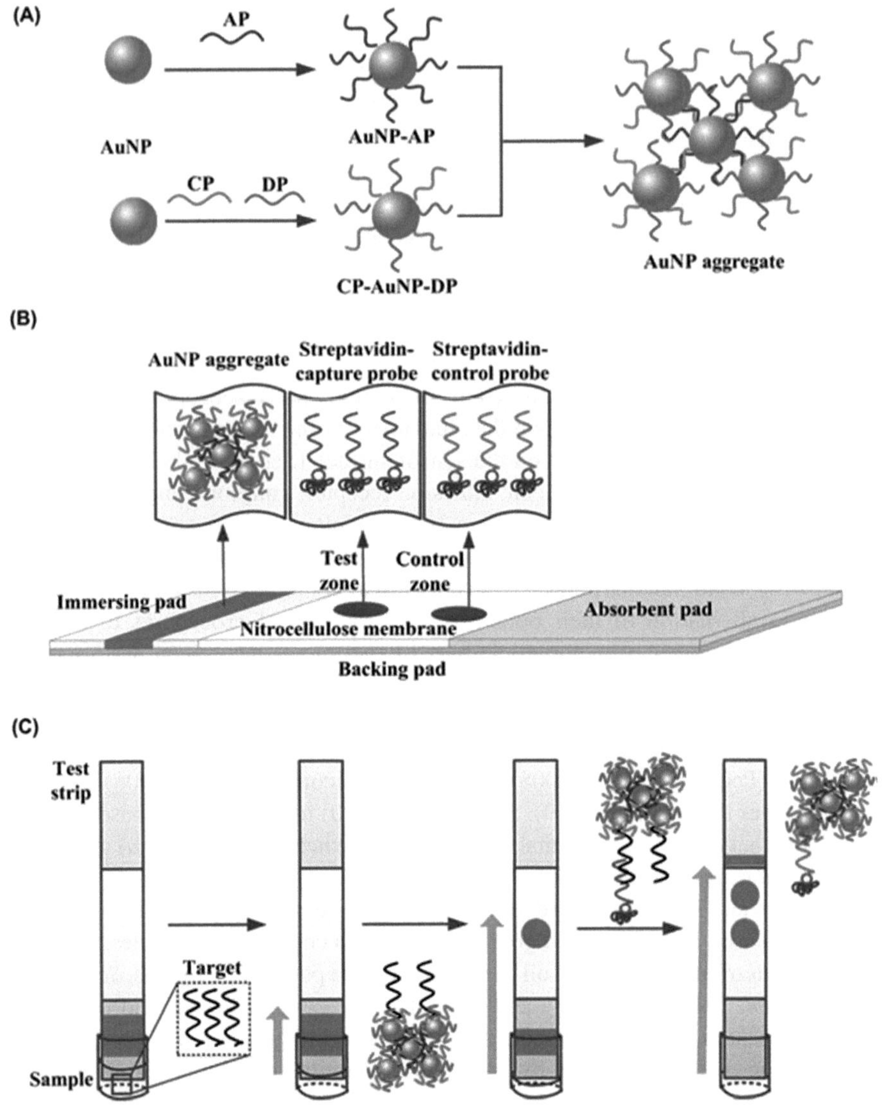

Fig. 8.4 Oligonucleotide-gold nanoparticle aggregation assay and detection strategy. (**a**) Preparation of oligonucleotide-linked gold nanoparticle (AuNP) aggregates. (**b**) Design of nucleic acid lateral flow (NALF) test strips. (**c**) Principle of the AuNP aggregates based NALF assay. *AP* amplification probe, *CP* complementary probe, *DP* detector probe. AP and CP are complementary to each other (Source: Hu et al. 2013)

gold nanoparticles (2-nm) at its 3′ terminal act as a nanobiotransducer in DNA hybridization. It produces a stronger fluorescence signal when hybridized to target DNA (Firrao et al. 2005).

The bio-barcode assay is another ultrasensitive method of amplification and detection of nucleic acids or proteins. They are characterized by their ultrahigh sensitivity and multiplexing capability for the detection of oligonucleotides or proteins. In case of oligonucleotide targets, Oligo-AuNP probes are hybridized to oligonucleotide-functionalized magnetic microparticle (MMP) probes using the target sequence as a linker. These complexes are then separated magnetically for subsequent release of the oligonucleotides from the Oligo-AuNP probes. These released biobarcodes are quantitatively analyzed by the scanometric assay (Fig. 8.5). The assay is also promising by allowing the quick detection of nucleic acids at high-zeptomolar levels (Nam et al. 2004) and protein targets at low-attomolar concentrations under optimized conditions (Goluch et al. 2006). Schwenkbier et al. (2015) developed a helicase-dependent isothermal amplification (HDA) in combination with on-chip hybridization for the detection of *Phytophthora* species. This approach allows efficient amplification of the yeast GTP-binding protein (Ypt1) target gene region at one constant temperature in a miniaturized heating device. The assay's specificity was determined by on-chip DNA hybridization and subsequent silver nanoparticle deposition. The silver deposits serve as stable endpoint signals that enable the visual as well as the electrical readout. These advancements, point to the direction of a near future on-site application of the combined techniques for a reliable detection of several kinds of plant pathogens.

8.2.3 Nanostructured Platforms Based Diagnostic System

Advancement of nanotechnology and biotechnology has prompted utilization of nanostructure as a novel sensing platforms, owing to their ultra-high surface area to volume ratio, size dependent electrical properties, and possibility of device miniaturization (Prieto-Simon et al. 2007; Sertova 2015). The principal application of such nanostructures is to decrease the time for pathogen detection. Nanomaterials, such as carbon nanomaterials (carbon nanotubes and graphene), nanowires, nanocomposites, and nanostructured metal oxide nanoparticles are playing an increasing role in the design of nanosensing platforms for pathogen and mycotoxin determination (Malhotra et al. 2014). Other types of nanostructure platforms are based on microfluidics systems, which can also be used to detect pathogens efficiently in real time and with high sensitivity (Baeummer 2004). A major advantage of such systems is their miniature format and their potential to detect compounds of interest in minute sample volumes with rapidity (García et al. 2010). Food spoilage due to fungal and bacterial microbes can be detected by several kinds of nanostructured platforms. For instance, Bhattacharya et al. (2007) designed an array of thousands of nanoparticles to be visualized in different colours in contact with food pathogens. Kaushik et al. (2009) fabricated nanoSiO2 and chitosan based nanobiocomposite film on an ITO substrate to co-immobilize r-IgGs and BSA for OTA (*Aspergillus ochraceus*) detection. They concluded that BSA/r-IgGs/CH-NanoSiO2/ITO immunoelectrode can be used for OTA detection with improved sensing characteristics.

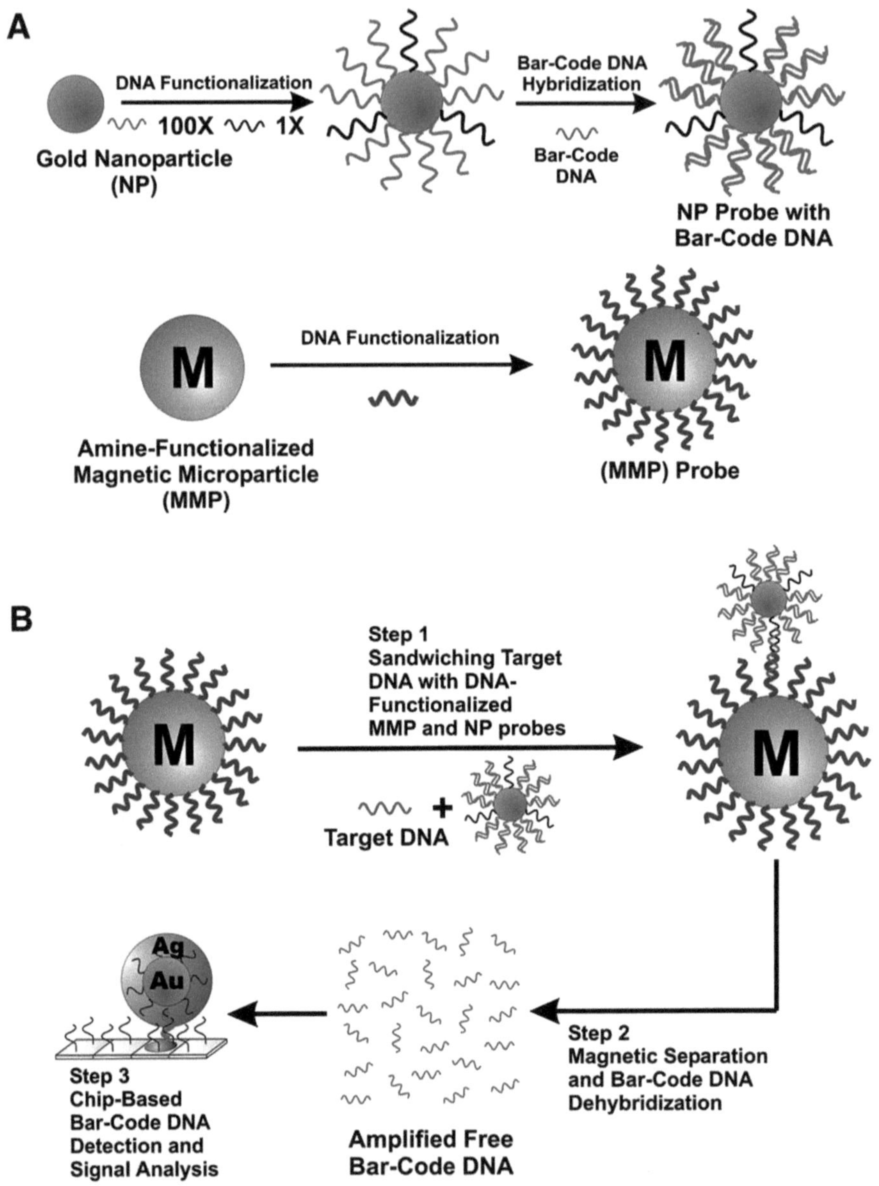

Fig. 8.5 A bio-barcode-amplification based assay for target DNA detection. (**a**) Preparation of gold nanoparticle and magnetic microparticle probes. Gold nanoparticles are functionalized with hybridized barcode DNA and capture strands for the target. Magnetic nanoparticles are functionalized with capture strands for the target. (**b**) Biobarcode based amplification (BCA) technique coupled with scanometric detection (Source: Nam et al. 2004)

Horseradish peroxidase (HRP) biosensors have been applied without any pre-treatment to determine OTA (*A. ochraceus* and *Penicillium viricatum*) in spiked beer samples and roasted coffee (Alonso-Lomilloa et al. 2010). Paniel et al. (2010) used magnetic nanoparticles to improvise electrochemical immunosensor for the detection of ultra-trace quantities of AFM1 (up to 0.01 ppb) produced by *A. flavus* in foodstuffs. Hervas et al. (2011) described a 'lab-on-chip' strategy integrating an electrokinetic magnetic bead-based electrochemical immunoassay on a microfluidic chip for the quick, sensitive and discriminatory quantification of zearalenone produced by *Fusarium* sp. Further, Panini et al. (2010) developed an immunosensor that utilized multi-wall carbon nanotubes and a continuous-flow system for the rapid and sensitive quantification of zearalenone in corn silage samples. Ansari et al. (2010) demonstrated that sol–gel derived Nano-ZnO film can be used for the immobilization of r-IgGs and BSA for blocking nonspecific binding sites of r-IgGs to detect OTA with a detection range, 0.006–0.01 nM/dm^3. Similarly, Kaushik et al. (2013) developed a nanostructured cerium oxide film based immunosensor for the detection of food-borne mycotoxins. Rabbit-immunoglobulin antibodies and BSA have been immobilized onto sol–gel derived nanostructured cerium oxide film synthesized onto an indium tin-oxide covered glass plate for the detection of ochratoxin-A.

Mak et al. (2010) reported an ultra-sensitive magnetic nanoparticle immunoassay for detecting more than one mycotoxin. Using the magnetic nanoparticle as the solid phase allowed a significantly increased surface area for the immobilization of the reactants and their uniform distribution throughout the whole volume of the reaction medium, thereby eliminating the diffusion limitations of traditional ELISA. The application of a magnetic field separated the reactants simply and rapidly, and facilitated the wash steps that are also required in traditional microplate-based ELISA (Fig. 8.6). Using these advantages, the MNP-based immunoassay scheme was developed and implemented in ELISA microplate wells for detection of aflatoxin-B1, zearalenone and HT-2 mycotoxins. The ozonization and adsorption efficiency of modified nanodiamonds to detect the content of aflatoxin-B1 has been examined by Puzyr et al. (2010). Recently, Actis et al. (2010) described ultrasensitive method for the detection of mycotoxins by STING (signal transduction by ion nano-gating) sensors, with a detection limit up to 100 fg ml^{-1}. Silica and clays are most efficient in combination with smaller sized water molecules and smaller mycotoxins such as aflatoxins and ochratoxins. However, clays are less efficient in binding the larger mycotoxins such as fumonisin and deoxynivalenol (vomitoxin) because the distance among clay layers is not sufficient to accommodate the larger molecules (Jaynes et al. 2007). By using nano-sized clay, the gap between the layers of clay has been prolonged ten times. As a result, the nanoclay can bind the whole family of mycotoxins.

A rapid enzyme-linked immunosorbent assay (ELISA) was improvised by using superparamagnetic nanoparticles (Radoi et al. 2008). The assay was effective for detecting aflatoxin M1 (AFM1) with a detection limit of 4–250 ng l^{-1} (Radoi et al. 2008). Cysteamine functionalized-gold nanoparticles (C-AuNP) along with aflatoxin B1 antibodies (aAFB1) were immobilized on a 4-mercaptobenzoic acid-based

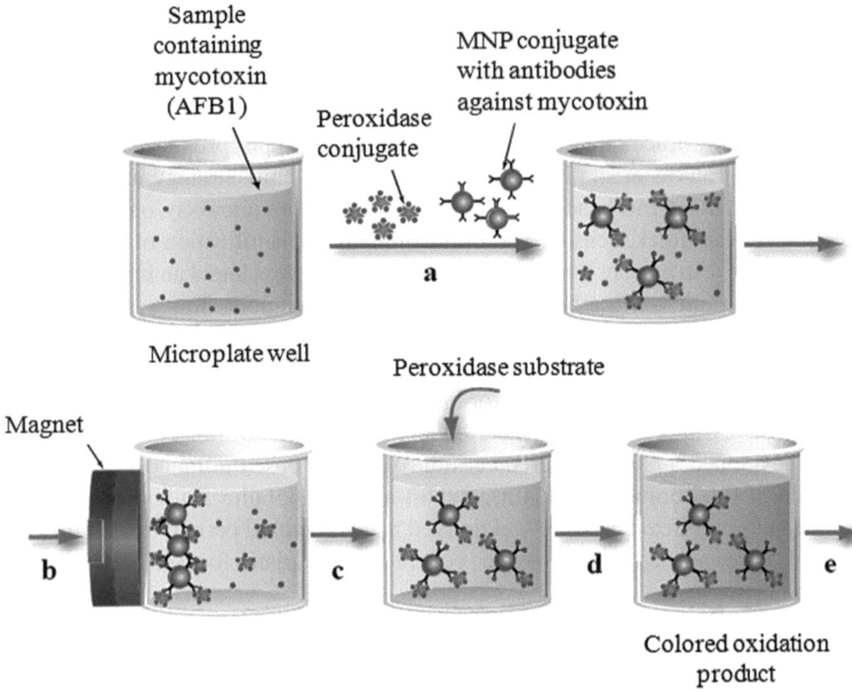

Fig. 8.6 Assay of mycotoxin (e.g. AFB1) detection using magnetic nanoparticles. (**a**) Free AFB1contained in the test sample competes with peroxidase-labeled AFB1 for binding sites on the antibodies adsorbed on the MNP surface; (**b**) A magnet separated magnetic nanoparticle (MNP) from unreacted components; (**c**) MNP are washed and, by removing the magnetic field, returned into solution; (**d**) the peroxidase substrate is added to the MNP suspension and; (**e**) development of colored oxidation product

self-collected monolayer on a gold electrode (MBA/Au) to fabricate a BSA/aAFB1-CAuNP/MBA/Au immunoelectrode. These electrodes were used to detect AFB1 in the range of 10–100 ng l^{-1} (Sharma et al. 2010). Recently, a moveable machine has been developed that can concurrently identify various bacterial, fungal toxins and pathogens in stored food (Yalcin and Otles 2010; Biswal et al. 2012). From all these reports, it seems that nanostructured platforms can be an exciting alternative to the conventional techniques for the detection of mycotoxins and pathogens spoiling food and agricultural crops.

8.2.4 *Nanofabrication Imaging*

Nanotechnology offers unique opportunities to precisely tune and control the chemical and physical properties of contrast materials in order to overcome problems of toxicity, useful imaging time, tissue specificity and signal strength. Nie (2013) reported that mesoscopic nanoparticles (5–100 nm diameter) have large surface

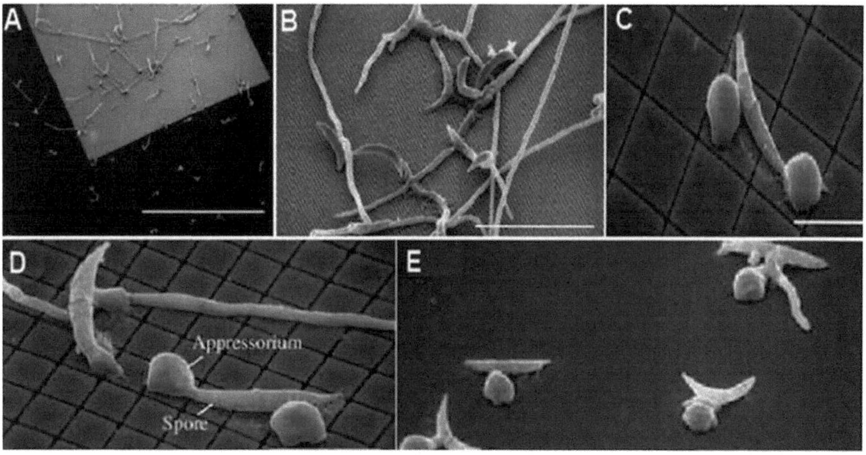

Fig. 8.7 Scanning electron micrographs (SEM) showing the fungus *Colletotrichum graminicola* grown on nanofabricated pillared arrays. When the individual pillars are very small (0.5 mm wide) and do not provide much surface contact (**a**, **b**), the spores of the fungus grow without forming 'appressoria'. When the pillars are wider (**c**, **d**) or when the surface is completely smooth (**e**), appressoria are formed quickly (Source: Mccandless 2005)

areas and ideal for conjugating functional groups in multiple pathogen diagnosis assays. Electron beam and photolithography techniques are also used to fabricate topographies that mimic leaf surface features as well as the internal plumbing of plants, and then nano-imaging technologies are used to study how pathogen invade and colonize the leaf tissue (Mccandless et al. 2005). Lithography was used to nanofabricate a pillared surface on silicon wafers. This lawn of miniature pillars (1.4 and 20 mm wide) was used to examine the movement across the surface by the fungus that mimicked some of the characteristics of the host plant. Images of the *Colletotrichum graminicola* crawling across the nanofabricated surface assisted the researchers to determine that the fungus needs to make a minimum contact (at least 4.5 mm) prior to initiation of appressoria formation (Fig. 8.7). To develop disease resistant cultivars, the infection process and behaviour of *Xylella fastidiosa* causing Pierce's disease inside grapevine xylem were studied using nanofabrication methods (Meng et al. 2005). The application of carbon-coated magnetic nanoparticles and microscopy methods at different levels of resolution to visualize the transport and deposition of nanoparticles inside the plant host was reported by González-Melendi et al. (2007). Further, Szeghalmi et al. (2007) investigated nanostructured surface-enhanced Raman scattering (SERS) substrates for imaging applications at high-spatial resolution (1 μm). They performed SERS imaging of dried fungal hyphae grown on commercially available nanostructured gold-coated substrates and concluded that this type of nanofabrication techniques offer a well-characterized and reproducible substrate for *in-situ* or *in-vivo* imaging studies of plant pathogen interactions. Rispail et al. (2014) evaluated the behavior of quantum dots and superparamagnetic nanoparticles on *Fusarium oxysporum* and indicated that both nanomaterials rapidly interacted with the fungal hypha labeling the presence of the pathogenic fungus, although, they showed differential behavior with respect to internalization. This work represents the first study on the behavior of quantum dots and superparamagnetic particles on fungal cells, and constitutes the first and essential step to address the feasibility of new nanotechnology-based systems for early detection and eventual control of pathogenic fungi.

8.2.5 Nanobiosensor Based Diagnostic System

Nanosensors with immobilized bioreceptor probes that are selective for target analyte molecules are termed as nano biosensors. They offer the advantages of being small, portable, sensitive with real-time monitoring, precise, quantitative, reliable, accurate, stable, reproducible and robust to identify potential and complex disease problems. At present, these systems were employed to detect and quantify minute amounts of contaminants such as viruses, bacteria, fungi, toxins and other biohazardous substances in the agriculture and food systems (Srinivasan and Tung 2015). Therefore, these nanosensors may have a huge impact on the precision farming methods (Rai and Ingle 2012). Moreover, early on-site detection of plant pathogens with portable nanobiosensors will enable to design the strategies to control the spread of diseases and will also help the study of disease epidemiology. These sensors can be linked to a GPS and distributed throughout the field for real-time monitoring of disease, soil conditions and crop health (Nezhad et al. 2014). The combination of biotechnological and nanotechnological approaches in bio-sensors can be used to construct equipment with increased sensitivity, allowing an earlier response to ecological changes and disease prevalence. Nanosensors will allow us to identify plant diseases before visible symptoms appear and thus facilitate their control. Precision farming will be improved by using nanosensors by providing precise data, helping growers to make better decisions, thus enhances agriculture production and productivity (Rai and Ingle 2012). Hashimoto et al. (2008) developed a new biosensor system for the rapid diagnosis of soil-borne diseases, consisting of two biosensors. The system was constructed using equal quantities of two different microbes, each individually immobilized on an electrode. Taking into consideration the particular optical properties of silver nanoparticles, the interaction between silver nanoparticles and sulphurazon-ethyl herbicide was investigated by Dubas and Pimpan (2008). They found that silver nanoparticles are sensitive to increase concentrations of herbicide in a solution and induced a variation in colour of the nanoparticles from yellow to orange red and finally to purple. This approach is useful for detecting contaminants, such as organic pollutants and microbial pathogens in water bodies and in the environment (Dubertret et al. 2001). Fluorescent silica nanoparticles (FSNP) combined with antibody molecules successfully detected plant pathogens such as *Xanthomonas axonopodis* pv. *vesicatoria* which

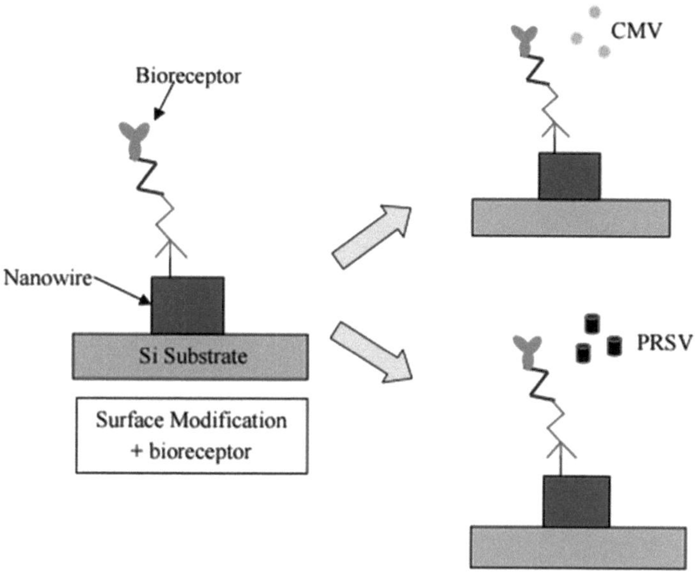

Fig. 8.8 Plant diseases detection using nanowire as biosensor transducer. Nanowire undergo surface modification with amino group solution and enzyme applied on the nanowire surface and thus, bioreceptor binds with nanowire structure which it used for grab plant viruses such as *Cucumber Mosaic Virus* (*CMV*) and *Papaya Ring Spot Virus* (Source: Ariffin et al. 2014)

causes bacterial spot disease in tomatoes and peppers (Yao et al. 2009). Copper oxide (CuO) nanoparticles and nanolayers were synthesized by sol–gel and spray pyrolysis methods, respectively. Both CuO nanoparticles and nanostructural layer biosensors were used for detecting *Aspergillus niger* fungi (Etefagh et al. 2013).

Furthermore, the highly ordered nanowires array combined with multiple biorecognition holds the promise of developing multiplexed nanobiosensors. Nanowire biosensors are a class of nanobiosensors, of which the major sensing components are made of nanowires coated by biological molecules such as DNA molecules, polypeptides, fibrin proteins, and bacteriophages. Since their surface properties are easily modified, nanowires can be decorated with virtually any potential chemical or biological molecular recognition unit, making the wires themselves analyte independent. The nanomaterials transduce the chemical binding event on their surface into a change in conductance of the nanowire in an extremely sensitive, real time and quantitative fashion. They will be very useful for high-throughput diagnosis and screening. By employing this concept, Ariffin et al. (2014) used nanowire as biosensor transducer for the detection of *Cucumber Mosaic Virus* (CMV) and *Papaya Ring Spot Virus* (PRSV) (Fig. 8.8). These results clear indicated that the nanowires are a good candidate material for fabricating nanoscale biosensors for making remote-controlled nanbiosensors for future applications in crop health-care testing, disease diagnostics and environmental monitoring.

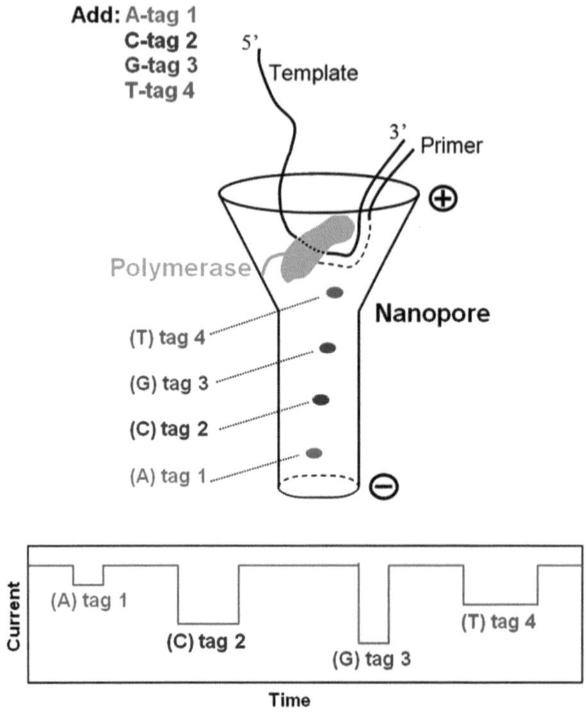

Fig. 8.9 Working flow of single molecule DNA sequencing by a nanopore with phosphate-tagged nucleotides. Each of the four nucleotides will carry a different tag. During sequencing by synthesis (SBS), these tags, attached via the terminal-phosphate of the nucleotide, will be released into the nanopore one at a time where they will produce unique current blockade signatures for sequence determination. A large array of such nanopores will lead to high throughput DNA sequencing (Source: Kumar et al. 2012)

8.2.6 Nanopore System

Nanopore systems are based on electronic detection of DNA sequence and have the potential of low sample preparation work, high speed, and low cost (Branton et al. 2008). Nucleotide identification using nanopore system is based on the measurement of conductivity changes across a lipid membrane while a DNA fragment is pulled through a nano-scale pore by an electric current. Conductivity changes are nucleotide-specific, enabling the identification of nucleotides as they cross the pore (Egan et al. 2012). The protein nanopore is inserted in a polymer bilayer membrane across the top of a microwell. Each microwell has a sensor chip that measures the ionic current as the single molecule passes through the nanopore. However, the speed at which the DNA strand travels through the nanopore is too fast for accurate identification (Clarke et al. 2009). It is worth mentioning that nanopore sequencing platform models have the potential to rapidly generate ultra long single molecule reads. Recently, Kumar et al. (2012) reported nanopore-based sequencing by synthesis (Nano-SBS) approach can accurately distinguish four DNA bases by detecting four different sized tags released from 5′-phosphate-modified nucleotides at the single molecule level for sequence determination.

The basic principle of the Nano-SBS strategy is described in Fig. 8.9. Another new sequencing technology developed by IBM and Roche together is 'DNA transistor' technology, which could potentially record the nucleotide sequence as the template is pulled through the nanopore sensor (Zhang et al. 2011). Recently, a portable DNA sequencing machine (MinION) was launched by U.K.-based Oxford Nanopore Technologies (Hayden 2015). This tool is able to sequence 10 kb of a single sense and anti-sense DNA strand and will make next-generation sequence (NGS) within the reach of many research groups. This system offers on the spot data, a special feature that can help scientists to quickly determine the cause of an epidemic outbreak, epidemiology of a disease or catalog the rare and exotic species. Further, it can discriminate between closely related bacteria, fungi and viruses, read complex portions of the genome, and differentiate between the two versions of a gene that are carried on each chromosome pair. Therefore, nanopore platform implemented within current diagnostic equipment has the potential of analyzing the entire genome in minutes instead of hours. From agricultural point of view, this technology can be applied to analyze plant and pathogen genomics and gene functions in addition to pathogens detection and prediction in agricultural crops.

8.2.7 Nanodiagonastic Kit Based Equipment System

Nanodiagonastic kit also known as 'lab in a box' refers to packing sophisticated measuring devices, reagents, power supply and other features that take up laboratory space into a parcel no larger or heavier than a briefcase (Khiyami et al. 2014). This type of a diagnostic kit can easily and quickly detect potential serious plant pathogens in fields, allowing experts to help farmers in prevention of disease epidemics from breaking out (Pimentel 2009; Nezhad 2014). For instance, 4mycosensor is a tetraplex competitive antibody-based assay in a dipstick format for the real-time detection of ZEA, T-2/HT-2, DON and FB1/FB2 mycotoxins on the same single strip for corn, wheat, oat and barley samples at or below their respective European maximum residue limits (MRLs) (Lattanzio et al. 2012). Nanodiagonastic kit based on immunoassay is fast, cheap, easy-to-use and suitable for the purpose of quick detection and screening of mycotoxins in cereals. However, there are many challenges which must be addressed before nanodiagonastic kit based equipment systems are truly ready for use in agriculture and allied sectors. These include the discovery and selection of effective antigen, antibody and nucleotide targets, which are required to improve the specificity of the diagnostic kits and permit strain differentiation. Furthermore, universal standards for the assessment of tests and levels of detection must be set so that studies of detection limit can be compared. In addition, for genomic target detection of a particular pathogen, strategies to simplify the purification and isolate genes of interest are vital. So far, the advances of nanotechnology have not been fully applied to pathogen disease detection in agricultural crops, nanotechnology can potentially address many of the challenges outlined earlier for effective on-site real time diagnostics of crop diseases.

8.3 Conclusion

Nanotechnology presents a wealth of potential tools for researchers involved in the detection, identification, and monitoring of plant pathogens. Indeed, the recent reports on the applications of portable diagnostic equipment, nanoparticle-based bio-barcoded DNA sensor, quantum dots, nanostructured platforms, nanoimaging and nanopore DNA sequencing tools have prompted virtually unqualified speculation as to the coming profusion of cheap, rapid, and accurate means of identifying and diagnosing complex disease problems. The promise for the development of portable handheld nano-devices for *in situ* field identification, has further whetted the appetite for nanodiagnostic technologies. Nanodiagnostics is an area of huge interest and future research will focus on the multimodality of nanoparticles. Research in this area is just at infancy stage for detecting plant pathogens and toxins in agricultural crops and various food commodities. There is a need to continue the pace for the optimization of nanodiagnostic assays to obtain an accurate signal for low levels of pathogens to solve plant disease-complex mysteries existing in the farmers' fields. The opportunities are endless for potential use of nanotechnology in disease diagnosis. The field is ever expanding and scientists are trying to keep up with the latest technologies that can be used to protect agricultural crops to achieve millennium nutrition and food security agenda.

Acknowledgments This work was supported by the Indian Council of Agriculture Research (ICAR) by a network project 'Application of Microorganisms in Agriculture and Allied Sectors' (AMAAS).

References

Actis P, Jejelowo O, Pourmand N (2010) Ultrasensitive mycotoxin detection by STING sensors. Biosens Bioelectron 26(2):333–337. doi:10.1016/j.bios.2010.08.016

Alghuthaymi MA, Almoammar H, Rai M, Said-Galiev E, Abd-Elsalam KA (2015) Myconanoparticles: synthesis and their role in phytopathogens management. Biotechnol Biotechnol Equip 29(2):221–236

Alonso-Lomilloa MA, Domınguez-Renedoa O, Ferreira-Gonc L, Arcos-Martıneza MJ (2010) Sensitive enzyme-biosensor based on screen-printed electrodes for ochratoxin A. Biosens Bioelectron 25:1333–1337

Alvarez AM (2004) Integrated approaches for detection of plant pathogenic bacteria and diagnosis of bacterial diseases. Annu Rev Plant Pathol 42:339–366

Ansari AA, Kaushik A, Pratima R, Solanki Malhotra BD (2010) Nanostructured zinc oxide platform for mycotoxin detection. Bioelectrochemistry 77(2):75–81

Ariffin SAB, Adam T, Hashim U, Faridah S, Zamri I, Uda MNA (2014) Plant diseases detection using nanowire as biosensor transducer. Adv Mater Res 832:113–117. doi:10.4028/www.scientific.net/AMR.832.113

Baeummer A (2004) Nanosensors identify pathogens in food. Food Technol 58:5155

Bakhori NM, Yusof NA, Abdullah AH, Hussein MZ (2013) Development of a fluorescence resonance energy transfer (FRET)-based DNA biosensor for detection of synthetic oligonucleotide of *Ganoderma boninense*. Biosensors 3(4):419–428

Bhattacharya S, Jang J, Yang L, Akin D, Bashir R (2007) Biomems and nanotechnology based approaches for rapid detection of biological entities. J Rapid Methods Autom Microbiol 15(1):132

Biswal SK, Nayak AK, Parida UK, Nayak PL (2012) Applications of nanotechnology in agriculture and food sciences. Int J Sci Innov Discov 2:21–36

Branton D et al (2008) The potential and challenges of nanopore sequencing. Nat Biotechnol 26(10):1146–1153

Chen G, Song F, Xiong X, Peng X (2013) Fluorescent nanosensors based on fluorescence resonance energy transfer (FRET). Ind Eng Chem Res 52:11228–11245. doi:10.1021/ie303485n

Clarke J, Wu H-C, Jayasinghe L, Patel A, Reid S, Bayley H (2009) Continuous base identification for single-molecule nanopore DNA sequencing. Nat Nanotechnol 4(4):265–270

Conde J, Dias JT, Grazú V, Moros M et al (2014) Revisiting 30 years of biofunctionalization and surface chemistry of inorganic nanoparticles for nanomedicine. Front Chem 2:48

Dameron CT, Reeser RN, Mehra RK, Kortan AR, Carroll PJ, Steigerwaldm ML, Brus LE, Winge DR (1989) Biosynthesis of cadmium sulphide quantum semiconductor crystallites. Nature 338(6216):596–597

De Boer SH, López MM (2012) New grower-friendly methods for plant pathogen monitoring. Ann Rev Phytopathol 50:197–218

Dubas ST, Pimpan V (2008) Green synthesis of silver nanoparticles for ammonia sensing. Talanta 76(1):29–33

Dubertret B, Calame M, Libchaber AJ (2001) Single-mismatch detection using gold-quenched fluorescent oligonucleotides. Nat Biotechnol 19(4):365–370

Edmundson MC, Capeness M, Horsfall L (2014) Exploring the potential of metallic nanoparticles within synthetic biology. New Biotechnol 31(6):572–578

Egan AN, Schlueter J, Spooner DM (2012) Applications of next-generation sequencing in plant biology. Am J Bot 99(2):175–185

Etefagh R, Azhir E, Shahtahmasebi N (2013) Synthesis of CuO nanoparticles and fabrication of nanostructural layer biosensors for detecting *Aspergillus niger* fungi. Sci Iran 20(3):1055–1058

Fan C, Wang S, Hong JW, Bazan GC et al (2003) Beyond superquenching: hyper-efficient energy transfer from conjugated polymers to gold nanoparticles. PNAS USA 100(11):6297–6301

Fang Y, Ramasamy RP (2015) Current and prospective methods for plant disease detection. Biosensors 4:537–561. doi:10.3390/bios5030537

Fang Y, Umasankar Y, Ramasamy RP (2014) Electrochemical detection of p-ethylguaiacol, a fungi infected fruit volatile using metal oxide nanoparticles. Analyst 139(15):3804–3810. doi:10.1039/c4an00384e

Firrao G, Moretti M, Ruiz-Rosquete M, Gobbi E, Locci R (2005) Nanobiotransducer for detecting flavescence doree phytoplasma. J Plant Pathol 87(2):101–107

Garcia F, Westfahl H, Schoenmaker J, Carvalho EJ et al (2010) Tailoring magnetic vortices in nanostructures. Appl Phys Lett 97:022501

Goluch ED, Nam JM, Georganopoulou DG, Chiesl TN et al (2006) A biobarcode assay for on-chip attomolar-sensitivity protein detection. Lab Chip 6(10):1293–1299

González-Melendi P, Fernandez-Pacheco R, Coronado MJ, Corredor E et al (2007) Nanoparticles as smart treatment delivery systems in plants: assessment of different techniques of microscopy for their visualization in plant tissues. Ann Bot 101(1):187–195

Hashimoto Y, Nakamura H, Koichi AK, Karube I (2008) A new diagnostic method for soil-borne disease using a microbial biosensor. Microbes Environ 23(1):35–39

Hayden EC (2015) Pint-sized DNA sequencer impresses first users. Nature 521:15–16

Hervás M, López MA, Escarpa A (2011) Integrated electrokinetic magnetic bead-based electrochemical immunoassay on microfluidic chips for reliable control of permitted levels of zearalenone in infant foods. Analyst 136(10):2131–2138

Hu J, Wang L, Li F, Han YL, Lin M, Lu TJ, Xu F (2013) Oligonucleotide-linked gold nanoparticle aggregates for enhanced sensitivity in lateral flow assays. Lab Chip 13:4352–4357. doi:10.1039/C3LC50672J

Jain K (2003) Nanodiagnostics: application of nanotechnology (NT) in molecular diagnostics. Expert Rev Mol Diagn 3(2):153–161

James C (2013) Polypyrrole nanoribbon based chemiresistive immunosensors for viral plant pathogen detection. Anal Methods 5:3497–3502

Jaynes WF, Zartman RE, Hudnall WH (2007) Aflatoxin B1 adsorption by clays from water and corn meal. Appl Clay Sci 36(13):197–205

Jeong J, Ju H, Noh J (2014) A review of detection methods for the plant viruses. Res Plant Dis 20(3):173–181. doi:10.5423/RPD.2014.20.3.173

Kashyap PL, Kaur S, Sanghera GS, Kang SS, Pannu PPS (2011) Novel methods for quarantine detection of Karnal bunt (*Tilletia indica*) of wheat. Elixir Agric 31:1873–1876

Kashyap PL, Kumar S, Gurjar MS, Singh A et al (2013a) Phytopathogenomics in plant disease management: a paradigm shift. In: Prasad D, Ray DP (eds) Biotechnological approaches in crop protection. Biotech Book Publishers, New Delhi, pp 241–262

Kashyap PL, Kumar S, Singh R, Kumar A et al (2013) LAMP for detection of plant pathogens. AGROBIOS pp 76–77

Kashyap PL, Kumar S, Srivastava AK, Sharma AK (2013b) Myconanotechnology in agriculture: a perspective. World J Microbiol Biotechnol 29(2):191–207

Kashyap PL, Xiang X, Heiden P (2015) Chitosan nanoparticle based delivery systems for sustainable agriculture. Int J Biol Macromol 77:36–51

Kaushik A, Solanki PR, Ansari AA, Ahmad S et al (2009) A Nanostructured cerium oxide film-based immunosensor for mycotoxin detection. Nanotechnology 20: Article ID: 055105

Kaushik A, Arya SK, Vasudev A, Bhansali S (2013) Recent advances in detection of ochratoxin-A. Open J Appl Biosens 2(1):1–11

Khan MR, Rizvi TF (2014) Nanotechnology: scope and application in plant disease management. Plant Pathol J 13:214–231

Khiyami MA, Almoammar H, Awad YM, Alghuthaym MA et al (2014) Plant pathogen nanodiagnostic techniques: forthcoming changes? Biotechnol Biotechnol Equip 28(5):775–785. doi:10.1080/13102818.2014.960739

Knudsen BR, Jepsen ML, Ho Y-P (2013) Quantum dot-based nanosensors for diagnosis via enzyme activity measurement. Expert Rev Mol Diagn 13(4):367–375. doi:10.1586/erm.13.17

Kumar S, Kashyap PL (2013) Expanding horizons of precision farming driven crop protection. In: Ram T, Lohan SK, Singh R, Singh P (eds) Precision farming: a new approach. Daya Publications, New Delhi, pp 192–212

Kumar S, Tao C, Chien M, Hellner B, Balijepalli A, Robertson JWF et al (2012) PEG-labeled nucleotides and Nanopore detection for single molecule DNA sequencing by synthesis. Sci Rep 2:684. doi:10.1038/srep00684

Kumar S, Singh R, Kashyap PL, Srivastava AK (2013) Rapid detection and quantification of *Alternaria solani* in tomato. Sci Hortic 151:184–189. doi:10.1016/j.scienta.2012.12.026

Lattanzio VMT, Nivarlet N, Lippolis V, Gatta SD et al (2012) Multiplex dipstick immunoassay for semi-quantitative determination of *Fusarium* mycotoxins in cereals. Anal Chim Acta 718:99–108

Lin H-Y, Huang C-H, Lu S-H, Kuo I-T, Chau L-K (2014) Direct detection of orchid viruses using nanorod-based fiber optic particle plasmon resonance immunosensor. Biosens Bioelectron 51:371–378

López MM, Bertolini E, Olmos A, Caruso P et al (2003) Innovative tools for detection of plant pathogenic viruses and bacteria. Int Microbiol 6:233–243

Mak AC, Osterfeld SJ, Yu H, Wang SX, Davis RW et al (2010) Sensitive giant magnetoresistive based immunoassay for multiplex mycotoxin detection. Biosens Bioelectron 25(7):1635–1639

Malhotra BD, Srivastava S, Ali MA, Singh C et al (2014) Nanomaterial-based biosensors for food toxin detection. Appl Biochem Biotechnol 174:880–896

Mann SK, Kashyap PL, Sanghera GS, Singh G, Singh S (2008) RNA interference: an eco-friendly tool for plant disease management. Transgenic Plant J 2(2):110–126

Martinelli F, Scalenghe R, Davino S, Panno S et al (2014) Advanced methods of plant disease detection: a review. Agron Sustain Dev 35(1):1–25. doi:10.1007/s13593-014-0246-1

Mccandless L (2005) Nanotechnology offers new insights into plant pathology. College of Agriculture and Life Sciences News, Cornell University, Ithaca, pp 17–18

McCartney HA, Foster SJ, Fraaije BA, Ward E (2003) Molecular diagnostics for fungal plant pathogens. Pest Manag Sci 59:129–142

Meng Y, Li Y, Galvani CD, Hao G et al (2005) Upstream migration of *Xylella fastidiosa* via pilus-driven twitching motility. J Bacteriol 187(16):5560–5567

Nam JM, Stoeva SI, Mirkin CA (2004) Bio-bar-code-based DNA detection with PCR-like sensitivity. J Am Chem Soc 126(19):5932–5933

Nie L (2013) Biomedical nanotechnology for optical molecular imaging, diagnostics, and therapeutics. JSM Nanotechnol Nanomed 1:1–2

Nezhad AS (2014) Future of portable devices for plant pathogen diagnosis. Lab Chip 14:2887–2904

Pal S, Ying W, Alocilja EC, Downes FP (2008) Sensitivity and specificity performance of a direct-charge transfer biosensor for detecting Bacillus cereus in selected food matrices. Biosyst Eng 99(4):461–468

Pan Z, Yang X-B, Li X, Andrade D et al (2010) Prediction of plant diseases through modeling and monitoring airborne pathogen dispersal. CAB reviews: perspectives in agriculture, veterinary science, nutrition and natural resources, 5, PAVSNNRD-09-00177R1

Paniel N, Radoi A, Marty JL (2010) Development of an electrochemical biosensor for the detection of aflatoxin M1 in milk. Sensors 10(10):9439–9448

Panini NV, Bertolino FA, Salinas E, Messina GA, Raba J (2010) Zearalenone determination in corn silage samples using an immunosensor in a continuous-flow/stopped-flow systems. Biochem Eng J 51(12):713

Pimentel D (2009) Invasive plants: their role in species extinctions and economic losses to agriculture in the USA. In: Management of invasive weeds, invading nature – Springer Series in invasion ecology. Springer, Dordrecht, pp 1–7

Poonam P, Deo N (2008) Current correlation functions for chemical sensors based on DNA decorated carbon nanotube. N Sensors Actuators B Chem 135(1):327–335

Prieto-Simon B, Noguer T, Campas M (2007) Emerging biotools for assessment of mycotoxins in the past decade. Trends Anal Chem 26:689–702

Puzyr AP, Burov AE, Bondar VS, Trusov YN et al (2010) Neutralization of aflatoxin b1 by ozone treatment and adsorption by nanodiamonds. Nanotechnol Russian 5:137–141

Rad F, Mohsenifar A, Tabatabaei M, Safarnejad MR et al (2012) Detection of *Candidatus Phytoplasma aurantifolia* with a quantum dots FRET-based biosensor. J Plant Pathol 94(3):525–534

Radoi A, Targa M, Prieto-Simon B, Marty JL (2008) Enzyme linked-nanoparticles for aflatoxin M1 detection. Talanta 77(1):138–143

Rai M, Ingle A (2012) Role of nanotechnology in agriculture with special reference to management of insect pests. Appl Microbiol Biotechnol 94(2):287–293

Rispail N, Matteis LD, Santos R, Miguel AS et al (2014) Quantum dot and superparamagnetic nanoparticle interaction with pathogenic fungi: internalization and toxicity profile. ACS Appl Mater Interfaces 6(12):9100–9110

Rosi NL, Mirkin CA (2005) Nanostructures in biodiagnostics. Chem Rev 105:1547–1562

Safarpour H, Safarnejad MR, Tabatabaei M, Mohsenifar A et al (2012) Development of a quantum dots FRET-based biosensor for efficient detection of *Polymyxa betae*. Can J Plant Pathol 34(4):507–515

Sankaran S, MishraA ER, Davis C (2010) A review of advanced techniques for detecting plant diseases. Comput Electron Agric 72:1–13. doi:10.1016/j.compag.2010.02.007
Savaliya R, Shah D, Singh R, Kumar A, Shanker R, Dhawan A, Singh S (2015) Nanotechnology in disease diagnostic techniques. Curr Drug Metab 16(8):645–661. doi:10.2174/13892002166 66150625121546
Schwenkbier L, Pollok S, König S, Urban M et al (2015) Towards on-site testing of *Phytophthora* species. Anal Methods 7:211–217
Sertova NM (2015) Application of nanotechnology in detection of mycotoxins and in agricultural sector. J Cent Eur Agric 16:117–130
Servin A, Elmer W, Mukherjee A, Torre-Roche RD et al (2015) A review of the use of engineered nanomaterials to suppress plant disease and enhance crop yield. J Nanopart Res 17:92
Sharma A, Matharu Z, Sumana G, Solanki PR et al (2010) Antibody immobilized cysteamine functionalized-gold nanoparticles for aflatoxin detection. Thin Solid Films 519(3):1213–1218
Singh S, Singh M, Agrawal VV, Kumar A (2010) An attempt to develop surface plasmon resonance based immunosensor for Karnal bunt (*Tilletia indica*) diagnosis based on the experience of nano-gold based lateral flow immune-dipstick test. Thin Solid Films 519:1156–1159
Singh R, Kumar S, Kashyap PL, Srivastava AK, Mishra S et al (2014) Identification and characterization of microsatellite from Alternaria brassicicola to assess cross-species transferability and utility as a diagnostic marker. Mol Biotechnol 56:1049–1059
Srinivasan B, Tung S (2015) Development and applications of portable biosensors. J Lab Autom 20(4):365–389
Stanisavljevic M, Son K, Vaculovicova M, Kizeka R, Adama V (2015) Quantum dots-fluorescence resonance energy transfer-based nanosensors and their application. Biosens Bioelectron 74(15):562–574
Sundelin T, Collinge DB, Lübeck M (2009) A cultivation independent, PCR-based protocol for the direct identification of plant pathogens in infected plant material. Eur J Plant Pathol 123(4):473–476
Szeghalmi A, Kaminskyj S, Rösch P, Popp J et al (2007) Time fluctuations and imaging in the SERS spectra of fungal hypha grown on nanostructured substrates. J Phys Chem B 111:12916–12924
Thaxton CS, Georganopoulou DG, Mirkin CA (2006) Gold nanoparticle probes for the detection of nucleic acid targets. Clin Chim Acta 363(1–2):120–126
Thind TS (2012) Fungicides in crop health security. Indian Phytopathol 65(2):109–115
Tothill IE (2011) Biosensors and nanomaterials and their application for mycotoxin determination. World Mycotoxin J 4(4):361–374
Upadhyayula VKK (2012) Functionalized gold nanoparticle supported sensory mechanisms applied in detection of chemical and biological threat agents: a review. Anal Chim Acta 715:1–18
Wang Z, Wei F, Liu SY, Xu Q, Huang JY et al (2010) Electrocatalytic oxidation of phytohormone salicylic acid at copper nanoparticles-modified gold electrode and its detection in oilseed rape infected with fungal pathogen *Sclerotinia sclerotiorum*. Talanta 80:1277–1281
Yadav A, Kon K, Kratosova G, Duran N et al (2015) Fungi as an efficient mycosystem for the synthesis of metal nanoparticles: progress and key aspects of research. Biotechnol Lett 37:2099–2120. doi:10.1007/s10529-015-1901-6
Yalcin B, Otles S (2010) Nanobiosensor and food pathogen interaction mechanisms. Electron J Environ Agric Food Chem 9:1257–1273
Yang H, Li H, Jiang X (2008) Detection of food borne pathogens using bioconjugated nanomaterials. Microfluid Nanofluid 5(5):571–583
Yao KS, Li SJ, Tzeng KC, Cheng TC et al (2009) Fluorescence silica nanoprobe as a biomarker for rapid detection of plant pathogens. Multi-Funct Mater Struct II (1–2) 79–82:513–516
Zhang J, Chiodini R, Badr A, Zhang GF (2011) The impact of next-generation sequencing on genomics. J Genet Genomics 38(3):95–109. doi:10.1016/j.jgg.2011.02.003
Zhao M-X, Zeng E-Z (2015) Application of functional quantum dot nanoparticles as fluorescence probes in cell labeling and tumor diagnostic imaging. Nanoscale Res Lett 10:171. doi:10.1186/s11671-015-0873-8

Chapter 9
Nanoparticle Toxicity in Water, Soil, Microbes, Plant and Animals

Naureen S. Khan, Ashwini K. Dixit, and Rajendra Mehta

Abstract Nanotechnology is defined as the design, synthesis and application of materials, devices whose size and shape have been engineered at the nanoscale. The building blocks of nanotechnology are nanoparticles. Nanoparticles are broadly classified as natural nanoparticles and anthropogenic nanoparticles. This chapter presents first the definitions, principles and applications of nanomaterials. Then the toxicity of nanomaterials on soil, water, food, microbes, plants and animals is detailed.

Keywords Nanoparticles • Toxicity • Water • Soil • Microbes • Bacteria • Plants • Animals

9.1 Introduction

The building blocks of nanotechnology are nanoparticles (Biswas and Wu 2005). The term nanoparticles is derived from the word 'nanos' meaning tiny. Particles that are measured in nanoscale level, i.e. 0.1–100 nm in diameter, are termed as nanoparticles (Royal society and Royal Academy of Engineering 2004). They have been present on earth and used by mankind for millions of years (Nowack and Bucheli 2007). Nanotechnology is defined as the design, synthesis and application of materials, devices whose size and shape have been engineered at the nanoscale (Buzea et al. 2007). It has bridged the gap among various other existing technologies of present scenario. It has mingled chemistry, physics, biology, engineering and technology, bioinformatics and biotechnology in such a manner that has offered a novel

N.S. Khan (✉) • A.K. Dixit (✉)
Department of Botany, Guru Ghasidas Vishwavidyalaya,
495009 Bilaspur, Chhattisgarh, India
e-mail: nicks30khan@gmail.com; dixitak@live.com

R. Mehta
Department of Rural Technology, Guru Ghasidas Vishwavidyalaya,
495009 Bilaspur, Chhattisgarh, India
e-mail: drrmehta21@gmail.com

S. Ranjan et al. (eds.), *Nanoscience in Food and Agriculture 2*, Sustainable Agriculture Reviews 21, DOI 10.1007/978-3-319-39306-3_9

concept of interdisciplinary research. Nanoparticles play a key role and are known as the backbone of this technology due to their distinguishing and incomparable properties. The small size offers a large surface area as compared to their bulk counter parts (Biswas and Wu 2005). The large surface to volume ratio, surface charge, geometry and state of existence decide their relative reactivity.

Nanoparticles are broadly classified as natural nanoparticles and manmade (anthropogenic) nanoparticles. Anthropogenic nanoparticles are further sub divided into unintentional and intentional/engineered nanoparticles. A variety of nanoparticles are introduced in the environment by various natural processes. Volcanic eruptions, forest fires, weathering phenomenon and soil erosion driven by water and wind are the sources of natural nanoparticles (Smita et al. 2012). Whereas, biological origin i.e. viral particles (Hogan et al. 2004), Pollen fragments (McMurry et al. 2000). Coal fired combustion systems and incinerators (Chang et al. 2004); automobiles and diesel powered vehicles (Kittelson 1998); welding processes industrial boilers, fire places, automobiles, diesel (Vincent and Clement 2000); trucks, and meat-cooking operations (Hildemann et al. 1991) are the sources of unintentional anthropogenic nanoparticles. The intentional/engineered nanoparticle i.e. carbon containing nanoparticles (fullerence, carbon nanotubes etc.) inorganic nanoparticles (metal nanoparticles like nano silver, nano gold, titanium dioxide nanoparticles, zinc oxide nanoparticles etc.), organic and inorganic nanohybrids (Sajid et al. 2015) are synthesized by following two approaches i.e. either top down or bottom up method.

The engineered nanoparticles have been exploited in every sector of society. They have become a compulsive content of today's smart products. Their rapidly increasing application in textiles, electronics, pharmaceutics, cosmetics and environmental remediation is noticeable and calls for their impact assessment (Royal society and Royal Academy of Engineering 2004). Table 9.1 corroborates the application of different nanoforms.

In the present review we have taken an attempt to commemorate the toxicity imposed by the nanoparticles on every sphere of life forms. Brief information about toxic effect of nanoparticles on the environment, microbes, animals and plants has been discussed in this article. The possible preventive measures and future challenges are also mentioned.

9.2 Nanotechnology, an Old Concept

The concept of nanotechnology was proposed early in 1959 by a physicist Richard Feynman in a lecture-"There's Plenty of Room at the Bottom" at a meeting of the American Physical Society and was honored by Nobel prize for this valuable contribution. Whereas, the term nanotechnology was coined by Norio Taniguchi in 1974. Unknowingly, the people were using the nanoparticles in their work place. Evidences of nanotechnology's existence in the middle ages are drawn from the stained glasses of various old churches. The glass windows of European cathedrals

Table 9.1 Application of nanoparticles

Nanoparticle	Sector	Use	Product	Reference
Silver (Ag)	Textile	As casuals	T-shirt, socks, sports clothing, shoe	Benn and Westerhoff (2008)
	Food packaging:	As antimicrobial agent	Blue moon goods fresh box silver	Cushen et al. (2012)
			Nanoparticle food storage container	Alfadul and Elneshwy (2010)
		As antimicrobial agent	Nano Care Technology Ltd. Antibacterial	Bouwmeester et al. (2007)
		As antimicrobial agent	Sunriver industrial nanosilver fresh food bag	Huang et al. (2011)
	Implants and medical devices	Dental hygeine	Toothpaste	Reidy et al. (2013)
		Treatment of eye condition		Reidy et al. (2013)
		Application in plastic		Reidy et al. (2013)
		To produce surgical meshes		Reidy et al. (2013)
		Vascular prosthesis		Reidy et al. (2013)
		Ventricular drainage catheters		Reidy et al. (2013)
		In surface coating of respirators		Cushen et al. (2012)
		Due to plasmonic property used in	Biosensor	Austin et al. (2011)
		In reducing secondary bacteremia	Acticoat	Reidy et al. (2013)
	Consumer products	In water treatment process/		Reidy et al. (2013)
		Silver impregnated (<100 nm)	Water filter	Nowack et al. (2011)
		Surface coatings, paints and washing machine		Reidy et al. (2013)
				Reidy et al. (2013)

(continued)

Table 9.1 (continued)

Nanoparticle	Sector	Use	Product	Reference
			Air filters	Reidy et al. (2013)
			Vaccum cleaners	Klaine et al. (2008)
			Phones	Klaine et al. (2008)
			Laptops	Klaine et al. (2008)
			Toys	Klaine et al. (2008)
			Baby products	Cerkez et al. (2012)
	Other products	In disinfection	Disinfecting spray	Wijnhoven et al. (2009)
		Imparting fragrance	Deodrants	Lorenz et al. (2011)
			Other cosmetics	Lorenz et al. (2011)
		As antifungal agent	Fungicide	Wright et al. (1999)
TiO_2	Domestic purpose	In cleaning items	Detergent	Kaida et al. (2004)
			Toothpaste	Kaida et al. (2004)
			Window panes	Remédios et al. (2012)
			Ceramic tiles	Remédios et al. (2012)
		Medical implants	Biosensor	Remédios et al. (2012)
	Commercial purpose	Displays	Television screen	Remédios et al. (2012)
			Computer moniters	Remédios et al. (2012)
		As photocatalyst:	Solar cells	Klaine et al. (2008)
			Batteries	Remédios et al. (2012)
	Food products			Remédios et al. (2012)
	Environmental remediation	As decontaminant		Esterkin et al. (2005)

(continued)

Table 9.1 (continued)

Nanoparticle	Sector	Use	Product	Reference
ZnO	Cosmetics	As UV light scattering additive	Sunscreen	Serpone, et al. (2007)
			Toothpaste	Serpone et al. (2007)
			Beauty products	Serpone et al. (2007)
		In electronic items		Song et al. (2010)
		In textiles		Dastjerdi and Montazer (2010)
		As essential ingredient in:	Antifouling paints	IPPIC (2012)
CuO	Electronics and technology		Lithium batteries	Sau et al. (2010)
Gold nanoparticles (Au)	Electonics	Used in conducting Inks/ Films		Klaine et al. (2008)
Iron oxide			Lipstick	Remédios et al. (2012)
		As detoxifying agent		Zhang (2003)
Alumina			Shampoo	Klaine et al. (2008)
CeO_2 cerium dioxide		As combustion catalyst	Diesel fuel	Park et al. (2008)
	Metallurgical and glass/ ceramic application		Solar cells	Serpone et al. (2007)
			Gas sensors	
			Oxygen pumps	Lin et al. (2006)
Carbon nanotubes	Commercial application	Polymer composites	Lubricants, machinable ceramics,	Remédios et al. (2012)
		Structural composites	Magnetic materials, military battle suits	Remédios et al. (2012)
		Conductive coatings		Remédios et al. (2012)
		Electromagnetic shielding		Remédios et al. (2012)
	Medical application	Orthopedic implants		Klaine et al. (2008)
		Automotive industries		Klaine et al. (2008)
	Others	In water purification system		Klaine et al. (2008)
		Adhesives		Klaine et al. (2008)

(continued)

Table 9.1 (continued)

Nanoparticle	Sector	Use	Product	Reference
Quantum dots	Electronics and technology		(Insulators Semiconductors Magnetic materials)	
			Solar cells	Royal society (2004)
			Photovoltaic cells	Royal society (2004)
			Security inks	Royal society (2004)
			Photonics	Royal society (2004)
			Telecommunication	Alivisatos et al. (2005)
	Medical application	In biological imaging	As markers	Hoet et al. (2004)
			As targeted therapeutics	Alivisatos et al. (2005)
Dendrimers	Medical devices	In drug delivery		Royal society (2004)
		As surface modifiers for enantioselective catalysis	Macrocapsules	Klaine et al. (2008)
			Nanolatex	Klaine et al. (2008)
			Coloured hlasses	Klaine et al. (2008)
			Chemical sensors	Klaine et al. (2008)
			Modified electrodes	Klaine et al. (2008)
			DNA transfecting agents	Klaine et al. (2008)
		As therapeutic agent	Hydrogels	Klaine et al. (2008)
			DNA chips	Klaine et al. (2008)
Nano composites	Vehicular application	Packaging auto- parts		Royal society (2004)
Carbon black			Plastic fillers	Hoet et al. (2004)
Fumed silica	Vehicular application	Used in car tyres and plastic fillers		Hoet et al. (2004)
Silica	Gene delivery			Ravi Kumar et al. (2004)

are live examples of this. The historical aspect of nanotechnology from the most primitive time to the middle age was commemorated by (Daniel and Astruc 2004). Due to very astonishing optical properties of nanoparticles they have been used since the fourth century, the prominent example of which is the dichroic glass of Lycurgus cup whereas, the use of nanomaterials in industrial sectors begins in twentieth century (Horikoshi and Serpone 2013). Use of silver nanoparticles as an antiseptic agent has been documented early in Rig veda. Various metals in nano form occupies a remarkable place in ayurvedic, sidhha and unani system of medicine due to their terrific potential in curing a wide range of diseases and abnormalities.

9.3 Distinguishing Properties of Nanoparticles

Before highlighting the hazardousness of nanoparticles the expected factors that are responsible for nanoparticle toxicity are size, shape, nature, reactivity, mobility, stability and surface chemical charge (Sajid et al. 2015). Property of the material changes in nanoparticle regime. Activity of the nanoparticles are strongly governed by their stability which is attained by the provision of ideal capping agents. Temperature, pressure and long term incubation governs the texture and consistency of the nano regime material due to the onset of aggregation or agglomeration among the nanopopulation.

9.4 Effect of Nanoparticles on the Environment

Environment harbors all the things that exist within itself. Any alteration in the environment gradually affects its state, constituents and vice-Versa. Environment consists of two factors broadly classified as biotic and abiotic factor. Nanoparticles that are expelled in the environment either naturally or by anthropogenic means, Spontaneously through natural phenomenon or by engineered sources imparts/leads to some changes in them. The inhalation of atmospheric nanoparticle has caused 60,000 deaths per year reported by US Environmental Protection Agency (EPA) and also these Nanoparticles directly get transferred in to the brain Oberdorster et al. (2004) and Raloff (2003). This is due to the large surface area which has resulted in direct generation of reactive oxygen species that are known as harmful oxyradicals, having the potential to attack DNA, proteins and membranes too as a result of which the cell gets injured Brown et al. (2001). In addition they show affinity towards the transition metals and organic chemical pollutants that enhances the level of toxicity to a greater extent Cheng et al. (2004) Route of entry of nanoparticles varies with the type of habitat. Figure 9.1 reflects the possible route of nanoparticle exposure to various life form in brief. In terrestrial organisms they may enter via inhalation or ingestion Brigger et al. (2002) whereas, in case of the aquatic animals they enter via direct passage across gill and other external surface epithelia as reported by Moore

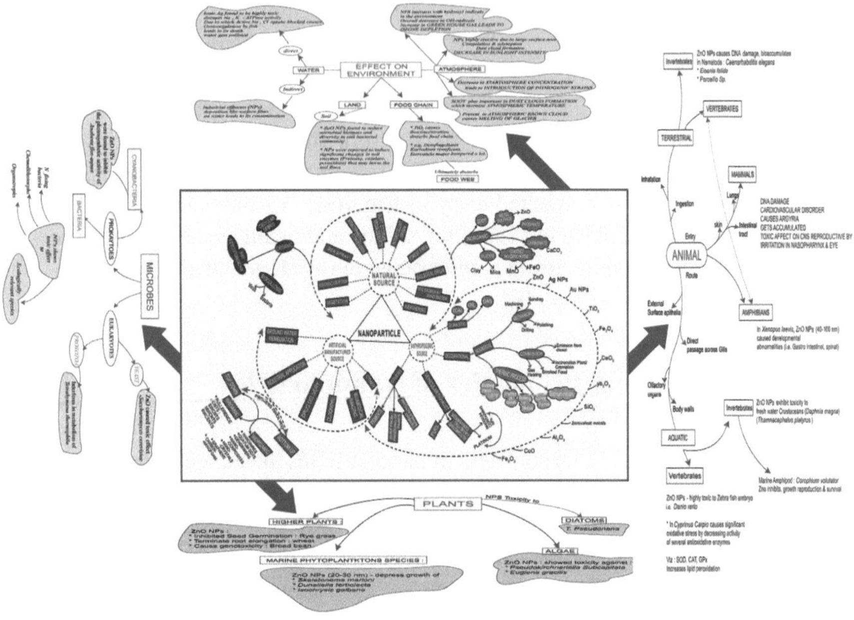

Fig. 9.1 Showing the sources of nanoparticles and their impact on various components of ecosystem

(2006). At cellular level engineered nanoparticles (single wall nanotubes) are responsible for swelling of the endoplasmic reticulum, it causes vacuolar changes. Additionally, high concentration of multiwalled nanotubes were found to be responsible for the degeneration, enlargement and rarefaction of macrophage cell's nucleus Jia et al. (2005).

In a nutshell, this can be clearly stated that nanoparticles (1–100 nm) can be life threatening. Source may vary but once entered in the environment they may harm every level of ecosystem.

9.4.1 Effect of Nanoparticles on Atmosphere

Accumulation of nanoparticle in the environment leads to generation of various environmentally toxic processes like dust cloud formation, Soot formation (Regional haze). These processes occur as a result of burning of organic waste materials found in nature that has been responsible for melting of Himalayan Glaciers Gustafsson et al. (2009). These are actually the naturally occurring nanoparticles that interfere with the natural resources. Incomplete combustion of fossil fuels increases risk of exposure to pathogenic species as suggested by Hua et al. (2007). According to Manning et al. (2005) others ill effects include abnormalities in level of environmental hydroxyl radical concentration. Wilson et al. (22007) reported the increase in ozone depletion mechanism by these tiny creatures that may be attributed due to change in tempaerature of stratospheric layer as reported by Tromp et al. (2003), Smita et al. (2012). According to Strawa et al. 2010 nanoparticles take part in the coarse material formation which is the major cause of global warming in present scenario.

9.4.2 *Effect on Soil*

The ultimate sink of every component in environment is soil and so as for nanoparticles. These tiny tools end up their journey in the soil or become dormant for some time. Soil is a natural reservoir for a wide variety of micro flora as it offer conducive environment for growth and survival of biological indicators. In turn they offer nutrients in the form of enzyme that take part in decomposition of soil content and improve the quality of soil by executing this cyclic process. In a broad sense their exists mutualism. The long term dormancy period of nanoparticles may evoke serious issues on the health of its inhabitant i.e. contribute in reducing the fertility of soil thereby causing an imbalance in the soil ecosystem (Kennedy and Smith 1995). Common ENPs include the metal ENPs (elemental Ag, Au, Fe etc.), fullerenes (grouping Buckminster fullerenes, CNTs, nanocones etc.), metal oxides (TiO_2, CuO, FeO_2, ZnO, Al_2O_3, SiO_2 etc.), complex compounds (Co–Zn–Fe oxide), quantum dots often coated with a polymer e.g. cadmium–selenide (CdSe) and organic polymers (dendrimers, polystyrene, etc.) Dinesh et al. 2012. Fortner et al. (2005) performed toxicity study of C_{60} aggregates on soil inhabiting *E. Coli* and *B. subtilis* bacteria. They found it to be toxic even at low concentration marked by reduced growth (>0.4 ppm). more over, this concentration has lowered the rate of aerobic respiration up to 4 ppm. Thus may interfere with soil development (pedogenesis) or behavior (e.g., crusting and dispersion) Klaine et al. (2008), Cameron et al. (1915).

9.4.3 *Effect of Nanoparticle on Water*

Domestic and Industrial wastes (may contain Nanoparticles in aggregate or colloidal form) expelled in the water bodies viz. ponds, rivers, sea etc. increases the threat of nanoparticle contamination. Discharges usually contains organic (protenaceous and polysaccharide from the microbes) and inorganic matter (heavy metal oxides) that make the water bodies unfit for use. It has been suggested that Ionic silver has found to be highly toxic that disrupts Na+K+− ATPase enzyme activity due to which active Na+ and Cl− uptake is blocked that causes Osmoregulation by fishes, leads to death of fishes thereby contaminate the water as studied by Tang and Wang (2004).

Nanoparticles may also accumulate on the surface microlayers of the oceans, presenting a route of aerosol exposure risk to marine birds and mammals, as well as the organisms living on the surface microlayer due to their hydrophobic nature. Kennedy et al. (2004), Simkiss K. (1990).

9.4.4 *Effect of Nanoparticles on Food Chain*

Starting from the primary producer (Plants) to that of the tertiary consumers (insects/animals). At every successive level nano-particles reflects their involvement by being entered in their metabolic pathway through active and passive transport. Confirmatory reports have been published by Holbrook et al. (2008), Lin et al. (2009) that shows their harmful effect in the food chain. Their small size offers large surface area that imparts them high potential to carry toxic materials i.e. such as lipophilic pollutants and heavy metals Baun et al. (2008) to enter the system.

9.4.5 *Effect of Nanoparticles on Agriculture Production*

About 70 % of India's economy is based on agricultural practices. The formation of dust clouds has found to be highly hazardous. Burning of wood for cooking and other purposes and the incomplete combustion of biomass give rise to natural nanoparticle. Their existence in environment has resulted in brown cloud formation which has remarkably lowered the sunlight intensity in south Asia. These pollution clouds have been indulged in the reduction of monsoon season in India. The productivity of chief crop plants like rice, wheat and soybean has immensely hampered as reported by UNEP 2002. Asian brown clouds are found to be loaded with large amount of soot and black carbon. These complexly degradable materials contributes to increase in melting of glaciers thereby blocking the availability of water to various glacier fed rivers as commemorated by Gustafsson et al. (2009). Water scarcity contributes in low yield of food crops.

9.5 Effect of Nanoparticles on Microbes

Nanoparticles not only affected abiotic factors rather have also been observed to pose harmful effects on microscopic creatures too ranging from the prokaryotes to Eukaryotes. For example – Akhavan and Ghaderi (2010) investigated the toxicity of graphene and graphene oxide damages the cell membrane of both the Gram-negative bacteria *Escherichia coli* and the Gram-positive bacteria *Staphylococcus aureus* by their sharp edges. A long descriptive about nanoparticles impact on micro flora has been expressed in Table 9.2.

Table 9.2 Harmful effect of nanoparticles on microorganisms

Test organism	Nanoparticle	Toxic effects	Reference
Bacterial *Sp.*	C60 (water suspension)	Antibacterial to broad range	Lyon et al. (2005, 2006), and Sayes et al. (2004)
Bacterial *Sp.*	C60 (encapsulated in polyvinylpyrrolidone)	Antibacterial to broad range	Kai et al. (2003)
Gm +veBacterial *Sp.*	Carboxyfullerence (malonic acid derivative)	Bactericidal effect	Mashino et al. (1999) and Tsao et al. (2002).
Gm +veBacterial *Sp.*	Hydroxylated fullerence	Bactericidal effect	Rozhkov et al. (2003)
Gm −veBacterial *Sp.* ***(E. Coli Sp.)***	C60 derivative with pyrrolidine groups	Inhibit its growth by interfering with energy metabolism	Mashino et al. (1999, 2003)
Gm +veBacterial *Sp.* ***(Mycobacteria Sp.)***	Other derivative of C60	Inhibit its growth	Babynin et al. (2002) and Bosi et al. (2000)
Gm −veBacterial *Sp.* ***(Salmonella typhimurium)***	Other derivative of C60: (carbon nanotube)	Antibacterial and antimutagenic effect	Babynin et al. (2002) and Bosi et al. (2000)
Gm −veBacterial *Sp.* ***(E. Coli Sp.)***	(SWCNT)	Antibacterial(cell membrane damage)	Kang et al. (2007) and Wei et al. (2007)
Bacterial *Sp.*	(MWCNT)	Cytotoxic in nature	Biswas and Wu (2005)
Gm −veBacterial *Sp.* ***(E. Coli Sp.)***	Metallic: quantum dots (uncoated)	Toxic causes oxidative damage to cell membrane	Kloepfer et al. (2005) and Hardman (2006)
Gm +veBacterial *Sp.* ***(Bacillus Subtilis)***	Metallic: quantum dots (uncoated)	Toxic causes oxidative damage to cell membrane	Kloepfer et al. (2005) and Hardman (2006)
Gm −veBacterial *Sp.* ***(E. Coli Sp.)***	Silver	Highly toxic (in environmental condition)	Chernousova and Epple (2013)
Bacterial *Sp.*	Silver	Bactericidal effect	Sondi and Salopek-Sondi (2004) and Morones et al. (2005)
Virus	Silver	Viricidal effect	Sondi and Salopek-Sondi (2004) and Morones et al. (2005)
Gm +veBacterial *Sp.* ***(Staphylococcus aureus)***	Gold	Low toxic	Nyberg et al. (2008) and Goodman et al. (2004)
Gm −veBacterial *Sp.* ***(E. Coli Sp.)***	Gold	Low toxic	Nyberg et al. (2008) and Goodman et al. (2004)
Gm −veBacterial *Sp.* ***(Shewanella oneidensis)***	Metal oxides: magnetite	Low toxic	De Windt et al. (2006)

(continued)

Table 9.2 (continued)

Test organism	Nanoparticle	Toxic effects	Reference
Gm –veBacterial *Sp.* ***(E. Coli Sp.)***	TiO_2	Accelerates solar disinfection by photocatalytic activity and (ROS)	Rincon and Pulgarin (2004) and Wolfrum et al. (2002)
Gm +veBacterial *Sp.* ***(Micrococcus luteus)***	TiO_2	Accelerates solar disinfection by photocatalytic activity and (ROS)	Rincon and Pulgarin (2004) and Wolfrum et al. (2002)
Gm +veBacterial *Sp.* ***(Bacillus Subtilis)***	TiO_2	Accelerates solar disinfection by photocatalytic activity and (ROS)	Rincon and Pulgarin (2004) and.Wolfrum et al. (2002)
Fungal Sp. ***(Aspergillus niger)***	TiO_2	Accelerates solar disinfection by photocatalytic activity and (ROS)	Rincon and Pulgarin (2004) and Wolfrum et al. (2002)
Gm +veBacterial *Sp.* ***(Bacillus Subtilis)***	MgO	Antibacterial effect	Huang et al. (2005)
Gm +veBacterial *Sp.* ***(Staphylococcus aureus)***	MgO	Antibacterial effect	Huang et al. (2005)
Gm –veBacterial *Sp.* ***(E. Coli Sp.)***	CeO_2	Antimicrobial effect	Thill et al. (2006)
Gm –veBacterial *Sp.* ***(E. Coli Sp.)***	Zno (uncoated)	Highly toxic	Bondarenko et al. (2012)
Gm +veBacterial *Sp.* ***(Bacillus Subtilis)***	Zno	Antibacterial activity	Sawai et al. (1995,1996)
Gm –veBacterial *Sp.* ***(E. Coli Sp.)***	Zno	Antibacterial activity	Sawai et al. (1995,1996)
Gm –veBacterial *Sp.* ***(E. Coli Sp.)***	ZnO (20 nm)	100 % mortality at 20 mg/ml of Zno	Jiang et al. (2009)
Gm –veBacterial *Sp.* ***(E. Coli Sp.)***	ZnO (10–30 nm)	Prevented growth at 500 mg/ml of ZnO	Premanathan et al. (2011)
Bacterial *Sp.*	SiO_2	Mild toxicity due to (ROS) production	Adams et al. (2006)
Gm –veBacterial *Sp.* ***(E. Coli Sp.)***	CuO (uncoated)	Highly toxic	Bondarenko et al. (2012)
Gm –veBacterial *Sp.*:-			
P. aeroginosa	CuO (80–160 nm)	Greatly affects the growth of these PGPR	Mahapatra et al. (2008)
Klebsiella puemoniae	CuO (80–160 nm)	Greatly affects the growth of these PGPR	Mahapatra et al. (2008)

(continued)

Table 9.2 (continued)

Test organism	Nanoparticle	Toxic effects	Reference
Salmonella Paratyphi	CuO (80–160 nm)	Greatly affects the growth of these PGPR	Mahapatra et al. (2008)
Shigella Strain	CuO (80–160 nm)	Greatly affects the growth of these PGPR	Mahapatra et al. (2008)
P. aeroginosa	Iron oxide nanoparticles	Reacts with peroxides present in environment,	Saliba et al. (2006) and Mishra and Kumar (2009)
		Generates free radical, which are highly toxic	
P. putida	Iron oxide nanoparticles	Reacts with peroxides present in environment,	Saliba et al. (2006) and Mishra and Kumar (2009)
		Generates free radical, which are highly toxic	
P. fluorescens	Iron oxide nanoparticles	Reacts with peroxides present in environment,	Saliba et al. (2006) and Mishra and Kumar (2009)
		Generates free radical, which are highly toxic	

9.5.1 *Effect of Nanoparticles on Nitrogen Fixing Bacteria (Cyanobacteria & Others)*

Cyanobacteria are integral part of producer community as they are the largest nitrogen fixer globallydue to the presence of nif gene and other ideal characteristics they are capable of fixing the atmospheric nitrogen themselves and serves as Biofertilizers thereby play a significant role in the growth of greenery. Due to the presence of oxygenic photosynthetic system they are capable of maintaining the equilibrium of earth atmosphere. These certain properties drag them in the list of essentials that vitally contribute for balancing the environment and nature. Brayner et al. (2010) performed the eco-toxicological investigation of ZnO Nanoparticles by taking Anabaena flos-aquae cyanobacteria and *Euglena gracilis* euglenoid microalgae as the experimental organism. ZnO Nanoparticles were synthesized by polyol process in which s trin-octylphosphine oxide (TOPO) and polyoxyethylene stearyl ether (Brij-76) were used as protective agents to control the particle size and shape. In the present investigation effect of ZnO nanoparticles on photosynthetic activity of the selected organisms was corroborated. In the case of Anabaena flos-aquae, addition of ZnO, ZnO-TOPO, and ZnO-Brij-76 induces stress which causes a progressive decline in the photosynthetic rate. Among these three different types of nanoparticles only ZnO-TOPO causes a fall in photosynthetic rate followed by cell death. However, In case of *Euglena gracilis*, all the nanoparticles were found to be indulged in cell death which was confirmed by doing the TEM analysis of their ultra thin sections. It reveals that polysaccharides produced by Anabaena flos-aquae avoid particle internalization after contact with ZnO and ZnO-Brij- 76 nanoparticles. Ma et al. (2013) suggested that nanoparticles have been found to interfere in the metabolic pathway of nitrogen fixing bacteria as well.

9.5.2 Effect of Nanoparticles on Plant Growth Promoting Rhizobacteria

Plant growth promoting rhizobacteria (PGPR) like *P. aeruginosa, P. putida, P. fluorescens, B. subtilis* and soil Nitrogen cycle bacteria viz., nitrifying bacteria and denitrifying bacteria have shown varying degrees of inhibition when exposed to ENPs in pure culture conditions or aqueous suspensions Mishra and Kumar (2009). Metal oxide nanoparticles of Cu (80–160 nm) were tested for antibacterial activity against plant growth promoting *Klebsiella pneumoniae, P. aeruginosa, Salmonella paratyphi* and *Shigella strains* Mahapatra et al. (2008). Iron and copper based nanoparticles are presumed to react with peroxides present in the environment generating free radicals known to be highly toxic to microorganisms like *P. aeruginosa* Saliba et al. (2006).

9.5.3 Effect of Nanoparticles on Eukaryotes

Likewise, the prokaryotes nanoparticles adversely affects the growth and metabolism of eukaryotes. Kasemets et al. (2009) has shown the toxic effect of ZnO nanoparticles on the unicellular Fungi i. e. yeast (*Saccharomyces cerevisiae*). Mortimer et al. (2010) suggested the toxicity of ZnO nanoparticles against free living Protozoan Sp. *(Tetrahymena thermophila).*

9.6 Effect of Nanoparticles on Plants

9.6.1 Effect of Nanoparticles on Algae & Diatom (Lower Plants)

Due to the adsorption of TiO_2 nanoparticles on the surface of algae, its weight becomes double as a result of this its floating capacity decreases and low surface gets exposed to sunlight leads to a gradual fall in photosynthetic rate as suggested by Navarro et al. (2008). In case of a green algae i.e. *Desmodesmus subspicatus*, TiO_2 nanoparticles were found to be toxic by Hund-Rinke and Simon (2006). ZnO nanoparticles showed toxicity against *Euglena gracilis* Euglinoid Brayner et al. (2010). Marine diatom *T. pseudonana* was found to be susceptible to ZnO nanoparticles and exhibit toxicity as reported by Miao et al. (2010).

9.6.2 Effect of Nanoparticles on Marine Phytoplanktons

Population growth studies on four marine phytoplanktons has been conducted by Miller et al. (2010) to assess the toxicity estimation of two metal oxide nanoparticle i.e. ZnO and TiO_2. Salty water bodies comprises a vast portion of earth and a repeated history of waste disposition has been in a common practice since ancient time, whether by accidental or intentional means. Marine phytoplanktons are adversely affected by engineered nanoparticles. In order to assess the sublethal effect of two selected metal NPs a modeling approach (Dynamic Energy Budget (DEB) framework) was employed in which They have found that ZnO nanoparticles depressed the growth rate of *Skeietonema marioni, Thalassiosira pseudonana, Dunaliella tertiolecta* and *Isochrysis galbana* by 50–70 %. Reproduction and early stage developmental study has been conducted by Nielsen et al. (2008). They have selected a marine brown algae *Fucus serratus* to test against manufactured Carbon black nanoparticle. Inference clearly states a great changes. Carbon black nanoparticles when subjected to the sperm suspension and embryo culture gets agglomerate thereby causing a decline in the fertilization success at a concentration of 100 mgl^{-1}.

9.7 Effect of Nanoparticles on Higher Plants

Nanoparticles are found to be indulged in the formation of reactive oxygen species in some higher plants explained by Navarro et al. (2008). Accumulation of nanoparticle takes place on the leaf lamina and on stomatal tissues causes an imbalance in the equilibrium inside outside the leaf and ultimately in plants Da Silva et al. (2006). Accumulation of nanoparticle on photosynthetic surfaces causes shading effects, i.e. reduced sun light availability and hence reduced photosynthetic rate Smita et al. (2012). Carbon nanotubes diminished rice yields and made wheat more prone to other pollutants Wild and Jones (2009) due to the asbestos-like behaviour of carbon nanotubes. Also, aluminium nanoparticle have been identified to inhibit root growth in various economically important plant species by interacting with root surface Canas et al. (2008), Yang and Watts (2005).

9.7.1 Phytotoxicity by Nanoparticles

Y. K. Mushtaq (2011) demonstrated The effects of Fe_3O_4, TiO_2, and carbon nanoparticles on cucumber plants. These nanoparticles exhibits the potential to negatively affect seed germination rate, root elongation, and germination index. Another study performed by Kumari et al. (2011) on the harsh effect of ZnO nanoparticles supports their toxic nature. Growing use of ZnO nanoparticles in the daily consumer products calls for an alert after their significant investigation. Workers have done a depth study by determining the Mitotic index (MI), micronuclei index (MN index), chromosomal aberration index, and lipid peroxidation of the root cells of *A. cepa* (grown hydroponically). They have treated the model organism with four different concentrations of ZnO nanoparticles dispersions (25, 50, 75, and 100 gml^{-1}). It was observed that MI decreases with the increase in concentration of ZnO nanoparticles to the contrary MN index and chromosomal index increases steadily. in the ZnO nanoparticles treated cells, the total count of micronucleated cell were higher than control set up which corroborates the genotoxic effect of ZnO nanoparticles on plant community.

9.7.2 Genotoxicity by Nanoparticles

Atha and collaborators (2012) reported for the first time that copper oxide Nanoparticles damaged DNA in some agricultural and grassland plants *(Raphanus sativus, Lolium perenne*, and *Lolium rigidum*). Mutation occurs in the crop plants that blocks their growth and results in gradual fall in productivity. Through this fraction of investigation they have clearified that metal nanoparticles acts as a carrier/mediator for DNA damage in all the living community including mammalian cell, plants and even the microscopic beings i.e. bacteria (Table 9.3).

Table 9.3 Harmful effect of nanoparticles on plants

Nanoparticle	Plant name	Harmful effect	Reference
	ALGAE:		
ZnO	***Pseudokirchneriella subcapitata***	Toxic interfere in metabolic pathway	Aruoja et al. (2009)
ZnO	***Anabaena flos-aquae***	Inhibit photosynthetic activity	Brayner et al. (2010)
Ag Nanoparticles	***unknown***	Highly toxic	Albright and Wilson (1974)
ZnO	***Euglena gracilis***	Cause cell death	Brayner et al. (2010)
ZnO (20 nm)	Marine diatom: ***T. pseudonana***	Found to be toxic	Miao et al. (2010)

(continued)

Table 9.3 (continued)

Nanoparticle	Plant name	Harmful effect	Reference
TiO_2	Unknown	Adsorbs on algal cell surface, affecting its ability to float.	Navarro et al. (2008)
		Reduced sunlight availability for photosynthesis	Navarro et al. (2008)
TiO_2	***Desmodesmus subpicatus***	Found to be toxic	Hund-Rinke and Simon (2006)
ZnO	***Allium cepa***	Exert cytotoxic & genotoxic effect	Kumari et al. (2011)
		Lipid peroxidation	Kumari et al. (2011)
		Decrease of the mitotic index	Kumari et al. (2011)
		Increase of the micronuclei	Kumari et al. (2011)
		Increase of chromosomal abberation index	Kumari et al. (2011)
ZnO	unknown	Penetrate plant system and may interfere with	Anita K. Patlolla (2013)
		intracellular components causing damage to cell division	Anita K. Patlolla (2013)
ZnO	***Triticum aestivum***	Makes it more vulnerable to other pollutants	Wild and Jones (2009)
		Reduced its biomass	Du et al. (2011)
		Due to dissolution of nanoparticles to Zn ions, facilitation of toxic Zn uptake occurs	Du et al. (2011)
ZnO 2000 mg/l	***Rye grass***	Inhibit germination of seed	Lin and Xing (2007)
		Terminate root elongation	Lin and Xing (2007)
ZnO	***Lepidium sativum*** (Garden cress)	Affects root elongation	Manzo et al. (2011)
ZnO	***Vicia faba*** (Broad bean)	Cause genotoxicity	Manzo et al. (2011)
ZnO (20–30 nm) at 1 mg/l	Marine phytoplankton species:		
	Skeletonema marioni	Depresses growth rate by 50–75 %	Miller et al. (2010)
	Thalassiosira pseudonana	Depresses growth rate by 50–75 %	Miller et al. (2010)
	Dunaliella tertiolecta	Depresses growth rate by 50–75 %	Miller et al. (2010)
	Isochrysis galbana	Depresses growth rate by 50–75 %	Miller et al. (2010)

(continued)

Table 9.3 (continued)

Nanoparticle	Plant name	Harmful effect	Reference
Carbon nanotubes	Economically important plants	Inhibit root elongation	Canas et al. (2008) and Yang and Watts (2005)
Aluminium nanoparticles	Economically important plants	Inhibit root elongation	Canas et al. (2008) and Yang and Watts (2005)
CNanoparticles	***Oryzae Sativa* (Rice)**	Diminishes yield	Wild and Jones (2009)
CuO	***Rhaphanus sativus***	Diminishes yield	Wild and Jones (2009)
CuO	***Lolium perenne***	Diminishes yield	Wild and Jones (2009)
CuO	***Lolium rigidum***	DNA damage, thus inhibits plant growth	Atha et al. (2012)
CuO (1000 mg/l)	***Cucurbita pepo***	Reduced emerging root length	Stampoulis et al. (2009)
Ag nanoparticles	***Cucurbita pepo***	Decrease plant biomass and transpiration	Remédios et al. (2012)
Yb_2o_3	Higher Plant Sp.	Affects root elongation	Ma et al. (2010)
	Cucumis sativus	Yb_2O_3 deposits were found in cytoplasm of root cells which are toxic in nature	Zhang et al. (2012)
CeO_2	Higher Plant Sp.	Affects root elongation	Ma et al. (2010)
La_2O_3	Higher Plant Sp.	Affects root elongation	Ma et al. (2010)
Gd_2O_3	Higher Plant Sp.	Affects root elongation	Ma et al. (2010)

9.8 Effect of Nanoparticles on Aquatic Organisms

To assess the merits and demerits of engineered nanoparticles several comparative studies has been done since time immemorial. These documentation provide a thorough and in depth study of the other face of a coin say the negative impacts of nanomaterials. Nanoparticles are being exploited and used in diversified applications. And being a multifunctional paradigm researchers have warned to use this versatile tool in an ecofriendly manner. Deleterious effects of nanoparticles are being experienced by atmosphere and land than how could the water bodies that covers 71 % of the earth surface will be ignorant and its inhabitant as well carbon nanotubes and fullerenes found to be potentially harmful especially to benthic organisms Velzeboer et al. (2011). Several researchers have excavated the harmful effects of nanomaterials by following different types of approaches. In recent years Blinova et al. (2010), have performed a comparative analysis of acute toxicity test of CuO and ZnO nanoparticles between crustaceans and protozoan population found in an artificially prepared fresh water ecosystem and a natural water body. In the study acute toxicity of crustaceans *Daphnia magna* and *Thamnocephalus platyurus* and protozoan *Tetrahymena thermophila* was determined.

9.9 Effect of Nanoparticles on Terrestrial Organisms

9.9.1 Effect of Nanoparticles on Terrestrial Invertebrates

Invertebrates cover a large group of animals ranging from top to bottom trophic levels of ecosystem. They are involved in various physiological phenomenon i.e. pollination, soil formation etc. Their occurrence differ from one habitat to the other and are considered as bioindicators (symbol of environmental change). Being an integral part of environment obviously this group of organisms encounters with the nanoparticles. In this section authors have tried to reflect the effect of nanoparticles on Annelids.

9.9.2 Effect of Nanoparticles on Annelids

Agricultural practices are the most important and widely used occupation cum source of income world wide. In order to earn the livelihood and growing demand of requirements one has to look for an effective and environmentally safe tool to reach the target of high productivity without compromising with the chemical based fertilizers and pesticides that has proved to be indulged in the formation of waste lands by successively decreasing the humus and fertility rate of the soil ultimately leading to a major cause of Bioaccumulation and biomagnification. Organic farming is the best alternative in which earthworms play the most significant role. Keeping in view the immense importance of earthworms in agriculture and zooming application of engineered nanoparticles in almost every sector of society studies has been conducted to see the effect of TiO_2 and ZnO nanoparticles by growing earthworms *Eisenia fetida* in an artificially created soil system by Hu et al. (2010). Distilled water at a concentration of, 0.1, 0.5, 1.0 or 5.0 g kg1 of Nanoparticles were prepared and earthworms were placed in that soil system for 7 days. By using Acute toxicity test and Comet assay concentration of Ti and Zn in earthworms was determined. It was observed that when doses of TiO_2 and ZnO were greater than 1.0 g kg^{-1} Zn accumulated in higher concentration than Ti and this leads to hyperaccumulation of Zn in the mitochondria of its gut cells which causes DNA damage. The cellulose activity was also inhibited at the dose level of 5.0 g kg^{1} of ZnO nanoparticles. Likewise, Khare et al. (2011) reported the toxicity of TiO_2 and ZnO nanoparticles less than 25 nm and 100 nm on the nematode (*Caenorhabditis elegans*). In this study toxicity of small and large sized nanoparticles of ZnO and TiO_2 were compared. It was found that small sized (25 nm) ZnO nanoparticles. were more toxic than that of the small (25 nm) and large sized (100 nm) TiO_2 nanoparticles.

9.9.3 Effect of Nanoparticles on Vertebrates

In the previous sections an attempt has been made to throw light upon the effect of nanoparticles on various abiotic and biotic factors of the environment. In this section the hazardous effect of inorganic nanoparticles i.e. ZnO has been demonstrated on two groups of vertebrates i.e. amphibians and mammals.

9.9.4 Effect of Nanoparticles on Amphibians

ZnO nanoparticles brought about developmental abnormalities in amphibians. Metallic nanoparticles have been found to cause deleterious effect up on encountered life forms. Among them ZnO, TiO_2, Fe_2O_3, and CuO nanomaterials (20–100 nm) were subjected to access their effect on amphibians in a study conducted by Nations et al. (2011a) on *Xenopus laevis* by employing the Frog Embryo Teratogenesis Assay Xenopus (FETAX) protocol. Experimental organism was subjected to dispersed nanomaterials at a concentration upto 1000 mg L^{-1} for TiO_2, Fe_2O_3, CuO, and ZnO. It was observed that inspite of causing mortality to the test specie these causes developmental abnormalities among them Viz.gastrointestinal, spinal, and other abnormalities. However, at a concentration of 10.3 mg L^{-1} of ZnO leads to total malformation. This study confers the harmful effects of ZnO nanoparticles. so its use and production needs to be checked for future perspective.

9.9.5 Effect of Nanoparticles on Mammals

Nanoparticles have been found to cause a number of serious problems such as cytotoxicity. Copper nanoparticles while incubated with zinc ions remains normal but when exposed to large sized ZnO nanoparticles. (at a non toxic concentration 6.25 μg/ml) its cytotoxic activity enhances gradually, also synergistic activity becomes strong. This is due to the presence of large sized ZnO nanoparticles which brings about the alteration in the human hepatoma cell line HepG2 Li et al. (2015) Table 9.4.

Table 9.4 Harmful effect of nanoparticles on animals

Test system	Nanoparticle	Harmful effect	Reference
Nematodes: (*Caenorhabditis elegans*)	TiO_2	Decrease in reproduction potential	Roh et al. (2009)
		Increased enzyme induction & protein formation	Roh et al. (2009)
Largemouth bass (*Micropterus salmoides*)	FullerenceC_{60}(0.5–1 mg/l)	Significant lipid peroxidation in brain	
		Total glutathione levels were marginally depleted in gills	Oberdörster (2004)
Daphnia magna	C_{60}water suspension (9 mg/l)	High rate of mortality has been noticed	Zhu et al. (2006)
Fathhead minnow	C_{60}water suspension (2.5–5 mg/l)	The peroxisomal lipid transport protein PMP70 was significantly reduced in the minnow	Oberdörster et al. (2006)
Zebrafish (*Danio rerio*)	C_{60}(prepared with benzene, THF, and acetone)	Suffered delayed embryo & larval development	
	(1.5 mg/l)	Decreased survival and hatching rates	Zhu et al. (2007)
Eukaryotic unicellular protozoan:	Carbon nanotubes	Observed pericardial edema	Zhu et al. (2006)
Stylonychia mytilus	MWCNT sonicated (1 mg/l)	Colocalization of CNTs within the mitochondria,	Zhu et al. (2006)
		Damage to micronucleus, macronucleus, and membrane	Gimbert et al. (2007)
Rainbow trout (*Oncorhynchus mykiss*)	Single-walled sonicated	Significant increases of $Na^{+}K^{+}$-adenosine triphosphatase	Gimbert et al. (2007)
		(ATPase) activity observed in the gills and intestine	
		Oxidative stress–linked effects observed	Gimbert et al. (2007)
Zebrafish (*Danio rerio*)	DWCNTs (240 mg/l)	Delay in hatching	Gulson and Wong (2006)
Zebrafish (*Danio rerio*)	Metallic copper (1.5 mg/l)	Gill injury and lethality observed	Seaman and Bertsch (2000)
		Acute toxicity observed	Seaman and Bertsch (2000)
Freshwater mussels (*Elliptio complanata*)	Quantum dots: cadmium telluride (CdTe)	Observed reduction in phagocytic activity	Karathanasis (1999)
		Lipid peroxidation increased at higher concentrations	Karathanasis (1999)
Mice	Graphene oxide [0–100 mg/L]	Down-regulation of genes, disturbance of cell energy metabolism, transduction	Wang et al. (2011)

(continued)

Table 9.4 (continued)

Test system	Nanoparticle	Harmful effect	Reference
Rats	SWCNT 1 or 5 mg kg71	Multifocal macrophage containing granulomas in lung	Warheit et al. (2004)
Guinea pigs	MWCNT 2.5 mg in 0.5 ml	Caused 'organising pneumonitis', pulmonary lesions	Grubek-Jaworska et al. (2006)
Mice	SWCNT 2-mg	SWCNT produced some activation of the histocompatibility complex, in CD4þ/CD8þT-cells	Koyama et al. (2006)
Rats	TiO_2 particles, rods or dots 1 or 5 mg	Produced transient inflammatory and cell injury effects	Warheit et al. (2006)
Rats	Ultrafine cadmium oxide 70 mg m^{-3}	Increased percentage of neutrophils	Warheit et al. (2006)
		Multifocal alveolar inflammation	Warheit et al. (2006)
		Showed elevated blood Cd.	Takenaka et al. (2004)
Rats	Ultrafine metallic nickel 0.15,–2.54 mg m^{-3}	Increase in pulmonary nickel,	Serita et al. (1999)
		Increase in lung weight	Serita et al. (1999)
		Accumulation of foamy alveolar macrophages (AM),	Serita et al. (1999)
		Generated AM indicating alveolar lipoproteinosis,	Serita et al. (1999)
		Acute calcification of the degenerated AM.	Serita et al. (1999)
Mouse spermatogonial stem cell line	Silver nanoparticles 5–100 mg ml71	Mitochondrial function are adversely affected,	
		Cells show increased LDH leakage	Braydich-Stolle et al. (2005)
Rat lung cells	Silver nanoparticles	Reduction in lung function and inflammatory lesions	Sung et al. (2008)
Sprague-Dawley rats		Silver nanoparticles accumulation in olfactory bulb and subsequent translocation to the brain	Kim et al. (2008a, b)
Rat liver cells	Silver nanoparticles	Cell leakage and reduction of mitochondrial function	Hussain et al. (2005)
Mouse fibroblast	Silver nanoparticles	50 µg/ml induced apoptosis to 43.4 % of cells	Arora et al. (2009)

(continued)

Table 9.4 (continued)

Test system	Nanoparticle	Harmful effect	Reference
HUMAN:			
Skin	Silver nanoparticles	Argyria-irreversible pigmentation of skin & argyrosis- pigmentation of eyes	Chen and Schluesener (2007)
Epithelium layer	Silver nanoparticles	Severe damage	Boosalis, et al. (1987)
Skin	Silver nanoparticles	Discolouration of the skin	Armitage et al. (1996) and Greene and Su (1987)
Eye	Silver nanoparticles	Severe irritation	Panyala et al. (2008)
Nose	Silver nanoparticles	Severe irritation	Panyala et al. (2008)
Brain	Silver nanoparticles	Neurotoxic damage	Cheng et al. (2004)
Male reproductive system	Silver nanoparticles	Cross blood-testes barrier,deposit in the testes, adversly effects sperm cells	Borm and Kreyling (2004)
Blood cells	Silver nanoparticles	Intensive toxic effect on PBMCs & inhibits phytohaemagglutinin-induced	Shin et al. (2007)
		Cytokine production	
Keratinocytes & fibroblasts	Acticoat, aquacel, Ag & contreet foam	Most significant cytotoxic effect	Panyala et al. (2008)
Cell	Ag nanoparticles >44.0 $\mu g\ ml^{-1}$	Necrotic in nature	Panyala et al. (2008)
Central nervous system (CNS)	Silver nanoparticles	Prolonged exposure leads to cerebral ataxia	Aaseth et al. (1981)
Respiratory system: (Lungs)	Silver nanoparticles	Produce surface radical and reactive oxygen specie, toxic to alveolar surface	Chen and Schluesener (2007)
(Lungs)	Silver nanoparticles	Induce oxidative stress in lungs epithelial cells	Limbach et al. (2007)
(Lungs)	Silver nanoparticles	Irritation in respiratory tract	Rosenman et al. (1987)
Brain	Silver nanoparticles	Manic depressive psychosis leads to ruptured aortic aneurysm, finally death	Dietl et al. (1984)
Skin	Silver nanoparticles	Cutaneous side effects: hypersensitivity reaction, allergic contact dermatitis, erythema multiforme	Fisher et al. (2003)

(continued)

Table 9.4 (continued)

Test system	Nanoparticle	Harmful effect	Reference
Lung	Pristine and MWCNTs	Alteration in its architecture, collapsed thick walled alveoli, presence of microhaemorrhagic foci	Coccini et al. (2013)
Human chest cavity	MWCNTs	Causes inflammation, production of scars/lesions, leads to mesothelioma	Poland et al. (2008)
Nucleus	C60 fullerence	Accumulate along nuclear membrane, penetrate and cause DNA damage	Porter et al. (2006)
Metabolic process in rats, human, mice	C60 fullerence	Reduce activity of hepatic enzymes i.e. glutathione	Iwata et al. (1998)
Blood and haematopoitic system	C60 fullerence	Triggered hemolysis of 40–50 % of cells	Bosi et al. (2004)
Human (HEK293) embryo kidney cell	SWCNTs – 0.78–200 $\mu g/ml^{-1}$	Inhibition of HEK293 cell proliferation, induced cell apoptosis, and decreased cellular adhesive ability	Cui et al. (2005)
Human bronchial epithelial cells(BEAS 2B)	TiO_2 nanoparticles (10 or 20 nm)	Cause DNA damage, lipid peroxidation and damage to nucleus	Gurr et al. (2005)
Human lung cell	Carbon nanotubes	Inflammation and oxidative stress	Pulskamp et al. (2007)
Human lung cell	Carbon black	Inflammation and oxidative stress	Monteiller et al. (2007)
Human lung cell	Silica	Inflammation and oxidative stress	Lin et al. (2006)
Human lung cell	Metal and metal oxides	Inflammation and oxidative stress	Limbach et al. (2007)
Human cancer cells	Graphene oxide [0–100 mg/L]	Oxidative stress	Zhang et al. (2010)

9.10 Conclusion

Nanoparticles are the most promising and widely exploited resources of present time. Their surface properties makes them unique, thus they are being used in every sector of society. To make and prove this technology as a boon we need to take some simple measures. By taking care while handling the engineered nanomaterials, by properlystudying the fate of the synthesized nanoparticle, also by implementing ecofriendly techniques that consumes minimum energy, labour and cost effective methods we can add beauty to our living world inspite of accelerating the pollution rate. Use of chemically engineered nanoparticles should be minimizes. Establishment of regulatory guidelines for nanoparticle synthesis is the urgent call, which should be followed by both the developed and developing nations. Now it's the need of time that without delaying the slogan of "GO Green" should be practiced not in words but in work. One thing is to be kept in mind i.e. the life cycle of engineered material should be well studied and the large scale synthesis of them should be executed on the basis of their fate in the environment. In the quest of innovation and advancement environmental issues are becoming more serious and uncontrolled. So, let's shake a leg towards the betterment of the earth in which we are living in.

References

Aaseth J, Olsen A, Halse J, Hovig T (1981) Argyria-tissue deposition of silver as selenide. Scand J Clin Lab Invest 41:247–251

Adams LK, Lyon DY, Alvarez PJJ (2006) Comparative ecotoxicity of nanoscale TiO_2, SiO_2, and ZnO water suspensions. Water Res 40:3527–3532

Akhavan O, Ghaderi E (2010) Toxicity of graphene and graphene oxide nano walls against bacteria. ACS Nano 4:5731–5736. doi:10.1021/nn101390x

Albright LJ, Wilson EM (1974) Sub-lethal effects of several metallic salt- organic compounds combinations upon hetrotrophic microflora of a natural water. Water Res 8:101–105

Alfadul SM, Elneshwy AA (2010) Use of nanotechnology in food processing, packaging and safety—review. Afr J Food Agric Nutr 10:2719–2739

Alivisatos AP, Gu W, Larabell C (2005) Quantum dots as cellular probes. Annu Rev Biomed Eng 7:55–76

Armitage SA, White MA, Wilson HK (1996) The determination of silver in whole blood and its application to biological monitoring of occupationally exposed groups. Ann Occup Hyg 40:331–338

Arora S, Jain J, Rajwade JM, Paknikar KM (2009) Interactions of silver nanoparticles with primary mouse fibroblasts and liver cells. Toxicol Appl Pharmacol 236:310–318

Aruoja V, Dubourguier HC, Kasemets K, Kahru A (2009) Toxicity of nanoparticles of CuO, ZnO and TiO_2 to microalgae Pseudokirchneriella subcapitata. Sci Total Environ 407(4):1461–1468

Atha DH, Wang H, Petersen EJ (2012) Copper oxide nanoparticle mediated DNA damage in terrestrial plant models. Environ Sci Technol 46(3):1819–1827

Austin LA, Kang B, Yen CW, El-Sayed MA (2011) Nuclear targeted silver nanospheres perturb the cancer cell cycle differently than those of nanogold. Bioconjug Chem 22:2324–2331

Babynin EV, Nuretdinov IA, Gubskaia VP, Barabanshchikov BI (2002) Study of mutagenic activity of fullerene and some of its derivatives using his? Reversions of *Salmonella typhimurium* as an example. Russ J Genet 38:359–363

Baun A, Hartmann NB, Grieger K, Kisk KO (2008) Ecotoxicity of engineered nanoparticles to aquatic invertebrates: a brief review and recommendations for future toxicity testing. Ecotoxicology 17(5):387–395. doi:10.1007/s10646-008-0208-y

Benn TM, Westerhoff P (2008) Nanoparticle silver released into water from commercially available sock fabrics. Environ Sci Technol 42:4133–4139

Biswas P, Wu C (2005) Nanoparticles and the environment. J Air Waste Manage Assoc 55:708–746

Blinova I, Ivask A, Heinlaan M, Mortimer M, Kahru A (2010) Ecotoxicity of nanoparticles of CuO and ZnO in natural water. Environ Pollut 158(1):41–47. doi:10.1016/j.envpol.2009.08.017

Bondarenko O, Ivask A, Käkinen A, Kahru A (2012) Sub-toxic effects of CuO nanoparticles on bacteria: kinetics, role of Cu ions and possible mechanisms of action. Environ Pollut 169:81–89

Boosalis MG, McCall JT, Ahrenhalz DH, Solem LH, McClain CJ (1987) Serum and urinary silver levels in thermal injury patients. Surgery 101:40–43
Borm PJ, Kreyling WJ (2004) Toxicological hazards of inhaled nanoparticles-potential implications for drug delivery. J Nanosci Nanotechnol 4:521–551
Bosi S, Da Ros T, Castellano S, Banfi E, Prato M (2000) Antimycobacterial activity of ionic fullerene derivatives. Bioorg Med Chem Lett 10:1043–1045
Bosi S, Feruglio L, Da Ros T, Spalluto G, Gregoretti B, Terdoslavich M, Decorti G, Passamonti S, Moro S, Prato M (2004) Hemolytic effects of water-soluble fullerene derivatives. J Med Chem 47:6711–6715
Bouwmeester H, Dekkers S, Noordam M, HagensW, Bulder A, de Heer C, ten Voordel S, Wijnhoven S, Sips A (nd) Health risks of application of nanotechnologies and nanoparticles within the food production chain. *RIKILT/RIVM* Report 2007.014
Braydich-Stolle L, Hussain S, Schlager JJ, Hofmann MC (2005) In vitro cytotoxicity of nanoparticles in mammalian germline stem cells. Toxicol Sci 88:412–419
Brayner R, Ferrari-Iliou R, Brivois N, Djediat S, Benedetti MF, Fievet F (2010) ZnO nanoparticles: synthesis, characterization, and ecotoxicological studies. Langmuir 26(9):6522–6528
Brigger I, Dubernet C, Couvreur P (2002) Nanoparticles in cancer therapy and diagnosis. Adv Drug Deliv Rev 54:631–51 PII: S0169-409X(02)00044-3
Brown DM, Wilson MR, MacNee W, Stone V, Donaldson K (2001) Size dependent proinflammatory effects of ultrafine polystyrene particles: a role for surface area and oxidative stress in the enhanced activity of ultrafines. Toxicol Appl Pharmacol 175:191–199. doi:10.1006/taap.2001.9240
Buzea C, Ivan I, Blandino P, Robbie K (2007) Nanomaterials and nanoparticles: sources and toxicity. Biointerphases 2(4):MR17–MR172. doi:10.1116/1.2815690
Cameron FK (1915) Soil colloids and the soil solution. J Phys Chem 19(1):1–13. doi: 10.1021/j150154a001
Canas JE, Long M, Nations S, Vadan R, Dai L, Luo M, Ambikapathi R, Lee EH, Olszyk D (2008) Effects of functionalized and nonfunctionalized single-walled carbon nanotubes on root elongation of selected crop species. Environ Toxicol Chem 27:1922–1931
Cerkez I, Kocer HB, Worley SD, Broughton RM, Huang TS (2012) Multifunctional cotton fabric: antimicrobial and durable press. J Appl Polym Sc 124(5):4230–4238. doi:10.1002/app.35402
Chang MCO, Chow JC, Watson JG, Hopke PK, Yi SM, England GC (2004) Measurement of ultrafine particle size distributions from coal-, oil-, and gas-fired stationary combustion sources. J Air Waste Manag Assoc 54:1494–1505
Chen X, Schluesener HJ (2007) Nanosilver: a nanoproduct in medical application. Toxicol Lett 176:1–12
Cheng D, Yang J, Zhao Y (2004) Antibacterial materials of silver nanoparticles application in medical appliances and appliances for daily use. Chin Med Equip J 4:26–32
Chernousova S, Epple M (2013) Silver as antibacterial agent: ion, nanoparticle, metal. Angew Chem. doi:10.1002/anie.201205923
Coccini T, Manzo L, Roda E (2013) Safety evaluation of engineered nanomaterials for health risk assessment: an experimental tiered testing approach using pristine and functionalized carbon nanotubes. ISRN Toxicol 2013, Article ID 825427, 13 pages
Cui DX, Tian FR, Ozkan CS, Wang M, Gao HJ (2005) Effect of single wall carbon nanotubes on human HEK293 cells. Toxicol Lett 155:73–85
Cushen M, Kerry J, Morris M, Cruz-Romero M, Cummins E (2012) Nanotechnologies in the food industry – recent developments, risks and regulation. Trends Food Sci Tech 24:30–46
Da Silva LC, Oliva MA, Azevedo AA, DeAraujo JM (2006) Responses of resting a plant species to pollution from an iron pelletization factory. Water Air Soil Pollut 175:241–256. doi:10.1007/s11270-006-9135-9
Daniel MC, Astruc D (2004) Gold nanoparticles: assembly, supramolecular chemistry, quantum-size-related properties and applications towards biology, catalysis and nanotechnology. Chem Rev 104:293–346. doi:10.1021/cr030698

Dastjerdi R, Montazer M (2010) A review on the application of inorganic nano-structured materials in the modification of textiles: focus on anti-microbial properties. Colloid Surf B79(1):5–18
De Windt W, Boon N, Van den Bulcke J, Rubberecht L, Prata F, Mast J, Hennebel T, Verstraete W (2006) Biological control of the size and reactivity of catalytic Pd(0) produced by *Shewanella oneidensis*. Anton Leeuw Int J Gen Mol Microbiol 90:377–389
Dietl HW, Anzil AP, Mehraein P (1984) Brain involvement in generalized argyria. Clin Neuropathol 3:32–38
Dinesh R, Anandaraj M, Srinivasan V, Hamza S (2012) Engineered nanoparticles in the soil and their potential implications to microbial activity. Geoderma 173–174:19–27
Du W, Sun Y, Ji R, Zhu J, Wu J, Guo H (2011) TiO_2 and ZnO nanoparticles negatively affect wheat growth and soil enzyme activities in agricultural soil. J Environ Monit 13(4):822–828
Esterkin CR, Negro AC, Alfano OM, Cassano AE (2005) Air pollution remediation in a fixed bed photocatalytic reactor coated with TiO_2. AIChE J 51:2298–2310. doi:10.1002/aic.10472
Fisher NM, Marsh E, Lazova R (2003) Scar-localized argyria secondary to silver sulfadiazine cream. J Am Acad Dermatol 49:730–732
Fortner JD, Lyon DY, Sayes CM, Boyd AM, Falkner JC, Hotze EM, Alemany LB, Tao YJ, Guo W, Ausman KD, Colvin VL, Huges JB (2005) C60 in water: nanocrystal formation and microbial response. Environ Sci Technol 39:4307–4316
Gimbert LJ, Hamon RE, Casey PS, Worsfold PJ (2007) Partitioning and stability of engineered ZnO nanoparticles in soil suspensions using field-flow fractionation. Environ Chem 4:8–10
Goodman CM, McCusker CD, Yilmaz T, Rotello VM (2004) Toxicity of gold nanoparticles functionalized with cationic and anionic side chains. Bioconjug Chem 15:897–900
Greene RM, Su WPD (1987) Argyria Am Fam Physician 36:151–154
Grubek-Jaworska H, Nejman P, Czuminska K, Przybylowski T, Huczko A, Lange H, Bystrzejewski M, Baranowski P, Chazan R (2006) Preliminary results on the pathogenic effects of intratracheal exposure to one-dimensional nanocarbons. Carbon 44:1057–1063
Gulson B, Wong H (2006) Stable isotope tracing—a way forward in nanotechnology. Environ Health Perspect 114:1486–1488
Gurr JR, Wang ASS, Chen CH, Jan KY (2005) Ultrafine titanium dioxide particles in the absence of photoactivation can induce oxidative damage to human bronchial epithelial cells. Toxicology 213:66–73
Gustafsson O, Kruså M, Zencak Z, Sheesley RJ, Granat L, Engström E, Praveen PS, Rao PS, Leck C, Rodhe H (2009) Brown clouds over South Asia: biomass or fossil fuel combustion? Science 323(5913):495–498
Hardman R (2006) A toxicologic review of quantum dots: toxicity depends on physicochemical and environmental factors. Environ Health Perspect 114:165–172
Hildemann LM, Markowski GR, Jones MC, Cass GR (1991) Submicrometer aerosol mass distributions of emissions from boilers, fireplaces, automobiles, diesel trucks, and meat-cooking operations. Aerosol Sci Technol 14:138–152
Hoet PH, Bruske-Hohlfeld I, Salata OV (2004) Nanoparticles known and unknown health risks. J Nanobiotechnol 2:12. doi:10.1186/1477-3155-2-12
Hogan CJ, Lee MH, Biswas P (2004) Capture of viral particles in soft X-ray-enhanced corona systems: charge distribution and transport characteristics. Aero Sci Technol 38:475–486
Holbrook RD, Murphy KE, Morrow JB, Cole KD (2008) Trophic transfer of nanoparticles in asimplified invertebrate food web. Nat Nanotechnol 335–355. doi:10.1038/nnano.2008.110
Horikoshi S, Serpone N (2013) Introduction to nanoparticles microwaves in nanoparticle synthesis. In: Satoshi H, Nick S (eds) Microwaves in nanoparticle synthesis: fundamentals and applications, 1st edn. Wiley-VCH Verlag GmbH & Co/KGaA, Berlin. ISBN ISBN-978-3-527- 64814–6
Hu CW, Li M, Cui YB, Li DS, Chen J, Yang LY (2010) Toxicological effects of TiO2 and ZnO nanoparticles in soil on earthworm Eiseniafetida. Soil Biol Biochem 42(4):586–591. doi:10.1016/j.soilbio.2009.12.007

Hua NP, Kobayashi F, Iwasaka Y, Shi G, Naganuma T (2007) Detailed identification of desert originated bacteria carried by Asian dust storms to Japan. Aerobiologia 23(4):291–298. doi:10.1007/s10453-007-9076-9

Huang L, Li DQ, Lin YJ, Wei M, Evans DG, Duan X (2005) Controllable preparation of Nano-MgO and investigation of its bactericidal properties. J Inorg Biochem 99:986–993

Huang Y, Chen S, Bing X, Gao C, Wang T, Yuan B (2011) Nanosilver migrated into food—simulating solutions from commercially available food fresh containers. Packag Technol Sci 24:291–297

Hund-Rinke K, Simon M (2006) Ecotoxic effect of photocatalytic active nanoparticles TiO_2 on algae and daphnids. Environ Sci Pollut Res 13:225–232

Hussain S, Hess K, Gearhart J, Geiss K, Schlager J (2005) Invitro toxicity of nanoparticles in BRL 3A rat liver cells. Toxicol In Vitro 19:975–983

IPPIC (International Paint and Printing Ink Council) (2012) http://www.ippic.org/site/assets/docs/Public%20AFWG/IPPIC%20%20Zinc%20oxide%20-%20regulatory%20status%20-final%20Draft%20March%201 %202012.pdf

Iwata N, Mukai T, Yamakoshi TN, Hara S, Yanase T, Shoji M, Endo T, Miyata N (1998) Effects of C_{60}, a fullerene, on the activities of glutathione s-transferase and glutathione-related enzymes in rodent and human livers. Fuller Sci Technol 6(2):213–226

Jia G, Wang HF, Yan L,Wang X, Pei RJ Yan T, Zhao YL, Guo XB (2005). Cytotoxicity of carbon nano materials:single wall nanotube, multi-wall nanotube and fullerence. Environ Sci Technol 39:1378–1383. doi:10.1021/es048729l

Jiang W, Mashayekhi H, Xing B (2009) Bacterialtoxicity comparison between nano and micro-scaled oxide particles. Environ Pollut 157(5):1619–1625

Kai Y, Komazawa Y, Miyajima A, Miyata N, Yamakoshi Y (2003) Fullerene as a novel photoinduced antibiotic. Fullerenes Nanotubes Carbon Nanostruct 11:79–87

Kaida T, Kobayashi K, Adachi M, Suzuki F (2004) Optical characteristics of titanium oxide interference film and the film laminated with oxides and their applications for cosmetics. J Cosmet Sci 55:219–220

Kang S, Pinault M, Pfefferle LD, Elimelech M (2007) Single walled carbon nanotubes exhibit strong antimicrobial activity. Langmuir 23:8670–8673

Karathanasis AD (1999) Subsurface migration of copper and zinc mediated by soil colloids. Soil Sci Soc Am J 63:830–838

Kasemets K, Ivask A, Dubourguier HC, Kahru A (2009) Toxicity of Nanoparticles of ZnO, CuO and TiO2 to yeast Sacchromyces cerevisiae. Toxicology In Vitro 23(6):1116–1122. pii/S0887233309001209

Kennedy AC, Smith KL (1995) Soil microbial diversity and the sustainability of agricultural soils. Plant Soil 170:75–86

Kennedy CB, Scott SD, Ferris FG (2004) Hydrothermal phase stabilization of 2-line Ferrihydrite by bacteria. Chem Geol 212(3–4):269–277. pii/S0009254104003158

Khare P, Sonane M, Pandey R, Ali S, Gupta KC, Satish A (2011) Adverse effects of TiO2 and ZnO nanoparticles in soil nematode, Caenorhabditis elegans. J Biomed Nanotechnol 7(1):116–117. http://dx.doi.org/10.1166/jbn.2011.1229

Kim Y, Kim J, Cho H, Rha D, Kim J, Park J, Choi B, Lim R, Chang H, Chung Y, Kwon I, Jeong J, Han B, Yu I (2008a) Twenty-eight-day oral toxicity, genotoxicity, and gender related tissue distribution of silver nanoparticles in Spra-gue-Dawley rats. Inhal Toxicol 20:575–583

Kim K, Sung W, Moon S, Choi J, Kim J, Lee D (2008b) Antifungal effect of silver nanoparticles on dermatophytes. J Microbiol Biotechnol 18:1482–1484

Kittelson DB (1998) Engines and nanoparticles: a review. J Aero Sci 29:575–588

Klaine SJ, Alvarez PJJ, Batley GE (2008) Nanomaterials in the environment: behavior, fate, bioavailability, and effects. Environ Toxicol Chem 27(9):1825–1851

Kloepfer JA, Mielke RE, Nadeau JL (2005) Uptake of CdSe and CdSe/ZnS quantum dots into bacteria via purine-dependent mechanisms. Appl Environ Microbiol 71:2548–2557

Koyama S, Endo M, Kim YA, Hayashi T, Yanagisawa T, Osaka K, Koyama H, Haniu H, Kuroiwa N (2006) Role of systemic T-cells and histopathological aspects after subcutaneous implantation of various carbon nanotubes in mice. Carbon 44:1079–1092

Kumari M, Khan SS, Pakrashi S, Mukherjee A, Chandrasekaran N (2011) Cytogenetic and genotoxic effects of zinc oxide nanoparticles on root cells of *Allium cepa*. J Hazard Mater 190(1–3):613–621

Li L, Cruz MLF, Connotly M, Conde E, Fernandez M, Schuster M, Navas JM (2015) The potentiation effect makes the difference: nontoxic concentration of ZnO nanoparticles enhance Cu nanoparticle toxicity In-vitro. Sci Total Environ 505:253–260. pii/S0048969714014582

Limbach LK, Wick P, Manser P, Grass RN, Bruinink A, Stark WJ (2007) Exposure of engineered nanoparticles to human lung epithelial cells: influence of chemical composition and catalytic activity on oxidative stress. Environ Sci Technol 41:4158–4163

Lin D, Xing B (2007) Phytotoxicity of nanoparticles: inhibition of seed germination and root growth. Environ Pollut 150(2):243–250

Lin W, Huang YW, Zhou XD, Ma Y (2006) Toxicity of cerium nanoparticles in human lung cancer cells. Int J Toxicol 25:451–457

Lin S, Reppert J, Hu Q, Hudson JS, Reid ML, Ratnikova TA, Rao AM, Luo H,Ke PC (2009) Uptake, translocation, and transmission of carbon nanomaterials in rice plants. Small 5:1128–1132. doi:10.1002/smll.200801556

Lorenz C, Hagendorfer H, von Goetz N, Kaegi R, Gehrig R, Ulrich A, Scheringer M, Hungerbühler K (2011) Nanosized aerosols from consumer sprays: experimental analysis and exposure modeling for four commercial products. J Nanopart Res 13:3377–3391

Lyon DY, Fortner JD, Sayes CM, Colvin VL, Hughes JB (2005) Bacterial cell association and antimicrobial activity of a C-60 water suspension. Environ Toxicol Chem 24:2757–2762

Lyon DY, Adams LK, Falkner JC, Alvarez PJJ (2006) Antibacterial activity of fullerene water suspensions: effects of preparation method and particle size. Environ Sci Technol 40:4360–4366

Ma Y, Kuang L, He X et al (2010) Effects of rare earth oxide nanoparticles on root elongation of plants. Chemosphere 78(3):273–279

Ma H, Williams PL, Diamond SA (2013) Ecotoxicity of manufactured ZnO nanoparticles - a review. Environ Pollut 172:76–85. pii/S0269749112003958

Mahapatra O, Bhagat M, Gopalakrishnan C, Arunachalam KD (2008) Ultrafine dispersed CuO nanoparticles and their antibacterial activity. J Exp Nanosci 3:185–193

Manning MR, Lowe DC, Moss RC, Bodeker GE, Allan W (2005) Short-term variations in the oxidizing power of the atmosphere. Nature 436:1001–1004. doi:10.1038/nature03900

Manzo S, Rocco A, Carotenuto R, De Luca Picione F, Miglietta M, Rametta G, Di Francia G (2011) Investigation of ZnO nanoparticles' ecotoxicological effects towards different soil organisms. Environ Sci Pollut Res 18(5):756–763

Mashino T, Okuda K, Hirota T, Hirobe M, Nagano T, Mochizuki M (1999) Inhibition of *E. coli* growth by fullerene derivatives and inhibition mechanism. Bioorg Med Chem Lett 9:2959–2962

Mashino T, Nishikawa D, Takahashi K, Usui N, Yamori T, Seki M, Endo T, Mochizuki M (2003) Antibacterial and antiproliferative activity of cationic fullerene derivatives. Bioorg Med Chem Lett 13:4395–4397

Mashino T, Usui N, Okuda K, Hirota T, Mochizuki M (2003) Respiratory chain inhibition by fullerene derivatives: hydrogen peroxide production caused by fullerene derivatives and a respiratory chain system. Bioorg Med Chem Lett 11:1433–1438

McMurry PH, Woo KS, Weber R, Chen DR, Pui DYH (2000) Size distributions of 3–10 nm atmospheric particles: implications for nucleation mechanisms. Philos Trans R Soc Lond Ser a-Math Phys Eng Sci 358:2625–2642

Miao AJ, Zhang XY, Luo Z, Chen CS, Chin WC, Santschi PH, Quigg A (2010) Zinc oxideeengineered nanoparticles: dissolution and toxicity to marine phytoplankton. Environ Toxicol Chem 29(12):2814–2822

Miller RJ, Lenihan HS, Muller EB, Tseng N, Hanna SK, Keller AA (2010) Impacts of metal oxide nanoparticles on marine phytoplankton. Environ Sci Technol 44(19):7329–7334
Mishra VK, Kumar A (2009) Impact of metal nanoparticles on plant growth promoting rhizobacteria. Dig J Nanomater Biostruct 4:587–592
Monteiller C, Tran L, MacNee W, Faux S, Jones A, Miller B, Donaldson K (2007) The proinflammatory effects of low-toxicity low-solubility particles, nanoparticles and fine particles, on epithelial cells in vitro: the role of surface area. Occup Environ Med 64:609–615
Moore MN (2006) Do nanoparticles present ecotoxicological risks for the health of the aquatic environment? Elsevier. doi:10.1016/j.envint.2006.06.014
Morones JR, Elechiguerra JL, Camacho A, Holt K, Kouri JB, Ramirez JT, Yacaman MJ (2005) The bactericidal effect of silver nanoparticles. Nanotechnology 16:2346–2353
Mortimer M, Kasemets K, Kahru A (2010) Toxicity of ZnO and CuO nanoparticles to ciliated protozoa Tetrahymena thermophila. Toxicology 269(2-3):182–189. doi:10.1016/j.tox.2009.07.007
Mushtaq YK (2011) Effect of nanoscale Fe3O4, TiO2 and carbon particles on cucumber seed germination. J Environ Sci Health A 46(14):1732–1735. doi:10.1080/10934529.2011.633403
Nations S, Wages M, Canas JE, Maul J, Theodorakis C, Cobb GP (2011a) Acute effect of Fe2O3, TiO2, ZnO and CuO nanomaterials on Xenopus 2 3 3 laevis. Chemosphere 83(8):1053–1061. pii/S0045653511001019
Navarro E, Baun A, Behra R, Hartmann NB, Filser J, Miao AJ, Quigg A, Santschi PH, Sigg L (2008) Environmental behavior and ecotoxicity of engineered nanoparticles to algae, plants, and fungi. Ecotoxicol 17:372–386
Nielsen HD, Berry LS, Stone V, Burridge TR Fernandes TF (2008) Interactions between carbon black nanoparticles and the brown algae Fucus serratus: inhibition of fertilization and zygotic development. Nanotoxicol 2:88–89. doi:10.1080/17435390802109185
Nowack B, Bucheli TD (2007) Review -occurrence, behavior and effects of nanoparticles in the environment. Environ Pollut 150:5–22. doi:10.1016/j.envpol.2007.06.006
Nowack B, Krug HF, Height M (2011) 120 years of nanosilver history: implications for policy makers. Environ Sci Technol 45(4):1177–1183
Nyberg L, Turco RF, Nies L (2008) Assessing the impact of nanomaterials on anaerobic microbial communities. Environ Sci Technol 42:1938–1943
Oberdorster E (2004) Manufactured nanomaterials (fullerences, C60) induce oxidative stress in the brain of juvenile largemouth bass. Environ Health Perspect 112:1058–1062. doi:10.1289/ehp.7021
Oberdörster E, Zhu S, Blickley TM, McClellan-Green P, Haasch ML (2006) Ecotoxicology of carbon-based engineered nanoparticles: effects of fullerene (C_{60}) on aquatic organisms. Carbon 44:1112–1120
Panyala NR, Pena-Mendez EM, Havel J (2008) Silver or silver nanoparticles: a hazardous threat to the environment and human health? Review. J Appl Biomed 6:117–129
Park B, Donaldson K, Duffin R, Tran L, Kelly F, Mudway I, Morin JP, Guest R, Jenkinson P, Samaras Z, Giannouli M, Kouridis H, Martin P (2008) Hazard and risk assessment of a nanoparticulate cerium oxide based diesel fuel additive—a case study. Inhal Toxicol 20:547–566
Patlolla AK (2013) Environmental toxicity monitoring of nanomaterials using vicia faba GENE-TOX assay. J Nanomed Nanotechol 4:e129. doi:10.4172/2157-7439.1000e129
Poland CA, Duffin R, Kinloch I et al (2008) Carbon nanotubes introduced intothe abdominal cavity of mice show asbestos like pathogenicity in a pilot study. Nat Nanotechnol 3:423–428
Porter AE, Muller K, Skepper J, Midgley P, Welland M (2006) Uptake of C_{60} by human monocyte macrophages, its localization and implications for toxicity: studied by high resolution electron microscopy and electron tomography. Acta Biomater 2:409–419
Premanathan M, Karthikeyan K, Jeyasubramanian K, Manivannan G (2011) Selective toxicity of ZnO nanoparticles toward Gram-positive bacteria and cancer cells by apoptosis through lipid peroxidation. Nanomed Nanotechnol Biol Med 7(2):184–192

Pulskamp K, Diabate S, Krug HF (2007) Carbon nanotubes show no sign of acute toxicity but induce intracellular reactive oxygen species in dependence on contaminants. Toxicol Lett 168:58–74

Raloff J (2003) Air sickness: how microscopic dust particles causes subtle but serious harm. Sci News 164(5):1–11. doi:10.2307/3982189

Ravi Kumar M, Hellermann G, Lockey RF, Mohapatra S (2004) Nanoparticle mediated gene delivery: state of the art. Expert Opin Biol Ther 4(8):1213–1224. doi:10.1517/14712598.4.8.1213

Reidy B, Haase A, Andreas L, Kenneth A, Dawson K, Lynch I (2013) Mechanisms of silver nanoparticle release, transformation and toxicity: a critical review of current knowledge and recommendations for future studies and applications. Materials 6:2295–2350

Remédios C, Rosário F, Bastos V (2012) Environmental nanoparticles interactions with plants: morphological, physiological and genotoxic aspects. Review Article. J Bot Article ID 751686, 8 pages

Rincon A, Pulgarin C (2004) Effect of pH, inorganic ions, organic matter and H_2O_2 on E. coli K12 photocatalytic inactivation by TiO_2 implications in solar water disinfection. Appl Catal B Environ 51:283–302

Roh J, Sim SJ, Yi J, Park K, Chung KH, Ryu D, Choi J (2009) Ecotoxicity of silver nanoparticles on the soil nematode *Caenorhabditis elegans* using functional ecotoxicogenomics. Environ Sci Technol 43:3933–3940

Rosenman KD, Seixas N, Jacobs I (1987) Potential nephrotoxic effects of exposure to silver. Br J Ind Med 44:267–272

Royal society and Royal Academy of Engineering (2004) Nanoscience and nanotechnologies: opportunities and uncertainties, RS policy document, 19/04. The Royal Society, London, p 113

Rozhkov SP, Goryunov AS, Sukhanova GA, Borisova AG, Rozhkova NN, Andrievsky GV (2003) Protein interaction with hydrated $C_{(60)}$ fullerene in aqueous solutions. Biochem BiophysRes Commun 303:562–566

Sajid M, Ilyas M, Basheer C, Tariq M, Daud M, Baig N, Shehzad F (2015) Impact of nanoparticles on human and environment: review of toxicity factors, exposures, control strategies, and future prospects. Environ Sci Pollut Res 22:4122–4143. doi:10.1007/s11356-014-3994-1

Saliba AM, de Assis MC, Nishi R, Raymond B, Marques EA, Lopes UG, Touqui L, Plotkowski MC (2006) Implications of oxidative stress in the cytotoxicity of *Pseudomonas aeruginosa* ExoU. Microbes Infect 8:450–459

Sau TK, Rogach AL, Jäckel F, Klar TA, Feldmann J (2010) Properties and applications of colloidal nonspherical noble metal nanoparticles. Adv Mater 22(16):1805–1825

Sawai J, Igarashi H, Hashimoto A, Kokugan T, Shimizu M (1995) Effect of ceramic powder slurry on spores of Bacillus subtilis. J Chem Eng Jpn 28:556–561

Sawai J, Igarashi H, Hashimoto A, Kokugan T, Shimizu M (1996) Effect of particle size and heating temperature of ceramic powders on antibacterial activity of their slurries. J Chem Eng Jpn 29:251–256

Sayes CM, Fortner JD, Guo W, Lyon D, Boyd AM, Ausman KD, Tao YJ, Sitharaman B, Wilson LJ, Hughes JB (2004) The differential cytotoxicity of water-soluble fullerences. Nano Lett 4:1881–1887

Seaman JC, Bertsch PM (2000) Selective colloid mobilization through surface-charge manipulation. Environ Sci Technol 34:3749–3755

Serita F, Kyono H, Seki Y (1999) Pulmonary clearance and lesions in rats after a single inhalation of ultrafine metallic nickel at dose levels comparable to the threshold limit value. Ind Health 37:353–363

Serpone N, Dondi D, Albini A (2007) Inorganic and organic UV filters: their role and efficacy in sunscreens and suncare products. Inorg Chim Acta 360:794–802

Shin SH, Ye MK, Kim HS, Kang HS (2007) The effects of nano-silver on the proliferation and cytokine expression by peripheral blood mononuclear cells. Int Immunopharmacol 7:1813–1821

Simkiss K (1990) Surface effects in ecotoxicology. Funct Ecol 4(3):303–308. StableURL:http://www.jstor.org/stable/2389590
Smita S, Gupta SK, Bartonova A, Dusinska M, Gutleb AC, Rahman Q (2012) Nanoparticles in the environment: assessment using the causal diagram approach. Environ Health 11:1–11
Sondi I, Salopek-Sondi B (2004) Silver nanoparticles as antimicrobial agent: a case study on *E. coli* as a model for Gram negative bacteria. J Colloid Interf Sci 275:177–182
Song W, Zhang J, Guo J, Zhang J, Ding F, Li L, Sun Z (2010) Role of the dissolved zinc ion and reactive oxygen species in cytotoxicity of ZnO nanoparticles. Toxicol Lett 199(3):389–397
Stampoulis D, Sinha SK, White JC (2009) Assay-dependent phytotoxicity of nanoparticles to plants. Environ Sci Technol 43(24):9473–9479
Strawa AW, Kirchstetter TW, Hallar AG, Ban-Weiss GA, McLaughlin JP, Harley RA, Lunden MM (2010) Optical and physical properties of primary on-road vehicle particle emissions and their implications for climate change. J Aerosol Sci 41: 36–50, No. 1,(January 2010), ISSN0021-8502
Sung J, Ji J, Yoon J, Kim D, Song M, Jeong J, Han B, Han J, Chung Y, Kim J, Kim T, Chang H, Lee E, Lee J, Yu I (2008) Lung function changes in Sprague-Dawley rats after prolonged inhalation exposure to silver nanoparticles. Inhal Toxicol 20:567–574
Takenaka S, Karg E, Kreyling W, Lentner B, Schulz H, Ziesenis A, Schramel P, Heyder J (2004) Fate and toxic effects of inhaled ultrafine cadmium oxide particles in the rat lung. Inhal Toxicol 16:83–92
Tang H, Wang D, Ge X (2004) Environmental nanopollutants (ENP) and aquatic micro interfacial processes. Water Sci Technol 50(12):103–109. ISSNPrint:0273–1223
The Royal Society (2004) Nanoscience and nanotechnologies: opportunities and uncertainties. The Royal Society & The Royal Academy of Engineering, London. ISBN 0 85403 604 0
Thill A, Spalla O, Chauvat F, Rose J, Auffan M, Flank AM (2006) Cytotoxicity of CeO_2 nanoparticles for Escherichia coli: a physico-chemical insight of the cytotoxicity mechanism. Environ Sci Technol 40:6151–6156
Tromp TK, Shia RL, Allen M, Eiler JM, Yung YL (2003) Potential environmental impact of a hydrogen economy on the stratosphere. Science 300:1740–1742. doi:10.1126/science.1085169
Tsao N, Luh TY, Chou CK, Chang TY, Wu JJ, Liu CC, Lei HY (2002) In vitro action of carboxyfullerene. J Antimicrob Chemother 49:641–649
UNEP Assessment Report (nd) The Asian brown cloud: climate and other environmental impacts. UNEP/DEWA/RS; 2002,02-3[http://www.rrcap.unep.org/abc/impactstudy/]
Velzeboer I, Kupryianchyk D, Peeters ETH, Koelmans AA (2011) Community effects of carbon nanotubes in aquatic sediments. Environ Int 37(6):1126–1130. pii/S0160412011000328
Vincent JH, Clement CF (2000) Ultrafine particles in workplace atmospheres. Philos Trans R Soc Lond Ser a-Math Phys Eng Sci 358:2673–2682
Wang K, Ruan J, Song H et al (2011) Biocompatibility of graphene oxide. Nanoscale Res Lett 6:1–8
Warheit DB, Laurence BR et al (2004) Comparative pulmonary toxicity assessment of single-wall carbon nanotubes in rats'. Toxicol Sci 77:117–125
Warheit DB, Webb TR, Sayes CM, Colvin VL, Reed KL (2006) Pulmonary instillation studieswith nanoscale TiO_2rods and dots in rats: toxicity is not dependent upon particle size and surface area. Toxicol Sci 91:227–236
Wei W, Sethuraman A, Jin C, Monteiro-Riviere NA, Narayan RJ (2007) Biological properties of carbon nanotubes. J Nanosci Nanotechnol 7:1284–1297
Wijnhoven SWP, Peijnenburg WJGM, Herberts CA, Hagens W, Oomen AG, Heugens EHV, Roszek B, Bisschops J, Gosens I, van de Meent D et al (2009) Nano-silver—a review of available data and knowledge gaps in human and environmental risk assessment. Nanotoxicology 3:109–138
Wild E, Jones KC (2009) Novel method for the direct visualization of in vivo nanomaterials and chemical interactions in plants. Environ Sci Technol 43:5290–5294

Wilson SR, Solomon KR, Tang X (2007) Changes in tropospheric composition and air quality due to stratospheric ozone depletion and climate change. Photochem Photobiol Sci 6:301. doi:10.1039/B700022G

Wolfrum EJ, Huang J, Blake DM, Maness PC, Huang Z, Fiest J, Jacoby WA (2002) Photocatalytic oxidation of bacteria, bacterial and fungal spores, and model biofilm components to carbon dioxide on titanium dioxide-coated surfaces. Environ Sci Technol 36:3412–3419

Wright JB, Lam K, Hansen D, Burrell RE (1999) Efficacy of topical silver against fungal burn wound pathogens. Am J Infect Control 27:344–350

Yang L, Watts DJ (2005) Particle surface characteristics may play an important role in phytotoxicity of alumina nanoparticles. Toxicol Lett 158:122–132

Zhang W (2003) Nanoscale iron particles for environmental remediation: an overview. J Nanopart Res 5(3):323–332. doi:10.1023/A:1025520116015

Zhang Y, Ali SF, Dervishi E et al (2010) Cytotoxicity effects of graphene and single-wall carbon nanotubes in neural phaeochromocytoma-derived pc12 cells. ACS Nano 4:3181–3186

Zhang P, Ma Y, Zhang Z et al (2012) Comparative toxicity of nanoparticulate/bulk Yb_2O_3 and $YbCl_3$ to cucumber (*Cucumis sativus*). Environ Sci Technol 46(3):1834–1841

Zhu Y, Zhao Q, Li Y, Cal X, Li W (2006) The interaction and toxicity of multi-walled carbon nanotubes with Stylonychia mytilus. J Nanosci Nanotechnol 6:1357–1364

Zhu X, Zhu L, Li Y, Duan Z, Chen W, Alvarez PJJ (2007) Developmental toxicity in zebrafish (Danio rerio) embryos after exposure to manufactured nanomaterials: buckminsterfullerene aggregates (nC60) and fullerol. Environ Toxicol Chem 26(5):976–979

Chapter 10
Nanoparticles in Water, Soils and Agriculture

El-Sayed Belal and Hassan El-Ramady

Abstract Nanotechnology is now present in the agricultural and food sectors with nanomaterials such as nanofertilizers and nanopesticides. These nanoparticles may contaminate the environment under certain conditions. The transport, mobility and sorption and bioavailability of nanoparticles is controlled by factors such as soil texture, clay content, pH, cation exchange capacity and soil organic matter. Here we review the fate of nanoparticles in agrosystems. Phytotoxicity of nanoparticles and nanofertilizers are highlighted.

Keywords Nanoparticles • Terrestrial environments • Soils • Water • Microbial community • Nanofertilizers • Phytotoxicity

10.1 Introduction

Nowadays, the agricultural researchers face a great challenge representing in producing the sufficient quantity and quality of food to feed the ever increasing global population without degrading the soil health and agroecosystems (Kashyap et al. 2015). Hence, agrinanotechnology has emerged as a technological advancement to overcome this challenge. Agrinanotechnology is gaining a relevant place in agriculture due to its use in devices like biosensors or for the application of new pesticides, genetic improvement of plants and animals, fertilizers, plant growth regulators and food additives. Furthermore, nanotechnology can help in protecting the environment through waste management and pollution control (Hong et al. 2013). On the other hand, novelty in agri-nanotechnology is rapidly entering almost every agricultural sector around the world. This surge in innovation is largely attributed to a steady increase in investments by both government and private sectors in research

E.-S. Belal
Agricultural Botany Department, Faculty of Agriculture, Kafr El Sheikh Uni, Kafr El-Sheikh, Egypt
e-mail: elsayedbelal1169@yahoo.com

H. El-Ramady (✉)
Soil and Water Department, Faculty of Agriculture, Kafrelsheikh Uni, Kafr El-Sheikh, Egypt
e-mail: ramady2000@gmail.com

S. Ranjan et al. (eds.), *Nanoscience in Food and Agriculture 2*, Sustainable Agriculture Reviews 21, DOI 10.1007/978-3-319-39306-3_10

and development (Scott and Chen 2013). Therefore, agrinanotechnology is an emerging technological platform for the next wave of development and transformation of agricultural systems. Furthermore, it is attracting a large-scale investment from different global corporations and governments around the world (Sandler and Kay 2006). As a result, agrinanotechnology is rapidly moving from the laboratory and onto the farm, supermarket shelves and the kitchen table (Scrinis and Lyons 2007). Agri-nanotechnology was and still a very rich field for research and development, so many researchers published their works related to the applications of nanotechnology in agriculture including food sector (Scrinis and Lyons 2007; Chaudhry et al. 2008; Sozer and Kokini 2009; Chen and Yada 2011; Mousavi and Rezaei 2011; Srilatha 2011; Ditta 2012; Prasad et al. 2012a; Mura et al. 2013; Tarafdar et al. 2013; Thul et al. 2013; Ngô and Van de Voorde 2014; Prasad et al. 2014; Sekhon 2014; Jampílek and Kráľová 2015; Kulkarni 2015; Siddiqui et al. 2015a, b; Shah et al. 2015; Aliofkhazraei 2016).

Therefore, this review is considered an important step in understanding the nanoparticles in agroecosystems including its different components such as soils, soil microbial community, water and plants. Phytotoxicity of nanoparticles as well as nanofertilizers will be also highlighted.

10.1.1 Nanoparticles in Agroecosystems

Nowadays, there is no field untouched by nanotechnology and its ground breaking scientific innovations including agriculture. Therefore, the use of nanotechnology in agriculture has been touched several fields including food industry, plant nanonutrition, plant protection and plant productivity (Thul et al. 2013). Concerning the effects of nanoparticles or nanomaterials on agroecosystems, it should be focused on the fate as well as behavior of nanoparticles on agricultural crops and possible toxic implications to plants, microorganisms naturally present in soil rhizosphere and generation of nano-wastes in agroecosystem (Fig. 10.1). Furthermore, the negative effects of nanomaterials (or nanoecotoxicology, which generated by free radicals leading to lipid peroxidation and DNA damage) on plant productivity, soil microorganisms as well as the environment must not be overlooked (Thul and Sarangi 2015). Hence, there is a need to predict the environmental effects of these nanoparticles in the foreseeable future (Thul et al. 2013).

There are several reports concerning the relationship between nanoparticles and effect of nanoparticles on the environment (Christian et al. 2008; Klaine et al. 2008; Bernhardt et al. 2010; Lin et al. 2010; Peralta-videa et al. 2011; Rico et al. 2011; Lowry et al. 2012; Maurer-Jones et al. 2013; Sekhon 2014; Sharma et al. 2015; Yang et al. 2015; Bottero 2016) or agroecosystems (Chinnamuthu and Boopathi 2009; Thul et al. 2013; Mura et al. 2013; Prasad et al. 2014; Judy and Bertsch 2014; Schwabe et al. 2015a; Van Aken 2015). It is well known that, several types of man-made nanoparticles produced by nano-manufacturing technologies. A side from the benefits, so far no information is elaborated on systematic investigation that raising

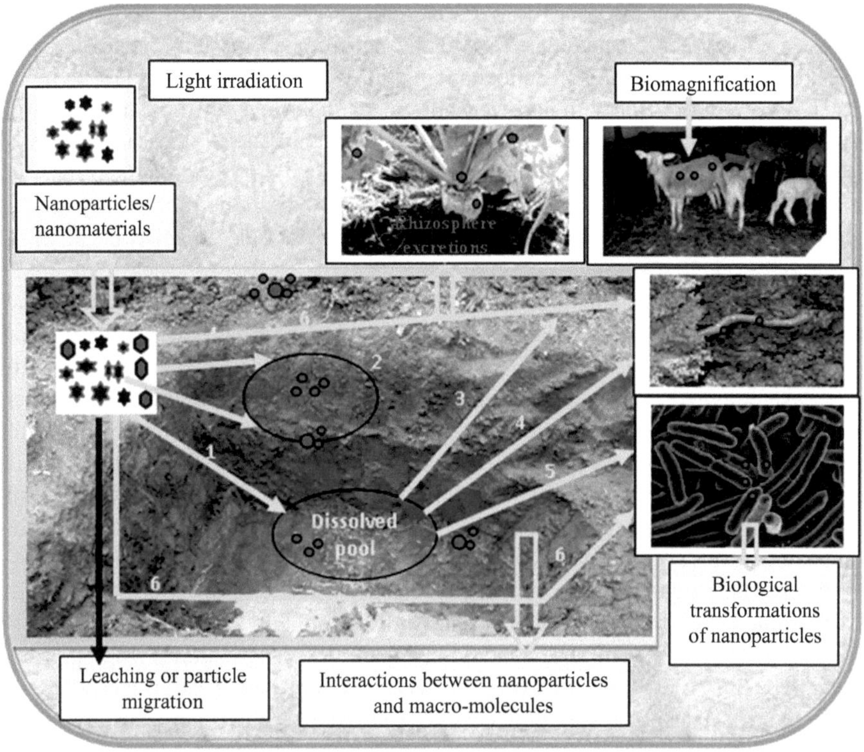

Fig. 10.1 Fate and sink of manufactured nanoparticles (MNPs) as well as chemical and biological processes in agroecosystem. Nanoproducts may release MNPs as individual nano-sized particles, and/or aggregated nano-scale particles. The aggregations may be dispersed by light irradiation, dissolved organic matter coatings and interactions with compounds from rhizosphere excretions. MNPs in soil may be bio-concentrated or bio-magnified through the food chain. Key processes in soil relating to transformation and potential risk from nanoparticles, where (*1*) dissolution, (*2*) sorption/aggregation, (*3*) plant bioaccumulation, (*4*) humans/animals accumulation and toxicity, (*5*) microbial toxicity and (*6*) direct particle uptake/toxicity and translocation. Nanoproducts can be directly applied for soils and thus enter groundwater and soils, whereas the incinerated nano-products during their disposal may enter the atmosphere and then be deposited on soils (Photos by El-Ramady except microorganism picture extracted from https://econature.wordpress.com/2.4.2015)
Source: Compiled from Klaine et al. (2008), Lin et al. (2010), Lowry et al. (2012), Pan and Xing (2012), Thul et al. (2013) and Aslani et al. (2014)

concerns about impacts of these nanoparticles on different environmental processes and organisms (Bernhardt et al. 2010). The ecosystems of plants as well as the environment of agricultural and/or forest crops are the main fields for these investigations. That means, plants and microorganisms have evolved themselves in presence of natural nanomaterials. Recently, the probability of plant exposure to these nanomaterials has increased to a greater extent with the ongoing increasing production and use of engineered nanomaterials for variety of products (Pan and Xing 2010; Thul et al. 2013).

Therefore, a growing concern over the potential for unintended, adverse consequences of engineered nanoparticles in the agroecosystems have generated new research initiatives focused on understanding the ecological effects of these nanoparticles. Furthermore, the fate and transport of nanoparticles in terrestrial environments including soils, waters, and sediments or about the biological impacts of these nanoparticles in natural environments, and the bulk of modern nanotoxicogical research should be focused on more than highly controlled laboratory studies with single species in simple media (Bernhardt et al. 2010; Ranjan et al. 2014; Dasgupta et al. 2015).

Therefore, nanoparticles exist in the environment naturally; however, natural nanoparticles are present at very low concentrations and have negligible impacts (Klaine et al. 2008; Remedios et al. 2012). In recent decades, fabricated nanoparticles have emerged and have been incorporated into a growing number of commercial products. Due to better understand the threat posed by nanoparticles and for more control their discharge into agroecosystem, both scientific and commercial efforts have become very large. Although, in many cases, the release of nanoparticles is unavoidable, the goal should be to minimize those releases of nanoparticles that could pose a significant risk to agroecosystem or to humans (Yadav et al. 2014).

It could be concluded that, nanoparticles have different effects on the environment. Concerning these different effects of nanoparticles on agroecosystems, it should be focused on the fate as well as behavior of nanoparticles on agricultural crops, possible toxic implications to plants, microorganisms naturally present in soil rhizosphere and generation of nano-wastes in agroecosystem. In general, these agroecosystems can include soils, waters, plants, microorganisms and animals as well as humans.

10.1.1.1 Nanoparticles in Soils

It is well known that, nanoparticles are not a human invention and naturally have existed from the beginning of the Earth's history (Handy et al. 2008). Soils, as natural nanoparticles, are a complex matrix containing mineral particles, colloids in pore water and there are concerns about adsorption and binding of pollutants within the matrix (Reid et al. 2000; Sharma et al. 2015). Furthermore, it is clear that exposure to natural nanoparticles vastly outweighs any anthropogenic production. Concerning the methods of nanoparticles formation, there are several mechanisms can create nanoparticles in the environment including geological or biological methods. Geological mechanisms include physicochemical weathering, authigenesis/neo-formation (e.g., in soils) and volcanic activity. These geological processes typically can produce inorganic particles. In regard to biological mechanisms, they can produce organic nanomolecules, although some organisms can produce mineral granules in cells (Handy et al. 2008).

The movement of nanoparticles in soils is not affected by gravitational settling but is guided by Brownian motion into the soil pores. Soils are composed micropores, i.e., network of humic materials and soil particles and macropores (Mura et al. 2013). Therefore, single nanoparticles can enter into the micropores and if they are sorbed on mobile colloids, the mobility is highly enhanced, while aggregates of nanoparticles remain in the macropores, whereas if sorbed on non-mobile particles their mobility is inhibited. The attachment between soil molecules and nanoparticles depends on the shape of nanoparticles and the collector and on the different properties that change the environment surrounding the particles. Therefore, soil conditions can enhance or inhibit nanoparticles mobility in soils (Fang et al. 2009). Ben-Moshe (2010) demonstrated that humic acids in soils and aquifers and the ionic strength of resident water can dramatically influence nanoparticle mobility of different types of metal oxides nanoparticles (like Fe_3O_4, TiO_2, CuO and ZnO), obtaining the highest mobility for nano-TiO_2. This may influence the monitoring of nanoparticle composition as well as the fate of soil nutrients, organic pollutants and contaminants (Mura et al. 2013).

There is a strong ascending interest in the use of nanoparticles in soils for different applications as reported in different papers (e.g., Boxall et al. 2007; Wilson et al. 2008; Theng and Yuan 2008; Zhan 2009; Mueller and Nowack 2008; He et al. 2011; Towell et al. 2011; Pan and Xing 2012; Collins et al. 2012; Priester et al. 2012; Jafar and Hamzeh 2013; Suppan 2013; Jośko and Oleszczuk 2013; Fernández et al. 2014; Jośko et al. 2014; Garner and Keller 2014; Conway et al. 2015; Watson et al. 2015; Schaumann et al. 2015a, b; Rabbani et al. 2016). However, researchers are still needed to develop new ways in monitoring possible nanoparticles toxicity (Karn et al. 2009; Grieger et al. 2010) and to evaluate the benefits or risks to organisms of interest in soil sciences including plants, bacteria, fungi and other living beings (Mura et al. 2013). It is also demonstrated that, negative effects of nanoparticles on soil bacterial communities, which were altered by nano-ZnO reducing the microbial biomass and diversity (Ge et al. 2011). Furthermore, soil enzyme activities were influenced by treatments with nano-ZnO. Due to immobilization and aggregation of nanoparticles in the soil showing phytotoxicity, it is also reported that biomass and root length decreased (Kim et al. 2011). Toxicity of nano-ZnO was also demonstrated on ryegrass and in maize where the germination is inhibited (Lin and Xing 2007). Other experiments carried out in plants showed that aluminum oxide nanoparticles and rare earth oxide nanoparticles can be toxic to a variety of crops (maize, soybeans, carrots, cabbage, cucumbers) as inhibitors of root elongation (Ma et al. 2010a).

Soil science is concerned with the science of all those materials we find in soils. For organisms this matrix can provide nutrition. In addition it is concerned with those microflauna and fauna that aid this process. This is a complex mix of chemicals and organisms some of which are organized at the nanolevel and some of which are not. The definition of nanotechnology has expanded from the initial discoveries of the capacity to move and locate atoms singularly to something much larger (Wilson et al. 2008). Soils are complex mixtures of solids (or particles) ranging from millimeter (mm) to nanometer (nm) in particle size. Using some developed techniques in nanotechnology like Transmission Electron Microscopy (TEM) and Atomic Force Microscopy (AFM), it could be possible to understand these structures of soils. These previous techniques can illustrate the organization of colloidal material in soils such as phyllosilicates, and humic acids as well as the discovery of new particles such as nanoparticles of iron oxides. Therefore, nanotechnology can offer further potential in identifying single cells, genes, individual DNA molecules, proteins, and other biological structures in soils (Wilson et al. 2008; Ranjan et al. 2015; Dasgupta et al. 2016c).

Concerning soil sciences, nanoscience is of crucial importance to these sciences because several natural compounds of the soils are nanoparticulate or have nanoscale features (Mura et al. 2013). Whereas at the nanoscale, the interactions between particle and other particle are either dominated by weak Van der Waals forces, stronger polar and electrostatic interactions, or covalent interactions. The details of nanoparticle–nanoparticle interaction forces and nanoparticle–fluid interactions are of key importance to describe physical and chemical processes, and the temporal evolution of free nanoparticles (Mura et al. 2013). On the other hand, different nanomaterials can be found in soil like nanominerals ranging from nanoparticle to mineral nanoparticles in nanosize but also exist at larger sizes (Maurice and Hochella 2008). It is reviewed about the natural inorganic nanoparticles including formation, fate and toxicity in the environment (Sharma et al. 2015). Furthermore, different nanomaterials also are present in soils including nanoscale aggregates of natural organic matter, bacterial appendages, organic and inorganic particles, clay minerals, amorphous substance and other nanominerals (Fe-minerals: hematite, goethite, magnetite; Mn-minerals and swelling clays: montmorillonite) (Mura et al. 2013). Manufactured or engineered nanoparticles can also be found in soils, whereas these nanoparticles can be released in the environment either intentionally in different forms including metal-based nanoparticles including metal oxides (e.g., nano-ZnO, -TiO_2, -CeO_2), zero valent metals (such as Au, Ag and Fe nanoparticles), and metal salts (e.g., nano silicates and ceramics) carbon derived nanomaterials (carbon nanotubes), quantum dots or semiconductor materials (e.g., CdSe, CdTe) and nanopolymers (e.g., polystyrene, and latex) or unintentionally by corrosion or combustion of products (Mura et al. 2013; Zhang et al. 2015).

Due to the dispersion of manufactured nanoparticles in soils, a change in size of their aggregates, suspension stability, transport and bioavailability can be noticed (de Santiago-Martín et al. 2015). Therefore, studies on colloidal processes of manufactured nanoparticles are essential to understand their fate and risk (Philippe and Schaumann 2014; Sharma et al. 2015). The suspension of these nanoparticles can be affected by soil conditions including ionic strength, the amount of dissolved organic matter and chemical and biological reactions (Nowack et al. 2012; Delay et al. 2015; Peng et al. 2015b; Zou et al. 2015; Li et al. 2016). Dissolved organic matter coats these nanoparticles and changes their surface properties, such as pore size distribution, specific surface area, characteristics of sorption to organic contaminants and aggregation status, as well as the mechanisms of toxic effects (Wang et al. 2011; Grillo et al. 2015; Li et al. 2016). It is demonstrated clearly and convincingly that dissolved organic matter increased the stability of metal-based manufactured nanoparticles in water suspension (Wang et al. 2011). In terrestrial environments, biological excretions such as exudates in the rhizosphere from roots and microorganisms, and contain macromolecules similar to those in dissolved organic matter (Grillo et al. 2015; Delay et al. 2015). These macro-molecules may form complex aggregates widely found in aqueous systems and are termed as biofilms (Lerner et al. 2012). The interactions between these macro-molecules and these nanoparticles may also result in the alteration of their aggregation as well as their transport. Due to the complexity and the heterogeneity of soil and sediment systems, they have hindered research on manufactured nanoparticle leaching and transport. The heterogeneity can be considered at two scales including both particles and systems. Concerning the particle scale, soil aggregates have large variability in their size and roughness, shape, composition and relative proportion of minerals, surface charge properties and size distribution and interconnectivity of pores (Jaisi and Elimelech 2009). Whereas at the system scale, natural soil columns usually contain open soil structures e.g., cracks, fissures, worm burrows and other open features (Jaisi and Elimelech 2009). Due to complexity, no single property can be applied as a general predictor of the transport and deposition of manufactured nanoparticles (Wang and Keller 2009). So, it is important to describe manufactured nanoparticle transport in soil columns quantitatively (Pan and Xing 2012).

Therefore, it could be concluded that, the environmental application and risk assessment of manufactured nanoparticles in soil greatly depend on our understanding of the interactions between these nanoparticles and different soil components. These engineered nanoparticles could be applied for soil remediation and then are released unintentionally through various other pathways to soils. Due to the complexity of soil system and the very early stage of nanoparticles research in soils, the understanding of nanoparticles behavior in this system is very limited. Nanoparticles can be formed naturally in different environments including soils, natural ores, microbial systems, aerosols, waters, and deepsea hydrothermal vents. Further studies are needed to evaluate different toxicity mechanisms of naturally formed nanoparticles.

10.1.1.2 Nanoparticles and Soil Microbial Community

From time to time, a large area of science attracts attention particularly when new developments occur between fields and new subject areas are born. Such processes are important, because they allow focusing of different disciplines on one area and this itself leads to new advancement. Over the last 20 years and particularly over the last ten, nanotechnology has evolved as an interdisciplinary area, which has attracted great interest (Wilson et al. 2008). Clearly the area of nanomaterial science is relevant to the analysis of soil structure and composition. However, nanobiology relevant to soil biology and instruments for detecting nano quantities of substances relevant to soil processes are also important as are various aspects of nanotechnology applications in environmental sciences (Wilson et al. 2008). It is reported that, engineered nanoparticles can be made of single elements like carbon or silver or a mixture of elements/molecules. These nanoparticles can be classified based on their chemical composition, size or morphology characteristics. It could be also described these nanoparticles into following groups including fullerenes, metal engineered nanoparticles (elemental Ag, Au, Fe, Se), oxides (TiO_2, CuO, FeO_2, ZnO, Al_2O_3, SiO_2, CeO_2), complex compounds like Co–Zn–Fe oxide, and quantum dots often coated with a polymer e.g., cadmium–selenide (CdSe) and organic polymers like polystyrene (Dinesh et al. 2012).

Microorganisms can be considered the drivers of global biogeochemical cycles. Whereas, these microorganisms are involved in the cycling of C, N, S as well as P. Due to their especial sensitive to environmental changes, the structure and abundance of the microorganism community may shift in response to foreign nanomaterials (Ge et al. 2011; Kumar et al. 2011). Because microorganisms help regulate and maintain overall ecosystem health and function (Janvier et al. 2007), changes in the microbial community will have a great effect on the entire ecosystem (Kanerva et al. 2008). Therefore, a better understanding of how microorganisms respond to nanoparticles and/or nanomaterials can help to address environmental and health concerns brought about by the manufacture and use of these nanomaterials (He et al. 2011). On the other hand, it is well documented that, several nanoparticles have already been reported to have anti-microbial properties and thus directly affect microorganisms (Shah and Belozerova 2009). There are no standard and proven methods for assessing the toxicity of nanoparticles on different soil microbial communities. To evaluate the effect of contaminants on the soil microbial community, various methods can be used including viability count, molecular-based methods, carbon utilization patterns and fatty acid methyl ester analysis. From a thorough literature review, it is reported that, the interaction between the nanoparticles and the bacterial cell leading to cytotoxicity has been hypothesized to involve two steps (Kumari et al. 2014). Concerning the first step, an oxidative damage by the nanoparticles is attached to the cell membrane, resulting in loss in membrane integrity without significant reduction in cell viability. Whereas, the second step involves leaking out of intracellular components, which is the leading cause of decreased viability and internalization of the nanoparticles, thereby causing damage to cell organelles including the nucleus (Klaine et al. 2008; Kumari et al. 2014).

Most of the microorganisms have developed effective molecular mechanisms and operated specific biochemical pathways to efflux, detoxify and accumulate the metals ions much before it was learnt by the plants. Furthermore, microorganisms are also capable to volatilize some of the metal ions to get rid of its acute toxicity (De Souza et al. 1999). Although microbes have developed resistance and avoidance mechanism, but more targeted studies are needed in regards to beneficial soil microbes such as N_2-fixing, phosphate solubilizers, arbuscular mycorrhizal fungi to establish the uptake mechanisms and consequences in soil and microbes (Thul and Sarangi 2015). It is noticed that, not only the size of nanoparticles but reduced length of shoot and root of wheat was observed in a dose-dependent manner (Dimpka et al. 2013). In another study, uptake of nano-ZnO causes damage of epidermal and cortical cells and transport from one cell to other through plasmodesmata (Lin and Xing 2008). Similarly, the evidence for the entrapment of nano-Ag by the cuticle, and penetration into the leaf tissue through stomata, and oxidation of nano-Ag and complexation of Ag^+ by thiol-containing molecules was reported by Larue et al. (2014). Furthermore, the cytotoxic and genotoxic impacts of nano-Ag were reported in root tips of onion (Kumari et al. 2009). Similar effects of chromosomal aberrations and DNA damage were also observed with TiO_2 (Pakrashi et al.

2014). Recently, nano-TiO_2 particles were reported to affect the molecular expression profiles of micro-RNAs (Frazier et al. 2014).

A range of studies has been reviewed concerning the interactions between nanoparticles and microorganisms to correlate the physico-chemical properties of engineered metal and metal oxide nanoparticles to their biological response (Shah and Belozerova 2009; Haensch and Emmerling 2010; Du et al. 2011; Kim et al. 2011; Rai and Duran 2011; Fajardo et al. 2012; Nogueira et al. 2012; Ge et al. 2012; Dinesh et al. 2012; Ansari and Husain 2012; Antisari et al. 2013; Pawlett et al. 2013; Holden et al. 2013; Tilston et al. 2013; Jośko et al. 2014; Holden et al. 2014a; Dimkpa 2014; García-Gómez et al. 2015; Simonin and Richaume 2015; Burke et al. 2015; Sillen et al. 2015; Van Aken 2015; Judy et al. 2015a; Xu et al. 2015; Aliofkhazraei 2016; Sirbu et al. 2016). Furthermore, it has been concluded that the species specific toxicity can be attributed to nanoparticle size and shape. However, the surface coating of the material, which can be altered significantly by environmental conditions, can ameliorate or promote microbial toxicity (Suresh et al. 2013; Thul et al. 2013).

From current literatures, it could be concluded that, there are several impacts of nanoparticles on soil microorganisms including the nitrogen cycle, soil enzyme activities, and processes involved in iron metabolism, phytohormones, and antibiotic production (Dimkpa 2014). These effects are negative or positive and the outcome being dependent on specific nanoparticles (type, size, surface charge, dose tested, specie of plant or microbe examined, and test media whether liquid, soil, agar, or other solid media). From these collectively published results, it could be suggested that, nanotechnology portends considerable, many negative, implications for soil microbes and, thus, agricultural processes that are microbially driven. Nonetheless, it could be observed that, the potential of plant and soil microbial also processes to mitigate the bioreactivity of nanoparticles (Dimkpa 2014). Although the roots of most terrestrial plants are associated with microbes, several studies of nanoparticle interactions with plants and microbes are generally conducted separately. The few studies in actual microbe/plant systems found effects of nanoparticles on the functioning of arbuscular mycorrhizal fungi, nitrogen fixation, as well as on the production of microbial siderophores in the plant rhizosphere. Therefore, it could be suggested that, a better understanding of the agroecological ramifications of nanoparticles would require more in-depth interactive studies in combined plant/microbe/nanoparticle systems (Dimkpa 2014).

Concerning soil microorganisms, there is a clear and specific effect of metal nanoparticles on soil microflora. For example, it is observed an impact of nano-TiO_2 on *Rhizobium*–legume symbiosis using garden peas and *Rhizobium* (*R. leguminosarum* bv. *viciae* 3 841). It is also found that, nano-TiO_2 exert morphological changes in bacterial cells. Furthermore, it was noticed that the interaction between these two organisms was disrupted in the form of root nodule development and the subsequent delay in onset of nitrogen fixation (Fan et al. 2014). The direct application of nanoparticles into soils or treated biosolids containing mobile nanoparticles may come in contacts with the soil microorganisms. These microorganisms are also efficient to adsorb and accumulate one or other form of nanomaterials, which in turn initiates the mobilization of nanomaterials through food chains and can alter communities comprising multiple populations (e.g., plant, fish, bacteria) within food webs (Holden et al. 2013; Thul and Sarangi 2015; Ranjan et al. 2014).

On the other hand, plants generally depend on soil bacteria and fungi to get their nutrients from the soils. A study finds that the popular microbicidal nano-Ag negatively impacts on the growth of plants and kills the soil microbes that sustain them (Zeliadt 2010). Not only microbes, but activity of several soil enzymes such as soil protease, catalase and peroxidase were found to be significantly reduced by nano-TiO_2 (Du et al. 2011). Moreover, inorganic SiO_2, ZnO and TiO_2 were found to exert toxic effect on bacteria. The toxicity of these elements further significantly enhanced in presence of light (Adams et al. 2006). A range of studies has been reviewed and focused on nanoparticles-microbial interactions to correlate the physicochemical properties of engineered metal and metal oxide nanoparticles as well as their biological response. Furthermore, it has been concluded that the species specific toxicity can be attributed to nanoparticles' size and shape. However, the surface coating of the material, which can be altered significantly by environmental conditions, can ameliorate or promote microbial toxicity (Suresh et al. 2013). Studies on ecologically relevant bacterial species such as *Escherichia coli, Bacillus subtilis, Pseudomonas putida*, and other have clearly indicated that nanoparticles can be taken up by microorganisms (Thul and Sarangi 2015).

Bacteria are essential elements of terrestrial and aquatic ecosystems as decomposers of organic matter and primary base for many food webs (Thul et al. 2013). These microorganisms are also efficient to adsorb and accumulate one or other form of nanomaterials, which in turn initiates the mobilization of nanomaterials through food chains and can alter communities comprising multiple populations like plant, bacteria and others within food webs (Holden et al. 2013). Due to plants depend on soil microorganisms including bacteria and fungi to get their nutrients from the air and soil, the antimicrobial and cellular toxicity effects of nanoparticles (e.g., silver, titanium dioxide and gold nanoparticles and nanoemulsions) could have impacts at the ecosystem level (Thul et al. 2013; Dasgupta et al. 2016a, b; Ranjan et al. 2016; Jain et al. 2016; Maddineni et al. 2016). In distinguished review, Hajipour et al. (2012) reported about antibacterial properties of nanoparticles. It is found that, the popular microbicidal silver nanomaterial negatively impacts the growth of plants as well as kills the soil microorganisms that sustain them (Zeliadt 2010). ZnO and TiO_2 nanoparticles significantly reduced activity for several soil enzymes including soil protease, catalase, and peroxidase, as well as significantly soil microbial communities (Du et al. 2011). It is also reported that, soil microorganisms, which are abundant and versatile catalysts, can sorb and disperse engineered nanomaterial agglomerates (Horst et al. 2010). It was observed that, the addition of nanoscale zerovalent iron leads to perturbation to soil bacterial community composition, and reduced the activity of chloroaromatic mineralizing microorganisms (Tilston et al. 2013). Moreover, inorganic TiO_2, SiO_2 and ZnO found to exert toxic effect on bacteria, the toxicity of these elements further significantly enhanced in presence of light (Adams et al. 2006).

The engineered nanoparticles was also found to significantly alter the bacterial communities in a dose-dependent manner, with some taxa increasing as a proportion of the community, but more taxa decreasing, indicating that effects mostly reduced diversity (Ge et al. 2011). In line to this, the uptake of manufactured nano-CeO_2 nanomaterials into roots and root nodules found to eliminate N_2 fixation potentials and impaired soybean growth (Priester et al. 2012). Likewise, most of the toxicity studies on ecologically relevant bacterial species such as *E. coli, Bacillus subtilis, Pseudomonas putida* (Załęska-Radziwiłł and Doskocz 2015) and other have been conducted for growth inhibition and cell viability clearly indicating that nanoparticles can be taken up by microbes, but more studies are needed in regards to beneficial soil microbes such as N_2 fixing, phosphate solubilizers, AM fungi to establish the uptake mechanisms and consequences of their accumulation in soil and microbes (Ge et al. 2011; Thul et al. 2013).

The mobility of the engineered nanoparticles in soils is dependent on their size, although it is the agglomerate size, not the primary size that is correlated with transportability. Many factors control the mobility of these nanoparticles in the soils, but size, charge, and agglomeration rate in the transport medium are predictive of these nanoparticles mobility in soils (Darlington et al. 2009). The existence and speciation of metal nanoparticles in soil solution and knowledge on the interaction between their active sites and soil solution or other ions are essential for a better understanding of the interactions between metal nanoparticles and microorganisms in soils. However, the solution chemistry of metal nanoparticles is quite limited and thermodynamic data such as solubility and reaction constants of nanoparticles are unavailable. On the long-term, data is required to know the effects of engineered nanoparticles on soil microbial populations in a range of soils with varying physicochemical characteristics and soils from different ecosystems (Dinesh et al. 2012).

It could be concluded that several new engineered nanoparticles as xenobiotics from both industrial and environmental applications are resulted from different human activities as by-products, which are already finding their way into soils. Therefore, the protection of soil microbial biomass and diversity is one of the major challenges for sustainable resource use because greater levels of microbial biomass and diversity mean greater nutrient turnover. A few studies have been conducted on the toxicity of engineered nanoparticles to soil organisms because of the complexity with which organisms are exposed to these nanoparticles within different soil phases. Further studies are needed to understand the entire effects of nanoparticles on different soil organisms under various environmental conditions through the identification of soil parameters controlling the bioavailability and toxicity of these nanoparticles.

10.1.1.3 Phytotoxicity of Nanoparticles

Several studies have addressed the uptake, translocation, bioaccumulation and phytotoxicity of nanoparticles by plant systems. Researchers have reported negative, positive and inconsequential effects from plants being exposed to nanoparticles. In general, nanoparticles-associated alteration of plant morphological features including roots, leaves and seed germination has been also reported. Unfortunately, very few studies have been conducted on the genetic response of plants that are exposed to nanoparticles (Yadav et al. 2014). In Fig. 10.1, it is presented an outline of the major possible pathways through which engineered nanoparticles may enter into agroecosystems to potentially cause toxicity to plants.

Nanotoxicology emerged in the early 2000s (Donaldson et al. 2004) and since this time it has been an exponential increase in the number of hazard and exposure studies concerned with evaluating nanomaterials safety (Johnston et al. 2013). Furthermore, a little was understood or known about risks of nanotechnology including also these following issues: basic toxicology of engineered nanomaterials, environmental fate and transport, reliable characterization strategies, occupational hazards, applicability of current governance schemes, and many other areas were in their infancy or largely unexplored (Hull and Bowman 2014). However, despite this progress there are still many challenges that exist when evaluating nanomaterial risks to human health and the environment, and therefore uncertainties surrounding the risks posed by nanomaterials (Johnston et al. 2013). Many key questions remained unanswered, making it difficult for the scientific community, policy makers and industry at large to make evidence-based decisions (Klaine et al. 2012). It is still too early to tell which specific action/s (or indeed inaction/s) will be taken, by whom, and when, and whether or not they will yield their desired outcomes (Hull and Bowman 2014).

The interactions between plants and engineered nanomaterials are of particular importance, as plants directly interact with soil, water, and the atmosphere, and serve as a potential pathway of these nanoparticles exposure for higher species through the food chain (Zhang et al. 2015; Singh et al. 2015b). It is worth to mention that, nanotechnology can support several applications to soil sciences (Lal 2007). Furthermore, the use of nanofertilizers can facilitate nutrient transport to the rhizosphere when needed and in more suitable amounts, thereby improving efficiency of use. Nanoparticles locked onto the roots can also enhance elemental uptake. The nanoscale delivery vehicles may be designed to fix the surrounding soil particles or organic matter (Johnston 2010), allowing more efficient release mechanisms. These features could promote the active substances to be taken up at a slower rate throughout crop growth, avoiding temporal overdoses, and minimizing input and wastes (Chen and Yada 2011). The slow release of nutrients into the environment could be achieved by using zeolites, which are a group of naturally occurring minerals with a honeycomb-like layered crystal structure (Mura et al. 2013).

A side from phytotoxicity, nanotoxicity is another concern of using nanomaterials as fertilizers. Liu and Lal (2015) reported that, additional research on the toxicity of newly developed nanofertilizers should be conducted to alleviate the public fears about the associated nanotoxicity. They listed two important points concerning nanotoxicity research including plants should have specific mechanisms in moderating these small particles and crops need only trivial amounts of nutrients to maintain normal physiological activities while extremely high concentrations of any nutrient would cause phytotoxicity. There are no enough researches concerning the influence of nanoparticles on terrestrial environments and phytotoxicity. Whereas and apart from detrimental effect upon direct contact of these nanoparticles, these can also diffuse into the intercellular space, the apoplast, and be adsorbed or incorporated into the membranes (Nowack and Bucheli 2007). Due to the negative surface charge of plant cells, these cells allow the apoplastic transport of negatively charged compounds into plants. Due to the existence of casparian strip, it poses a barrier to the apoplastic flow and transport, and only symplastic transport is possible into the xylem. However, this barrier is not perfect and compounds can enter the xylem through holes or damaged cells without ever crossing a cell membrane and be further transported to the shoots. Therefore this process is found to be a dominant process for the uptake of metal complexes with chelators such as EDTA and their translocation to the shoots (Tandy et al. 2006). This indicates that negatively charged nanoparticles could enter the apoplast of the root cortex and eventually also the xylem, but are not taken up by the cells (Thul and Sarangi 2015).

It is also worth to mention that, the phytotoxicity due to biomagnification, bioaccumulation and biotransformation of engineered nanoparticles in food crops is still not well understood (Thul and Sarangi 2015). Few studies have been reported on the accumulation of engineered nanomaterials in crop plants such as rape, radish, lettuce, corn, and cucumber (Rico et al. 2011). The carbon-based fullerenes (C_{70} and fullerols $C_{60}(OH)_{20}$) and most of the metal-based nanomaterials (zinc oxide, titanium dioxide, cerium oxide, magnetite, gold, silver, iron, and copper) were reported to be accumulated in plants (Rico et al. 2011). Moreover, accumulated nanomaterials in the plants can be the part of biological food chain. As a part, positive effects of metal-based nanomaterial on plant encouraged for some crops, on the other hand, significant negative effects were also observed, such as reduced root growth, shoot length and germination (Thul et al. 2013; Thul and Sarangi 2015).

Therefore, nanoparticles can be involved in natural processes including soil development and cycling of nutrients but can also act as vehicles of contaminant transport, alter the bioavailability of substances and hence their toxicity. Particles in the nanoscale or colloidal form are abundant in all environmental compartments. These nanophases are composed of natural organic matter including humic substances, biota itself (viruses, bacteria including pathogens), inorganic particles (clays, oxides or carbonates) and engineering nanoparticles (Tourinho et al. 2012; Mura et al. 2013). Therefore, it has still further question whether nanoparticles taken up into plants having toxic effects will be consumed by animals and humans? Hence, assessing the environmental risks requires a clear understanding concerning nanoparticles ecotoxicity, biomobility, bioavailability, reactivity and persistency in agroecosystems. Nanomaterials can indirectly cause membrane damage by generating reactive oxygen species (ROS), which can oxidize double bonds on fatty acid tails of membrane phospholipids in a process known as lipid peroxidation. This process may further leads to membrane permeability and fluidity, making cells more susceptible to osmotic stress and failure to nutrient uptake (Navarro et al. 2008). Peroxidized fatty acids can trigger reactions that generate other free radicals, leading to more cell membrane and DNA damage (Thul et al. 2013).

Many metal oxide nanoparticles and carbon-based materials are being screened for phytotoxicity as well as environmental toxicity or ecotoxicity (Fubini et al. 2007; Handy et al. 2008; Navarro et al. 2008; Kahru et al. 2008; Stampoulis et al. 2009; Ruffini and Cremonini 2009; Kahru and Dubourguier 2010; Ma et al. 2010b;

Peralta-Videa et al. 2011; Rico et al. 2011; Dietz and Herth 2011; Mondal et al. 2011; Dimkpa et al. 2012; Smita et al. 2012; Suriyaprabha et al. 2012a; Batley et al. 2013; Maurer-Jones et al. 2013; Holden et al. 2014b; Husen and Siddiqi 2014; Magdolenova et al. 2014; Li et al. 2015; Arruda et al. 2015; Watson et al. 2015; Bhattacharyya et al. 2015; Bour et al. 2015; Jorio 2016). Several metal and metal oxide nanoparticles have been involved in many studies in relation to its phytotoxicity as follows:

1. Aluminum oxide, Al_2O_3 for wheat (Yanık and Vardar 2015), microalgae (Sadiq et al. 2011) and bacteria (Załęska-Radziwiłł and Doskocz 2015).
2. Cerium dioxide, CeO_2 (Rico et al. 2013a,b; Schwabe et al. 2014; Hong et al. 2014; Hawthorne et al. 2014; Zhao et al. 2014; Majumdar et al. 2014; Ma et al. 2015; Rui et al. 2015; Schwabe et al. 2015b; Wang et al. 2015b).
3. Copper oxide, CuO (Shaw and Hossain 2013; Lalau et al. 2014; Nair and Chung 2014; Perreault et al. 2014; Ouda 2014; Shi et al. 2014; Peng et al. 2015a, b; Stewart et al. 2015; Dimkpa et al. 2015; Wang et al. 2015a; Anjum et al. 2015).
4. Gold, Au (Beattiew and Haverkamp 2011; Sabo-Attwood et al. 2012; Koelmel et al. 2013; Zhai et al. 2014; Ramesh et al. 2014; Taylor et al. 2014; Kern 2015).
5. Iron oxide, Fe_xO_y or FeOOH (Ghafariyan et al. 2013; Pardha-Saradhi et al. 2014; Binns 2014; Kim et al. 2015; Burke et al. 2015; Martínez-Fernández et al. 2015; Libralato et al. 2015; Bombin et al. 2015; Canivet et al. 2015).
6. Molybdenum, Mo (Aubert et al. 2012; Taran et al. 2014; Kanneganti and Talasila 2014).
7. Nickel oxide, NiO (Faisal et al. 2013; Oukarroum et al. 2015; Antisari et al. 2015).
8. Selenium (Domokos-Szabolcsy 2011; Domokos-Szabolcsy et al. 2012, 2014; El-Ramady 2014; El-Ramady et al. 2014a, b, c, 2015a, b, c, 2016).
9. Silicon, SiO_2 (Adams et al. 2006; Li et al. 2012; Suriyaprabha et al. 2012a,b; Siddiqui et al. 2014; Siddiqui and Al-Whaibi 2014; Kalteh et al. 2014; Abdul Qados and Moftah 2015; Janmohammadi et al. 2015).
10. Silver, Ag (Kaveh et al. 2013; De La Torre-Roche et al. 2013; Pokhrel and Dubey 2013; Prakash et al. 2014; Geisler-Lee et al. 2014; Ouda 2014; Larue et al. 2014; Boenigk et al. 2014; Sharma et al. 2014; Sillen et al. 2015; Kohan-Baghkheirati and Geisler-Lee 2015; Gorczyca et al. 2015; Doolette et al. 2015; Judy et al. 2015b; Almutairi and Alharbi 2015; Singh et al. 2015a).
11. Titanium dioxide, TiO_2 (Lei et al. 2008; Menard et al. 2011; Foltete et al. 2011; Larue et al. 2012; Feizi et al. 2012; Gao et al. 2013; Larue et al. 2014; Castiglione et al. 2014; Burke et al. 2015; Chen et al. 2015; Wang et al. 2015b).
12. Zinc oxide, ZnO (Lin and Xing 2008; Ghodake et al. 2011; Prasad et al. 2012a, b; Ma et al. 2013; Sedghi et al. 2013; Zhao et al. 2013a,b; Pokhrel and Dubey 2013; Raliya and Tarafdar 2013; de la Rosa et al. 2013; Mukherjee et al. 2014; Ramesh et al. 2014; Zhao et al. 2014; Helaly et al. 2014; Raskar and Laware 2014; Liu et al. 2015a; Zhang et al. 2015; Chen et al. 2015; Stewart et al. 2015; Dimkpa et al. 2015; Watson et al. 2015).

The toxic action of metal and metal oxide nanoparticles can potentially involve at least three distinct mechanisms (Brunner et al. 2006). These mechanisms include: (1) particles may release toxic substances into exposure media, e.g., free Ag ions from silver particles. (2) Surface interactions with media may produce toxic substances, such as chemical radicals or reactive oxygen species (ROS). (3) Particle or their surfaces may interact directly with, and disrupt biological targets, like carbon nanotubes interaction with membranes or intercalation with DNA (Ma et al. 2013). However, there are several knowledge gaps need to be filled to gain a thorough understanding on nanoparticles ecotoxicity for risk assessment and management for several material. Concerning ZnO nanoparticles, it is reported that there three reasons concluded by Ma et al. (2013). First, there is a significant lack of characterization for ZnO nanoparticles and the exposure system in ecotoxicity studies conducted thus far. Second, tools and techniques are needed to differentiate between particle induced toxicity and dissolved Zn^{2+} effects. Third, there is no sufficient data on chronic effects from long term and low concentration exposure, which may be more representative for real environmental exposure. Based on these knowledge gaps, future ecotoxicity study of ZnO nanoparticles should focus on the long-term effects of ZnO nanoparticles at low exposure levels as well as bioaccumulation and trophic transfer of these nanoparticles in aquatic and terrestrial environments.

Concerning factors affecting nanoparticle phytotoxicity, it is reported from several literatures that, the uptake, translocation, and accumulation of nanoparticles depends on the species of plant, and the size, type, chemical composition, fictionalizations, and stability of the nanoparticles in the system (Larue et al. 2012; Yadav et al. 2014). That means, there are factors related to nanoparticles, plants and environment, whereas the first one includes particle size and shape, surface characteristics and concentration (reactive surface, surface coating, concentration, solubility), dispersion medium, and stabilizers. About the factors related to plants, they include stomatal opening, membrane transporters, root surface receptors, physical state of plant and finally the environmental factors includes radiation or sun light, moisture, living organisms and the components of soil/water/air (Yadav et al. 2014).

Several researchers studied the interactions of nanoparticles with plants. They have shown different degrees of phenotoxicity and genotoxicity as well as phytotoxicity. It is clear to notice that, most of the studied nanomaterials affect plants at different growth stages. For instance, silicon nanoparticles completely inhibit zucchini seed germination (Stampoulis et al. 2009), whereas Siddiqui and Al-Whaibi (2014) suggested that nanoparticles-SiO_2 could be used as a fertilizer for the crop improvement (up up to 8 g L^{-1} nanoparticles-SiO_2 for tomato). The same trend was observed by Suriyaprabha et al. (2012a, b) for maize and basil (*Ocimum basilicum*) under salinity stress (Kalteh et al. 2014). CeO_2 nanoparticles reduce seed germination in corn, tomato, cucumber, and spinach (Lopez-Moreno et al. 2010). Root development is also affected by engineered nanomaterials. For example, TiO_2 nanoparticles reduce root growth in onions, wheat, and tobacco (Larue et al. 2012; Feizi et al. 2012) and Cu nanoparticles reduce root growth in zucchini (Stampoulis et al. 2009) and onion (Nagaonkar et al. 2015). More studies were carried out concerning the phytotoxicity of copper oxide nanoparticles on *Landoltia punctuate*, *Arabidopsis thaliana* and *Lemna gibba* (Lalau et al 2014; Nair and Chung 2014; Perreault et al. 2014).

Other rare earth element nanoparticles like La_2O_3, Gd_2O_3, and Yb_2O_3 inhibit root growth in cabbage, cucumber, lettuce, radish, rape, tomato, and wheat (Ma et al. 2010a), whereas more physiological impacts of applied CeO_2 nanoparticles on cucumber, rice and barley were observed (Zhao et al. 2013b; Rico et al. 2013a,b; Hong et al. 2014; Zhao et al. 2014; Rico et al. 2015). Co and Zn nanoparticles inhibit root growth in onions and tomato (Ghodake et al. 2011; Faisal et al. 2013) and ZnO nanoparticles completely block root growth in garlic (Shaymurat et al. 2012). Other engineered nanomaterials like ZnO nanoparticles reduce biomass production in zucchini (Stampoulis et al. 2009), where more phytotoxicity of ZnO nanoparticles on different crops including oilseed rape (Kouhi et al. 2015a, b), maize (Liu et al. 2015a), onion (Raskar and Laware 2014; Taranath et al. 2015), peanut (Prasad et al. 2012b), sunflower (Asadzade et al. 2015), soybean (Sedghi et al. 2013), wheat (Ramesh et al. 2014; Watson et al. 2015) and these effects may be vary with soil properties. Ag and Cu nanoparticles reduce biomass production in squash (Musante and White 2012), while TiO_2 and ZnO nanoparticles reduce biomass production in wheat (Du et al. 2011). Reports also indicate that nanoparticles can exhibit genotoxic effects. TiO_2 nanoparticles have shown to produce DNA damage in onions, tobacco, maize and beans (Ghosh et al. 2010; Castiglione et al. 2011). In addition, Lopez-Moreno et al. (2010) reported that CeO_2 and ZnO nanoparticles cause the appearance of new bands in the genomic DNA of soybean. A review by Rico et al. (2011) highlighted that the uptake, bioaccumulation, biotransformation, and risks of nanoparticles for food crops are still not well understood. Nowadays, several researchers worldwide are performing research to answer questions about the interaction of engineered nanomaterials with agricultural plants (Hong et al. 2013).

Therefore, across the major compartments of water (surface water and waste water treatment plant effluent) and solid media (including soil, sediments, and biosolids), and across major receptors of aquatic organisms, microorganisms, complex communities (in soils, sediments or wastewater) and terrestrial plants, there are some overlaps in modeled and measured engineered nanoparticle concentrations versus those administered in toxicity studies. Furthermore, there are great uncertainties regarding engineered nanoparticles bioavailability and the effective concentrations that cause toxicity by a particular mode of action within a receptor. To maximize environmental realism, hazard assessment should, as closely as possible, be conducted to include ecologically relevant receptors and exposure conditions. So, extremely low to rather high engineered nanoparticles concentrations could be routinely administered in toxicity assessments (Holden et al. 2014b).

Therefore, it could be concluded that, several further questions are still need to answer including assess the environmental risks resulted from nanoparticles ecotoxicity, biomobility, bioavailability, reactivity and persistency in agroecosystems.

10.1.1.4 Nanoparticles in Water

Nanoparticles can be categorized into natural, incidental, and fabricated (Farre et al. 2011). Nanoparticles have existed in nature from the beginning of the Earth's history and they are still found in the environment in the form of volcanic dust, lunar dust, soil mineral composites, soil colloids, etc. (Lee and Richards 2004). Concerning the incidental nanoparticles, also they can be defined as anthropogenic nanoparticle wastes, are produced as a result of human activities such as industrial processes, welding fumes, coal combustion, vehicle exhaust, etc. (Smita et al. 2012). In contrast, fabricated nanoparticles are designed and produced to achieve specific physicochemical properties targeted towards unique applications and these nanoparticles comprise four major types including carbon-based nanomaterials, metal-based nanomaterials, dendrimers and nanocomposites (Yadav et al. 2014).

Fabricated or manufactured or engineered particles show some complex colloid and aggregation chemistry, which is likely to be affected by some factors including particle size, shape, surface area and surface charge, as well as the adsorption properties of the material (Handy et al. 2008). A side from these engineered nanoparticles, it could be considered the natural freshwater and other natural waters as a source for natural nanoparticles. Concerning the natural freshwater, it contains very complex colloid materials including inorganic minerals and organic matter such as humic substances (Lead and Wilkinson 2006; Sharma et al. 2015). Whereas, the other natural waters, nanoparticles were found in many types of these natural waters including the oceans, groundwater, surface waters, atmospheric water, and even treated drinking water (Wigginton et al. 2007). In aquatic systems, colloid is the generic term applied to particles in the 1 nm to 1 μm size range. The natural nanomaterial fraction has been identified as being of particular concern because of the changes that occur in this size range, although the most important size range in terms of environmental processes is not well defined. Aquatic colloids comprise macromolecular organic materials, such as humic and fulvic acids, proteins, and peptides, as well as colloidal inorganic species, typically hydrous iron and manganese oxides. Their small size and large surface area per unit mass make them important binding phases for both organic and inorganic contaminants (Klaine et al. 2008).

Naturally occurring and anthropogenic incidental nanomaterials play an important and often completely overlooked role in regulating the behavior of contaminants in complex aquatic systems. This statement is applicable to both manmade aquatic systems, such as nanomaterials in water treatment plants and distribution systems (Ma et al. 2014), and natural aquatic systems, such as rivers and groundwater (Johnson et al. 2014). What has been generally missing is an example of a major, acute aqueous contaminant spill where the central role of naturally occurring and incidental nanomaterials has been clearly delineated (Yang et al. 2015). Due to a heavy application of nanofertilizers, significant amounts of several nutrients (mainly N and P) are transported into surface and groundwater bodies, disrupting aquatic ecosystems and threatening health of human as well as aquatic life. Therefore, an urgent and practically essential research direction is to develop highly efficient and environmentally-friendly nanofertilizers (including macronutrients and micronutrients) to replace the conventional fertilizers and to ensure the sustainable food production while protecting the environment. Thus, development of nanofertilizers is a high priority in fertilizer research, whereas it could be considered these nanofertilizers promising plant nanonutrients (Liu and Lal 2015).

Concerning the environmental behavior and impact of nanoparticles under aquatic environments, it is crucial to be aware that nanoparticles might undergo significant alterations in the liquid phase. There are many distinguished reviews about the behavior and effects of nanoparticles in aquatic systems (Moore 2006; Wigginton et al. 2007; Zhang et al. 2008; Pradeep 2009; Sharma 2009; Brar et al. 2010; Fabrega et al. 2011; Arvidsson et al. 2011; Delay and Frimmel 2012; Batley et al. 2013; He et al. 2014; Corsi et al. 2014; Canesi et al. 2015; Zou et al. 2015; Zhang et al. 2015), as well as some studies (Ferry et al. 2009; von der Kammer et al. 2010; Ma et al. 2014; Johnson et al. 2014; Kumari et al. 2014; Mallevre et al. 2014; Yang et al. 2015) and some distinguished books (Savage et al. 2009; Frimmel and Niessner 2010; Street et al. 2014).

Regarding particles in aquatic systems, research and discussion in the last decade has mainly focused on synthetically tailored nanoparticles, mostly called manufactured or engineered nanoparticles. Owing to their outstanding properties, these nanoparticles are widely and beneficially used in industrial, technical, medical, pharmaceutical, cosmetic and life science applications (Loeffler 2005). In particular, engineered nanoparticles are promising for improving environmental quality when utilized, e.g., in water treatment processes (Wiesner and Bottero 2007) or in remediation (Joo and Cheng 2006). However, there have been early warnings on the environmental risks associated with the widespread use and production of these particles (RS-RAE 2004) as engineered nanoparticles are likely to be emitted into environmental systems and have been shown to be of ecotoxicological relevance (Fent 2010). The scientific community in this field of research to a large extent still seems to walk in a terra incognita, whereas this terra incognita is not entirely unknown (Delay and Frimmel 2012).

Understanding the behavior and role of these nanoparticles in aquatic systems is a task of enormous complexity, including understanding the manifold physical–chemical properties and taking into account that there is a large suite of geogenic, biogenic and anthropogenic these nanoparticles (Delay and Frimmel 2012). It is reported that, nanoparticles in aquatic systems can be formed by (1) natural abiotic or biotic (Lenz et al. 2011) processes, (2) of natural origin but induced by human activity such as mining, (3) engineered nanoparticles that enter aquatic systems incidentally, or (4) engineered nanoparticles that are intentionally added to these systems, e.g., $Fe^{(0)}$ nanoparticles for groundwater remediation (Kirschling et al. 2010) and TiO_2 nanoparticles in water treatment (Delay and Frimmel 2012). On the other hand, it is reported about the occurrence of nanoparticles in aquatic systems, whereas engineered nanoparticles are likely to be an infinitesimal fraction of nanoparticles in the environment compared with natural nanoparticles (Delay and Frimmel 2012).

Both naturally occurring and anthropogenic incidental nanomaterials play an important and often completely overlooked role in regulating the behavior of contaminants in complex aquatic systems. This statement is applicable to both man-made aquatic systems (nanomaterials in water treatment plants and distribution systems) and natural aquatic systems (rivers and groundwater). By applying nano-science perspectives, it could be established that nanoparticles are an important component of the contaminant mixture within the aquatic systems. Due to their transport, bioavailability toxicity, and ecological impact, nanoparticle as a toxin depends not only on its concentration but also on its form. Therefore, it is pointed out that, the importance of examining the role of nanoscale contaminants as an important component of both existing and future aquatic ecosystem pollution scenarios (Yang et al. 2015). Hence, nanoparticles have been employed to improve the quality of soils, air and water using the nanoremediation (Mohmood et al. 2013; Wang and Chen 2015; Stuart and Compton 2015; Gil-Díaz et al. 2016). Nanoremediation is the process in which the application of reactive nanomaterials can enhance both catalysis and chemical reduction of the contaminants resulting in and transformation detoxification of pollutants (Kuppusamy et al. 2016). This process has become one of the main foci of research and development with great potential for protecting the environment from pollution using different nanomaterials like zeolites, metal oxides, carbon nanotubes, noble metals and titanium dioxide (Liu et al. 2015b; Kuppusamy et al. 2016).

It could be concluded that, the movement and translocation of nanoparticles within waters is an emerging issue. Otherwise, different effects of nanoparticles have been addressed as well as using nanoremediation in waste waters, ground water and contaminated waters. Different effects of nanoparticles on waters have been observed and at the same time using some nanoparticles (like nano–zero iron) or nanomaterials to transform and degrade contaminants in soils and water have been successfully established.

10.1.1.5 Nanofertilizers

The increasing anthropogenic demands has jeopardized natural resources and exacerbated soil and environmental degradation. Due to nature does not recognize wastes, where a new life from every death emerges through a meticulous recycling of essential elements contained in the so-called wastes (Lal 2015). Therefore, the importance of nexuses and inter-connectivity is also documented by a close relationship between soil security, water security, energy security, climate security, economic security and political security. Indeed, both economic and political securities are closely linked with food security on the one hand and security of natural resources on the other (Fig. 10.2). Therefore, the co-productivity generated by the anthropogenic use of primary resources (soil, water, climate) and secondary inputs (fertilizers, irrigation, tillage, amendments) must be optimized. Furthermore, understanding and judiciously managing the water-soil-waste nexus for food security is

Fig. 10.2 A unique management for rice straw in Kafrelsheikh Uni. (*photos from 1 to 3*), where Prof. Belal and his co-workers have already extracted several products from this rice straw using special microbial strains. These products include cellulose A, B, and C (*Photo 1*), from lift to right: humic acid, bioethanol, silica, lignin, compost, alternative soil and rice straw (*Photo 2*) cellulose (*Photo 3*), and algae (*photo 4*). Production a biogas from farm yard manure (*Photos from 5 to 8*), where biogas slurry enriched with algae can be obtained (*Photos 5 and 6*) and biogas can be noticed from the ignition in Photo 7 and dried biogas slurry (*Photo 8*) (Photos by El-Ramady)

important to achieving the sustainable use of natural resources, improving the environment and sustaining ecosystem functions and services, enhancing human well-being (Lal 2015).

Nanoscience or nanotechnology is leading to the development of a range of inexpensive nanotechnology applications for enhanced plant growth and protection.

Therefore, nanoparticles and nanocapsules provide an efficient means to distribute pesticides and fertilizers in a controlled fashion with high site specificity thus reducing collateral damage (Nair et al. 2010). Due to using of pesticides and fertilizers in improving food production, it leads to an uncontrolled release of undesired substances into the environment (Sekhon 2014). Nowadays, nanotechnology represents a promising approach in improving the agricultural production and in remediation contaminated soil and groundwater. Concerning the current applications of nanotechnologies in agro-environmental studies, researchers reported that, it should paid attention to the fate of nanomaterials once introduced in both water and soils (Sekhon 2014). In frame of nanotechnology, the use of nanoparticles have already improved the quality of the environment and helped detect and remediate polluted sites (Schaumann et al. 2015b; de Santiago-Martín et al. 2015). However, it is reported about only a small number of nanomaterials demonstrated potential toxic effects (Mura et al. 2013).

Nanofertilizers could be defined as the nanoparticles or nanomaterials which can provide plants with nutrients or the materials which increase the efficiency of the activities of conventional fertilizers. Application of nanofertilizers for traditional fertilizer is beneficial because these nanofertilizers have the ability to release nutrients into soils steadily and in a controlled way preventing water pollution (Naderi and Danesh-Shahraki 2013; Singh et al. 2015b). Because of development and application of nanofertilizers are still at initial stages, there are a few researches or systemic studies on the effects and advantages of applying micronutrient nanofertilizers under field conditions (Liu and Lal 2015). However practically, nanotechnology permits broad advantages in agricultural research, such as disease prevention and treatment in plants using various nano-cides and nutrient management of agriculture field using nanofertilizers (Gonzalez-Melendi et al. 2008; De Rosa et al. 2010; Corradini et al. 2010; Ghormade et al. 2011; Priester et al. 2012; Naderi and Danesh-Shahraki 2013; El-Kereti et al. 2013; Delfani et al. 2014; Liu and Lal 2014; Tarafdar et al. 2014; Asadzade et al. 2015; Thul and Sarangi 2015; Liu and Lal 2015; Solanki et al. 2015; Monreal et al. 2015; Subramanian et al. 2015; Mastronardi et al. 2015; Soliman et al. 2015; Rai et al. 2015; Rameshaiah et al. 2015; Sharonova et al. 2015; Yatim et al. 2015). Hence, it has been reported that, nanofertilizers have several roles in sustainable agriculture (Naderi and Danesh-Shahraki 2013; El-Ramady 2014; Sekhon 2014). Thus, micronutrient-containing nanoparticles could potentially improve plant growth by supplying nutrients (Liu and Lal 2015).

Because of scarce water resources and limited additional arable lands in the world, a significant increase in agricultural fertilizer application is needed to achieve the required massive increase in global food production. Therefore, for maintaining the current levels of grain production, serious environmental issues globally have caused due to the applications of diverse conventional fertilizers at high rates and for a long period in the agricultural sector (Liu and Lal 2015). Recently, it is reported about development and application of new types of fertilizers using innovative nanotechnology. This new generation of nanofertilizers is one of the potentially effective options of significantly enhancing the global agricultural productions needed to meet the future demands of the growing population (Liu and Lal 2015). It is also reviewed about some engineered nanomaterials (as nanofertilizers), which can enhance plant-growth in certain concentration ranges. These nanofertilizers could be used in agriculture to increase agronomic yields of crops and/or minimize environmental pollution (Liu and Lal 2015).

Due to many issues for the use of traditional chemical fertilizers, low use efficiency is the prominent one, which not only increases the cost of production but also causes several environmental pollutions (Wilson et al. 2008). Therefore, there is a crucial need for nanofertilizers, which have proved to be another landmark in the history of crop production through nanotechnology. Furthermore, nanofertilizers could solve this problem due to their nano-size and as nanomaterials with large surface area (Ditta et al. 2015). These nanofertilizers could be also utilized as nanocoatings, e.g., sulfur nano-coating ($\leq$100 nm layer), ensuring their controlled release, surface protection and ultimately boosting up their use efficiency (Brady and Weil 1996; Ditta et al. 2015). Nevertheless, nanofertilizers have been proved more efficient compared to the conventional or ordinary fertilizers because these nanoparticles reduce nitrogen loss due to emissions, leaching and long-term incorporation by soil microorganisms (Liu et al. 2006). Moreover, controlled release fertilizers may also improve soil by decreasing toxic effects associated with over-application of traditional chemical fertilizers (Suman et al. 2010). It is also reported that, there are some reports about the use of nano-encapsulated slow-release fertilizers (De Rosa et al. 2010). It has been also used the biodegradable, polymeric chitosan nanoparticles (nearly 78 nm) for controlled release of the NPK fertilizer sources such as urea, calcium phosphate and potassium chloride (Corradini et al. 2010). It is reported also about some nanomaterials like kaolin and polymeric biocompatible nanoparticles, which could be also utilized for this purpose (Wilson et al. 2008; Ditta et al. 2015).

Due to the excessive applications of nitrogen and phosphorus fertilizers, they affect the ground water and also lead to eutrophication in aquatic ecosystems. Moreover, due to the low of fertilizer use efficiency of conventional fertilizers (30–35 %, 18–20 %, and 35–40 % for N, P, and K, respectively; Subramanian et al. 2015), food production will have to be much more efficient than ever before (Chinnamuthu and Boopathi 2009). According to this and limited availability of land and water resources, development of agriculture can be achieved exclusively through increasing productivity by effective use of modern technologies. Among these, nanotechnology has the potential to revolutionize the agricultural systems, biomedicine, environmental engineering, safety and security, water resources, energy conversion, and numerous other areas (Baruah and Dutta 2009; Naderi and Danesh-Shahraki 2013). The changing climate, sustainable use of natural resources, environmental factors, urbanization, accumulation of pesticides and over use fertilizers are the most important problems of modern agriculture. New techniques and methods have been used in order to avoid the detrimental effects of these factors. The nanomaterial is one of the new technologies that into almost all areas of our lives and being to be used in agriculture production. The researchers indicate many of the potential benefits of nanotechnology (Siddiqui et al. 2015b). Therefore, it is reported that, one of the most notable characteristics of nanoscale metals and metal oxides is the greatly enhanced availability to plants and translocation within these plants. Consequently, the use of nanoparticle-based micronutrient formulations (nano-plant nutrition) may offer a highly effective novel platform for crop disease suppression and yield enhancement through more targeted and strategic nutrition-based promotion of host resistance (Servin et al. 2015).

Although fertilizers are very important for plant growth and development, most of the applied fertilizers are rendered unavailable to plants due to many factors, such as leaching, degradation by photolysis, hydrolysis, and decomposition. Hence, it is necessary to minimize nutrient losses in fertilization, and to increase the crop yield through the exploitation of new applications with the help of nanotechnology and nanomaterials. Concerning nanofertilizers and their roles in sustainable agriculture, it was and still one of the most important issues in the agricultural sector. Due to the importance of fertilizers derived from nanotechnology, they have started to attract attention in agriculture. Whereas, there are insufficient studies on fertilizers produced with nanotechnology, although it is well known that, these nanofertilizers have a significant impact in agricultural production (Siddiqui et al. 2015b). Nanofertilizers or nano-encapsulated nutrients might have properties that are effective to crops, released the nutrients on-demand, controlled release of chemicals fertilizers that regulate plant growth and enhanced target activity (De Rosa et al. 2010; Nair et al. 2010). It is well known that, higher plants have a remarkable ability to develop mechanism to perform better under suitable and unsuitable conditions. Concerning these higher plants, scientists/researchers want to develop new techniques that could be suitable for plants to support their native functions. Nanofertilizers or nanoparticles have unique physicochemical properties and the potential to boost the plant metabolism (Giraldo et al. 2014; Siddiqui et al. 2015b).

On the other hand, it is reported that, engineered nanoparticles have the ability to inter into plants cells and leaves, and also can transport DNA and chemicals into plant cells. This area of research offers new possibilities in plant biotechnology to target specific genes manipulation and expression in the specific cells of the plants (Siddiqui et al. 2015b). Research in the field of nanotechnology is required to discover the novel applications to target specific delivery of chemicals, proteins, nucleotides for genetic transformation of crops (Scrinis and Lyons 2007). Nanotechnology has large potential to provide an opportunity for the researchers of plant science and other fields, to develop new tools for incorporation of nanoparticles into plants that could augment existing functions and add new ones (Cossins 2014; Siddiqui et al. 2015b).

In frame of sustainable agriculture, applying innovative nanotechnology in agriculture (including fertilizer development) is regarded as one of the promising approaches to significantly increase crop production (Lal 2008). Concerned by the low efficiency (merely 30–50 %) of the conventional fertilizers and few management options to enhance the rates, it is also urged application of nanotechnology to fertilizer research and development (De Rosa et al. 2010). Compared with the conventional fertilizers, nanofertilizers are expected to achieve the following benefits including enhancement the efficiency of fertilizer use, a significant improvement of crops' growth and yields, reduction nutrient losses and/or minimization the adverse environmental impacts (Liu and Lal 2015). Although several of these agricultural experts have been interested in development and application of nanomaterial related fertilizers (Ghormade et al. 2011; Khot et al. 2012), the directly-related research is lacking. However, some recent research in nanotechnology has demonstrated the promising perspective of nanofertilizer development and application (Liu and Lal 2015). Accordingly, nanofertilizers are either nanomaterials which can supply one or more nutrients to the plants and enhance their growth and yields, or those which facilitate to improve the performance of conventional fertilizers, but do not directly provide crops with nutrients (Liu and Lal 2015).

Therefore, the nanofertilizers have the following properties:

1. The nutrient carriers of nano-dimensions ranging from 30 to 40 nm and capable of holding bountiful of nutrient ions due to their high surface area as well as exploiting unique properties of these nanoparticles (Subramanian et al. 2015),
2. These nanofertilizers intended to improve the nutrient use efficiencies due to release nutrients slowly and steadily for more than 30 days which may assist in improving the nutrient use efficiency without any associated ill-effects and
3. These nanofertilizers can be synthesized by fortifying nutrients singly or in combinations onto the adsorbents with nano-dimension and designed to deliver slowly over a long period of time, the loss of nutrients is substantially reduced the environmental safety (Subramanian et al. 2015).

Nano-fertilizers and nanocomposites can be used in improving the nutrient use efficiency by preventing the nutrient ions from either getting fixed or lost in the environment as well as controlling the release of nutrients from the fertilizer granules (Table 10.1; Subramanian et al. 2008). Extensive studies had been undertaken to characterize nitrogen-nanofertilizers (Subramanian and Sharmila Rahale 2013; Mohanraj 2013; Manikandan and Subramanian 2014), phosphate-nanofertilizers (Bansiwal et al. 2006; Adhikari 2011; Behnassi et al. 2011), potassium-nanofertilizers (Subramanian and Sharmila Rahale 2012), sulfur-nanofertilizers (Selva Preetha et al. 2014; Thirunavukkarasu 2014) and zinc-nanofertilizers (Subramanian and Sharmila Rahale 2012), as reviewed by Subramanian et al. (2015).

Nanofertilizers can be classified as macronutrient nanofertilizers and micronutrient nanofertilizers. Concerning macronutrient nanofertilizers, they are chemically comprised of one or more macronutrient elements such as N, P, K, Mg, and Ca, which are being able to supply these essential macronutrients to plants. In general, these macronutrient fertilizers (mainly N and P fertilizers) are used in large quantities for increasing production of food, fiber, and other essential commodities (Liu and Lal 2015). On the other hand, micronutrient nanofertilizers include Fe, Mn, Zn, Cu, and Mo among others. It is reported that, micronutrient nanofertilizers may enhance the bioavailability of previous nutrients to plants even under the worst-cases or conditions (Liu and Lal 2015). A side from nanofertilizers, there is the nanomaterial-enhanced fertilizers as well as other nanoparticulate fertilizer carriers. These nanomaterial-enhanced fertilizers can be defined as those nanomaterials, which, when augmented with plant nutrient(s), can increase plant uptake efficiency of the nutrient(s) and/or reduce the adverse impacts of conventional fertilizer

Table 10.1 Comparison between nano-fertilizers and conventional fertilizers

Properties	Nano-fertilizers	Conventional fertilizers
Solubility and dispersion of mineral micro-nutrients	Nano-fertilizers may improve solubility and dispersion of insoluble nutrients in soil, reduce soil absorption and fixation as well as increase the bioavailability	Less bioavailability to plants due to large particle size and less solubility
Nutrient uptake efficiency	Nano-fertilizers may increase fertilizer efficiency and uptake ratio of the soil nutrients in crop production as well as save fertilizer resource	Bulk composite is not available for roots and decrease efficiency
Controlled release modes	Both release rate and release pattern of nutrients for water soluble fertilizers might be precisely controlled through encapsulation in envelope forms of semipermeable membranes coated by resin-polymer, waxes and sulfur	Excess release of fertilizers may produce toxicity and destroy ecological balance of soil
Effective duration of nutrient release	Nano-fertilizers can extend effective duration of nutrient supply of fertilizers into soil	Used by the plants at the time of delivery, the rest is converted into insoluble salts in the soil
Loss rate of fertilizer nutrients	Nanostructured formulation can reduce loss rate of fertilizer nutrients into soil by leaching and/or leaking	High loss rate by leaching, rain off and drift

Source: Cui et al. (2010) and Solanki et al. (2015)

application, but those nanomaterials neither contain nor provide the targeted nutrient(s). The most important example of this type is nutrient-augmented-zeolites (Liu and Lal 2015). It is worth to mention that, there are some new nanoparticulate plant-growth enhancers with unclear mechanisms. Whereas, there are several reports documenting that application of some other types of nanoparticles could also enhance plant growth to some degree (under low concentration), despite the fact that these particles did not contain any essential plant nutrients (Liu and Lal 2015). Typical examples of this type include nanoparticles of SiO_2 (Roldugin et al. 2015; Asadzade et al. 2015), selenium (Zhang et al. 2015; Chen et al. 2015), copper (Jain et al. 2016), iron (Delfani et al. 2014; Bakhtiari et al. 2015; Soliman et al. 2015), zinc (El-Kereti et al. 2013; Tarafdar et al. 2014; Asadzade et al. 2015; Soliman et al. 2015), phosphorus (Liu and Lal 2014; Sharonova et al. 2015) and carbon nanotubes (Yatim et al. 2015).

Therefore, it could be concluded that, nanofertilizers could be categorized into four categories including macronutrients nanofertilizers, micronutrient nanofertilizers, nutrient-loaded nanofertilizers, and plant-growth-enhancing nanomaterials. Each category can be discussed with reference to nanomaterials' chemical composition, particle size, concentrations applied, benefited plant species, plant incubation methods, and plant-growth enhancement aspects and the rates. It is reported that, the use of nanofertilizers (especially biological ones) have some benefits including the high nutrients use efficiency, reduces soil toxicity, minimizes the potential negative effects associated with over dosage and reduces the frequency of the application. Hence, nanotechnology can be considered a high potential for achieving sustainable agriculture, especially in developing countries.

10.1.2 Towards Nanotechnology for Sustainable Agriculture

Environmental pollution can be considered an emerging issue including pollution from air, water and soil. This pollution can be expected when the level of chemicals, metal (ions), smoke, bacteria, viruses or pathogens increases beyond some tolerance limit for living animals. Furthermore, this environmental pollution is an age-old problem and began when man started to cut wood for houses, learning to make fire and other activities. The environmental pollution also started when human began to establish his settlements and then flourished various civilizations, the pollution problems mainly water contamination appeared. However, the population on the Earth was not large until probably the nineteenth century and pollution problems were not severe (Kulkarni 2015). There is a major threat of fossil fuel getting depleted to an alarming level on which we are dependent for our energy resources for all our activities. It should be not continue with use of fossil fuel which contaminates air, as still it is cheaper than any other alternative energy. This in turn also gives rise to global warming i.e. possibility of increasing the overall temperature of the earth, resulting into melting of large icebergs on poles (Kulkarni 2015).

Excessive and inappropriate use of fertilizers and pesticides has increased nutrients and toxins in groundwater and surface waters, incurring health and water purification costs, and decreasing fishery and recreational opportunities (Mukhopadhyay 2014). Furthermore, different agricultural practices, which degrade soil quality contribute to eutrophication of aquatic habitats and may necessitate the expense of increased irrigation, fertilization and energy to maintain productivity on degraded soils. These previous practices can also kill the beneficial insects and other wildlife. Moreover, intensive of tillage, irrigation and fertilizer dressing have also caused more extensive damage to the carbon profile in soils comparing with the early agrarian practices (Knorr et al. 2005; Mukhopadhyay 2014). It is well known that, nanotechnology applications in agriculture can be successful if the natural processes are simulated in greater scientific sophistication/articulation for successful implementation. For instance, the target might be to make soils more capable to improve efficient nutrient use for greater productivity and better environmental security. Therefore, nutrient management in frame of nanotechnology must rely on some important parameters including (1) nutrient-ions must be present in the plant available forms in soil system and (2) nutrient transport in soil-plant systems relies on ion exchange, adsorption/desorption and solubility/precipitation reactions, and (3) nanomaterials must facilitate processes that would ensure availability of nutrients to plants in the rate and manner that plants demand (Mukhopadhyay 2014).

There are several modern strategies or approaches of nanotechnology can be used for the management of water, pesticides, fertilizers, sensors, limitations in the use of chemical pesticides and potential of nanomaterials in sustainable agriculture management (Prasad et al. 2014). Several publications focused on the sustainable agriculture under the umbrella of nanotechnology and effects of nanoparticles on the terrestrial environments (Chen and Yada 2011; Anandaraj et al. 2011; Ditta 2012; Doong et al. 2013; Shamim and Sharma 2013; Mura et al. 2013; Prasad et al. 2014; Mukhopadhyay 2014; El-Beyrouthya and El Azzi 2014; Takeuchi et al. 2014; de Oliveira et al. 2014; Hu et al. 2014; Sekhon 2014; Kulkarni 2015; Siddiqui et al. 2015a; Shah et al. 2015; Ditta et al. 2015; Thul and Sarangi 2015; Patil et al. 2016; Salamanca-Buentello and Daar 2016). From these studies, it is clear to notice that, nanotechnology will play an increasingly important role in agriculture. Furthermore, studies over the past decade showed that nanotechnology has the potential to revolutionize agriculture with the use of following fields: biosensors, plant growth regulators, food additives, genetic improvement of plants and animals, smart delivery systems for drugs, pesticides and fertilizers and nanopesticides (Hong et al. 2013). To prevent the possible adverse effects from the use of nanotechnology in agriculture, studies on issues such as the fate and transport of the nanomaterials in the ecosystem, uptake and accumulation in plants and animals, and evaluation of the toxicity of nanomaterials are needed to perform. Risk assessment studies should also be performed before the application of nanoproducts to the agricultural field, and the effects should be monitored (Hong et al. 2013).

The opportunity for application of nanotechnology in agriculture is prodigious. It is also well recognized that, the adoption of some new nanotechnologies is crucial issue in frame of sustainable development (Mukhopadhyay 2014). It is thought that, agro-nanotechnology might take a few decades to move from laboratory to land, especially since it has to avoid the pitfalls experienced with biotechnology. For this to happen, sustained funding and understanding on the part of policy planners and science administrators as well as nano-ethics, along with reasonable expectations, would be crucial for this nascent field to blossom. Application of nanotechnology is essential, given the millions of people worldwide who continue to lack access to health care, education, safe water, reliable sources of energy, and other basic human development needs (Mukhopadhyay 2014; Salamanca-Buentello and Daar 2016).

Therefore, it could be concluded that, potential applications of nanotechnology need to be reassessed in the light of sustainability, taking into account the economic (Shapira and Youtie 2015), societal (Roure 2016), ethical (Salamanca-Buentello and Daar 2016) and environmental factors (Ditta et al. 2015; Bottero 2016) and interdependencies. That means, the sustainability of nanotechnology based products needs should not just during the manufacturing phase but should be considered over the entire life cycle. Therefore, it should be taken into account the whole product life cycle for the technology assessment and it is required at an early stage of development to achieve the correct balance between cost effectiveness and the environmental impacts, as reviewed by Rickerby (2013). Furthermore, life cycle analysis is expected to provide valuable insights in this respect. Due to the standard methods for risk assessment, they may be inadequate to identify the particular hazards associated with nanomaterials as well as nano-specific risk assessment tools should be developed. Appropriate recovery and recycling strategies in frame of sustainability are also needed to limit dispersion of nanomaterials in the environment. Measures must also be adopted to minimize the health and environmental risks due to release of nanomaterials at each stage of the product life cycle from the production phase, during use, through to final disposal or recycling. In the absence of reliable data and ethics, however, worst case scenarios must be assumed for risk management of nanotechnology in frame of sustainable agriculture.

10.2 Conclusion

Nowadays, there is no field untouched by nanotechnology and its ground breaking scientific innovations including agricultural sector. Therefore, the use of nanotechnology in agriculture has been touched several fields including food industry, plant nano-nutrition, plant protection, plant productivity, nanofertilizers and nanopesticides. Concerning the effects of nanoparticles on agroecosystems, the fate and behavior of these nanoparticles on agricultural crops and possible toxic implications to plants and microorganisms naturally present in soil rhizosphere and generation of nano-wastes in agroecosystem are emerging issues. Furthermore, the negative effects of nanoparticles (or nanoecotoxicology, which generated by free radicals leading to lipid peroxidation and DNA damage) on plant productivity, soil microorganisms, soil enzyme activities as well as the entire environment must not be overlooked. Hence, there is a need to predict the environmental effects of these nanoparticles in the foreseeable future towards the sustainability of agriculture.

Acknowledgements Authors thank the outstanding contribution of STDF research teams (Science and Technology Development Fund, Egypt) and MBMF/DLR (the Federal Ministry of Education and Research of the Federal Republic of Germany), (Project ID 5310) for their help. Great support from this German-Egyptian Research Fund (GERF) is gratefully acknowledged.

References

Abdul Qados AMS, Moftah AE (2015) Influence of silicon and nano-silicon on germination, growth and yield of faba bean (*Vicia faba* L.) under salt stress conditions. doi:10.9734/AJEA/2015/14109

Adams LK, Lyon DY, Alvarez PJ (2006) Comparative eco-toxicity of nanoscale TiO_2, SiO_2, and ZnO water suspensions. Water Res 40:3527–3532

Adhikari T (2011) Nano-particle research in soil science: micronutrients. In: Proceedings of the national symposium on 'applications of clay science: agriculture environment and industry', NBSS & LUP, Nagpur, 18–19 February 2011, pp 74–75

Aliofkhazraei M (2016) Handbook of nanoparticles. Springer International Publishing, Cham. doi:10.1007/978-3-319-15338-4

Almutairi ZM, Alharbi A (2015) Effect of silver nanoparticles on seed germination of crop plants. J Adv Agric 4(1):283–288

Anandaraj M, Dinesh R, Srinivasan V, Harnza S (2011) Nanotechnology in agriculture: the use of novel materials and environmental issues. Bot 59–61:22–34

Anjum NA, Adam V, Kizek R, Duarte AC, Pereira E, Iqbal M, Lukatkin AS, Ahmad I (2015) Nanoscale copper in the soil–plant system – toxicity and underlying potential mechanisms. Environ Res 138:306–325, http://dx.doi.org/10.1016/j.envres.2015.02.019

Ansari SA, Husain Q (2012) Potential applications of enzymes immobilized on/in nano materials: a review. Biotechnol Adv 30(3):512–523. doi:10.1016/j.biotechadv.2011.09.005

Antisari LV, Carbone S, Gatti A, Vianello G, Nannipieri P (2013) Toxicity of metal oxide (CeO_2, Fe_3O_4, SnO_2) engineered nanoparticles on soil microbial biomass and their distribution in soil. Soil Biol Biochem 60:87–94, http://dx.doi.org/10.1016/j.soilbio.2013.01.016

Antisari LV, Carbone S, Gatti A, Vianello G, Nannipieri P (2015) Uptake and translocation of metals and nutrients in tomato grown in soil polluted with metal oxide (CeO_2, Fe_3O_4, SnO_2, TiO_2) or metallic (Ag, Co, Ni) engineered nanoparticles. Environ Sci Pollut Res. doi:10.1007/s11356-014-3509-0

Arruda SCC, Silva ALD, Galazzi RM, Azevedo RA, Arruda MAZ (2015) Nanoparticles applied to plant science: a review. Talanta. http://dx.doi.org/10.1016/j.talanta.2014.08.050

Arvidsson R, Molander S, Sandén BA, Hassellöv M (2011) Challenges in exposure modeling of nanoparticles in aquatic environments, human and ecological risk assessment. Int J 17(1):245–262. doi:10.1080/10807039.2011.538639

Asadzade N, Moosavi SG, Seghatoleslam MJ (2015) Effect of low irrigation and Zn and SiO_2 nano-fertilizers and conventional fertilizers on morphophysiological traits and seed yield of sunflower. Biol Forum – Int J 7(1):357–364, ISSN: 0975-1130

Aslani F, Bagheri S, Julkapli NM, Juraimi AS, Hashemi FSG, Baghdadi A (2014) Effects of Engineered Nanomaterials on Plants Growth: an Overview. Sci World J, http://dx.doi.org/10.1155/2014/641759

Aubert T, Burel A, Esnault MA, Cordier S, Grasset F, Cabello-Hurtado F (2012) Root uptake and phytotoxicity of nanosized molybdenum octahedral clusters. J Hazard Mater 219–220:111–118

Bakhtiari M, Moaveni P, Sani B (2015) The effect of iron nanoparticles spraying time and concentration on wheat. Biol Forum – Int J 7(1), ISSN: 0975-1130

Bansiwal AK, Rayalu SS, Labhasetwar NK, Juwarkar AA, Devotta S (2006) Surfactant-modified zeolite as a slow release fertilizer for phosphorus. J Agric Food Chem 54:4773–4779

Baruah S, Dutta J (2009) Nanotechnology applications in sensing and pollution degradation in agriculture. Env Chem Lett J 7:191–204

Batley GE, Kirby JK, McLaughlin MJ (2013) Fate and risks of nanomaterials in aquatic and terrestrial environments. Acc Chem Res 46:854–862

Beattiew IR, Haverkamp RG (2011) Silver and gold nanoparticles in plants: sites for the reduction to metal. Metallomics 3:628–632. doi:10.1039/c1mt00044f

Behnassi M, Shahid AS, D'Silva J (2011) Sustainable agricultural development. Springer Science Business Media, London, pp 171–184

Ben-Moshe T (2010) Transport of metal oxide nanoparticles in saturated porous media. Chemosphere 81:387–393

Bernhardt ES, Colman BP, Hochella MF Jr, Cardinale BJ, Nisbet RM et al (2010) An ecological perspective on nanomaterial impacts in the environment. J Environ Qual 39:1954–1965

Bhattacharyya A, Mohammad F, Naika HR, Epid STT, Reddy J, Prakasham RS (2015) Nanoparticles and their impact on plants. Res J Nanosci Nanotechnol 5(2):27–43. doi:10.3923/rjnn.2015.27.43

Binns C (2014) Nanomagnetism: fundamentals and applications, vol 6, Frontiers of nanoscience. Elsevier Ltd., Amsterdam

Bodale I, Oprisan M, Stan C, Tufescu F, Racuciu M, Creanga D, Balasoiu M (2016) Nanotechnological application based on $CoFe_2O_4$ nanoparticles and electromagnetic exposure on agrotechnical plant growth. In: Sontea V Tiginyanu I (eds) 3rd International conference on nanotechnologies and biomedical engineering, IFMBE Proceedings 55, Springer Science+Business Media Singapore, pp 153 – 156. doi:10.1007/978-981-287-736-9_37

Boenigk J, Beisser D, Zimmermann S, Bock C, Jakobi J, Grabner D, Groβmann L, Rahmann S, Barcikowski S, Sures B (2014) Effects of silver nitrate and silver nanoparticles on a planktonic community: general trends after short-term exposure. PLoS ONE 9:e95340. doi:10.1371/journal.pone.00953

Bombin S, LeFebvre M, Sherwood J, Xu Y, Bao Y, Ramonell KM (2015) Developmental and peproductive effects of iron oxide nanoparticles in *Arabidopsis thaliana*. Int J Mol Sci 16(10):24174–24193. doi:10.3390/ijms161024174

Bottero J-Y (2016) Environmental risks of nanotechnology: a new challenge? In: Nanosciences and nanotechnology, Lourtioz J-M et al. (eds) Springer International Publishing, Cham, pp 287 – 311. doi:10.1007/978-3-319-19360-1_13

Bour A, Mouchet F, Silvestre J, Gauthier L, Pinelli E (2015) Environmentally relevant approaches to assess nanoparticles ecotoxicity: a review. J Hazard Mater, http://dx.doi.org/10.1016/j.jhazmat.2014.10.021

Boxall AB, Tiede K, Chaudhry Q (2007) Engineered nanomaterials in soils and water: how do they behave and could they pose a risk to human health? Nanomedicine (Lond) 2(6):919–927

Brady NC, Weil RR (1996) The nature and properties of soils. Prentice-Hall Inc., Upper Saddle River

Brar SK, Verma M, Tyagi RD, Surampalli RY (2010) Engineered nanoparticles in wastewater and wastewater sludge – evidence and impacts. Waste Manag 30:504–520. doi:10.1016/j.wasman.2009.10.012

Brunner TJ, Wick P, Manser P, Spohn P, Grass RN, Limbach LK, Bruinink A, Stark WJ (2006) *In vitro* cytotoxicity of oxide nanoparticles: comparison to asbestos, silica, and the effect of particle solubility. Environ Sci Technol 40(14):4374–4381

Burke DJ, Pietrasiak N, Situ SF, Abenojar EC, Porche M, Kraj P, Lakliang Y, Samia ACS (2015) Iron oxide and titanium dioxide nanoparticle effects on plant performance and root associated microbes. Int J Mol Sci 16(10):23630–23650. doi:10.3390/ijms161023630

Canesi L, Ciacci C, Balbi T (2015) Interactive effects of nanoparticles with other contaminants in aquatic organisms: friend or foe? Mar Environ Res, http://dx.doi.org/10.1016/j.marenvres.2015.03.010

Canivet L, Dubot P, Garçon G, Denayer F-O (2015) Effects of engineered iron nanoparticles on the bryophyte, *Physcomitrella patens* (Hedw.) Bruch & Schimp, after foliar exposure. Ecotoxicol Environ Saf 113:499–505, http://dx.doi.org/10.1016/j.ecoenv.2014.12.035

Castiglione MR, Giorgetti L, Geri C, Cremonini R (2011) The effects of nano-TiO_2 on seed germination, development and mitosis of root tip cells of *Vicia narbonensis* L. and *Zea mays* L. J Nanopart Res 13(6):2443–2449

Castiglione MR, Giorgetti L, Cremonini R, Bottega S, Spanò C (2014) Impact of TiO_2 nanoparticles on *Vicia narbonensis* L.: potential toxicity effects. Protoplasma 251:1471–1479

Chaudhry Q, Scotter M, Blackburn J, Ross B, Boxall A, Castle L, Aitken R, Watkins R (2008) Applications and implications of nanotechnologies for the food sector. Food Addit Contam: Part A 25(3):241–258. doi:10.1080/02652030701744538

Chen H, Yada R (2011) Nanotechnologies in agriculture: new tools for sustainable development. Trends Food Sci Technol 22:585–594. doi:10.1016/j.tifs.2011.09.004

Chen J, Liu X, Wang C, Yin S-S, Li X-L, Hu W-J, Simon M, Shen Z-J, Xia Q, Chu C-C, Peng X-X (2015) Nitric oxide ameliorates zinc oxide nanoparticles-induced phytotoxicity in rice seedlings. J Hazard Mater 297:173–182, http://dx.doi.org/10.1016/j.jhazmat.2015.04.077

Chinnamuthu CR, Boopathi PM (2009) Nanotechnology and agroecosystem. Madras Agric J 96(1–6):17–31

Christian P, Von der Kammer F, Baalousha M, Hofmann T (2008) Nanoparticles: structure, properties, preparation and behaviour in environmental media. Ecotoxicology 17(5):326–343

Collins D, Luxton T, Kumar N, Shah S, Walker VK et al (2012) Assessing the impact of copper and zinc oxide nanoparticles on soil: a field study. PLoS ONE 7(8):e42663. doi:10.1371/journal.pone.0042663

Conway JR, Beaulieu AL, Beaulieu NL, Mazer SJ, Keller AA (2015) Environmental stresses increase photosynthetic disruption by metal oxide nanomaterials in a soil-grown plant. ACS Nano. doi:10.1021/acsnano.5b03091

Corradini E, De Moura M, Mattoso L (2010) A preliminary study of the incorporation of NPK fertilizer into chitosan nanoparticles. Express Polym Lett 4(8):509–515

Corsi I, Cherr GN, Lenihan HS, Labille J, Hassellov M, Canesi L, Dondero F, Frenzilli G, Hristozov D, Puntes V, Torre CD, Pinsino A, Libralato G, Marcomini A, Sabbioni E, Matranga V (2014) Common strategies and technologies for the ecosafety assessment and design of nanomaterials entering the marine environment. ACS Nano 8(10):9694–9709. doi:10.1021/nn504684k

Cossins D (2014) Next generation: nanoparticles augment plant functions. The incorporation of synthetic nanoparticles into plants can enhance photosynthesis and transform leaves into biochemical sensors. The scientist, news & opinion, March 16. http://www.the-scientist.com/?articles.view/articleNo/39440/title/Next-Generation–Nanoparticles-Augment-Plant-Functions/

Cui HX, Sun CJ, Liu Q, Jiang J, Gu W (2010) Applications of nanotechnology in agrochemical formulation, perspectives, challenges and strategies. In: International conference on Nanoagri, Sao Pedro, Brazil, pp 28–33

Darlington TK, Neigh AM, Spencer MT, Guyen OTN, Oldenburg SJ (2009) Nanoparticle characteristics affecting environmental fate and transport through soil. Environ Toxicol Chem 28:1191–1199

Dasgupta N, Shivendu R, Deepa M, Chidambaram R, Ashutosh K, Rishi S (2015) Nanotechnology in agro-food: from the field to plate. Food Res Int 69:381–400

Dasgupta N, Shivendu R, Shraddha M, Ashutosh K, Chidambaram R (2016a) Fabrication of food grade vitamin E nanoemulsion by low energy approach: characterization and its application. Int J Food Prop 19(3):700–708. doi:10.1080/10942912.2015.1042587

Dasgupta N, Shivendu R, Bhavapriya R, Venkatraman M, Chidambaram R, Avadhani GS, Ashutosh K (2016b) Thermal co-reduction approach to vary size of silver nanoparticle: its microbial and cellular toxicology. Environ Sci Pollut Res 23:4149–4163. doi:10.1007/s11356-015-4570-z

Dasgupta N, Shivendu R, Patra D, Srivastava P, Kumar A, Ramalingam C (2016c) Bovine serum albumin interacts with silver nanoparticles with a "side-on" or "end on" conformation. Chem Biol Int 253:100–111. doi:10.1016/j.cbi.2016.05.018

de la Rosa G, Lopez-Moreno ML, de Haro D, Botez CE, Peralta-Videa JR, Gardea-Torresdey JL (2013) Effects of ZnO nanoparticles in alfalfa, tomato, and cucumber at the germination stage: root development and X-ray absorption spectroscopy studies. Pure Appl Chem 85(12):2161–2174

De La Torre-Roche R, Hawthorne J, Musante C, Xing B, Newman LA, Ma X, White JC (2013) Impact of Ag nanoparticle exposure on p, p'-DDE bioaccumulation by Cucurbita pepo (zucchini) and Glycine max (soybean). Environ Sci Technol 47(2):718–725

De Oliveira JL, Campos EVR, Bakshi M, Abhilash PC, Fraceto LF (2014) Application of nanotechnology for the encapsulation of botanical insecticides for sustainable agriculture: prospects and promises. Biotechnol Adv 32(8):1550–1561. doi:10.1016/j.biotechadv.2014.10.010

De Rosa MC, Monreal C, Schnitzer M, Walsh R, Sultan Y (2010) Nanotechnology in fertilizers. Nat Nanotechnol 5:91

de Santiago-Martín A, Constantin B, Guesdon G, Kagambega N, Raymond S, Cloutier RG (2015) Bioavailability of engineered nanoparticles in soil systems. J Hazard Toxic Radioact Waste. doi:10.1061/(ASCE)HZ.2153-5515.0000263

De Souza MP, Pilon-Smits EAH, Terry N (1999) The physiology and biochemistry of selenium volatilization by plants. In: Raskin I, Ensley BD (eds) Phytoremediation of toxic metals: using plants to clean up the environment. Wiley, New York, pp 171–188

Delay M, Frimmel FH (2012) Nanoparticles in aquatic systems. Anal Bioanal Chem 402:583–592. doi:10.1007/s00216-011-5443-z

Delay M, Schwegmann H, Frimmel FH (2015) Nanoparticles and refractory organic matter: interactions and consequences. J Environ Chem Eng Vol. 3 (Issue 4, Part B): 2997–3004. http://dx.doi.org/10.1016/j.jece.2015.02.012
Delfani M, Firouzabadi MB, Farrokhi N, Makarian H (2014) Some physiological responses of black-eyed pea to iron and magnesium nanofertilizers. Commun Soil Sci Plant Anal 45:530–540
Dietz KJ, Herth S (2011) Plant nanotoxicology. Trends Plant Sci 16(11):582–589. doi:10.1016/j.tplants.2011.08.003
Dimkpa CO (2014) Can nanotechnology deliver the promised benefits without negatively impacting soil microbial life? J Basic Microbiol 2014(54):1–16. doi:10.1002/jobm.201400298
Dimkpa CO, McLean JE, Britt DW, Anderson AJ (2012) Bioactivity and biomodification of Ag, ZnO and CuO nanoparticles with relevance to plant performance in agriculture. Ind Biotechnol 8:344–357
Dimkpa CO, McLean JE, Britt DW, Anderson AJ (2015) Nano-CuO and interaction with nano-ZnO or soil bacterium provide evidence for the interference of nanoparticles in metal nutrition of plants. Ecotoxicology. doi:10.1007/s10646-014-1364-x
Dimpka CO, Latta ED, McLean JE et al (2013) Fate of CuO and ZnO nano- and microparticles in the plant environment. Environ Sci Technol 47:4734–4742
Dinesh R, Anandaraj M, Srinivasan V, Hamza S (2012) Engineered nanoparticles in the soil and their potential implications to microbial activity. Geoderma 173–174:19–27. doi:10.1016/j.geoderma.2011.12.018
Ditta A (2012) How helpful is nanotechnology in agriculture? Adv Nat Sci: Nanosci Nanotechnol 3(3):033002. doi:10.1088/2043-6262/3/3/033002
Ditta A, Arshad M, Ibrahim M (2015) Nanoparticles in sustainable agricultural crop production: applications and perspectives. In: Siddiqui MH, Al-Whaibi MH, Mohammad F (eds) Nanotechnology and plant sciences: nanoparticles and their impact on plants. Springer International Publishing, Cham, pp 55–75. doi:10.1007/978-3-319-14502-4
Domokos-Szabolcsy E (2011) Biological effect and fortification possibilities of inorganic selenium forms in higher plants. PhD dissertation, Debrecen University
Domokos-Szabolcsy E, Marton L, Sztrik A, Babka B, Prokisch J, Fari M (2012) Accumulation of red elemental selenium nanoparticles and their biological effects in *Nicotinia tabacum*. Plant Growth Regul 68:525–531. doi:10.1007/s10725-012-9735-x
Domokos-Szabolcsy E, Abdalla N, Alshaal T, Sztrik A, Márton L, El-Ramady H (2014) *In vitro* comparative study of two *Arundo donax* L. ecotypes' selenium tolerance. Int J Hortic Sci 20(3–4):119–122, ISSN 1585-0404
Donaldson K, Stone V, Tran CL, Kreyling W, Borm PJ (2004) Nanotoxicology. Occup Environ Med 61:727–728
Doolette CL, McLaughlin MJ, Kirby JK, Navarro DA (2015) Bioavailability of silver and silver sulfide nanoparticles to lettuce (*Lactuca sativa*): effect of agricultural amendments on plant uptake. J Hazard Mater, http://dx.doi.org/10.1016/j.jhazmat.2015.08.012
Doong R, Sharma VK, Kim H (2013) Interactions of nanomaterials with emerging environmental contaminants, vol 1150, ACS symposium series. American chemical society, Washington, DC
Du W, Sun Y, Ji R, Zhu J, Wu J et al (2011) TiO_2 and ZnO nanoparticles negatively affect wheat growth and soil enzyme activities in agricultural soil. J Environ Monit 13:822–828
El Beyrouthya M, El Azzi D (2014) Nanotechnologies: novel solutions for sustainable agriculture. Adv Crop Sci Technol 2:e118. doi:10.4172/2329-8863.1000e118
El-Kereti MA, El-feky SA, Khater MS, Osman YA, El-sherbini EA (2013) ZnO nanofertilizer and He Ne laser irradiation for promoting growth and yield of sweet basil plant. Recent Pat Food Nutr Agric 5(3):169–181
El-Ramady H (2014) Integrated nutrient management and postharvest of crops. *In*: Sustainable agriculture reviews, Lichtfouse E (ed) Sustainable agriculture reviews, Vol. 13, doi:10.1007/978-3-319-00915-5_8, Springer International Publishing, Cham, pp 163–274

El-Ramady H, Domokos-Szabolcsy É, Abdalla NA, Alshaal TA, Shalaby TA, Sztrik A, Prokisch J, Fári M (2014a) Selenium and nano-selenium in agroecosystems. Environ Chem Lett 12(4):495–510. doi:10.1007/s10311-014-0476-0

El-Ramady HR, Abdalla NA, Alshaal TA, Elhawat N, Domokos-Szabolcsy É, Prokisch J, Sztrik A, Fári M (2014b) Nano-selenium: from *in vitro* to micro farm experiments. The international conference "*Biogeochemical processes at air-soil-water interfaces and environmental protection*" for the European society for soil conservation, Imola–Ravenna, 23–26 June 2014. doi:10.13140/2.1.2260.4481

El-Ramady HR, Abdalla NA, Alshaal TA, Elhawat N, Domokos-Szabolcsy É, Prokisch J, Sztrik A, Fári M (2014c) Nano-selenium: from *in vitro* to micro farm experiments. The international conference "biogeochemical processes at air-soil-water interfaces and environmental protection" for the European society for soil conservation, Imola–Ravenna, 23–26 June 2014. doi:10.13140/2.1.2260.4481

El-Ramady H, Abdalla N, Alshaal T, El-Henawy A, Faizy SE-DA, Shams MS, Shalaby T, Bayoumi Y, Elhawat N, Shehata S, Sztrik A, Prokisch J, Fári M, Pilon-Smits EA, Domokos-Szabolcsy É (2015a) Selenium and its role in higher plants. In: Lichtfouse E et al. (eds) Environmental chemistry for a sustainable world, vol 7. Springer science + business media B.V, pp 235–296. doi:10.1007/978-3-319-19276-5_6

El-Ramady H, Domokos-Szabolcsy É, Shalaby TA, Prokisch J, and Fári M (2015b) Selenium in agriculture: water, air, soil, plants, food, animals and nanoselenium. In: Lichtfouse E (ed) Environmental chemistry for a sustainable world, vol 5. (CO_2 sequestration, biofuels and depollution). Springer, Berlin, pp 153–232. doi:10.1007/978-3-319-11906-9_5

El-Ramady H, Abdalla N, Taha HS, Alshaal T, El-Henawy A, Faizy SE-DA, Shams MS, Youssef SM, Shalaby T, Bayoumi Y, Elhawat N, Shehata S, Sztrik A, Prokisch J, Fári M, Domokos-Szabolcsy É, Pilon-Smits EA, Selmar D, Haneklaus S and Schnug E (2015c) Selenium and nano-selenium in plant nutrition. Environ Chem Lett. doi:10.1007/s10311-015-0535-1

El-Ramady H, Alshaal T, Abdalla N, Prokisch J, Sztrik A, Fári M, Domokos-Szabolcsy É (2016) Selenium and nano-selenium biofortified sprouts using micro-farm systems. In: Global advances in selenium research from theory to application. Bañuelos GS, Lin Z-Q, Guilherme LRG, dos Reis AR (eds) Proceedings of the 4th international conference on selenium in the environment and human health, Sao Paulo, 18–21 October 2015. CRC, Taylor & Francis Group, London, pp 189–190

Fabrega J, Luoma SN, Tyler CR, Galloway TS, Lead JR (2011) Silver nanoparticles: behaviour and effects in the aquatic environment. Environ Int 37:517–531. doi:10.1016/j.envint.2010.10.012

Faisal M, Saquib Q, Alatar AA et al (2013) Phytotoxic hazards of NiO-nanoparticles in tomato: a study on mechanism of cell death. J Hazard Mater 250–251:318–332

Fajardo C, Ortiz LT, Rodriguez-Membibre ML, Nande M, Lobo MC, Martin M (2012) Assessing the impact of zero-valent iron (ZVI) nanotechnology on soil microbial structure and functionality: a molecular approach. Chemosphere 86:802–808. doi:10.1016/j.chemosphere.2011.11.041

Fan R, Huang YC, Grusak MA, Huang CP, Sherrier DJ (2014) Effects of nano-TiO_2 on the agronomically-relevant *Rhizobium*–legume symbiosis. Sci Total Environ 466–467:503–512. doi:10.1016/j.scitotenv.2013.07.032

Fang J, Shan XQ, Wen B, Lin JM, Owens G (2009) Stability of titania nanoparticles in soil suspensions and transport in saturated homogeneous soil columns. Environ Pollut 157:1101–1109

Farre M, Sanchis J, Barcelo D (2011) Analysis and assessment of the occurrence, the fate and the behavior of nanomaterials in the environment. Trends Anal Chem 30(3):517–527

Feizi H, Moghaddam PR, Shahtahmassebi N, Fotovat A (2012) Impact of bulk and nanosized titanium dioxide (TiO_2) on wheat seed germination and seedling growth. Biol Trace Elem Res 146:101–106

Fent K (2010) Ecotoxicology of engineered nanoparticles. In: Frimmel FH, Niessner R (eds) Nanoparticles in the water cycle: properties, analysis and environmental relevance. Springer, Berlin, pp 183–206

Fernández MD, Alonso-Blázquez MN, García-Gómez C, Babin M (2014) Evaluation of zinc oxide nanoparticle toxicity in sludge products applied to agricultural soil using multispecies soil systems. Sci Total Environ 497–498:688–696, http://dx.doi.org/10.1016/j.scitotenv.2014.07.085

Ferry JL, Craig P, Hexel C, Sisco P, Frey R, Pennington PL, Fulton MH, Scott IG, Decho AW, Kashiwada S, Murphy CJ, Shaw TJ (2009) Transfer of gold nanoparticles from the water column to the estuarine food web. Nat Nanotechnol 4:441–444. doi:10.1038/NNANO.2009.157

Foltete AS, Masfaraud JF, Bigorgne E, Nahmani J, Chaurand P, Botta C, Labille J, Rose J, Férard JF, Cotelle S (2011) Environmental impact of sunscreen nanomaterials: ecotoxicity and gentoxicity of altered TiO_2 nanocomposites on *Vicia faba*. Environ Pollut 159:2515–2522. doi:10.1016/j.envpol.2011.06.020

Frazier TP, Burklew CE, Zhang B (2014) Titanium dioxide nanoparticles affect the growth and microRNA expression of tobacco (*Nicotiana tabacum*). Funct Integr Genomics 14(1):75–83

Frimmel FH, Niessner R (2010) Nanoparticles in the water cycle properties, analysis and environmental relevance. Springer, Berlin/Heidelberg. doi:10.1007/978-3-642-10318-6

Fubini B, Fenoglio I, Tomatis M (2007) Physicochemical characteristics of nanoparticles that determine potential toxicity. In: Monteiro-Riviere NA, Tran CL (eds) Nanotoxicology: characterization, dosing, and health effects. Informa Healthcare, New York, pp 59–70

Gao J, Xu G, Qian H, Liu P, Zhao P, Hu Y (2013) Effects of nano-TiO_2 on photosynthetic characteristics of *Ulmus elongata* seedlings. Environ Pollut 176:63–70. doi:10.1016/j.envpol.2013.01.027

García-Gómez C, Babin M, Obrador A, Álvarez JM, Fernández MD (2015) Integrating ecotoxicity and chemical approaches to compare the effects of ZnO nanoparticles, ZnO bulk, and $ZnCl_2$ on plants and microorganisms in a natural soil. Environ Sci Pollut Res. doi:10.1007/s11356-015-4867-y

Garner KL, Keller AA (2014) Emerging patterns for engineered nanomaterials in the environment: a review of fate and toxicity studies. J Nanopart Res 16:2503. doi:10.1007/s11051-014-2503-2

Ge Y, Schimel JP, Holden PA (2011) Evidence for negative effects of TiO_2 and ZnO nanoparticles on soil bacterial communities. Environ Sci Technol 45:1659–1664

Ge Y, Schimel JP, Holden PA (2012) Identification of soil bacteria susceptible to TiO_2 and ZnO nanoparticles. Appl Environ Microbiol 78(18):6749–6758. doi:10.1128/AEM.00941-12

Geisler-Lee J, Brooks M, Gerfen JR, Wang Q, Fotis C, Sparer A, Ma X, Berg RH, Geisler M (2014) Reproductive toxicity and life history study of silver nanoparticle effect, uptake and transport in *Arabidopsis thaliana*. Nanomaterials 4:301–318. doi:10.3390/nano4020301

Ghafariyan MH, Malakouti MJ, Dadpour MR et al (2013) Effects of magnetite nanoparticles on soybean chlorophyll. Environ Sci Technol 47:10645–10652

Ghodake G, Seo YD, Lee DS (2011) Hazardous phytotoxic nature of cobalt and zinc oxide nanoparticles assessed using *Allium cepa*. J Hazard Mater 186:952–955

Ghormade V, Deshpande MV, Paknikar KM (2011) Perspectives for nano-biotechnology enabled protection and nutrition of plants. Biotechnol Adv 29:792–803

Ghosh M, Bandyopadhyay M, Mukherjee A (2010) Genotoxicity of titanium dioxide (TiO_2) nanoparticles at two trophic levels: plant and human lymphocytes. Chemosphere 81(10):1253–1262. doi:10.1016/j.chemosphere.2010.09.022

Gil-Díaz M, Gonzalez A, Alonso J, Lobo MC (2016) Evaluation of the stability of a nanoremediation strategy using barley plants. J Environ Manag 165:150e158, http://dx.doi.org/10.1016/j.jenvman.2015.09.032

Giraldo JP, Landry MP, Faltermeier SM, McNicholas TP, Iverson NM, Boghossian AA, Reuel NF, Hilmer AJ, Sen F, Brew JA, Strano MS (2014) Plant nanobionics approach to augment photosynthesis and biochemical sensing. Nat Mater 13(4):400–408. doi:10.1038/nmat3890

Gonzalez-Melendi P, Fernandez-Pacheco R, Coronado MJ, Corredor E, Testillano PS, Risueno MC et al (2008) Nanoparticles as smart treatment-delivery systems in plants: assessment of different techniques of microscopy for their visualization in plant tissues. Ann Bot Lond 101:187–195

Gorczyca A, Pociecha E, Kasprowicz M, Niemiec M (2015) Effect of nanosilver in wheat seedlings and *Fusarium culmorum* culture systems. Eur J Plant Pathol 142:251–261

Grieger KD, Fjordbøge A, Hartmann NB, Eriksson E, Bjerg PL, Baun A (2010) Environmental benefits and risks of zero-valent iron nanoparticles (nZVI) for in situ remediation: risk mitigation or trade-off? J Contam Hydrol 118:165–183

Grillo R, Rosa AH, Fraceto LF (2015) Engineered nanoparticles and organic matter: a review of the state-of-the-art. Chemosphere 119:608–619, http://dx.doi.org/10.1016/j.chemosphere.2014.07.049

Haensch M, Emmerling C (2010) Effects of silver nanoparticles on the microbiota and enzyme activity in soil. J Plant Nutr Soil Sci 173:554–558. doi:10.1002/jpln.200900358

Hajipour MJ, Fromm KM, Ashkarran AA, de Aberasturi DJ, de Larramendi IR, Rojo T, Serpooshan V, Parak WJ, Mahmoudi M (2012) Antibacterial properties of nanoparticles. Trends biotechnol 30(10):499–511. d. http://dx.doi.org/10.1016/j.tibtech.2012.06.004

Handy RD, von der Kammer F, Lead JR, Richard Owen MH, Crane M (2008) The ecotoxicology and chemistry of manufactured nanoparticles. Ecotoxicology 17:287–314. doi:10.1007/s10646-008-0199-8

Hawthorne J, De la Torre Roche R, Xing B, Newman LA, Ma X, Majumdar S, Gardea-Torresdey J, White JC (2014) Particle-size dependent accumulation and trophic transfer of cerium oxide through a terrestrial food chain. Environ Sci Technol 48(22):13102–13109, dx.doi.org/10.1021/es503792f

He S, Feng Y, Ren H, Zhang Y, Gu N, Lin X (2011) The impact of iron oxide magnetic nanoparticles on the soil bacterial community. J Soils Sediments 11:1408–1417. doi:10.1007/s11368-011-0415-7

He X, Aker WG, Leszczynski J, Hwang H-M (2014) Using a holistic approach to assess the impact of engineered nanomaterials inducing toxicity in aquatic systems. J Food Drug Anal 22:128–146, http://dx.doi.org/10.1016/j.jfda.2014.01.011

Helaly MN, El-Metwally MA, El-Hoseiny H, Omar SA, El-Sheery NI (2014) Effect of nanoparticles on biological contamination of *in vitro* cultures and organogenic regeneration of banana. Aust J Crop Sci 8:612–624

Holden PA, Nisbet RM, Lenihan HS, Miller RJ, Cherr GN et al (2013) Ecological nanotoxicology: integrating nanomaterial hazard considerations across the subcellular, population, community, and ecosystems levels. Acc Chem Res 46:813–822

Holden PA, Klaessig F, Turco RF, Priester J, Rico CM, Arias HA, Mortimer M, Pacpaco K, Gardea-Torresdey JL (2014a) Evaluation of exposure concentrations used in assessing manufactured nanomaterial environmental hazards: are they relevant? Environ sci technol 48(18):10541–10551. doi:10.1021/es502440s

Holden PA, Schimel JP, Godwin HA (2014b) Five reasons to use bacteria when assessing manufactured nanomaterial environmental hazards and fates. Curr Opin Biotechnol 27:73–78. doi:10.1016/j.copbio.2013.11.008

Hong J, Peralta-Videa JR, Gardea-Torresdey JL (2013) Nanomaterials in agricultural production: benefits and possible threats? In: Shamim N, Sharma VK (eds) Sustainable nanotechnology and the environment: advances and achievements, vol 1124, ACS symposium series. American Chemical Society, Washington, DC, pp 73–90. doi:10.1021/bk-2013-1124.ch001

Hong J, Peralta-Videa JR, Rico CM et al (2014) Evidence of translocation and physiological impacts of foliar applied CeO_2 nanoparticles on cucumber (*Cucumis sativus*) plants. Environ Sci Technol 48(8):4376–4385

Horst AM, Neal AC, Mielke RE, Sislian PR, Suh WH et al (2010) Dispersion of TiO_2 nanoparticle agglomerates by *Pseudomonas aeruginosa*. Appl Environ Microbiol 76:7292–7298

Hu YH, Burghaus U, Qiao S (2014) Nanotechnology for sustainable energy, ACS symposium series 1140. American Chemical Society, Washington DC, United States

Hull MS, Bowman DM (2014) Nanotechnology environmental health and safety risks, regulation and management, 2nd edn. William Andrew Inc/Elsevier Inc, New York

Husen A, Siddiqi KS (2014) Phytosynthesis of nanoparticles: concept, controversy and application. Nano Res Lett 9:229

Jain A, Shivendu R, Nandita D, Chidambaram R (2016) Nanomaterials in food and agriculture: an overview on their safety concerns and regulatory issues. Crit Rev Food Sci. doi:10.1080/10408398.2016.1160363

Jafar G, Hamzeh G (2013) Ecotoxicity of nanomaterials in soil. Ann Biol Res 4(1):86–92

Jain R, Dominic D, Jordan N, Rene ER, Weiss S, van Hullebusch ED, Hübner R, Lens PNL (2016) Preferential adsorption of Cu in a multi-metal mixture onto biogenic elemental selenium nanoparticles. Chem Eng J 284:917–925

Jaisi DP, Elimelech M (2009) Single-walled carbon nanotubes exhibit limited transport in soil columns. Environ Sci Technol 43:9161–9166

Jampílek J, Kráľová K (2015) Application of nanotechnology in agriculture and food industry, its prospects and risks. Ecol Chem Eng S 22(3):321–361. doi:10.1515/eces-2015-0018

Janmohammadi M, Sabaghnia N, Ahadnezhad A (2015) Impact of silicon dioxide nanoparticles on seedling early growth of lentil (*Lens culinaris* Medik) genotypes with various origins. Agric For 61(3):19–33. doi:10.17707/AgricultForest.61.3.02

Janvier C, Villeneuve F, Alabouvette C, Edel-Hermann V, Mateille T, Steinberg C (2007) Soil health through soil disease suppression: which strategy from descriptors to indicators? Soil Biol Biochem 39:1–23

Johnson CA, Freyer G, Fabisch M, Caraballo MA, Ksel K, Hochella MF (2014) Observations and assessment of iron oxide and green rust nanoparticles in metal-polluted mine drainage within a steep redox gradient. Environ Chem 11(4):377

Johnston CT (2010) Probing the nanoscale architecture of clay minerals. Clay Miner 45:245–279

Johnston H, Pojana G, Zuin S, Jacobsen NR, Moeller P, Loft S, Semmler-Behnke M, McGuiness C, Balharry D, Marcomini A, Wallin H, Kreyling W, Donaldson K, Tran L, Stone V (2013) Engineered nanomaterial risk. Lessons learnt from completed nanotoxicology studies: potential solutions to current and future challenges. Crit Rev Toxicol 43(1):1–20. doi:10.3109/10408444.2012.738187

Joo SH, Cheng IF (2006) Nanotechnology for environmental remediation, Modern inorganic chemistry. Springer, New York

Jorio A (2016) Bioengineering applications of carbon nanostructures. Nanomedicine and nanotoxicology series. Zucolotto V (ed) Springer international publishing, Cham, doi:10.1007/978-3-319-25907-9

Jośko I, Oleszczuk P (2013) Influence of soil type and environmental conditions on the ZnO, TiO_2 and Ni nanoparticles phytotoxicity. Chemosphere 92:91–99

Jośko I, Oleszczuk P, Futa B (2014) The effect of inorganic nanoparticles (ZnO, Cr_2O_3, CuO and Ni) and their bulk counterparts on enzyme activities in different soils. Geoderma 232–234:528–537, http://dx.doi.org/10.1016/j.geoderma.2014.06.012

Judy JD, Bertsch PM (2014) Bioavailability, toxicity, and fate of manufactured nanomaterials in terrestrial ecosystems. In: Sparks D (ed) Adv Agron, Vol. 123, pp:1–64. Elsevier Inc. http://dx.doi.org/10.1016/B978-0-12-420225-2.00001-7

Judy JD, McNear DH Jr, Chen C, Lewis RW, Tsyusko OV, Bertsch PM, Rao W, Stegemeier J, Lowry GV, McGrath SP, Durenkamp M, Unrine JM (2015a) Nanomaterials in biosolids inhibit nodulation, shift microbial community composition, and result in increased metal uptake relative to bulk/dissolved metals. Environ Sci Technol 49(14):8751–8758. doi:10.1021/acs.est.5b01208

Judy JD, Kirby JK, Creamer C, McLaughlin MJ, Fiebiger C, Wright C, Cavagnaro TR, Bertsch PM (2015b) Effects of silver sulfide nanomaterials on mycorrhizal colonization of tomato plants and soil microbial communities in biosolid-amended soil. Environ Pollut 206:256–263, http://dx.doi.org/10.1016/j.envpol.2015.07.002

Kahru A, Dubourguier H-C (2010) From ecotoxicology to nanoecotoxicology. Toxicology 269:105–119. doi:10.1016/j.tox.2009.08.016

Kahru A, Dubourguier H-C, Blinova I, Ivask A, Kasemets K (2008) Biotests and biosensors for ecotoxicology of metal oxide nanoparticles: a minireview. Sensors 8:5153–5170. doi:10.3390/s8085153

Kalteh M, Alipour ZT, Ashraf S, Aliabadi MM, Nosratabadi AF (2014) Effect of silica nanoparticles on basil (*Ocimum basilicum*) under salinity stress. J Chem Health Risks 4:49–55

Kanerva T, Palojarvi A, Ramo K, Manninen S (2008) Changes in soil microbial community structure under elevated tropospheric O_3 and CO_2. Soil Biol Biochem 40:2502–2510

Kanneganti A, Talasila M (2014) MoO_3 nanoparticles: synthesis, characterization and its hindering effect on germination of *Vigna unguiculata* seeds. Int. J Eng Res Appl 4(7) (Version 3):116–120. ISSN: 2248-9622

Karn B, Kuiken T, Otto M (2009) Nanotechnology and in situ remediation: a review of the benefits and potential risks. Environ Health Perspect 117:1813–1831

Kashyap PL, Xiang X, Heiden P (2015) Chitosan nanoparticle based delivery systems for sustainable agriculture. Int J Biol Macromol, http://dx.doi.org/10.1016/j.ijbiomac.2015.02.039

Kaveh R, Li Y-S, Ranjbar S, Tehrani R, Brueck CL, Van Aken B (2013) Changes in *Arabidopsis thaliana* gene expression in response to silver nanoparticles and silver ions. Environ Sci Technol 47:10637–10644. doi:10.1021/es402209w

Kern ME (2015) Interaction of gold nanomaterials with the edible food crop, *Helianthus annuus* (Common sunflower). MS (Master of science) thesis, University of Iowa. http://ir.uiowa.edu/etd/1657

Khot RL, Sankaran S, Maja JM, Ehsani R, Schuster EW (2012) Applications of nanomaterials in agricultural production and crop protection: a review. Crop Prot 35:64–70. doi:10.1016/j.cropro.2012.01.007

Kim S, Kim J, Lee I (2011) Effects of Zn and ZnO nanoparticles and Zn^{2+} on soil enzyme activity and bioaccumulation of Zn in *Cucumis sativus*. Chem Ecol 27(1):49–55. doi:10.1080/02757540.2010.529074

Kim J-H, Oh Y, Yoon H, Hwang I, Chang Y-S (2015) Iron nanoparticle-induced activation of plasma membrane H^+-ATPase promotes stomatal opening in *Arabidopsis thaliana*. Environ Sci Technol 49(2):1113–1119

Kirschling TL, Gregory KB, Minkley JEG, Lowry GV, Tilton RD (2010) Impact of nanoscale zero valent iron on geochemistry and microbial populations in trichloroethylene contaminated aquifer materials. Environ Sci Technol 44(9):3474–3480

Klaine SJ, Alvarez PJJ, Batley GE, Fernandes TF, Handy RD, Lyon DY, Mahendra S, McLaughlin MJ, Lead JR (2008) Nanomaterials in the environment: behavior, fate, bioavailability, and effects. Environ Toxicol Chem 27(9):1825–1851. doi:10.1897/08-090.1

Klaine SJ, Koelmans AA, Horne N, Carley S, Handy RD, Kapustka L, Nowack B, von der Kammer F (2012) Paradigms to assess the environmental impact of manufactured nanomaterials. Environ Toxicol Chem 31(1):3–14

Knorr W, Prentice IC, House JI, Holland EA (2005) Long-term sensitivity of soil carbon turnover to warming. Nature 433:298–302

Koelmel J, Leland T, Wang H, Amarasiriwardena D, Xing B (2013) Investigation of gold nanoparticles uptake and their tissue level distribution in rice plants by laser ablation-inductively coupled-mass spectrometry. Environ Pollut 174:222–228, http://dx.doi.org/10.1016/j.envpol.2012.11.026

Kohan-Baghkheirati E, Geisler-Lee J (2015) Gene expression, protein function and pathways of *Arabidopsis thaliana* responding to silver nanoparticles in comparison to silver ions, cold, salt, drought, and heat. Nanomaterials 5:436–467

Kouhi SMM, Lahouti M, Ganjeali A, Entezari MH (2015a) Long-term exposure of rapeseed (*Brassica napus* L.) to ZnO nanoparticles: anatomical and ultrastructural responses. Environ Sci Pollut Res Int 22(14):10733–10743. doi:10.1007/s11356-015-4306-0

Kouhi SMM, Lahouti M, Ganjeali A, Entezari MH (2015b) Comparative effects of ZnO nanoparticles, ZnO bulk particles, and Zn^{2+} on *Brassica napus* after long-term exposure: changes in growth, biochemical compound. Water Air Soil Pollut 226:364. doi:10.1007/s11270-015-2628-7

Kulkarni SK (2015) Nanotechnology: principles and practices, 3rd edn. Springer, Capital Publishing Company, Cham. doi:10.1007/978-3-319-09171-6

Kumar N, Shah V, Walker VK (2011) Perturbation of an arctic soil microbial community by metal nanoparticles. J Hazard Mater 190(1–3):816–822

Kumari M, Mukherjee A, Chandrasekaran N (2009) Genotoxicity of silver nanoparticles in *Allium cepa*. Sci Total Environ 407(19):5243–5246

Kumari J, Kumar D, Mathur A, Naseer A, Kumar RR, Chandrasekaran PT, Chaudhuri G, Pulimi M, Raichur AM, Babu S, Chandrasekaran N, Nagarajan R, Mukherjee A (2014) Cytotoxicity of TiO_2 nanoparticles towards fresh water sediment microorganisms at low exposure concentrations. Environ Res 135:333–345. doi:10.1016/j.envres.2014.09.025

Kuppusamy S, Palanisami T, Megharaj M, Venkateswarlu K, Naidu R (2016) In-situ remediation approaches for the management of contaminated sites: a comprehensive overview. In: de Voogt P (ed) Reviews of environmental contamination and toxicology, vol 236. Springer International Publishing, Cham, pp 1–115. doi:10.1007/978-3-319-20013-2_1

Lal R (2007) Soil science and the carbon civilization. Soil Sci Soc Am J 71:1425–1437

Lal R (2008) Promise and limitations of soils to minimize climate change. J Soil Water Conserv 63:113A–118A

Lal R (2015) The nexus approach to managing water, soil and waste under changing climate and growing demands on natural resources. In: Kurian M, Ardakanian R (eds) Governing the nexus: water, soil and waste resources considering global change. Springer, Cham, pp 39–61. doi:10.1007/978-3-319-05747-7_3

Lalau CM, Mohedano RA, Schmidt EC, Bouzon ZL, Ouriques LC, dos Santos RW, da Costa CH, Vicentini DS, Matias WG (2014) Toxicological effects of copper oxide nanoparticles on the growth rate, photosynthetic pigment content, and cell morphology of the duckweed *Landoltia punctate*. Protoplasma. doi:10.1007/s00709-014-0671-7

Larue C, Laurette J, Herlin-Boime N, Khodja H, Fayard B, Flank A, Brisset F, Carriere M (2012) Accumulation, translocation and impact of TiO_2 nanoparticles in wheat (*Triticum aestivum*): influence of diameter and crystal phase. Sci Total Environ 431:197–208

Larue C, Castillo-Michel H, Sobanska S, Cécillon L, Bureau S, Barthès V, Ouerdane L, Carrière M, Sarret G (2014) Foliar exposure of the crop *Lactuca sativa* to silver nanoparticles: evidence for internalization and changes in Ag speciation. J Hazard Mater 264:98–106. doi:10.1016/j.jhazmat.2013.10.053

Lead JR, Wilkinson KJ (2006) Aquatic colloids and nanoparticles: current knowledge and future trends. Environ Chem 3:159–171

Lee SH, Richards RJ (2004) Montserrat volcanic ash induces lymph node granuloma and delayed lung inflammation. Toxicology 195:155–165

Lei Z, Mingyu S, Xiao W, Chao L, Chunxiang Q, Liang C, Hao H, Xiaoqing L, Fashui H (2008) Antioxidant stress is promoted by nanoanatase in spinach chloroplasts under UV-B radiation. Biol Trace Elem Res 121:69–79

Lenz M, Kolvenbach B, Gygax B, Moes S, Corvini PFX (2011) Shedding light on selenium biomineralization: proteins associated with bionanominerals. Appl Environ Microbiol 77(13):4676–4680. doi:10.1128/AEM.01713-10

Lerner RN, Lu Q, Zeng H, Liu Y (2012) The effects of biofilms on the transport of stabilized zerovalent iron nanoparticles in saturated porous media. Water Res 46:975–985

Li B, Tao G, Xie Y, Cai X (2012) Physiological effects under the condition of spraying nano- SiO_2 onto the *Indocalamus barbatus* McClure leaves. J Nanjing For Univ (Natural Science Edition) 36:161–164

Li K-E, Chang Z-Y, Shen C-X, Yao N (2015) Toxicity of nanomaterials to plants. In: Siddiqui MH, Al-Whaibi MH, Mohammad F (eds) Nanotechnology and plant sciences nanoparticles and their impact on plants. Springer International Publishing, Cham, pp 101–123. doi:10.1007/978-3-319-14502-6

Li S, Ma H, Wallis LK, Etterson MA, Riley B, Hoff DJ, Diamond SA (2016) Impact of natural organic matter on particle behavior and phototoxicity of titanium dioxide nanoparticles. Sci Total Environ 542:324–333, http://dx.doi.org/10.1016/j.scitotenv.2015.09.141

Libralato G, Costa Devoti A, Zanella M, Sabbioni E, Mičetić I, Manodori L, Pigozzo A, Manenti S, Groppi F, Ghirardini AV (2015) Phytotoxicity of ionic, micro- and nano-sized iron in three plant species. Ecotoxicol Environ Saf, http://dx.doi.org/10.1016/j.ecoenv.2015.07.024

Lin D, Xing B (2007) Phytotoxicity of nanoparticles: inhibition of seed germination and root growth. Environ Pollut 150:243–250

Lin D, Xing B (2008) Root uptake and phytotoxicity of ZnO nanoparticles. Environ Sci Technol 42:5580–5585

Lin D, Tian X, Wu F, Xing B (2010) Fate and transport of engineered nanomaterials in the environment. J Environ Qual 39:1–13. doi:10.2134/jeq2009.0423

Liu R, Lal R (2014) Synthetic apatite nanoparticles as a phosphorus fertilizer for soybean (*Glycine max*). Sci Rep 4:5686–5691

Liu R, Lal R (2015) Potentials of engineered nanoparticles as fertilizers for increasing agronomic productions. Sci Total Environ 514:131–139, http://dx.doi.org/10.1016/j.scitotenv.2015.01.104

Liu F, Wen LX, Li ZZ, Yu W, Sun HY, Chen JF (2006) Porous hollow silica nanoparticles as controlled delivery system for water soluble pesticide. Mat Res Bull 41:2268–2275

Liu W, Tian S, Zhao X, Xie W, Gong Y, Zhao D (2015a) Application of stabilized nanoparticles for in situ remediation of metal-contaminated soil and groundwater: a critical review. Curr Pollut Rep 1:280–291. doi:10.1007/s40726-015-0017-x

Liu X, Wang F, Shi Z, Tong R, Shi X (2015b) Bioavailability of Zn in ZnO nanoparticle-spiked soil and the implications to maize plants. J Nanopart Res 17:175. doi:10.1007/s11051-015-2989-2

Loeffler J (2005) Nanomaterial roadmap 2015, overview on promising nanomaterials for industrial applications. Steinbeis-Europa-Zentrum, Karlsruhe

Lopez-Moreno ML, de la Rosa G, Hernandez-Viezcas JA, Peralta-Videa JR, Gardea-Torresdey JL (2010) X-ray Absorption Spectroscopy (XAS) corroboration of the uptake and storage of CeO_2 nanoparticles and assessment of their differential toxicity in four edible plant species. J Agric Food Chem 58(6):3689–3693. doi:10.1021/Jf904472e

Lowry GV, Gregory KB, Apte SC, Lead JR (2012) Transformations of nanomaterials in the environment. Environ Sci Technol 46:6893–6899. doi:10.1021/es300839e

Ma Y, Kuang L, HeX BW, Ding Y, Zhang Z, Zhao Y, Chai Z (2010a) Effects of rare earth oxide nanoparticles on root elongation of plants. Chemosphere 78:273–279

Ma X, Geiser-Lee J, Deng Y, Kolmakov A (2010b) Interactions between engineered nanoparticles (ENPs) and plants: phytotoxicity, uptake and accumulation. Sci Total Environ 408(16):3053–3061

Ma H, Williams PL, Diamond SA (2013) Ecotoxicity of manufactured ZnO nanoparticles: a review. Environ Pollut 172:76–85, http://dx.doi.org/10.1016/j.envpol.2012.08.011

Ma R, Levard C, Judy JD, Unrine JM, Durenkamp M, Martin B, Jefferson B, Lowry GV (2014) Fate of zinc oxide and silver nanoparticles in a pilot wastewater treatment plant and in processed biosolids. Environ Sci Technol 48(1):104–112

Ma Y, Zhang P, Zhang Z, He X, Li Y, Zhang J, Zheng L, Chu S, Yang K, Zhao Y Chai Z (2015) Origin of the different phytotoxicity and biotransformation of cerium and lanthanum oxide nanoparticles in cucumber. Nanotoxicology, pp:1–9, ISSN: 1743-5390

Maddineni SB, Badal KM, Shivendu R, Nandita D (2015) Diastase assisted green synthesis of size-controllable gold nanoparticles. RSC Adv 5:26727–26733. doi:10.1039/C5RA03117F

Magdolenova Z, Collins A, Kumar A, Dhawan A, Stone V, Dusinska M (2014) Mechanisms of genotoxicity: a review of *in vitro* and *in vivo* studies with engineered nanoparticles. Nanotoxicology 8(3):233–278. doi:10.3109/17435390.2013.773464

Majumdar S, Peralta-Videa JR, Bandyopadhyay S, Castillo-Michel H, Hernandez-Viezcas J-A, Sahi S, Gardea-Torresdey JL (2014) Exposure of cerium oxide nanoparticles to kidney bean shows disturbance in the plant defense mechanisms. J Hazard Mater 278:279–287

Mallevre F, Fernandes TF, Aspray TJ (2014) Silver, zinc oxide and titanium dioxide nanoparticle ecotoxicity to bioluminescent *Pseudomonas putida* in laboratory medium and artificial wastewater. Environ Pollut 195:218–225, http://dx.doi.org/10.1016/j.envpol.2014.09.002

Manikandan A, Subramanian KS (2014) Fabrication and characterisation of nanoporous zeolite based N fertilizer. Afr J Agric Res 9:276–284

Martínez-Fernández D, Barroso D, Komárek M (2015) Root water transport of *Helianthus annuus* L. under iron oxide nanoparticle exposure. Environ Sci Pollut Res. doi:10.1007/s11356-015-5423-5

Mastronardi E, Tsae P, Zhang X, Monreal C, DeRosa MC (2015) Strategic role of nanotechnology in fertilizers: potential and limitations. In: Rai M et al (eds) Nanotechnologies in food and agriculture. Springer International Publishing, Cham, pp 25–67. doi:10.1007/978-3-319-14024-7_2

Maurer-Jones MA, Mousavi MPS, Chen LD, Bühlmann P, Haynes CL (2013) Characterization of silver ion dissolution from silver nanoparticles using fluorous-phase ion-selective electrodes and assessment of resultant toxicity to *Shewanella oneidensis*. Chem Sci 4:2564–2572. doi:10.1039/C3SC50320H

Maurice PA, Hochella MF (2008) Nanoscale particles and processes: a new dimension in soil science. Adv Agron 100:123–138

Menard A, Drobne D, Jemec A (2011) Ecotoxicity of nanosized TiO_2: review of *in vivo* data. Environ Pollut 159:677–684. doi:10.1016/j.envpol.2010.11.027

Mohanraj J (2013) Effect of nano-zeolite on nitrogen dynamics and green house gas emission in rice soil eco system. M. Tech. (Ag.) Thesis, TNAU, Coimbatore, India

Mohmood I, Lopes CB, Lopes I, Ahmad I, Duarte AC, Pereira E (2013) Nanoscale materials and their use in water contaminants removal—a review. Environ Sci Pollut Res 20:1239–1260. doi:10.1007/s11356-012-1415-x

Mondal AM, Mondal S, Samanta S, Mallick S (2011) Synthesis of ecofriendly silver nanoparticle from plant latex used as an important taxonomic tool for phylogenetic interrelationship. Adv Biores 2(1):122–133, ISSN 0976-4585

Monreal CM, DeRosa M, Mallubhotla SC, Bindraban PS, Dimkpa C (2015) Nanotechnologies for increasing the crop use efficiency of fertilizer-micronutrients. Biol Fertil Soils. doi:10.1007/s00374-015-1073-5

Moore MN (2006) Do nanoparticles present ecotoxicological risks for the health of the aquatic environment? Environ Int 32:967–976. doi:10.1016/j.envint.2006.06.014

Mousavi SR, Rezaei M (2011) Nanotechnology in agriculture and food production. J Appl Environ Biol Sci 1(10):414–419

Mueller NC, Nowack B (2008) Exposure modeling of engineered nanoparticles in the environment. Environ Sci Technol 42(12):4447–4453. doi:10.1021/es7029637

Mukherjee A, Peralta-Videa JR, Bandyopadhyay S, Rico CM, Zhao L, Gardea-Torresdey JL (2014) Physiological effects of nanoparticulate ZnO in green peas (*Pisum sativum* L.) cultivated in soil. Metallomics 6:132–138. doi:10.1039/c3mt00064h

Mukhopadhyay SS (2014) Nanotechnology in agriculture: prospects and constraints. Nanotechnology. Sci Appl 7:63–71

Mura S, Seddaiu G, Bacchini F, Roggero PP, Greppi GF (2013) Advances of nanotechnology in agro-environmental studies. Ital J Agron 8:127–140

Musante C, White JC (2012) Toxicity of silver and copper to *Cucurbita pepo*: differential effects of nano and bulk-size particles. Environ Toxicol 27:510–517

Naderi MR, Danesh-Shahraki A (2013) Nanofertilizers and their roles in sustainable agriculture. Int J Agric Crop Sci 5(19):2229–2232

Nagaonkar D, Shende S, Rai M (2015) Biosynthesis of copper nanoparticles and its effect on actively dividing cells of mitosis in *Allium cepa*. Biotechnol Prog 31(2):557–565. doi:10.1002/btpr.2040

Nair PMG, Chung IM (2014) Impact of copper oxide nanoparticles exposure on *Arabidopsis thaliana* growth, root system development, root lignification, and molecular level changes. Environ Sci Pollut Res. doi:10.1007/s11356-014-3210-3

Nair R, Varghese SH, Nair BG, Maekawa T, Yoshida Y, Kumar DS (2010) Nanoparticulate material delivery to plants. Plant Sci 179:154–163

Navarro E, Baun A, Behra R, Hartmann NB, Filser J et al (2008) Environmental behavior and ecotoxicity of engineered nanoparticles to algae, plants, and fungi. Ecotoxicolgy 17:372–386

Ngô C, Van de Voorde MH (2014) Nanotechnologies in agriculture and food. In: Ngô C, Van de Voorde MH (eds) Nanotechnology in a nutshell. Springer, New York, pp 233–247

Nogueira V, Lopes I, Rocha-Santos T, Santos AL, Rasteiro GM, Antunes F, Gonçalves F, Soares AMVM, Cunha A, Almeida A, Gomes NNCM, Pereira R (2012) Impact of organic and inorganic nanomaterials in the soil microbial community structure. Sci Total Environ 424:344–350. doi:10.1016/j.scitotenv.2012.02.041

Nowack B, Bucheli TD (2007) Occurrence, behavior and effects of nanoparticles in the environment. Environ Pollut 150:5–22. doi:10.1016/j.envpol.2007.06.006

Nowack B, Ranville JF, Diamond S, Gallego-Urrea JA, Metcalfe C, Rose J et al (2012) Potential scenarios for nanomaterial release and subsequent alteration in the environment. Environ Toxicol Chem 31:50–59

Ouda SM (2014) Antifungal activity of silver and copper nanoparticles on two plant pathogens, *Alternaria alternata* and *Botrytis cinerea*. Res J Microbiol 9:34–42. doi:10.3923/jm.2014.34.42

Oukarroum A, Barhoumi L, Samadani M, Dewez D (2015) Toxic effects of nickel oxide bulk and nanoparticles on the aquatic plant *Lemna gibba* L. Biomed Res Int, Article ID 501326, http://dx.doi.org/10.1155/2015/501326

Pakrashi S, Jain N, Dalai S et al (2014) *In vivo* genotoxicity assessment of titanium dioxide nanoparticles by *Allium cepa* root tip assay at high exposure concentrations. PLoS ONE 9(2):e87789

Pan B, Xing B (2010) Manufactured nanoparticles and their sorption of organic chemicals. Adv Agron 108:137–181

Pan B, Xing B (2012) Applications and implications of manufactured nanoparticles in soils: a review. Eur J Soil Sci. doi:10.1111/j.1365-2389.2012.01475.x

Pardha-Saradhi P, Yamal G, Peddisetty T, Sharmila P, Singh J, Nagarajan R, Rao KS (2014) Plants fabricate Fe-nanocomplexes at root surface to counter and phytostabilize excess ionic Fe. Biometals 27(1):97–114. doi:10.1007/s10534-013-9690-7

Patil SS, Shedbalkar UU, Truskewycz A, Chopade BA, Ball AS (2016) Nanoparticles for environmental clean-up: a review of potential risks and emerging solutions. Environ Technol Innov 5:10–21

Pawlett M, Ritz K, Dorey RA, Rocks S, Ramsden J, Harris JA (2013) The impact of zero-valent iron nanoparticles upon soil microbial communities is context dependent. Environ Sci Pollut Res 20(2):1041–1049. doi:10.1007/s11356-012-1196-2

Peng C, Duan D, Xu C, Chen Y, Sun L, Zhang H, Yuan X, Zheng L, Yang Y, Yang J, Zhen X, Chen Y, Shi J (2015a) Translocation and biotransformation of CuO nanoparticles in rice (*Oryza sativa* L.) plants. Environ Pollut 197:99–107, http://dx.doi.org/10.1016/j.envpol.2014.12.008

Peng C, Zhang H, Fang H, Xu C, Huang H, Wang Y, Sun L, Yuan X, Chen Y, Shi J (2015b) Natural organic matter-induced alleviation of the phytotoxicity to rice (*Oryza sativa* L.) caused by CuO nanoparticles. Environ Toxicol Chem 34(9):1996–2003. doi:10.1002/etc.3016

Peralta-Videa JR, Zhao L, Lopez-Moreno ML, de la Rosad G, Hong J, Gardea-Torresdey JL (2011) Nanomaterials and the environment: a review for the biennium 2008–2010. J Hazard Mater 186:1–15. doi:10.1016/j.jhazmat.2010.11.020

Perreault F, Popovic R, Dewez D (2014) Different toxicity mechanisms between bare and polymer-coated copper oxide nanoparticles in *Lemna gibba*. Environ Pollut 185:219–227. doi:10.1016/j.envpol.2013.10.027

Philippe A, Schaumann GE (2014) Interactions of dissolved organic matter with natural and engineered inorganic colloids: a review. Environ Sci Technol 48(16):8946–8962. doi:10.1021/es502342r

Pokhrel LR, Dubey B (2013) Evaluation of developmental responses of two crop plants exposed to silver and zinc oxide nanoparticles. Sci Total Environ 452–453:321–332. doi:10.1016/j.scitotenv.2013.02.059

Pradeep TA (2009) Noble metal nanoparticles for water purification: a critical review. Thin Solid Films 517:6441–6478. doi:10.1016/j.tsf.2009.03.195

Prakash M, Nair G, Chung M (2014) Assessment of silver nanoparticle-induced physiological and molecular changes in *Arabidopsis thaliana*. Environ Sci Pollut Res 21:8858–8869

Prasad R, Bagde US, Varma A (2012a) Intellectual property rights and agricultural biotechnology: an overview. Afr J Biotechnol 11(73):13746–13752

Prasad TNVKV, Sudhakar P, Sreenivasulu Y, Latha P, Munaswamy V, Reddy KR, Sreeprasad TS, Sajanlal PR, Pradeep T (2012b) Effect of nanoscale zinc oxide particles on the germination, growth and yield of peanut. J Plant Nutr 35(6):905–927

Prasad R, Kumar V, Prasad KS (2014) Nanotechnology in sustainable agriculture: present concerns and future aspects. Afr J Biotechnol 13(6):705–713. doi:10.5897/AJBX2013.13554

Priester JH, Ge Y, Mielke RE, Horst AM, Moritz SC et al (2012) Soybean susceptibility to manufactured nanomaterials with evidence for food quality and soil fertility interruption. Proc Natl Acad Sci U S A 109:E2451–E2456

Rabbani MM, Ahmed I, Park S-J (2016) Application of nanotechnology to remediate contaminated soils. In: Hasegawa H, Rahman Mofizur IM, Azizur Rahman M (eds) Environmental remediation technologies for metal-contaminated soils. Springer, Tokyo, pp 219–229. doi:10.1007/978-4-431-55759-3_10

Rai M, Duran N (2011) Metal nanoparticles in microbiology. Springer, Berlin/Heidelberg. doi:10.1007/978-3-642-18312-6

Rai M, Ribeiro C, Mattoso L, Duran N (2015) Nanotechnologies in food and agriculture. Springer, Cham/Heidelberg/New York/Dordrecht/London. doi:10.1007/978-3-319-14024-7

Raliya R, Tarafdar JC (2013) ZnO nanoparticle biosynthesis and its effect on phosphorous-mobilizing enzyme secretion and gum contents in cluster bean (*Cyamopsis tetragonoloba* L.). Agric Res 2(1):48–57. doi:10.1007/s40003-012-0049-z

Ramesh M, Palanisamy K, Babu K, Sharma NK (2014) Effects of bulk & nano-titanium dioxide and zinc oxide on physio-morphological changes in *Triticum aestivum* Linn. J Glob Biosci 3:415–422

Rameshaiah N, Pallavi J, Shabnam S (2015) Nanofertilizers and nanosensors – an attempt for developing smart agriculture. Int J Eng Res Gen Sci 3(1):2091–2730

Ranjan S, Nandita D, Arkadyuti RC, Melvin SS, Chidambaram R, Rishi S, Ashutosh K (2014) Nanoscience and nanotechnologies in food industries: opportunities and research trends. J Nanopart Res 16(6):2464. doi:10.1007/s11051-014-2464-5

Ranjan S, Nandita D, Sudandiradoss C, Ramalingam C, Ashutosh K (2015) A novel approach to evaluate titanium dioxide nanoparticle-protein interaction through docking: an insight into the mechanism of action. Proc Natl Acad Sci, India Sect B Biol Sci. doi:10.1007/s40011-015-0673-z

Ranjan S, Nandita D, Bhavapriya R, Ganesh SA, Chidambaram R, Ashutosh K (2016) Microwave-irradiation-assisted hybrid chemical approach for titanium dioxide nanoparticle synthesis: microbial and cytotoxicological evaluation. Env Sci Pollut Res. doi:10.1007/s11356-016-6440-8

Raskar SV, Laware SL (2014) Effect of zinc oxide nanoparticles on cytology and seed germination in onion. Int J Curr Microbiol App Sci 3:467–473

Reid BJ, Jones KC, Semple KT (2000) Bioavailability of persistent organic pollutants in soils and sediments—a perspective on mechanisms, consequences and assessment. Environ Pollut 108:103–112

Remedios C, Rosario F, Bastos V (2012) Environmental nanoparticles interactions with plants: morphological, physiological, and genotoxic aspects. J Bot 2012:1–8, http://dx.doi.org/10.1155/2012/751686

Rickerby DG (2013) Nanotechnology for more sustainable manufacturing: opportunities and risks. In: Shamim N, Sharma VK (eds) Sustainable nanotechnology and the environment: advances and achievements, vol 1124, ACS symposium. American Chemical Society, Washington, DC, pp 91–105

Rico CM, Majumdar S, Duarte-Gardea M, Peralta-Videa JR, Gardea-Torresdey JL (2011) Interaction of nanoparticles with edible plants and their possible implications in the food chain. J Agric Food Chem 59:3485–3498

Rico CM, Hong J, Morales MI, Zhao L, Barrios AC, Zhang JY, Peralta-Videa JR, Gardea-Torresdey JL (2013a) Effect of cerium oxide nanoparticles on rice: a study involving the antioxidant defense system and *in vivo* fluorescence imaging. Environ Sci Technol 47:5635–5642. doi:10.1021/es401032m

Rico CM, Morales MI, McCreary R, Castillo-Michel H, Barrios AC, Hong J, Tafoya A, Lee WY, Varela-Ramirez A, Peralta-Videa JR, Gardea-Torresdey JL (2013b) Cerium oxide nanoparticles modify the antioxidative stress enzyme activities and macromolecule composition in rice seedlings. Environ Sci Technol 47:14110–14118. doi:10.1021/es4033887

Rico CM, Barrios AC, Tan W, Rubenecia R, Lee SC, Varela-Ramirez A, Peralta-Videa JR, Gardea-Torresdey JL (2015) Physiological and biochemical response of soil-grown barley (*Hordeum vulgare* L.) to cerium oxide nanoparticles. Environ Sci Pollut Res Int 14:10551–10558. doi:10.1007/s11356-015-4243-y

Roldugin VI, Fedotov MA, Folmanis GE, Kovalenko LV, Tananaevc IG (2015) Formation of aqueous colloidal solutions of selenium and silicon by laser ablation. Dokl Phys Chem, 463 (Part 1), 161–164. doi:10.1134/S0012501615070064

Roure F (2016) Societal approach to nanoscience and nanotechnology: when technology reflects and shapes society. In: Lourtioz J-M et al (eds) Nanosciences and nanotechnology. Springer International Publishing, Cham, pp 357–404. doi:10.1007/978-3-319-19360-1_17

RS-RAE, The royal society and royal academy of engineering (2004). Nanoscience and nanotechnologies: opportunities and uncertainties. http://www.nanotec.org.uk/finalReport.htm

Ruffini MC, Cremonini R (2009) Nanoparticles and higher plants. Caryologia: Int J Cytol Cytosystematics Cytogenet 62(2):161–165. doi:10.1080/00087114.2004.10589681

Rui Y, Zhang P, Zhang Y, Ma Y, He X, Gui X, Li Y, Zhang J, Zheng L, Chu S, Guo Z, Chai Z, Zhao Y, Zhang Z (2015) Transformation of ceria nanoparticles in cucumber plants is influenced by phosphate. Environ Pollut 198:8–14

Sabo-Attwood T, Unrine JM, Stone JW, Murphy CJ, Ghoshroy S, Blom D, Bertsch PM, Newman LA (2012) Uptake, distribution and toxicity of gold nanoparticles in tobacco (*Nicotiana xanthi*) seedlings. Nanotoxicology 6(4):353–360. doi:10.3109/17435390.2011.579631

Sadiq IM, Pakrashi S, Chandrasekaran N, Mukherjee A (2011) Studies on toxicity of aluminum oxide (Al_2O_3) nanoparticles to microalgae species: *Scenedesmus sp.* and *Chlorella sp.* J Nanopart Res 13:3287–3299. doi:10.1007/s11051-011-0243-0

Salamanca-Buentello F, Daar AS (2016) Dust of wonder, dust of doom: a landscape of nanotechnology, nanoethics, and sustainable development. In: Bagheri A et al (eds) Global bioethics: the impact of the UNESCO international bioethics committee, vol 5, Advancing global bioethics. Springer International Publishing, Cham, pp 101–123. doi:10.1007/978-3-319-22650-7_10

Sandler R, Kay WD (2006) The national nanotechnology initiative and the social good. J Law, Med Ethics 34(4):675–681

Savage N, Diallo M, Duncan J, Street A, Sustich R (2009) Nanotechnology applications for clean water. Elsevier Inc., William Andrew Inc, New York

Schaumann GE, Philippe A, Bundschuh M, Metreveli G, Klitzke S, Rakcheev D, Grün A, Kumahor SK, Kühn M, Baumann T, Lang F, Manz W, Schulz R, Vogel H-J (2015a) Understanding the fate and biological effects of Ag- and TiO_2-nanoparticles in the environment: the quest for advanced analytics and interdisciplinary concepts. Sci Total Environ 535:3–19, http://dx.doi.org/10.1016/j.scitotenv.2014.10.035

Schaumann GE, Baumann T, Lang F, Metreveli G, Vogel H-J (2015b) Engineered nanoparticles in soils and waters. Sci Total Environ 535:1–2, http://dx.doi.org/10.1016/j.scitotenv.2015.06.006

Schwabe F, Schulin R, Rupper P, Rotzetter A, Stark W, Nowack B (2014) Dissolution and transformation of cerium oxide nanoparticles in plant growth media. J Nanopart Res 16:2668. doi:10.1007/s11051-014-2668-8

Schwabe F, Zhai G, Kern M, Turner A, Schnoor JL, Wiesner MR (2015a) Barriers, pathways and processes for uptake, translocation and accumulation of nanomaterials in plants – critical review. Nanotoxicology 11:1–22. doi:10.3109/17435390.2015.1048326

Schwabe F, Tanner S, Schulin R, Rotzetter A, Stark W, von Quadt A, Nowack B (2015b) Dissolved cerium contributes to uptake of Ce in the presence of differently sized CeO_2 -nanoparticles by three crop plants. Metallomics 7:466–477

Scott N, Chen H (2013) Nanoscale science and engineering for agriculture and food systems. Ind Biotechnol 9(1):17–18

Scrinis G, Lyons K (2007) The emerging nano-corporate paradigm: nanotechnology and the transformation of nature, food and agri-food systems. J Sociol Food Agric 15(2):22–44, ISSN: 0798-1759

Sedghi M, Hadi M, Toluie SG (2013) Effect of nano zinc oxide on the germination of soybean seeds under drought stress. Ann West Uni Timis oara ser Biol XVI(2):73–78

Sekhon BS (2014) Nanotechnology in agri-food production: an overview. Nanotechnol Sci Appl 7:31–53, http://dx.doi.org/10.2147/NSA.S39406

Selva Preetha P, Subramanian KS, Sharmila Rahale C (2014) Sorption characteristics of nanozeolite based slow release sulphur fertilizer. Int J Dev Res 4:225–228

Servin A, Elmer W, Mukherjee A, De la Torre-Roche R, Hamdi H, White JC, Bindraban P, Dimkpa C (2015) A review of the use of engineered nanomaterials to suppress plant disease and enhance crop yield. J Nanopart Res 17:92. doi:10.1007/s11051-015-2907-7

Shah V, Belozerova I (2009) Influence of metal nanoparticles on the soil microbial community and germination of lettuce seeds. Water Air Soil Pollut 197:143–148

Shah MA, Bhat MA, Davim JP (2015) Nanotechnology applications for improvements in energy efficiency and environmental management. IGI Global, Hershey

Shamim N, Sharma VK (2013) Sustainable nanotechnology and the environment: advances and achievements, vol 1124, ACS symposium. American Chemical Society, Washington, DC. doi:10.1021/bk-2013-1124.ch001

Shapira P, Youtie J (2015) The economic contributions of nanotechnology to green and sustainable growth. In: Basiuk VA, Basiuk EV (eds) Green processes for nanotechnology. Springer International Publishing, Cham, pp 409–434. doi:10.1007/978-3-319-15461-9_15

Sharma VK (2009) Aggregation and toxicity of titanium dioxide nanoparticles in aquatic environment—a review. J Environ Sci Health A Tox Hazard Subst Environ Eng 44(14):1485–1495, http://dx.doi.org/10.1080/10934520903263231

Sharma VK, Siskova KM, Zboril R, Gardea-Torresdey L (2014) Organic-coated silver nanoparticles in biological and environmental conditions: fate, stability and toxicity. Adv Colloid Interf Sci 204:15–34, http://dx.doi.org/10.1016/j.cis.2013.12.002

Sharma VK, Filip J, Zboril R, Varma RS (2015) Natural inorganic nanoparticles – formation, fate, and toxicity in the environment. Chem Soc Rev. doi:10.1039/c5cs00236b

Sharonova NL, Yapparov AK, Khisamutdinov NS, Ezhkova AM, Yapparov IA, Ezhkov VO, Degtyareva IA, Babynin EV (2015) Nanostructured water-phosphorite suspension is a new promising fertilizer. Nanotechnol Russ 10(7–8):651–661. doi:10.1134/S1995078015040187

Shaw AK, Hossain Z (2013) Impact of nano-CuO stress on rice (*Oryza sativa* L.) seedlings. Chemosphere 93(6):906–915

Shaymurat T, Gu J, Xu C, Yang Z, Zhao Q, liu Y, Liu Y (2012) Phytotoxic and genotoxic effects of ZnO nanoparticles on garlic (*Allium sativum* L.): a morphological study. Nanotoxicology 6(3):241–248. doi:10.3109/17435390.2011.570462

Shi J, Peng C, Yang Y, Yang J, Zhang H, Yuan X, Chen Y, Hu T (2014) Phytotoxicity and accumulation of copper oxide nanoparticles to the Cu-tolerant plant *Elsholtzia splendens*. Nanotoxicology 8(2):179–188. doi:10.3109/17435390.2013.766768

Siddiqui MH, Al-Whaibi MH (2014) Role of nano-SiO_2 in germination of tomato (*Lycopersicum esculentum* seeds mill.). Saudi J Biol Sci 21:13–17

Siddiqui MH, Al-Whaibi MH, Faisal M, Al Sahli AA (2014) Nano-silicon dioxide mitigates the adverse effects of salt stress on *Cucurbita pepo* L. Environ Toxicol Chem 33(11):2429–2437. doi:10.1002/etc.2697

Siddiqui MH, Al-Whaibi MH, Mohammad F (2015a) Nanotechnology and plant sciences nanoparticles and their impact on plants. Springer International Publishing, Cham. doi:10.1007/978-3-319-14502-0

Siddiqui MH, Al-Whaibi MH, Firoz M, Al-Khaishany MY (2015b) Role of nanoparticles in plants. In: Siddiqui MH, Al-Whaibi MH, Mohammad F (eds) Nanotechnology and plant sciences nanoparticles and their impact on plants. Springer International Publishing, Cham, pp 19–32. doi:10.1007/978-3-319-14502-2

Sillen WMA, Thijs S, Abbamondi GR, Janssen J, Weyens N, White JC, Vangronsveld J (2015) Effects of silver nanoparticles on soil microorganisms and maize biomass are linked in the rhizosphere. Soil Biol Biochem 91:14–22, http://dx.doi.org/10.1016/j.soilbio.2015.08.019

Simonin M, Richaume A (2015) Impact of engineered nanoparticles on the activity, abundance, and diversity of soil microbial communities: a review. Environ Sci Pollut Res 22:13710–13723. doi:10.1007/s11356-015-4171-x

Singh A, Singh NB, Hussain I, Singh H, Singh SC (2015a) Plant-nanoparticle interaction: an approach to improve agricultural practices and plant productivity. Int J Pharm Sci Invent 4(8):25–40, ISSN: 2319 – 6718

Singh R, Shedbalkar UU, Wadhwani SA, Chopade BA (2015b) Bacteriagenic silver nanoparticles: synthesis, mechanism, and applications. Appl Microbiol Biotechnol. doi:10.1007/s00253-015-6622-1

Sirbu T, Maslobrod SN, Yu. A. Mirgorod, Borodina VG, Borsch NA, Ageeva LS (2016) Influence of dispersed solutions of copper, silver, bismuth and zinc oxide nanoparticles on growth and catalase activity of *Penicillium Funiculosum*. In: Sontea V, Tiginyanu I (eds) 3rd international conference on nanotechnologies and biomedical engineering, IFMBE Proceedings 55, Springer Science+Business Media Singapore, pp:271–274. doi:10.1007/978-981-287-736-9_66

Smita S, Gupta SK, Bartonova A, Dusinska M, Gutleb AC, Rahman Q (2012) Nanoparticles in the environment: assessment using the causal diagram approach. Environ Health 11(Suppl 1):S13. doi:10.1186/1476-069X-11-S1-S13

Solanki P, Bhargava A, Chhipa H, Jain N, Panwar J (2015) Nano-fertilizers and their smart delivery system. In: Rai M et al (eds) Nanotechnologies in food and agriculture. Springer International Publishing, Cham, pp 81–101. doi:10.1007/978-3-319-14024-7_3

Soliman AS, El-feky SA, Darwish E (2015) Alleviation of salt stress on *Moringa peregrina* using foliar application of nanofertilizers. J Hortic For 7(2):36–47. doi:10.5897/JHF2014.0379

Sozer N, Kokini JL (2009) Nanotechnology and its applications in the food sector. Trends Biotechnol 27(2):82–89. doi:10.1016/j.tibtech.2008.10.010

Srilatha B (2011) Nanotechnology in agriculture. J Nanomedic Nanotechnol 2(7):1–5, http://dx.doi.org/10.4172/2157-7439.1000123

Stampoulis D, Sinha SK, White JC (2009) Assay – dependent phytotoxicity of nanoparticles to plants. Environ Sci Technol 43:9473–9479

Stewart J, Hansen T, McLean JE, McManus P, Das S, Britt DW, Anderson AJ, Dimkpa CO (2015) Salts affect the interaction of ZnO or CuO nanoparticles with wheat. Environ Toxicol Chem 34:2116–2125. doi:10.1002/etc.v34.9

Street A, Sustich R, Duncan J, Savage N (2014) Nanotechnology applications for clean water: solutions for improving water quality, 2nd edn. Elsevier Inc/William Andrew Inc, New York

Stuart EJE, Compton RG (2015) Nanoparticles-emerging contaminants. In: Moretto LM, Kalcher K (eds) Environmental analysis by electrochemical sensors and biosensors. Springer Science+Business Media, New York, pp 855–878. doi:10.1007/978-1-4939-1301-5_8

Subramanian KS, Sharmila Rahale C (2012) Ball milled nanosized zeolite loaded with zinc sulfate: a putative slow release Zn fertilizer. Int J Innov Hortic 1:33–40

Subramanian KS, Sharmila Rahale C (2013) Nano-fertilizers – synthesis, characterization and application. In: Adhikari T, Subba R (eds) Nanotechnology in soil science and plant nutrition. New India Publishing Agency, New Delhi

Subramanian KS, Paulraj C, Natarajan S (2008) Nanotechnological approaches in nutrient management. In: Chinnamuthu CR, Chandrasekaran B, Ramasamy C (eds) Nanotechnology applications in agriculture. TNAU technical bulletin. TNAU, Coimbatore, pp 37–42

Subramanian KS, Manikandan A, Thirunavukkarasu M, Rahale CS (2015) Nano-fertilizers for balanced crop nutrition. In: Ra M et al (eds) Nanotechnologies in food and agriculture. Springer International Publishing, Cham, pp 69–80. doi:10.1007/978-3-319-14024-7_3

Suman PR, Jain VK, Varma A (2010) Role of nanomaterials in symbiotic fungus growth enhancement. Curr Sci 99:1189–1191

Suppan S (2013) Nanomaterials in soil: our future food chain? The institute for agriculture and trade policy, Published by IATP (www.iatp.org).

Suresh AK, Pelletier DA, Doktycz MJ (2013) Relating nanomaterial properties and microbial toxicity. Nanoscale 5:463–474

Suriyaprabha R, Karunakaran G, Yuvakkumar R, Prabu P, Rajendran V, Kannan N (2012a) Growth and physiological responses of maize (*Zea mays* L.) to porous silica nanoparticles in soil. J Nanopart Res 14:1294–1296

Suriyaprabha R, Karunakaran G, Yuvakkumar R, Rajendran V, Kannan N (2012b) Silica nanoparticles for increased silica availability in maize (*Zea mays* L.) seeds under hydroponic conditions. Curr Nanosci 8:1–7

Takeuchi MT, Kojima M, Luetzow M (2014) State of the art on the initiatives and activities relevant to risk assessment and risk management of nanotechnologies in the food and agriculture sectors. Food Res Int. doi:10.1016/j.foodres.2014.03.022

Tandy S, Schulin R, Nowack B (2006) Uptake of metals during chelant-assisted phytoextraction with EDDS related to the solubilized metal concentration. Environ Sci Technol 40:2753–2758

Tarafdar JC, Sharma S, Raliya R (2013) Nanotechnology: interdisciplinary science of applications. Afr J Biotechnol 12(3):219–226

Tarafdar JC, Raliya R, Mahawar H, Rathore I (2014) Development of zinc nanofertilizer to enhance crop production in pearl millet (*Pennisetum americanum*). Agric Res 3(3):257–262

Taran NY, Gonchar OM, Lopatko KG, Batsmanova LM, Patyka MV, Volkogon MV (2014) The effect of colloidal solution of molybdenum nanoparticles on the microbial composition in rhizosphere of *Cicer arietinum* L. Nanoscale Res Lett 9:289

Taranath TC, Patil BN, Santosh TU, Sharath BS (2015) Cytotoxicity of zinc nanoparticles fabricated by *Justicia adhatoda* L. on root tips of *Allium cepa* L.—a model approach. Environ Sci Pollut Res Int (11):8611–8617. doi:10.1007/s11356-014-4043-9

Taylor AF, Rylott EL, Anderson CWN, Bruce NC (2014) Investigating the toxicity, uptake, nanoparticle formation and genetic response of plants to gold. PLoS ONE 9(4):e93793. doi:10.1371/journal.pone.0093793

Theng BKG, Yuan G (2008) Nanoparticles in the soil environment. Elements 4:395–399

Thirunavukkarasu M (2014) Synthesis and evaluation of sulphur nano-fertilizers for groundnut. Ph.D. Thesis submitted to tamil nadu agricultural university, Coimbatore, India

Thul ST, Sarangi BK (2015) Implications of nanotechnology on plant productivity and its rhizospheric environment. In: Siddiqui MH, Al-Whaibi MH, Mohammad F (eds) Nanotechnology and plant sciences: nanoparticles and their impact on plants. Springer International Publishing, Cham, pp 37–54. doi:10.1007/978-3-319-14502-3

Thul ST, Sarangi BK, Pandey RA (2013) Nanotechnology in agroecosystem: implications on plant productivity and its soil environment. Expert Opin Environ Biol 2:1. doi:10.4172/2325-9655.1000101

Tilston EL, Collins CD, Mitchell GR, Princivalle J, Shaw LJ (2013) Nanoscale zero valent iron alters soil bacterial community structure and inhibits chloroaromatic biodegradation potential in Aroclor 1242-contaminated soil. Environ Pollut 173:38–46

Tourinho PS, van Gestel CAM, Lofts S, Svendesen C, Soares AMVM, Loureiro S (2012) Metal-based nanoparticles in soil: fate, behavior and effects on soil invertebrates. Environ Toxicol Chem 31:1679–1692

Towell MG, Browne LA, Paton GI, Semple KT (2011) Impact of carbon nanomaterials on the behaviour of ^{14}C-phenanthrene and ^{14}C-benzo-[a] pyrene in soil. Environ Pollut 159:706–715. doi:10.1016/j.envpol.2010.11.040

Van Aken B (2015) Gene expression changes in plants and microorganisms exposed to nanomaterials. Curr Opin Biotechnol 33:206–219. doi:10.1016/j.copbio.2015.03.005

von der Kammer F, Ottofuelling S, Hofmann T (2010) Assessment of the physico-chemical behavior of titanium dioxide nanoparticles in aquatic environments using multi-dimensional parameter testing. Environ Pollut 158:3472–3481. doi:10.1016/j.envpol.2010.05.007

Wang D, Chen Y (2015) Critical review of the influences of nanoparticles on biological wastewater treatment and sludge digestion. Crit Rev Biotechnol. doi:10.3109/07388551.2015.1049509

Wang P, Keller AA (2009) Natural and engineered nano and colloidal transport: role of zeta potential in prediction of particle deposition. Langmuir 25:6856–6862

Wang Z, Li J, Zhao J, Xing B (2011) Toxicity and internalization of CuO nanoparticles to prokaryotic alga microcystis aeruginosa as affected by dissolved organic matter. Environ Sci Technol 45:6032–6040

Wang S, Liu H, Zhang Y, Xin H (2015a) The effect of CuO NPs on reactive oxygen species and cell cycle gene expression in roots of rice. Environ Toxicol Chem. doi:10.1002/etc.2826

Wang Y, Peng C, Fang H, Sun L, Zhang H, Feng J, Duan D, Liu T, Shi J (2015b) Mitigation of Cu(II) phytotoxicity to rice (*Oryza sativa*) in the presence of TiO_2 and CeO_2 nanoparticles combined with humic acid. Environ Toxicol Chem 34:1588–1596

Watson JL, Fang T, Dimpka CO, Britt DW, McLean JE, Jacobson A, Anderson AJ (2015) The phytotoxicity of ZnO nanoparticles on wheat varies with soil properties. Biometals 28(1):101–112. doi:10.1007/s10534-014-9806-8

Wiesner MR, Bottero J-Y (2007) Environmental nanotechnology- applications and impacts of nanomaterials. McGraw-Hill, New York

Wigginton NS, Haus KL, Hochella MF (2007) Aquatic environmental nanoparticles. J Environ Monit 9:1306–1316

Wilson MA, Tran NH, Milev AS, Kannangara GSK, Volk H, Lu GQM (2008) Nanomaterials in soils. Geoderma 146:291–302. doi:10.1016/j.geoderma.2008.06.004

Xu C, Peng C, Sun L, Zhang S, Huang H, Chen Y, Shi J (2015) Distinctive effects of TiO_2 and CuO nanoparticles on soil microbes and their community structures in flooded paddy soil. Soil Biol Biochem, http://dx.doi.org/10.1016/j.soilbio.2015.03.011

Yadav T, Mungray AA, Mungray AK (2014) Fabricated nanoparticles: current status and potential phytotoxic threats. In: Whitacre DM (ed) Reviews of environmental contamination and toxicology, vol 230. Springer International Publishing, Cham, pp 83–110. doi:10.1007/978-3-319-04411-8_4

Yang Y, Colman BP, Bernhardt ES, Hochella MF (2015) Importance of a nanoscience approach in the understanding of major aqueous contamination scenarios: case study from a recent coal ash spill. doi:10.1021/es505662q

Yanık F, Vardar F (2015) Toxic effects of aluminum oxide (Al_2O_3) nanoparticles on root growth and development in *Triticum aestivum*. Water Air Soil Pollut 226:296. doi:10.1007/s11270-015-2566-4

Yatim NM, Shaaban A, Dimin MF, Yusof F (2015) Statistical evaluation of the production of urea fertilizer-multiwalled carbon nanotubes using plackett burman experimental design. Procedia – Soc Behav Sci 195:315–323

Załęska-Radziwiłł M, Doskocz N (2015) DNA changes in *Pseudomonas putida* induced by aluminum oxide nanoparticles using RAPD analysis. Desalin Water Treat. doi:10.1080/19443994.2014.996015

Zeliadt N (2010) Silver beware: antimicrobial nanoparticles in soil may harm plant life. Sci Am. http://www.scientificamerican.com/article/silver-beware-antimicrobial-nanoparticles-in-soil-may-harm-plant-life/3.4.2015

Zhai G, Walters KS, Peate DW, Alvarez PJJ, Schnoor JL (2014) Transport of gold nanoparticles through plasmodesmata and precipitation of gold ions in woody poplar. Environ Sci Technol Lett 1(2):146–151

Zhan J (2009) Multifunctional colloidal particles for *in situ* remediation of chlorinated hydrocarbons. Environ Sci Technol 43:8616–8621

Zhang Y, Chen Y, Westerhoff P, Hristovski K, Crittenden JC (2008) Stability of commercial metal oxide nanoparticles in water. Water res 42:2204–2212. doi:10.1016/j.watres.2007.11.036

Zhang P, Ma Y, Zhang Z (2015) Interactions between engineered nanomaterials and plants: phytotoxicity, uptake, translocation, and biotransformation. In: Siddiqui MH, Al-Whaibi MH, Mohammad F (eds) Nanotechnology and plant sciences nanoparticles and their impact on plants. Springer International Publishing, Cham, pp 77–99. doi:10.1007/978-3-319-14502-5

Zhao L, Hernandez-Viezcas JA, Peralta-Videa JR, Bandyopadhyay S, Peng B, Munoz B, Keller AA, Gardea-Torresdey JL (2013a) ZnO nanoparticle fate in soil and zinc bioaccumulation in corn plants (*Zea mays*) influenced by alginate. Environ Sci: Processes Impacts 15:260–266. doi:10.1039/C2EM30610G

Zhao L, Sun Y, Hernandez-Viezcas JA, Servin AD, Hong J, Genhua N, Peralta-Videa JR, Duarte-Gardea M, Gardea-Torresdey JL (2013b) Influence of CeO_2 and ZnO nanoparticles on cucumber physiological markers and bioaccumulation of Ce and Zn: a life cycle study. J Agric Food Chem 61:11945–11951, dx.doi.org/10.1021/jf404328e

Zhao L, Peralta-Videa JR, Rico CM, Sun Y, Niu G, Servin A, Nunez JE, Duarte-Gardea M, Gardea-Torresdey JL (2014) CeO_2 and ZnO nanoparticles change the nutritional qualities of cucumber (*Cucumis sativus*). J Agric Food Chem 62:2752–2759. doi:10.1021/jf405476u

Zou X, Shi J, Zhang H (2015) Morphological evolution and reconstruction of silver nanoparticles in aquatic environments: the roles of natural organic matter and light irradiation. J Hazard Mater 292:61–69, http://dx.doi.org/10.1016/j.jhazmat.2015.03.005

Index

S. Ranjan et al. (eds.), *Nanoscience in Food and Agriculture 2*, Sustainable Agriculture Reviews 21, DOI 10.1007/978-3-319-39306-3

C

V

W

X

MIX
Papier aus verantwortungsvollen Quellen
Paper from responsible sources
FSC® C105338

If you have any concerns about our products,
you can contact us on
ProductSafety@springernature.com

In case Publisher is established outside the EU,
the EU authorized representative is:
Springer Nature Customer Service Center GmbH
Europaplatz 3, 69115 Heidelberg, Germany

Printed by Libri Plureos GmbH
in Hamburg, Germany